Power Systems

Murari Mohan Saha • Jan Izykowski
Eugeniusz Rosolowski

Fault Location on Power Networks

Springer

Murari Mohan Saha, Dr.
Värmlandsvägen 11
SE-722 44 Västerås
Sweden
murari-saha@tele2.se

Jan Izykowski, Prof.
Eugeniusz Rosolowski, Prof.

Wroclaw University of Technology
Faculty of Electrical Engineering
Wybrzeze S. Wyspianskiego 27
50-370 Wroclaw
Poland

jan.izykowski@pwr.wroc.pl
eugeniusz.rosolowski@pwr.wroc.pl

ISSN 1612-1287
e-ISSN 1860-4676
ISBN 978-1-4471-2525-9
e-ISBN 978-1-84882-886-5
DOI 10.1007/978-1-84882-886-5
Springer London Dordrecht Heidelberg New York

British Library Cataloguing in Publication Data
A catalogue record for this book is available from the British Library

Softcover reprint of the hardcover 1st edition 2009

Cover design: deblik, Berlin, Germany

Printed on acid-free paper

Springer is part of Springer Science+Business Media (www.springer.com)

Preface

Electric power systems, which are growing in size and complexity, will be always exposed to failures of their components. In the case of a failure, the faulty element should be disconnected from the rest of the sound system in order to minimize the damage of the faulty element and to remove the emergency situation for the entire system. This action should be taken fast and accurately and is accomplished by a set of automatic protective relaying devices. At the same time, when a fault occurs on a line (distribution or transmission), it is very important for the utility to identify the fault location as quickly as possible for improving the service reliability. If a fault location cannot be identified quickly and this produces prolonged line outage during a period of peak load, severe economic losses may occur and reliability of service may be questioned. All these circumstances have raised the great importance of fault-location research studies and thus the problem has attracted widespread attention among researchers in power-system technology in recent years.

Basic algorithms used in fault locators are intended to make distance to fault calculation as accurate as possible. The fault locator is mainly associated with protection relays. Distance relays for transmission-line protection provide some indication of the general area where a fault occurred, but they are not designed to pinpoint the location. Moreover, both the tasks: line protection and fault location are fulfilled by processing the same current and voltage signals that are obtained from the instrument transformers and recorded at the substation. Fault-location estimation is a desirable feature in any protection scheme. Locating the fault on the transmission line accelerates line restoration and maintains system stability. That is why these two subjects are closely related to each other. There are, however, different demands formulated for protection and fault location. The last function should be made most precisely and with great accuracy. Distance to fault is estimated off-line from the recorded data. On the other hand – the relaying function is made on-line as fast as possible.

Different algorithms were developed in order to get a better estimation of fault distance depending on the extracted data from one or both ends of the transmission line. Most of the research done to date, has been aimed at finding the locations of transmission-line faults. This is mainly because of the impact of transmission-line faults on the power systems and the time required to physically check the lines is much larger than the faults in the sub-transmission and distribution systems. Of late, the location of faults on sub-transmission and distribution systems has started receiving some attention as many utilities are operating in a deregulated environment and are competing with each other to increase the availability of power supply to the customers. A fault location in the distribution system is not an easy job due to its high complexity and difficulty caused by non-homogeneity of line, fault resistance, load uncertainty, and phase unbalance.

The material is organized in 9 chapters. It starts with the introduction of basic concept and characteristic of methods of fault location in Chap. 1, which is followed by a general presentation of network configurations and models in Chap. 2. A separate chapter, 3, is dedicated to the nature of power-line faults, models and analyses. The algorithms for fault detection, fault-direction discrimination and determination of a type of fault are mentioned in the same Chap. 3 as well. Since a fault locator is a microprocessor device, its performance is determined primarily by an algorithm running on adequate hardware. From this perspective, the signal-processing methods for fault location are presented in detail in Chap. 4. A separate chapter, 5, is included that describes the measurement chains of fault locators. The main part of this book covers description of fault-location methods based on different approaches. In Chap. 6, it describes one-end impedance-based fault-location algorithms. It continues in Chap. 7 with two-end and multi-end fault-location algorithms. A separate chapter, 8, is dedicated to describe in brief different fault-location methods in distribution networks. Finally, Chap. 9 describes intelligence-methods application with respect to fault location.

Intended to link the design and application perspectives, written by authors from both university and relay manufacturer domains, this book is aimed at the audience of application, design and R&D engineers in protective relaying and automation as well as at university graduate and continuous-education students.

Distinctive Feature of the Book

The underlying assumption of this work is that fault location cannot be truly understood, applied, set, tested and analyzed without a deep and detailed knowledge of their interiors. Consequently, the design perspective organizes the included material. Unlike many other books, this one does not direct the reader to manufacturers' documentation, but instead, it tends to gather detailed information for both better understanding and comparison. It is intended to provide information enabling one to reproduce complete algorithms of a fault locator at least in their basic forms. The authors believe that this will help the reader to understand all aspects of fault location.

Västerås, Sweden
Wrocław, Poland
October 2009

Murari Mohan Saha
Jan Iżykowski
Eugeniusz Rosołowski

Contents

1 Fault Location – Basic Concept and Characteristic of Methods

Abstract This chapter begins with explaining the aim of fault location and its importance. It also describes the role of fault locators in comparison to protective relays. Then, a general division of fault-location techniques is presented. Usage of different input signals of fault locators is considered. The fault-location error is defined and the sources of possible errors are characterized. The basics of the traveling-wave and high-frequency methods are briefly presented.

1.1 Introduction

The rapid growth of electric power systems observed over the last several decades has resulted in a large increase in the number of transmission and distribution lines in operation in the world. At the same time, free marketing and deregulation introduced all over the world imposes more and more restrictive requirements on providing a continuous and good-quality power supply, without a significant increase in the cost of the energy being delivered. Terms such as the continuity of power supply, dependability and reliability play very important roles for contemporary power systems. As a result of the imposed restrictive requirements, an increased demand on the high quality of power-system protection and control devices together with their supplementary equipments has appeared as of prime importance. Among different capabilities of these devices, the fault-location function is considered as very important [44, 106, 109, 157, 258, 292, 352].

Transmission and distribution lines experience faults that are caused by storms, lightning, snow, freezing rain, insulation breakdown and, short circuits caused by birds and other external objects. In most cases, electrical faults manifest in mechanical damage, which must be repaired before returning the line to service. The restoration can be expedited if the location of the fault is either known or can be estimated with reasonable accuracy.

Fault locators provide estimates for both sustained and transient faults. Generally, transient faults cause minor damage that is not easily visible on inspection. Fault locators help identify these locations for early repairs to prevent recurrence and consequent major damages.

The subject of fault location has been of considerable interest to electric power utility engineers and researchers for many years. Most of the research done to date, has been aimed at finding the locations of transmission-line faults. This is mainly because of the impact of transmission-line faults on the power systems and the time required to physically check the lines is much larger than the faults in the sub-transmission and distribution systems. Of late, the location of faults on sub-transmission and distribution systems has started to receive some attention as many utilities are operating in a deregulated environment and are competing with each other to increase the availability of power supply to the customers.

In modern society the customers are more sensitive to the outages. Therefore, more efficient methods for fault location, supply restoration and high-quality customer service, which reduce the overall costs, are required. Fast location of the faulted section in distribution networks results in minimizing of inconvenience caused to the affected customers. This is becoming more important as there is an increasing emphasis placed on quality and reliability of supply and, therefore, fault location is considered to be one of the first functions to be integrated into modern substation control system [265].

1.2 Aim of Fault Location and Its Importance

Fault location is a process aimed at locating the occurred fault with the highest possibly accuracy. *A fault locator* is mainly the supplementary protection equipment, which apply the *fault-location algorithms* for estimating the distance to fault. When locating faults on the line consisting of more than one section, i.e., in the case of a three-terminal or multi-terminal line, the faulted section has to be identified and a fault on this section has to be located.

A fault-location function can be implemented into [109, 258]:

- microprocessor-based protective relays;
- digital fault recorders (DFRs);
- stand-alone fault locators; and
- post-fault analysis programs.

Including the fault-location function as an additional function of microprocessor-based relays is commonly used in practice. In this case high computational capability and communication with remote sites of modern relays are utilized at little, or almost no additional cost. Also, digital fault recorders enable easy and not costly incorporation of the fault-location function. In turn, stand-alone fault locators are applied in the case of using sophisticated fault-location algorithms and under the condition that higher cost of the implementation is accepted. Yet the other possibility is related to post-fault analysis programs [120] with included fault-location algorithms. Such programs are used mainly for verification of operation of protective relays.

Transmission and distribution lines experience temporary and permanent faults. Temporary faults, which are the most dominant faults on overhead lines, are self-cleared. In consequence, the power-supply continuity is not permanently affected, which is advantageous. In turn, after the permanent fault occurrence, the related protective relaying equipment enables the associated circuit breakers to de-energize the faulted sections. If a given line is taken out of service, the connected loads are not supplied or, if possible, the other lines are forced to supply the loads supplied by the tripped line. It is also possible that a series of cascading trips can happen, taking out of service successively larger and larger parts of the system. In some unfavorable cases this can lead even to blackouts of large power systems, as has happened recently in some countries. Contemporary power systems operate closer and closer to their operating limits. Therefore, in order to avoid blackouts special care ought to be paid in equipping power systems with protection and control devices, as well as in their settings.

In the case of permanent faults, the restoration of power supply can be done after the maintenance crew finishes the repair of the damage caused by the fault. For this purpose, the fault position has to be known; otherwise the whole line has to be inspected to find the damage place. Thus, it is important that the location of a fault is either known or can be estimated with reasonably high accuracy. This allows saving money and time for the inspection and repair, as well as to provide a better service due to the possibility of faster restoration of power supply. This also enables the blackouts to be avoided.

Temporary faults are self-cleared and do not affect permanently the supply continuity, however, the location of such faults is also important. In this case the fault location can help to pinpoint the weak spots on the line. As a result, the plans of maintenance schedules can be fixed for avoiding further problems in the future.

Even when helicopters are immediately available for patrol following unsuccessful reclosing, fault locators perform a valuable service. Trouble cannot always be found with a routine patrol with no indication of where the fault occurred. For example, tree growth could reduce clearances, resulting in a flashover during severe conductor sagging. By the time the patrol arrives, the conductors have cooled, increasing the clearance to the tree. The weak spot is not obvious [62].

The importance of fault locators is more obvious where foot patrols are relied upon, particularly on long lines, in rough terrain. Also, they can help where maintenance jurisdiction is divided between different companies or divisions within a company.

Fault locators are valuable even where the line has been restored either automatically or non-automatically. In this category are faults caused by cranes swinging into the line, brushfires, damaged insulators and vandalism. The locator allows rapid arrival at the site before the evidence is removed or the 'trail becomes cold'. Also, the knowledge that repeating faults are occurring in the same area can be valuable in detecting the cause. Weak spots that are not obvious may be found because a more through inspection can be focused in the limited area defined by the fault locator.

Because of economical reason, feeder-dedicated fault locators can hardly be applied in distribution networks. However, substantial monitoring using digital fault recorders (DFR) is a common utility practice in most countries. Moreover, a fault-recording function is available in new installed digital relays. Under these circumstances, low-cost fault location has become feasible [281].

1.3 Fault Locators *Versus* Protective Relays

Fault locators and protective relays are closely related, however, there are some important differences between them. These differences can be considered as related to the following features [258]:

- accuracy of fault location;
- speed of determining the fault position;
- speed of transmitting data from remote site;
- used data window; and
- digital filtering of input signals and complexity of calculations.

Fault locators are used for pinpointing the fault position accurately and not only for indication of the general area (defined by a protective zone) where a fault occurred – which is the case for protective relays.

Both the measurement and decision making of protective relays are performed in an on-line regime. High speed of operation of protective relays appears as a crucial requirement imposed on them. This is so since in order to prevent spreading out the fault effects; the faulted line has to be switched off as quickly as possible. Therefore, high-speed measuring algorithms are applied in contemporary protective relays. Use of high-speed operating circuit breakers is also of prime importance. Fault-clearing time is an important consideration in the selection of protective relays and requirements for relaying speed must be carefully determined. If the relaying is too slow, system instability, excessive equipment damage, and adverse effects on customer service may result. On the other hand, faster protection tends to compromise relay system security and selectivity. Requirement for fast clearing of faults demands that the decision for tripping transmission lines has to be made in short time, even faster than in one cycle of the fundamental frequency (20 ms for the systems operating at 50 Hz). In contrast, the calculations of fault locators are performed in an off-line mode since the results of these calculations (position of the fault and in the case of some algorithms also the involved fault resistance) are for human users. This implies that the fault-location speed of calculations can be measured in seconds or even minutes.

Low-speed data communications or supervisory control and data acquisition (SCADA) can be applied for fault-location purposes, which differs from communication used by protective relays.

The best data window segment from the whole available window can be selected for fault location to reduce errors. This is so since the computations are performed in an off-line regime and searching for the best data window can be easily applied. The fault interval lasts from a fault incipience up to a fault clearing by a circuit breaker, and usually this takes around three fundamental frequency cycles, which is wider than required for fault location.

In the case of the protective relays the required high speed imposes that the applied calculations have not to be too complex and too time consuming. In contrast, fault-location calculations do not have such limitations. As for example, more accurate phasor calculation for fault location, including rejection of DC components, can be applied. Also, the models of the power line and the fault in fault-location algorithms are usually more advanced than for relaying.

Among different types of relays commonly used for protecting power lines, distance relays [38, 39, 219, 335, 349] are the most related to fault locators. These relays are designated for fast and reliable indication of the general area where a fault occurred. If the fault is recognized as occurring within the pre-defined protective zone, then a trip signal to the corresponding circuit breaker is sent immediately. In consequence, the fault becomes isolated quickly, which minimizes the impact of a fault on a power network.

Distance relays have multiple protection zones to provide back capability. The relay that detects the fault in the 1st-zone is designed to trip first. Generally, a pair of distance relays is used to protect a two-terminal line. Usually, they can communicate with each other, forming a pilot relaying. As a result of exchanging information between the distance relays from the line terminals, they both could trip at the 1st-zone setting.

Operation of a distance relay may be significantly influenced by the combined effect of load and fault resistance, which is known as the *reactance effect* [38, 39, 62, 330]. The distance relay may misoperate for a forward external fault, or may not operate for an internal fault if the value of the fault resistance is too large. The value of the fault resistance may be particularly large for ground faults, which are the most frequent faults on overhead lines.

Influence of fault resistance on measurement performed by distance relays is explained with Figs 1.1 and 1.2. The explanation is performed in relation to the single-phase case, which can be easy extended to the three-phase case, while considering different fault types [38, 39, 62, 330]. Figure 1.1 shows the circuit diagram of the transmission network experiencing a fault (F), involving a resistance (R_F), on a homogeneous single-phase line A–B. There is a line impedance ($\underline{Z}_L$) between bus A and B. For the sake of keeping simplicity of considerations, the lumped-line model with neglected shunt capacitances is taken into account here. Parts of the network behind the local (A) and remote (B) terminals are replaced by the Thevenin equivalents containing electromotive forces (emfs) and equivalent source impedances [11].

The fault loop seen from the bus A can be described with the following formula:

$$\underline{U}_A - d\underline{Z}_L\underline{I}_A - R_F\underline{I}_F = 0 \qquad (1.1)$$

where:
d – distance from the bus A to fault point F, expressed in relative units (p.u.),
$\underline{U}_A$, $\underline{I}_A$ – voltage and current from the measurement point (here considered as measured at the bus A),
$\underline{I}_F$ – total fault current (flowing through the fault-path resistance), which for the assumed line model (neglected shunt capacitances) equals:

$$\underline{I}_F = \underline{I}_A + \underline{I}_B \tag{1.2}$$

The formula (1.1) can be written as follows:

$$\underline{Z}_A = \frac{\underline{U}_A}{\underline{I}_A} = d\underline{Z}_L + R_F \frac{\underline{I}_F}{\underline{I}_A} \tag{1.3}$$

where:
$\underline{Z}_A$ – fault-loop impedance determined as the ratio of the voltage and current, both acquired at the measurement point A.

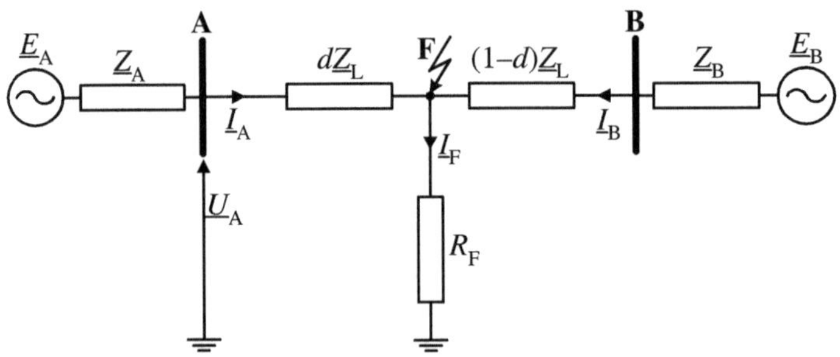

Fig. 1.1 Circuit diagram of transmission network with line A–B affected by fault (F) involving fault resistance R_F

From (1.3) it is seen that the measured impedance $\underline{Z}_A$ is a strict measure of the distance to the fault (d), only if the fault resistance is equal to zero or is very low and can be neglected. Otherwise, the fault resistance R_F is seen in general as some impedance:

$$\underline{R}_F^{\#} = R_F \frac{\underline{I}_A + \underline{I}_B}{\underline{I}_A} \tag{1.4}$$

Depending on the currents at both line ends ($\underline{I}_A$, $\underline{I}_B$), the fault resistance R_F can be seen as (1.4):

- pure resistance (Fig. 1.2a);
- resistance and capacitive reactance (Fig. 1.2b);
- resistance and inductive reactance (Fig. 1.2c).

In the last two cases (Fig. 1.2b and Fig. 1.2c) there is a contribution of the reactance (capacitive or inductive) in the seen resistance, and therefore it is called as the reactance effect [62, 69, 129, 215].

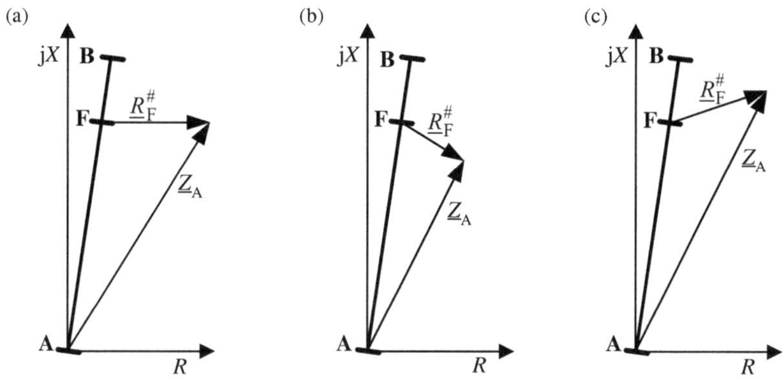

Fig. 1.2 Influence of remote infeed on one-end fault-loop impedance measurement – fault resistance is seen as: (a) pure resistance, (b) resistance and capacitive reactance, and (c) resistance and inductive reactance

1.4 General Division of Fault-location Techniques

In a natural way a fault location can be done by foot patrols or by patrols equipped with different transportation means and binoculars. Such means of faulted-line inspection is considered as time consuming. The primitive method of fault location consisted of visual inspection as is mentioned in [303]. Also, calls from witnesses of damages on the power line, or customer calls, can provide the required knowledge about the fault position. However, such primitive ways do not satisfy the requirements imposed on fault location.

Valuable information on fault location can be obtained also from fault indicators, installed either in substations or on poles (or towers) along the transmission or distribution line [312]. Additional use of a radio link allows use of the information from indicators even during inclement weather. The other, unconventional fault-location system for monitoring transients of induced radiation from power-system arcing faults, using both VLF and VHF reception, has been tested in the experimental installation [322].

In spite of various attempts to different unconventional techniques, *automatic fault location* is still considered as the most widely used. It is based on determining the physical location of a fault by processing the voltage and current waveform values. Automatic fault location can be classified into the following main categories:

1. technique based on fundamental-frequency currents and voltages, mainly on impedance measurement;
2. technique based on traveling-wave phenomenon;
3. technique based on high-frequency components of currents and voltages generated by faults; and
4. knowledge-based approaches.

Making use of the fundamental-frequency voltages and currents at the line terminal (or terminals), together with the line parameters, appears as the simplest way for determining the fault location. It is mainly considered that the calculated impedance of the faulted-line segment is a measure of the distance to fault. The methods belonging to this category are simple and economical for implementing. Depending on the utilized input signals of the fault locator, these methods – as applied to the two-terminal line, can be further classified. Performing such classification one has to take into account an availability of measurements: whether from one or both ends, and also whether complete measurements (voltage and current) or incomplete measurements (voltage or current) from a particular line end are utilized. Similarly, different availability of the fault-locator input signals could be distinguished in application to three-terminal and multi-terminal lines.

Various fault-location methods, with acceptable accuracy for most of the practical applications, have been developed using one-end impedance techniques. A major advantage for these techniques is that communication means are not needed and simple implementation into digital protective relays or digital fault recorders is possible. However, the fault-location algorithms will be more accurate, if more information about the system is available. Therefore, if communication channels are at the disposal, then the two-terminal fault-location methods may be used. Only low-speed communications are necessary for this application. If necessary, the data could be retrieved manually for estimation of the fault location. The two-end technique offers improved fault-location determination, without any assumptions and information regarding the external networks such as impedances of the equivalent sources. In this way, if the two-terminal technique can be applied, the compensation for the reactance effect becomes immaterial.

The technique based on fundamental frequency currents and voltages, which exhibits numerous advantages and being the most popular in real applications is taken into detailed consideration in the next chapters of this book.

Traveling-wave methods consider the voltage and current waves, traveling at the speed of light from the fault towards the line terminals. These methods are considered as very accurate, however, also as complex and costly for application, as requiring high sampling frequency [197, 312]. The basics of traveling-wave methods are briefly presented in Sect. 1.7.

The technique based on high-frequency components of currents and voltages generated by faults, which travel between the fault and the line terminals, is also not widely used. This method is considered expensive and complex, since use of specially tuned filters for measuring high-frequency components is required [21,

198]. The basic principle of the technique based on high-frequency components of currents and voltages generated by faults is described in Sect. 1.8.

Recently, a lot of research efforts have been focused on fault-location techniques both in transmission and distribution networks using knowledge-based (artificial intelligence) methods, such as artificial neural networks, fuzzy-sets theory and expert systems [6, 127, 128, 195, 275]. Selected examples of artificial intelligence application are presented in Chap. 9.

1.5 Input Signals of Fault Locators

1.5.1 Pre-fault, Fault and Post-fault Data

Figure 1.3 presents example waveforms of three-phase voltage recorded under the sample single phase-to-earth fault. Concerning the position of time intervals with respect to this fault incipience and its clearance (achieved as a result of the protective relay operation and switching off the line by the circuit breaker) one can distinguish the following time intervals:

- pre-fault interval: lasting from the beginning of the registration up to the detected fault-incipience instant (Fig. 1.3: Fault incipience);
- fault interval: lasting from the fault incipience instant (tflt_incipience) up to the detected fault-clearance instant (Fig. 1.3: Fault clearance); and
- post-fault interval: lasting from the fault-clearance instant (Fig. 1.3: Fault clearance) up to the end of the recorded event.

According to the kind of time interval, one can distinguish:

- pre-fault quantities – signals recorded within the pre-fault interval;
- fault quantities – signals recorded within the fault interval; and
- post-fault quantities – signals recorded within the post-fault interval.

However, there is no uniform usage of this nomenclature within the open literature of the fault-location issue. Sometimes, instead of using: 'fault interval' and 'fault quantities', the terms: 'post-fault interval', 'post-fault quantities' are utilized. This can be explained in such a case, the prefix 'post-' has a meaning 'after the fault (incipience)' and not 'after the fault (clearance)'.

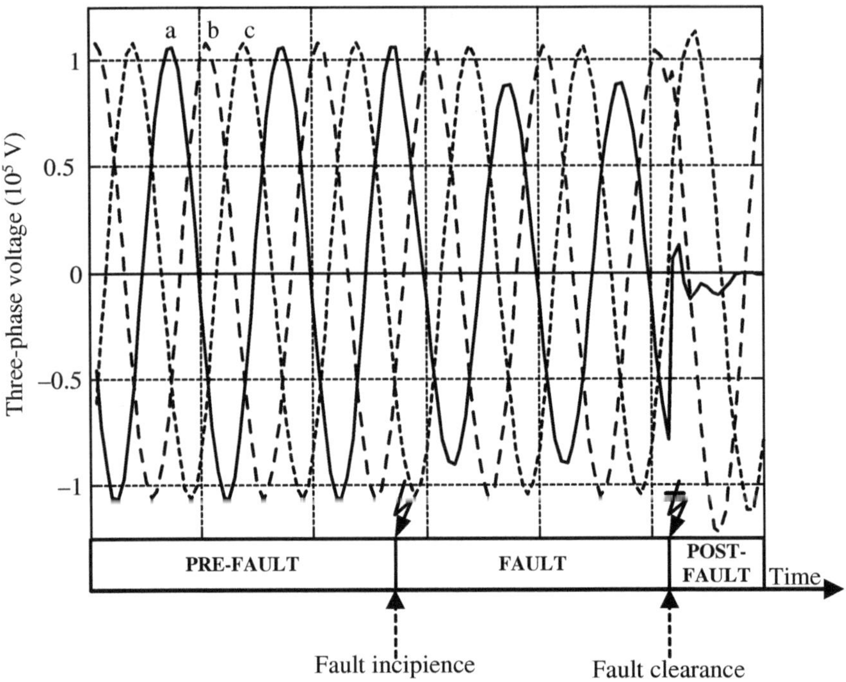

Fig. 1.3 Specification of time intervals according to position with respect to instances of fault incipience and its clearance

Mostly, the fault quantities (voltage and current) are utilized for fault location. However, there are also many fault-location approaches, in which the pre-fault quantities are additionally included as the fault-locator input signals. However, sometimes, usage of the pre-fault measurements is treated as the drawback of the fault-location method. This is so, since in some cases the pre-fault quantities could be not recorded or they do not exist, as for example in the case of the current during some intervals of the automatic reclosure process. Also, the pre-fault quantities can be not of pure sinusoidal shape, due to the appearance of the fault symptoms just before its occurrence. Also, in some hardware solutions, measurement of pre-fault (load) currents is accomplished with lower accuracy than for much higher fault currents. Therefore, if it is possible, usually the usage of pre-fault measurements is avoided.

Rather rare usage of the post-fault quantities for the fault-location purpose is observed. One such technique is presented in [304].

1.5.2 Use of Different Input Signals for Fault Location

Varieties of fault-location algorithms have been developed so far. The majority of them are based on an impedance principle, making use of the fundamental-frequency voltages and currents. Depending on the availability of the fault-locator input signals they can be categorized as:

- one-end algorithms [52, 62, 121, 155, 309, 343–345];
- two-end algorithms [24, 38, 39, 80, 123, 135, 164, 225, 234, 259, 270, 271, 341]; and
- multi-end algorithms [4, 319, 320].

One-end impedance-based fault-location algorithms estimate a distance to fault with use of voltages and currents acquired at a particular end of the line. Such a technique [62] is simple and does not require communication means with the remote end. Therefore, it is attractive and is commonly incorporated into the microprocessor-based protective relays.

Two-end algorithms process signals from both terminals of the line and thus a larger amount of information is utilized. Therefore, the performance of the two-end algorithms is generally superior in comparison to the one-end approaches. Different input signals are used for two-end fault locators, as for example: complete currents and voltages from the line terminals [80, 164, 225] or quantities from impedance relays at the line terminals [38, 39, 259, 270]. In [23, 24, 341] only voltages from both line terminals are utilized, which results in ensuring complete immunity to saturation of current transformers (CTs). Taking into account that in real-life situations the CTs saturation can not happen simultaneously at both line terminals it is also possible to use voltages from both ends and additionally currents from that end at which there is no CTs saturation [123, 271], which also ensures immunity to CTs saturation. Yet another utilization of two-end measurements has been proposed in [319, 320], where exchanging a minimal amount of information between the line terminals over a protection channel is considered.

Digital measurements at different line terminals can be performed synchronously if the satellite GPS (global positioning system) is available. A synchronized measurement system requires that the measurements taken at different substations include, in addition to magnitude, the phase-angle data with respect to an arbitrary but common reference. Phase information is obtained from knowledge of the absolute time at which the measurements were obtained (time tagging). The time for all measurements must be synchronized with a time reference that must be the same for all local systems. This time reference is obtained from the GPS [41, 65]. The ability of GPS to provide a time reference signal, synchronized at widely separated locations has been widely recognized as having great potential in power-system applications [41, 107, 236, 329].

The satellite GPS [41, 107, 236] is presently considered as the most common synchronization means. This system is owned and operated by the US Department of Defense, but civilian users also have access. There are 24 satellites (in 2007 there were 30 active satellites) that are positioned in such a way that four or more of them are observable at every location on the earth. Each satellite contains a highly accurate clock. Satellites maintain the so-called coordinated universal time with an accuracy of ±0.5 μs.

In order to show how high accuracy is assured let us relate the GPS accuracy to a single cycle (T1) for the fundamental frequency of 50 Hz. The accuracy of ±0.5 μs is equivalent to: $(\pm 1/40000)\times 360° = \pm 0.009°$, which is a very small fraction of the angle corresponding to the fundamental frequency cycle: 360°.

If no GPS signal is received during 8 h, the time drift of the backing up crystal oscillator in the application reported [41] does not exceed 20 ms.

A schematic diagram for two-end synchronized fault-location arrangement is shown in Fig. 1.4.

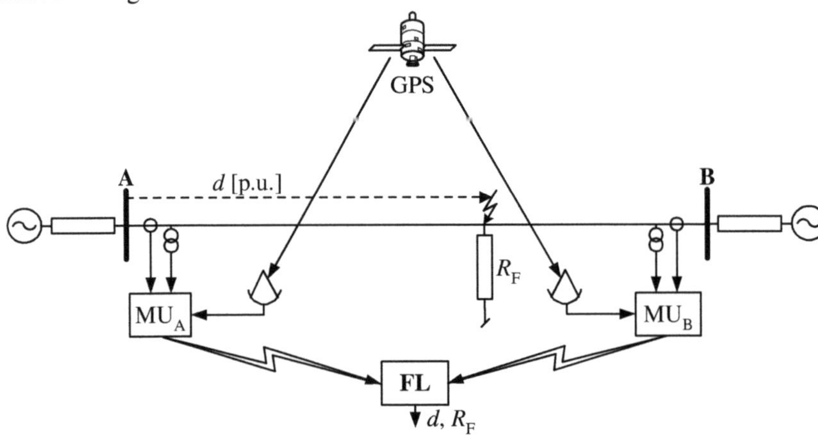

Fig. 1.4 Schematic diagram of two-end synchronized fault location with using GPS

Synchronized two-end measurements allow for simple and accurate fault location. Such algorithms were introduced in [23, 24, 89, 132, 133, 135, 163, 164, 190–192]. In another application [310], a Web-based fault-locating system, which uses GPS synchronized sampling and Internet technology for multi-terminal fault location is described. The system was put to practical use after intensive simulation test cases for various types of line and after trial use for one year in an actual power system. The fault-location system has been applied in many transmission systems on single and double circuit lines and many fault-location results have been obtained. The results show good performance with location errors within several hundred meters for solidly earthed systems.

When the GPS is not available (Fig. 1.5) or also if the signal from GPS is lost, the measurements from the line terminals are performed asynchronously [116, 120, 225, 259, 341]. Such measurements do not have a common time reference. The sampling instants (marked in Fig. 1.6 by small circles) at the A and B line ends do not coincide. This is so since the A/D converters from both line ends are not controlled by the GPS. As a result, there is a certain random shift (ΔT_{B-A}) between the sampling instants of the A/D converters from both ends.

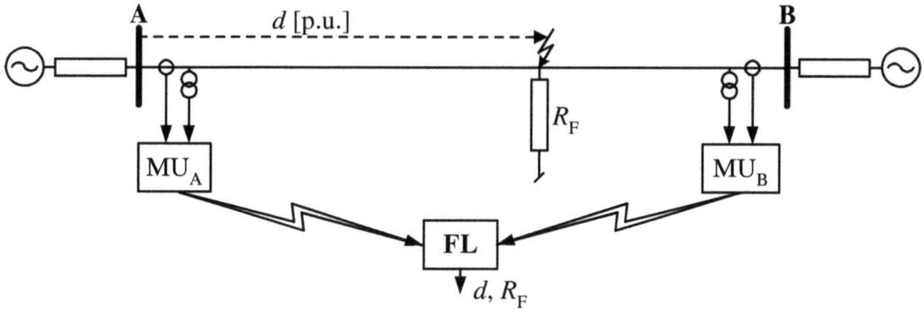

Fig. 1.5 Schematic diagram of two-end unsynchronized fault location

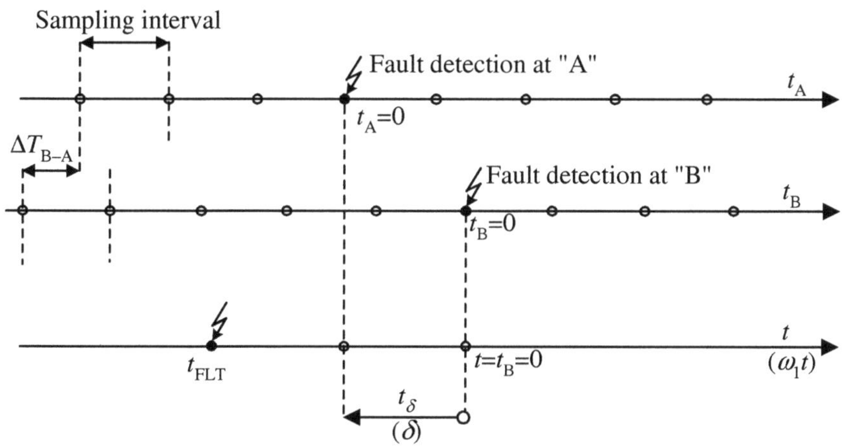

Fig. 1.6 Illustration of the need of phase alignment in the case of using two-end unsynchronized measurements

Moreover, the instant, at which the fault is detected, is usually considered as the time stamp: t_A=0 (at the terminal A) and t_B=0 (at the terminal B), which in the general case also do not coincide. In consequence, the measurements from both line ends do not have a common time reference. In order to assure such common base, one has to take the measurements from the particular terminal as the base (for example from the terminal B), as will be assumed in further considerations), while for the other terminal (the terminal A) to introduce the respective alignment. When formulating the fault-location algorithm in terms of phasors of the measured quantities, such alignment is done by multiplying all the unsynchronized (the superscript: unsynchr.) phasors from the terminal A by the synchronization operator: exp(jδ). As for example for the positive-sequence voltage we apply:

$$\underline{U}_{A1}^{\text{synchr.}} = \underline{U}_{A1}^{\text{unsynchr.}} e^{j\delta} \tag{1.5}$$

where: δ – unknown synchronization angle.

In general, this angle can be:

- eliminated by mathematical manipulations, for example as in [341];
- calculated from the pre-fault quantities, for example as in [4]; and
- calculated with processing the fault quantities, for example as in [116, 120, 225].

1.6 Fault-location Errors

1.6.1 Accuracy of Fault Location

In [106], as well as in other numerous references, the following definition for the fault-location error is given:

> Percentage error in fault-location estimate based on the total line length: (error) = (instrument reading – exact distance to the fault) divided by (total line length).

This definition can be written down as the following formula:

$$\text{error}'(\%) = \frac{d - d_{\text{exact}}}{\ell} 100\% \tag{1.6}$$

where: d, d_{exact} – estimated and exact distance to the fault (in km or in relative units: p.u.), ℓ – total line length (in km, or if relative units are used: $\ell = 1\,\text{p.u.}$).

When performing the statistical evaluation of the accuracy for the particular fault-location method, different measures for the fault-location error are determined. For example, the maximum, average, standard deviation values are determined for the evaluation purposes. It is characteristic that the absolute value is usually taken for the nominator from the definition formula (1.6), and thus obtaining [132]:

$$\text{error}(\%) = \frac{|d - d_{\text{exact}}|}{\ell} 100\% \tag{1.7}$$

Note that usage of (1.7) assures that, when for example the average error is determined for a given population of the evaluation tests, the errors having identical magnitude but different signs do not compensate each other.

Performing the fault-location accuracy evaluation, different factors affecting the accuracy are taken into account. The main factors commonly considered as changeable during the evaluation analysis are:

- fault position (location);
- fault type;

- fault resistance including presence of an arc;
- level of pre-fault power flow and its direction,
- strength of equivalent sources behind the line terminals;
- line imbalance due to lack of transposition;
- inaccuracy in providing impedance data for the overhead line (or cable);
- inaccuracy in providing impedance data for the vicinity of the considered overhead line (or underground cable), as for example the possible mismatch with respect to the source impedances (if they are involved in the evaluated fault-location algorithm) is considered;
- presence and status of series and shunt devices in the line, as for example installations of the banks of series compensating capacitors equipped with metal oxide varistors (MOVs);
- fault-inception angle;
- identification of a fault, in terms of the correctness and accuracy of fault-incipience detection, fault-clarification detection, fault-type classification;
- transient and steady errors of instrument voltage and current transformers, including the possibility of CT saturation;
- frequency response of voltage measurement chains; and
- accuracy of A/D conversion, etc.

1.6.2 Factors Influencing Fault-location Accuracy

Different factors affect the accuracy of fault-location methods. In general, without specifying the fault-location method, the main factors can be listed as follows:

1. Inaccurate compensation for the reactance effect in the case of fault-location algorithms using one-end measurements. This is so if the vicinity of the line is inaccurately represented in the algorithm, i.e., in the case of providing impedances of the equivalent sources behind the line terminals, which do not match the actual strength of the sources.
2. Inaccurate fault-type (faulted phases) identification for fault-locating algorithms based on considering the natural fault loops (phase-to-earth or phase-to-phase loops), similarly as applied in distance relays.
3. Inaccurate line parameters, which do not match the actual parameters. Note that, even if the geometry of line conductors is accurately taken for calculating the line impedances, the total line length could be known with some error.
4. Uncertainty about the line parameters, particularly for the zero-sequence impedance. It is often difficult to obtain the accurate zero-sequence impedance for the line. This is so since this impedance is affected by a soil resistivity, which may be variable under the whole line route, and it is also dependent on weather conditions.

5. Inaccurate compensation for the mutual effects on the zero-sequence components. This takes place if the current required for compensating for the mutual coupling is for some reasons unavailable.
6. Insufficient accuracy of the line model, i.e., if untransposed lines are represented as being transposed, and line shunt capacitance is not considered.
7. Presence of shunt reactors and capacitors or series capacitor compensating devices.
8. Load-flow unbalance.
9. Errors of current and voltage instrument transformers and unfaithful reproduction of the primary signals due to their limited bandwidth.
10. Insufficient sampling frequency and bit resolution of A/D system.

To improve the fault-location estimation, it is important to eliminate, or at least to reduce errors possible for the considered method. Note that a particular factor affecting fault-location accuracy has to be considered strictly in relation to the analyzed method. If this factor appears important, then the means of its elimination or minimization can be considered, as for example as in the cases discussed below.

Various fault-location methods, with acceptable accuracy for most of the practical applications, have been developed using one-end impedance techniques (Chap. 6). A major advantage for these techniques is that communication means are not needed and simple implementation into digital protective relays or digital fault recorders is possible. However, the fault-location algorithms will be more accurate, if more information about the system is available. Therefore, if communication channels are available, then the two-terminal fault-location methods (Chap. 7) may be used. The two-end technique offers improved fault-location determination, without any assumptions and information regarding the external networks such as impedances of the equivalent sources. In this way, if the two-terminal technique can be applied, the compensation for the reactance effect becomes immaterial.

In turn, considering the influence of saturation of CTs on fault-location accuracy, one can arrive at the conclusion that complete elimination of this source of errors can be achieved by using such a set of measurements that the currents from the saturated CTs are not used for fault-distance estimation [123, 271]. Otherwise, the sophisticated compensation algorithms aimed at faithful reproduction of the CT primary current have to be applied.

Fault location in distribution networks creates new problems comparing with the same task in transmission lines. In transmission networks each transmission line may be equipped with its own fault locator (FL). In such a case, the FL algorithm is a numerical procedure, which converts voltage and current, given in a digital form, into a single number being a distance to a fault. In contrast, in distribution networks, FLs are usually assumed to be of a centralized type, i.e., they measure the quantities common for the whole substation (busbar voltages and transformer currents) that makes accurate fault location more difficult [265, 278].

Methods have also been proposed in the past for estimating fault locations on radial transmission lines [301]. These methods, when used for distribution lines are prone to errors because of non-homogeneity of lines, presence of 'laterals' and load taps. A method using fundamental-frequency voltages and currents has been proposed for rural distribution feeders [79]. It does not consider the dynamic nature of the loads. Performance of the technique in situations where cables are used could also be an issue.

1.7 Traveling-wave Method

1.7.1 Introduction

Nowadays, increased competition due to the deregulation of the power industry and the desire to provide better services to the customers having more and more sensitive loads, make the development of faster and reliable fault-location methods for transmission and distribution systems an important issue.

The development of fault-location techniques using digital data is important to speed the restoration of service and to pinpoint the trouble areas, following a fault. Methods of locating faults on power networks can be classified into two fundamental categories: techniques based on power-frequency components, and the other utilizing the higher-frequency components of the transient fault signals. The latter is referred to as the traveling-wave or ultra-high-speed fault-location method, due to its use of traveling-wave theory and shorter sampling windows [198].

Traveling-wave theory has long been studied for the purpose of fault detection and location in transmission lines. The essential idea behind these methods is based on the correlation between the forward and backward traveling waves along the line. The principle of the fault-location techniques is based on the successive identification of the fault, initiated by traveling high-frequency voltage/current signal present where the locator is installed. In particular, the first and few subsequent signals are used to identify the fault position. The propagation time of the high-frequency components is also used to determine the fault position. The technique has proved to be immune to power-frequency phenomena such as power swings and current transformer (CT) saturation, and insensitive to fault type, fault resistance, fault-inception angle, and source parameters of the system [296].

Traveling-wave methods of fault location for both underground cables and overhead lines have been reported since 1931 [71, 181]. During the 1940s the methods were widely adopted for use on de-energized underground cables. For many years, utilities have recognized traveling-wave methods as a way to overcome the inaccuracies and limitations of traditional reactance methods for fault location [258]. Alternative traveling-wave fault-location methods were extensively reported during the 1950s [181, 303]. Traveling-wave systems provided better accuracy than anything else previously available, but were gradually abandoned due to reliability and maintenance problems resulting in a loss of both interest and confidence in the techniques.

Recently, traveling-wave methods have re-emerged as an alternative for fault location [13, 21, 63, 104, 125, 157, 197, 199, 296]. This is mostly due to utilities' desire for faster, more accurate fault location, coupled with improvements in data acquisition: GPS time synchronization and communication systems. Generally, the latest traveling-wave fault-location systems for overhead lines can be divided into three modes [312]:

- single-ended mode relying on the fault to produce traveling-wave transients;
- double-ended mode relying on the fault to produce traveling-wave transients; and
- single-ended mode employing traveling-wave transients produced by circuit-breaker operations.

Among the limitations of the traveling-wave methods, the requirement of high sampling rate is frequently stated. Other stated problems include the uncertainty in the choice of sampling window and problems of distinguishing between traveling waves reflected from the fault and from the remote end of the line.

In distribution systems, fault-location methods based on the information provided by a digital fault recorder (DFR) are not so common. Several new techniques have been proposed in recent years to address the problem [3, 79, 127, 128]. Nevertheless, the case of a feeder having multiple laterals, with measurements normally available only at the substation where multiple possibilities of fault location exist for a given recording has not been fully resolved. The essential problem of such a case is that a number of possible locations, the same electrical distance from the substation, can be found for a recorded signal at the substation end. Some attempts to address the problem of multiple solutions have been made using knowledge-based approaches, or integrating the information available from a substation DFR with the known feeder configuration as well as the protective coordination scheme used for the feeder. It is possible, however, that the information needed by these schemes, may not be accurate or even not available at all at the substation [198]. Recent developments in transducer technology enabled high sampling rate recording of transient signals during faults. The availability of such broad bandwidth sampling capability facilitates better and efficient use of traveling-wave-based methods for fault analysis.

Another significant development in parallel with the advances in the transducer technology is the newly introduced signal-processing tools such as the discrete wavelet transform (DWT), which allows the analysis of sampled signals with localized transients [197–199, 229]. Wavelet transforms have been applied to power-system problems in the area of transient analysis for automatic feature detection, incipient failure detection and power-quality assessment. The attractive feature of wavelet transform in analyzing the traveling waves due to fault is the automatic adjustment of window width of the wavelets depending upon the duration of the transient under study. This is accomplished by time dilation of the chosen mother wavelet. As a result, accurate information on the arrival time of signals traveling at different speeds along the faulted line can be captured. This information is then used to calculate the distance to the fault point along the monitored line.

Fault-location methods using traveling waves are independent of the network configuration and devices installed in the network. These techniques are very accurate, but require high sampling rate and their implementation is more costly than implementation of impedance-based techniques.

1.7.2 Traveling-wave Fault-location Theory

When a fault occurs along a transmission line, the voltage and current transients will travel towards the line terminals. These transients will continue to bounce back and forth between the fault point and the two terminals for the faulted line until the post-fault steady state is reached. Evolution of the terminal bus transients can be constructed using the well-known lattice diagram method [19].

Considering a single-phase lossless transmission line of length ℓ, connected between buses A and B, with characteristic impedance Z_c and traveling-wave velocity of v. If a fault occurs at a distance x from bus A, this will appear as an abrupt injection at the fault point. This injection will travel like a surge along the line in both directions between the fault point and two terminals until the post-fault steady state is reached. A lattice diagram illustrating the reflection and refraction of traveling waves initiated by the fault transients is shown in Fig. 1.7.

The voltage and current at any point x obey the partial differential equations:

$$\begin{aligned} \frac{\partial e}{\partial x} &= -L' \frac{\partial i}{\partial t} \\ \frac{\partial i}{\partial x} &= -C' \frac{\partial e}{\partial t} \end{aligned} \tag{1.8}$$

where: L' and C' are the inductance and capacitance of the line per unit length.

Resistance is assumed as to be negligible. The solutions of these equations are:

$$\begin{aligned} e(x,t) &= e_f(x - vt) + e_r(x + vt) \\ i(x,t) &= \frac{1}{Z_c} e_f(x - vt) - \frac{1}{Z_c} e_r(x + vt) \end{aligned} \tag{1.9}$$

where: $Z_c = \sqrt{\frac{L'}{C'}}$ is the characteristic impedance of the transmission line and $v = \sqrt{\frac{1}{L'C'}}$ is the velocity of propagation.

Forward (e_f for voltage and similar for current and i_f) and reverse (e_r and i_r) waves, as shown in Fig. 1.7, leave the disturbed area x traveling in different directions at v, which is a little less than the speed of light, toward the transmission-line ends. Transmission-line ends represent a discontinuity or impedance change where some of the wave's energy will reflect back to the disturbance. The remaining energy will travel to other power-system elements or transmission lines. The lattice diagram in Fig. 1.7 illustrates the multiple waves (represented by subscripts 2 and 3) generated at line ends. Wave amplitudes are represented by reflection coefficients k_A and k_B, which are determined by characteristic impedance ratios at the discontinuities. τ_A and τ_B represent the travel time from the fault to the discontinuity.

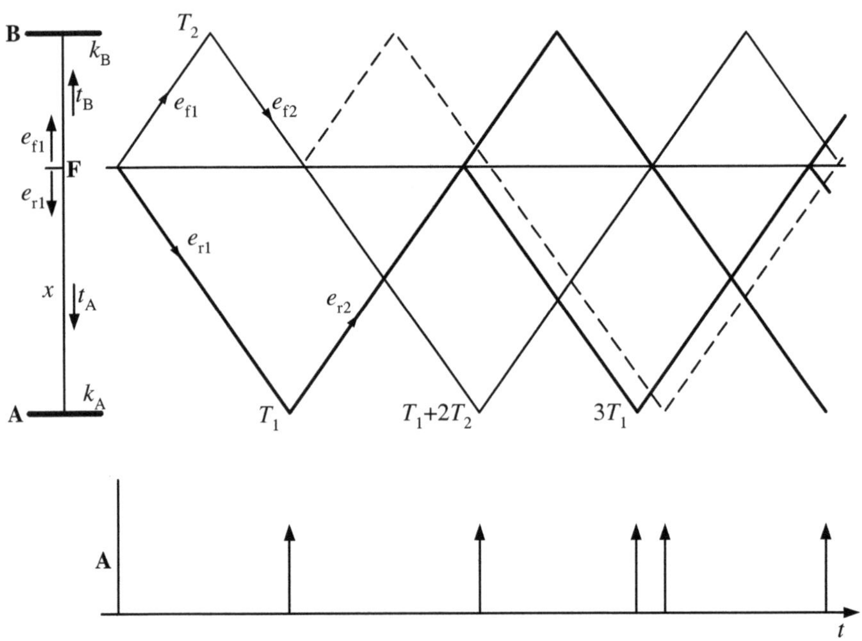

Fig. 1.7 Traveling voltage and current waves: lattice diagram for a fault at distance x from A

With the GPS technology, τ_A and τ_B can be determined very precisely. By knowing the length (ℓ) of the line and the time of arrival difference ($\tau_A - \tau_B$), one can calculate the distance (x) to the fault from substation A by:

$$x = \frac{\ell - c(\tau_A - \tau_B)}{2} \tag{1.10}$$

where: c – the wave propagation of 299.79 m/s.

Construction of the lattice diagram becomes computationally difficult if the attenuation and distortion of the signals are taken into account as they travel along the line. On the other hand, time-frequency resolution of the transient signals can be used to determine the travel times of these transients between the fault point and the line terminals. In three-phase transmission lines, if losses are taken into account, there are three modes of propagation, therefore for the analysis of the traveling-wave effect, phase values must be converted into modal values.

1.7.3 Data and Equipment Required

Traveling-wave methods typically rely on accurate timing information. This is required in order to provide accurate fault-location results.

Data Required

The traveling-wave method relies on calculation of time for the line disturbance to reach the end of the line. Essentially, when a disturbance occurs, very accurate time tagging must be done. Since the wave moves at the speed of light, by comparing the wave arrival time difference at each end, one can determine the distance to the source of the disturbance. This requires extremely accurate timing for calculation of fault location. Either voltage or current wave data can be used. The voltage portion of the traveling waveform tends to be reduced as the result of buses with lower impedance. On the other hand, the current waveform tends to double as the result of a constant current source.

The first data requirement is a standard time reference at both receiving terminals. Then, some method of distinguishing which waveform must be used is required. The appropriate current waveform (or time tag in the voltage method) must be known to accurately calculate a fault location. From this point, a fault location can be calculated based on the precise wave-arrival times on each end of the line. The main piece of data that must be known is the very precise time that the traveling wave reaches each end of the line. At this point, it is merely a matter of calculation.

Equipment Required

The following equipment is necessary to locate faults using the traveling-wave method:

- A very accurate time-stamping device (GPS) on both ends of the line.

- An appropriate sensor to detect the voltage or current, depending on the parameter used. In the case of the current wave, normal relaying accuracy CTs are used. The secondary CT circuits then have the current pulses detected by clamp-on inductive sensors. In the case of detecting voltage pulses, capacitive potential transformers are utilized.
- A communications circuit is required to transmit the time-stamped data back to a central location.
- A computer capable of retrieving the remote data, distinguishing the appropriate waveform for the fault-location calculation, and providing the appropriate calculations to the fault.

Accuracy Limitations

1. Assumptions made in determining fault location

   The traveling waveform travels at the speed of light.
   - Discontinuities in the electrical system produce wave reflections. Each discontinuity can be used as calibration for the timing of the wave arrival at the receiving end of the line. Since the velocity of propagation is constant, the distance can be calculated quite accurately.
   - Time stamping must be very precise to make the system work.
   - The two-terminal method allows timing from the initiation of the short circuit, hence, reflected waves are not involved.

2. Accuracy

   - GPS-based traveling-wave fault-locating systems where time-stamp information is provided from both ends of the faulted transmission lines has proven to be very accurate. Operating results have shown an accuracy on the performance of ±300 m, even for long lines.
   - Wave-detection error due to interpretation of the transient is a major form of error. This error results from many transients and/or reflected transients appearing the same. This is especially true of lightning strikes. Lightning storms with multiple strokes can cause major confusion in terms of which transient was associated with which fault.
   - Stronger buses tend to dampen voltage transients. The result is lower fault-location accuracy.
   - The GPS system is the time-measuring standard. Hence, any errors in this system are reflected into the ability to accurately locate faults. The Department of Defense of USA intentionally builds a small amount of uncertainty into the system.
   - Current and voltage transformers provide reasonable reproduction of transients.
   - When utilizing the one-terminal form of the traveling-wave method, analysis of the waveforms has to be more sophisticated. Potentially, signature analysis may be required because the transients are more complex.

1.7.4 Classifications of Traveling-wave Methods

In the 1950s, overhead line-fault locators were classified as Types A, B, C, and D according to their mode of operation – each type being associated with one particular implementation or design, and all being based on voltage rather than current waves [106]. The latest traveling-wave systems, which generally use current waves, operate simultaneously in Type-A and Type-D mode and also include a new Type-E mode. Type-A and Type-D modes rely on the fault to produce one or more traveling-wave transients from which the fault location can be determined, whereas Type E uses transients produced by circuit-breaker operations.

Type A is a single-ended mode in which the flashover at the fault point launches two waves that travel in opposite directions away from the fault. The effective impedances at the line terminals are assumed to be lower than the line surge impedance so that significant reflections are produced that then travel back along the faulty line to the fault point. If the fault arc still exists, and also presents an effective resistance lower than the surge impedance of the line, then any waves arriving at the fault will be almost totally reflected back to the line terminals. Thus, if the transients appearing at either end of the line are captured, they can be analyzed to determine the fault position.

The assumptions of near-total reflection at the fault point are not always satisfied, and in some instances the fault arc may extinguish prematurely, making analysis of the transients more difficult. By adopting Type-D operation, in which the difference in the times of first arrival of the two fault-generated waves at both line terminals are determined, we avoid problems associated with the behavior of the fault arc. Also, reflections from other discontinuities between the two line terminals, such as branches, tapped loads, cable sections, and even intermediate substations become unimportant.

Type D requires both an accurate method of time synchronization and an easy means of bringing the measurements from the two terminals to a common point so that the fault position can be determined. Using small, non-directional aerials, GPS now provides time synchronization accuracies of better than 1 μs over the entire surface of the Earth, 24 h per day, which equates to a fault-location accuracy of $\pm$152.4 m, irrespective of line length. While calculating a Type-D result requires communication with the data acquisition units in two or more substations, it does not have to be 'on-line' and can, therefore, use any communications channel.

Type E is a single-ended mode that uses the transients created when a line is re-energized by closing a circuit breaker. The method is equivalent to the impulse current method of fault location widely used on underground cables. Type E can be used to locate permanent faults, including open-circuit conductors, where little or no fault current is flowing, and additionally can be used to measure the electrical length of healthy lines.

1.8 High-frequency Methods

1.8.1 Introduction

Contemporary methods for fault location on overhead lines and underground cables can be classified as two fundamental types: methods based on the measurement of post-fault line impedance and methods based on the measurement of the fault-generated traveling-wave component. There has been considerable research effort into the development of impedance-based methods for fault location. However, like any other power frequency-based measurement methods, they suffer from limitations due to fault-path resistance, line loading and source parameters, etc. As a result, the accuracy attained in fault location is rather limited.

Theoretically, the pattern of the fault-generated traveling wave contains information about the fault location that can be used to accurately locate the fault. However, present traveling-wave-based fault-location methods exhibit shortcomings; a fault will not generate many traveling-wave components when it occurs at a voltage inception close to zero degree; for a close-up fault, the time difference between the arrival of an incident wave and the arrival of its reflection from the bus bar will be so short that the waves are unlikely to be detected separately. This could make the interpretation of the information available in the first few milliseconds after the arrival of the first wave front virtually impossible. Where the measurement involves voltage signals, then the bandwidth limitation of the capacitive voltage transformer (CVT) can be a serious impediment [21].

An approach to transmission-line protection has been developed based on the detection of fault-generated high-frequency transient signals, and the research shows that the technique can be applied to achieve very high accuracy in fault location [20]. The method has been shown to be immune to power frequency phenomena such as power swings and CT saturation. Similar techniques have been developed [199], where traveling-wave phenomena were used for fault detection on 345-kV transmission lines. The use of wavelet transforms to extract the high-frequency transients was introduced. Two basic systems were described; one using recordings from both ends of the line and synchronized using global positioning satellite receivers, and the second using recordings made at one end of the line.

A technique developed and described in [21] concerns determining the fault location on distribution overhead lines and cable systems. It relies on detecting fault-generated high-frequency signals and avoids the problem of identifying multiple reflections from bus-bars and the fault point, which restrict the use of other traveling-wave techniques. The low fault-inception angle is no longer a problem, since the high-frequency signals associated with the fault arc do not vary with the point on the wave at which the fault occurs. In the scheme described [21], a high-speed sampling system is used to capture the fault-generated high-frequency transients. To achieve the high accuracy inherent in the system, higher sampling rates are used.

A single-ended fault-location technique for radial distribution feeders based on the information provided by the measurements and the feeder configuration has been described in [198]. The proposed method first identifies the fault path, based on the traveling-wave information provided by the high-frequency components of the recorded fault-transient signals. It then calculates the exact fault location along the identified path based on the power-frequency signals. The multi-phase transient signals are first decomposed into modal components. Then, the modal signals are decomposed into their wavelet components and the corresponding wavelet coefficients are obtained. These wavelet coefficients are used to extract the relevant signal features, which are subsequently used to identify the branch or path where the fault is located. Finally, the fault distance from the main substation is calculated using the information provided by the power-frequency signal.

The technique based on high-frequency components of currents and voltages generated by faults, which travel between the fault and the line terminals, is still also not widely used. This method is considered as expensive and complex, since use of specially tuned filters for measuring high-frequency components is required.

1.8.2 Basic Principle and Fault-locator Design

A sudden change in system voltage on a power line or cable will generate a wide-band signal, which covers the entire frequency range [21]. The initial values of these waves are dependent, among other factors, on the fault position on the line, the fault-path resistance and, the most important of all, the instance of fault occurrence. These different frequency components propagate away from the fault point in both directions. In time, these signals reach other discontinuities on the line/cable and are reflected back towards the fault point. In the frequency domain, the magnitude of the individual signal components decreases as the frequency rises and traveling speed increases.

The principle of the fault-location method is based on the successive identification arrival of traveling high-frequency voltage signals at the bus-bar where the locator is installed. In particular, the timing of the first received and subsequent signals referenced to that first signal are used to identify the fault position.

Fault-locator Unit

The three-phase voltages are monitored using high-voltage coupling capacitors. The signals from these are then digitized for processing. Modal mixing transforms are used to extract the aerial-mode and ground-mode signals. This process provides inherent filtering. Digital band-pass filters then extract the high-frequency components used for the fault location. In practice, the accuracy of the technique is mostly affected by noise interference, which arises mainly from two sources: noise on the power line/cable, such as corona and partial discharge; and background noise coupled into the equipment. The prevalent anti-noise techniques can be employed to reduce the effects of background noise. Moreover, the interference noise can be effectively reduced by controlling the gain of the input signals to maximize the accuracy of the fault location and reduce the effect of noise interference.

2 Network Configurations and Models

Abstract The objective of this chapter is to review the basic configurations of power networks. In particular, transmission and distribution lines are considered. The first ones are usually built as overhead lines. Different configurations of such lines are described in detail: single- and double-circuit lines, multi-terminal and tapped lines, traditional and series-compensated transmission lines. Lumped- and distributed-parameter models of the lines are presented. Basic features related to distribution networks are presented.

2.1 Introduction

Specific features of lines used in contemporary power systems depend on many factors, such as their construction (overhead or cable, materials used and line geometry), voltage level (generally: high voltage or medium voltage), network configurations (single-circuit, double-circuit, two- or multi-terminal lines, tapped lines), reactance compensation (uncompensated or series-compensated) and others. All these details should be taken into considerations when the network mathematical models for steady-state and transients are formulated. These models, together with the respective measurement procedures, which are introduced in Chap. 4, constitute the basis for the fault-location algorithms. This chapter gives an overview of different types of lines and their models from the fault-location point of view.

2.2 Overhead Lines

Fault location in transmission networks is based on considering the flow of a fault current. Depending on the availability of measurements for the fault locator, the flow of a fault current within the faulted line itself or also in its vicinity is considered. A particular fault-location method has to be considered in strict relation to the configuration of the power network and its model.

2.2.1 Single-circuit Overhead Lines

Single-circuit three-phase overhead lines are the simplest means for transmitting power energy from the generation center to the consumption region. A schematic diagram of a power network with a single-circuit overhead line is presented in Fig. 2.1a [11]. The line is marked with a graphic symbol typical of an impedance description. Moreover, the description $\underline{Z}_L$, corresponding to a general indication of the line impedance is used. The line ends are denoted here by letters: A and B. The fault occurring on the line is marked with a common graphic symbol for the fault and letter F. The vicinity of the line A–B under consideration is represented by the external network. Assuming linearity of the whole circuit, the external network can be equivalented [11], as shown in Fig. 2.1b. The obtained equivalent of the external network in a general case consists of:

- two equivalent sources behind the line terminals A, B – consisting of the emfs ($\underline{E}_A$, $\underline{E}_B$) and source impedances ($\underline{Z}_A$, $\underline{Z}_B$); and
- extra link ($\underline{Z}_E$) between the line terminals A, B.

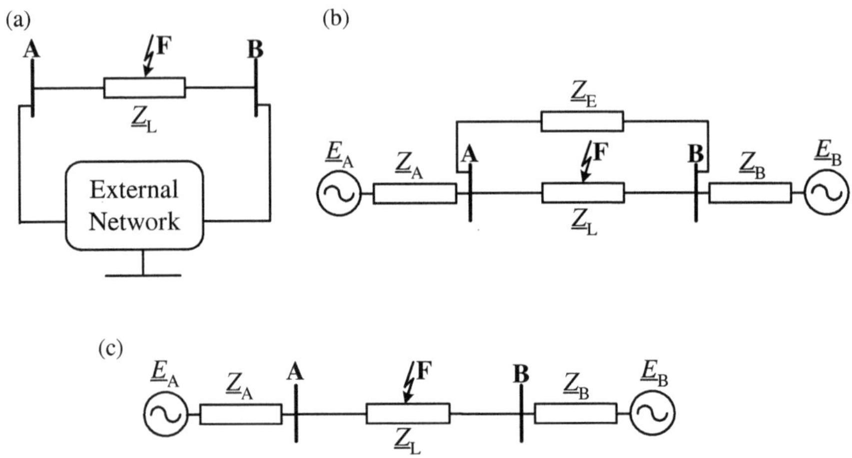

Fig. 2.1 Transmission network with single-circuit overhead line: (a) generic scheme, (b) general equivalent scheme, and (c) simplified equivalent scheme with the line being the only connection between buses A, B

Since the load and generation in a power network as well as the network topology undergo changes, the equivalent network of the line external network also changes and is not fixed. As a result, the source impedances ($\underline{Z}_A$, $\underline{Z}_B$) are considered in the fault-location process to be the uncertain parameters. Therefore, the fault-location algorithms, which do not require that the source impedances be known, are generally more accurate than the algorithms for which this impedance data is used as the input data. The one-end fault-location algorithms require setting the source impedances and due to dynamic changes of the network it is difficult to provide the actual values of these impedances. Fortunately, in many applications it is sufficient to provide the representative values of the source impedances, which are obtained for the most typical conditions of the network operation. Possible mismatch between the provided representative source impedances and the actual parameters in many applications does not cause considerable errors in fault location. This is so especially in the case of strong sources, which is the case when the source impedance is much smaller than the line impedance.

If the line ($\underline{Z}_L$) considered is the only connection between the buses A, B, then the extra link ($\underline{Z}_E$) does not exist, and there are only equivalent sources, as shown in Fig. 2.1c. This is the well-known double-machine network.

2.2.2 Double-circuit Lines

Both fault location and protective relaying for double-circuit lines (also called parallel lines) are dealt with in numerous references [7, 29, 68, 76, 90, 106, 109, 113, 114, 121, 122, 131, 140, 155, 204, 205, 218, 286, 293, 294, 299, 344]. Such lines are basically constructed due to constraints in obtaining new right-of-ways and are very common in power networks. For such lines the two three-phase transmission circuits are arranged on the same tower or follow on adjacent towers the same right-of-way. The circuits may be either of the same or different voltage level. Also, more than two three-phase circuits can be arranged in such a way (multi-circuit lines) [350].

Due to the nearness of both circuits of a double-circuit line, they are mutually magnetically coupled. The magnetic coupling is related to the effect of a current flowing in one circuit, which influences the voltage profile in the other circuit, and *vice versa*. This means that the voltage profile of a given circuit is not entirely dependent on the current flowing in this circuit.

The mutual coupling effect can be expressed in terms of various inter-circuit mutual impedances. Using the symmetrical components approach to the line description, the positive-, negative- and zero-sequence mutual impedances are considered. The positive- and negative-sequence mutual impedances are usually a small fraction of the positive-, negative-sequence self-impedances and therefore are usually neglected in the analysis. In contrast, the zero-sequence mutual impedance ($\underline{Z}_{0m}$) is of relatively high value and thus cannot be ignored in the analysis of single phase-to-ground faults. The mutual coupling of double-circuit lines for the zero-sequence is thus important for the fault location based on considering the natural fault loops [109].

Different configurations of double-circuit lines [11, 109, 350] are met in power networks. Figure 2.2 presents a general configuration of a power network with a double-circuit overhead line terminated on both sides at the separate buses. The line circuits are denoted by $\underline{Z}_{LI}$, $\underline{Z}_{LII}$ and their mutual coupling for the zero-sequence by $\underline{Z}_{0m}$. The vicinity of the line circuits is represented with:

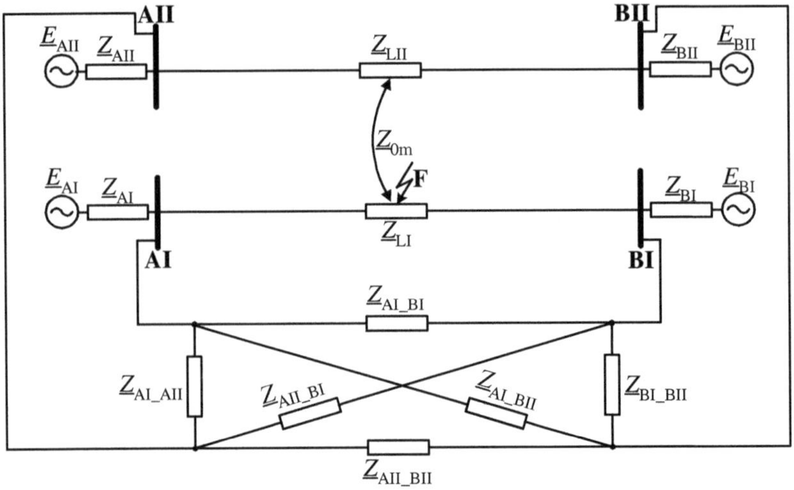

Fig. 2.2 Schematic diagram of power network with double-circuit overhead line terminated at both ends at separate buses

- equivalent source behind the line terminal AI (emf: $\underline{E}_{AI}$, impedance: $\underline{Z}_{AI}$);
- equivalent source behind the line terminal AII (emf: $\underline{E}_{AII}$, impedance: $\underline{Z}_{AII}$);
- equivalent source behind the line terminal BI (emf: $\underline{E}_{BI}$, impedance: $\underline{Z}_{BI}$);
- equivalent source behind the line terminal BII (emf: $\underline{E}_{BII}$, impedance: $\underline{Z}_{BII}$); and
- links between the line terminals AI, AII, BI, BII in the form of a complete tetragonal of impedances: $\underline{Z}_{AI\_AII}$, $\underline{Z}_{AI\_BI}$, $\underline{Z}_{AI\_BII}$, $\underline{Z}_{AII\_BI}$, $\underline{Z}_{AII\_BII}$, $\underline{Z}_{BI\_BII}$.

Figure 2.3 presents the classical case of the network with two line circuits connected at both ends to the common buses. This scheme is obtained from the general scheme of Fig. 2.2, considering the following:

- equivalent source ($\underline{E}_A$, $\underline{Z}_A$) obtained as the resultant for parallel connection of the sources: ($\underline{E}_{AI}$, $\underline{Z}_{AI}$) and ($\underline{E}_{AII}$, $\underline{Z}_{AII}$);
- equivalent source ($\underline{E}_A$, $\underline{Z}_B$) obtained as the resultant for parallel connection of the sources: ($\underline{E}_{BI}$, $\underline{Z}_{BI}$) and ($\underline{E}_{BII}$, $\underline{Z}_{BII}$); and
- extra link ($\underline{Z}_E$) obtained as the resultant for parallel connection of the following impedances: $\underline{Z}_{AI\_BI}$, $\underline{Z}_{AI\_BII}$, $\underline{Z}_{AII\_BI}$, $\underline{Z}_{AII\_BII}$.

The extra link shown in the network of Fig. 2.3 is not always present, especially in high-voltage networks that are not highly interconnected. Operating conditions of a double-circuit line could change due to different reasons, such as load dispatch, forced outage, scheduled maintenance, etc. The mutual coupling of double-circuit lines depends on the mode of operation of the healthy circuit ($\underline{Z}_{LII}$), which is in parallel to the faulted-line circuit ($\underline{Z}_{LI}$) considered. In order to present these modes, the status of circuit breakers and also grounding connectors of the healthy parallel line has to be considered [201].

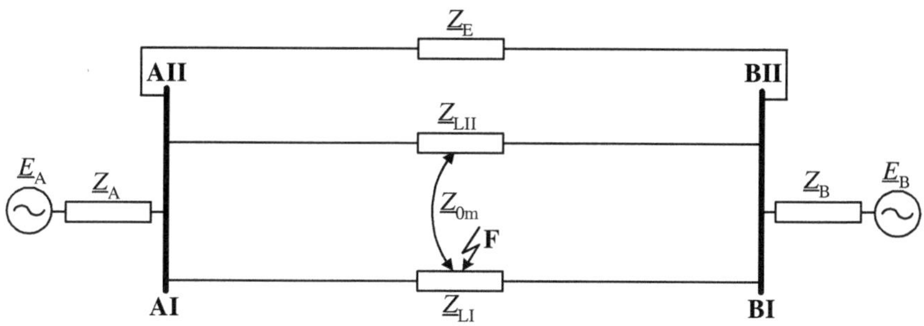

Fig. 2.3 Schematic diagram of power network with double-circuit overhead line terminated at both ends at common buses

Figure 2.4 presents two modes for which the mutual coupling of parallel lines is of interest. In the case of Fig. 2.4a the parallel line is in operation, which is the normal operating mode. The mutual coupling of parallel lines also exists if the parallel line is switched off and grounded at both ends [114] (Fig. 2.4b).

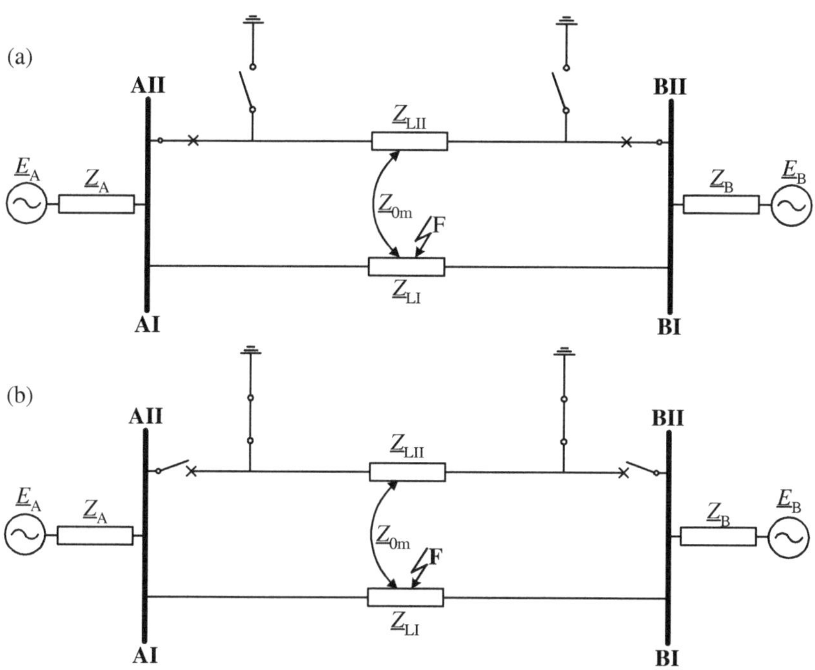

Fig. 2.4 Double-circuit overhead-line modes with mutual coupling of parallel lines: (a) both lines in operation, and (b) parallel line is switched off and grounded at both ends

Figure 2.5 presents three cases for which there is a discontinuity for the current flow in the healthy parallel line, and therefore there is no mutual coupling between the lines.

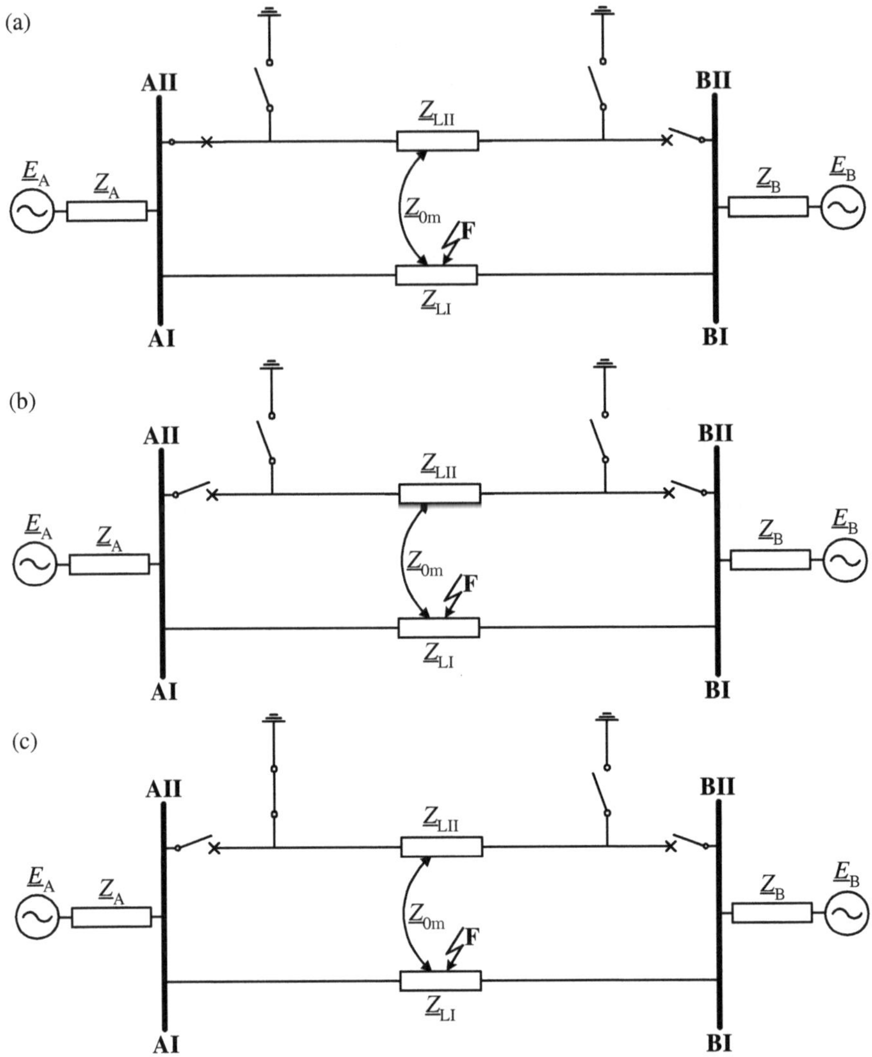

Fig. 2.5 Double-circuit overhead-line modes with no mutual coupling of parallel lines: (a) parallel line is switched off at one end (BII) and not grounded, (b) parallel line is switched off at both ends and not grounded, and (c) parallel line is switched off at both ends and grounded only at one end

In some cases [113, 350], the line circuits may run in parallel only for a part of the route. The circuits for this part are mutually coupled, while for the remaining part of the route, they are hung on different towers and are terminated at distant substations. This is illustrated in Fig. 2.6.

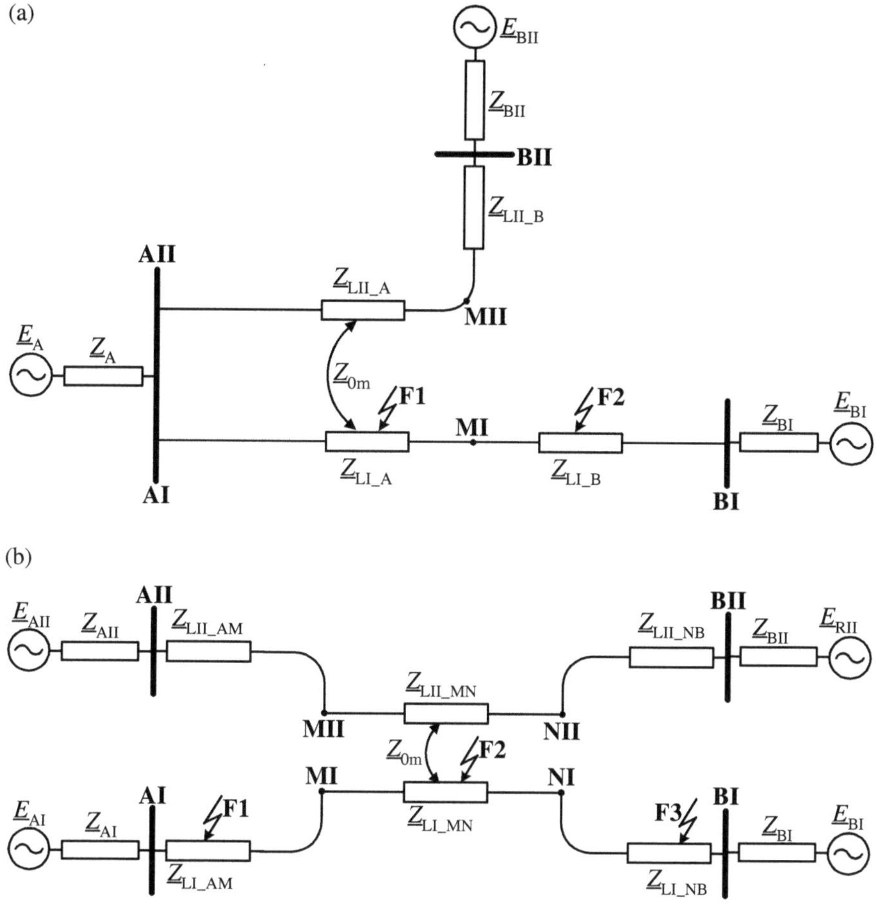

Fig. 2.6 Examples of power networks containing partially parallel line circuits with mutual coupling for the line sections: (a) $\underline{Z}_{LI\_A}$, $\underline{Z}_{LII\_A}$, and (b) $\underline{Z}_{LI\_MN}$, $\underline{Z}_{LII\_MN}$

Figure 2.6 presents two examples of power networks with partially parallel circuits. The need for taking into account the mutual coupling effect depends on the fault position: Fig. 2.6a – faults F1, F2; Fig. 2.6b – faults F1, F2, F3. Considering the fault loop between bus AI and fault point F1 in the network of Fig. 2.6a, the mutual coupling has to be taken into account along the whole distance. By contrast, as regards the fault loop between bus AI and the fault point F2, the mutual coupling has to be considered for the distance between bus AI and point MI, and not for the remaining part (MI–F2).

2.2.3 Multi-terminal and Tapped Lines

It is for economical or environmental-protection reasons that use is made of multi-terminal and tapped lines [350]. Lines having three or more terminals with substantial generation behind each are named multi-terminal lines [350]. Depending on the number of terminals we can distinguish three-terminal lines having three terminals, four-terminal lines having four terminals, and so on.

Tapped lines are those having three or more terminals with substantial power generation behind, at a maximum at two of them [350]. The number of taps per line varies from one to even more than ten. The taps themselves feed only loads, which means that they are terminated by the passive networks, while at the remaining terminals there are active networks (with power generation) [81, 350].

Examples of power-network configurations with single-circuit three-terminal line are shown in Fig. 2.7. In the case of using double-circuit lines, typical configurations are as shown in Fig. 2.8.

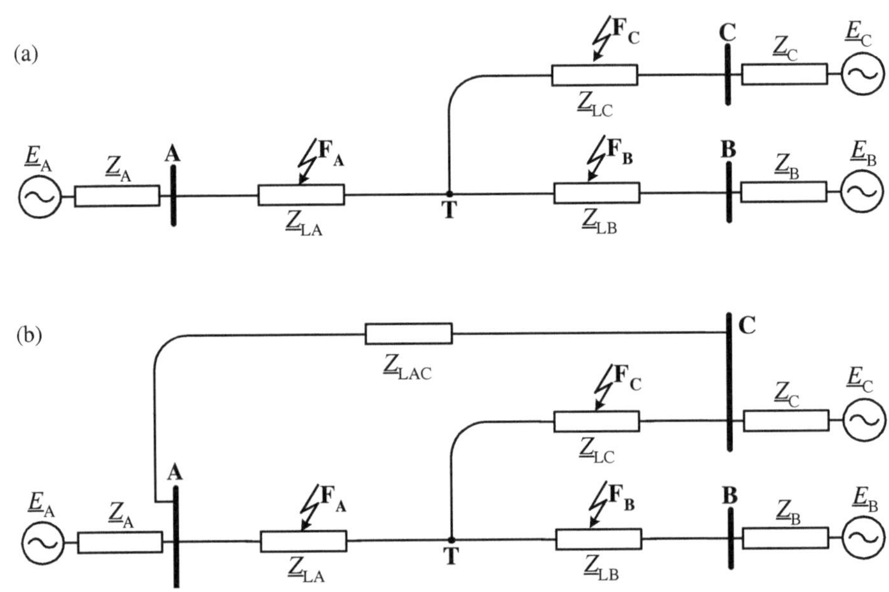

Fig. 2.7 Examples of power-network configurations with single-circuit three-terminal line: (a) basic teed network, and (b) teed network with extra link between two substations

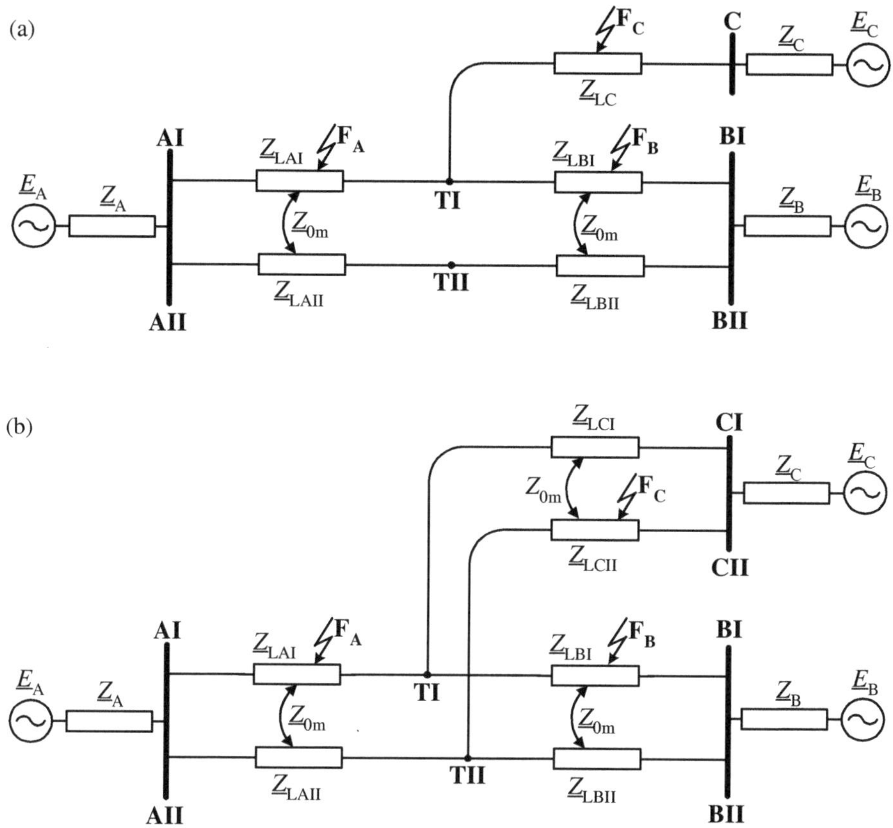

Fig. 2.8 Examples of power-network configurations with parallel three-terminal line: (a) two line sections are of double-circuit type, and (b) all three line sections are double circuits

Figure 2.9 presents typical configurations of power networks with tapped line supplying load in two different ways [350]: via a transformer connected to the tap point through a circuit breaker (Fig. 2.9a) and additionally with overhead line section ($\underline{Z}_{LC}$), Fig. 2.9b.

Fault location on multi-terminal and tapped lines relies on determining the following:

- identifying the line section at which the fault (F_A or F_B or F_C) occurred; and
- determining the distance to fault for the faulted section, usually measured from the respective bus (A, B or C) towards the fault point (F_A or F_B or F_C).

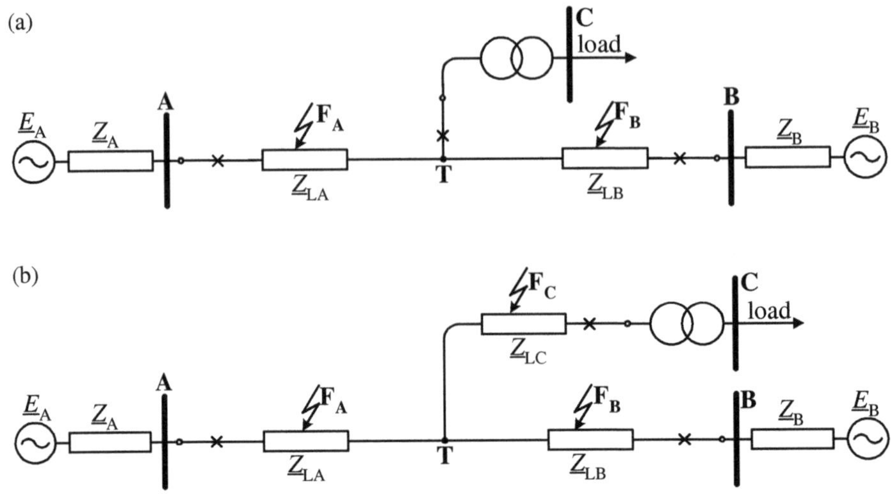

Fig. 2.9 Typical configurations of power networks with tapped line supplying load through: (a) transformer, and (b) overhead line ($\underline{Z}_{LC}$) and transformer

2.2.4 Overhead Line and Cable Composite Networks

In Figs. 2.10 and 2.11, examples of configurations of overhead line and cable composite networks [214, 308] are presented.

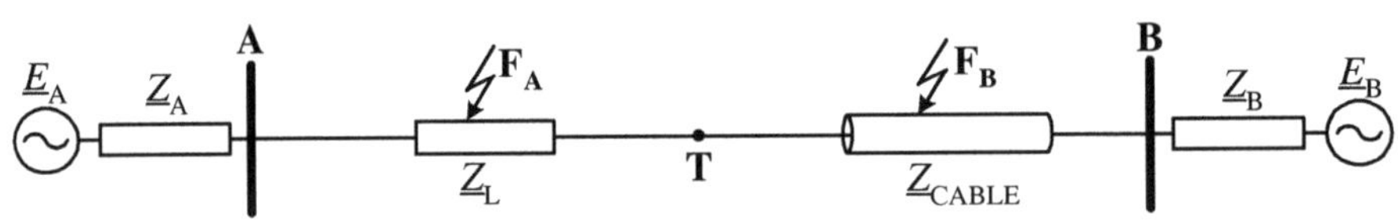

Fig. 2.10 Overhead line in series connection with cable

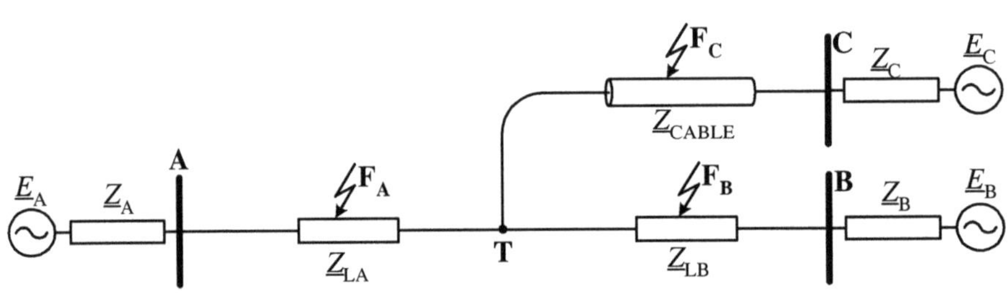

Fig. 2.11 Overhead line tapped with cable

Fault location in such networks is considered to be a difficult task due to large differences in parameters of the line and cable. Moreover, the problem of cable parameters changing, especially changes in its relative permittivity occurring with aging, has to be accounted for [308].

2.3 Models of Overhead Lines

Models of overhead lines are considered in strict relation to a particular application. Among different applications, the line models are considered in relation to the following:

- representing a faulted line in the fault-location algorithm; and
- simulating faults for generating the fault data, which is used in evaluation of the fault-location algorithms under study [56, 57, 203, 207].

In this section, the representation of a faulted line in fault-location algorithms is addressed. Assumption of the line model is a starting point for derivation of the fault-location algorithms.

In turn, the line models adopted in simulation of line faults are widely described in reference manuals (theory books) of the well-known simulation transients programs, such as ATP-EMTP [56] and others, and therefore are not considered here.

Overhead-line parameters are calculated using supporting routines available in the simulation programs. Also, an on-line measurement of transmission-line impedance performed either during normal operation or during faults is used in practice. In general, there are two types of line models:

- lumped-parameter models; and
- distributed-parameter models.

Lumped-parameter models represent a line by lumped elements, whose parameters are calculated at a single frequency, predominantly the fundamental power frequency. Using these models, steady-state calculations for fault location or transient simulations in the neighborhood of the frequency considered can be performed.

As opposed to the lumped-parameter models, the distributed-parameter line models are used for more accurate representation of the line. Two categories of distributed-parameter line models can be distinguished:

- constant-parameter model; and
- frequency-dependent parameter model.

Series parameters: resistance (R), inductance (L), and shunt parameters: capacitance (C) and conductance (G) characterize the line. Usually, line conductance, which accounts for the leakage currents along the insulators and in the air, can be neglected, except at very low frequencies. Shunt capacitance can usually be assumed as frequency independent. In turn, series resistance and inductance can be considered to be frequency dependent. However, this is considered for the simulation [56, 200] and rather not for representing a line in the fault-location algorithm.

2.3.1 Lumped-parameter Models

In the simplest lumped-parameter model of an overhead line, only the series resistance (R_L) and reactance (X_L) are included (Fig. 2.12). Such a model is considered adequate for representing a short line, usually less than 80 km long [84].

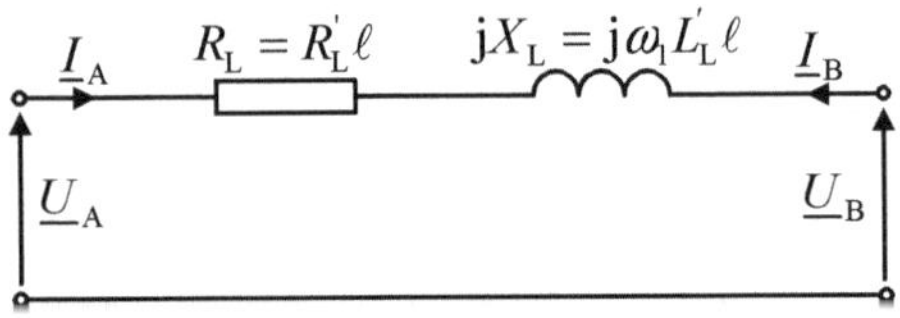

Fig. 2.12 Model of short unfaulted overhead line

In Fig. 2.12, the following signals and parameters are used:
$\underline{U}_A$, $\underline{U}_B$ – voltage from A and B line ends,
$\underline{I}_A$, $\underline{I}_B$ – current from A and B line ends,
R'_L, L'_L – line resistance and inductance per unit length,
ℓ – line length,
ω_1 – angular fundamental frequency.

The circuit of Fig. 2.12 applies either to single-phase or completely transposed three-phase lines operating under balanced conditions. For a completely transposed three-phase line and balanced conditions, line resistance and inductance are considered for the positive-sequence. In turn, $\underline{U}_A$, $\underline{U}_B$ are the positive-sequence line-to-neutral voltages; $\underline{I}_A$, $\underline{I}_B$ are the positive-sequence line currents.

Under unbalanced conditions, mainly under faults, a three-phase line representation has to be considered. Figure 2.13 presents a faulted line together with the equivalent sources behind the line terminals A and B. A fault (occurring at a point marked by F) divides the line into two segments:

- A–F of the relative length d (p.u.),
- F–B of the relative length $(1–d)$ (p.u.).

All signals (voltages and currents) in the circuit of Fig. 2.13 are three-phase (note that particular phases are marked in subscripts with letters: a, b, c), and thus are represented by 3×1 column vectors, as for example, for the sending end voltage $\mathbf{U}_A$ we have:

$$\mathbf{U}_A = \begin{bmatrix} \underline{U}_{Aa} \\ \underline{U}_{Ab} \\ \underline{U}_{Ac} \end{bmatrix} \tag{2.1}$$

All impedances are described by 3×3 matrices, as for example, for the line impedance:

$$\mathbf{Z}_{\mathrm{L}} = \begin{bmatrix} \underline{Z}_{\mathrm{Laa}} & \underline{Z}_{\mathrm{Lab}} & \underline{Z}_{\mathrm{Lac}} \\ \underline{Z}_{\mathrm{Lba}} & \underline{Z}_{\mathrm{Lbb}} & \underline{Z}_{\mathrm{Lbc}} \\ \underline{Z}_{\mathrm{Lca}} & \underline{Z}_{\mathrm{Lcb}} & \underline{Z}_{\mathrm{Lcc}} \end{bmatrix} \tag{2.2}$$

where:
diagonal elements present the self-impedances of the phase conductors and off-diagonal elements present the mutual impedances between two phase conductors, for which the following is satisfied: $\underline{Z}_{\mathrm{Lba}} = \underline{Z}_{\mathrm{Lab}}$, $\underline{Z}_{\mathrm{Lca}} = \underline{Z}_{\mathrm{Lac}}$, $\underline{Z}_{\mathrm{Lcb}} = \underline{Z}_{\mathrm{Lbc}}$.
Note that the line in Fig. 2.13 is represented with only series parameters, while shunt parameters are here neglected.

At the fault point there is a three-phase fault model marked by $\underline{Z}_{\mathrm{F}}$, while $\underline{I}_{\mathrm{F}}$, $\underline{U}_{\mathrm{F}}$ denote the total fault current and voltage at the fault, respectively. Detailed considerations for the fault models are presented in Chap. 3.

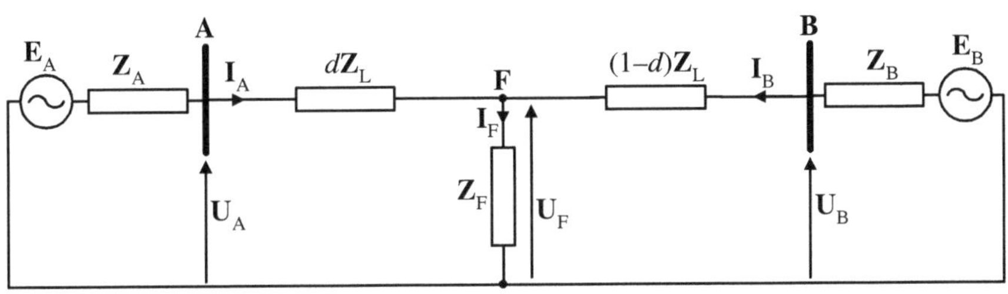

Fig. 2.13 Circuit diagram of three-phase faulted line using matrices for presenting components (sources and impedances) and column matrices for signals

The self- and mutual impedances and admittances of each phase of overhead lines are determined by line geometry and they are not identical for all phases. In general, the line-impedance matrix $\mathbf{Z}_{\mathrm{L}}$ is not symmetrical. For a symmetrical impedance matrix the diagonal elements are equal and the off-diagonal elements are equal, too. This is satisfied if the line is completely transposed. A complete transposition (Fig. 2.14) is achieved by exchanging the conductor positions along the line in such a way that each phase (a, b and c) occupies each position for one-third of the line length.

Fig. 2.14 Completely transposed section of three-phase line

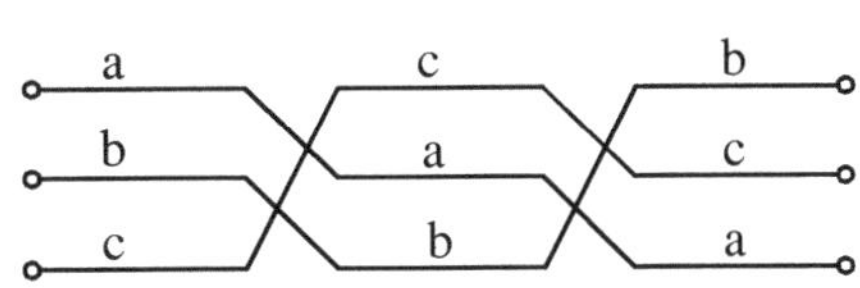

For a completely transposed three-phase line the impedance matrix is symmetrical as follows:

$$\mathbf{Z}_\text{L} = \begin{bmatrix} \underline{Z}_\text{Ls} & \underline{Z}_\text{Lm} & \underline{Z}_\text{Lm} \\ \underline{Z}_\text{Lm} & \underline{Z}_\text{Ls} & \underline{Z}_\text{Lm} \\ \underline{Z}_\text{Lm} & \underline{Z}_\text{Lm} & \underline{Z}_\text{Ls} \end{bmatrix} \tag{2.3}$$

where in the last position of the subscript for the matrix impedance elements the character of the impedance is denoted by s – self-impedance of the phase conductor and by m – mutual impedance between phase conductors.

If the transposition technique is not applied, the impedance matrix is no longer a symmetrical one; however, the following simplification is sometimes made. It relies on using the averaged value for the diagonal and the average for the off-diagonal elements. In this case, additional ramifications with respect to accuracy of the calculation results occur. Applying such simplification one gets:

$$\underline{Z}_\text{Ls} = \frac{1}{3}(\underline{Z}_\text{Laa} + \underline{Z}_\text{Lbb} + \underline{Z}_\text{Lcc}) \tag{2.4}$$

$$\underline{Z}_\text{Lm} = \frac{1}{3}(\underline{Z}_\text{Lab} + \underline{Z}_\text{Lbc} + \underline{Z}_\text{Lca}) \tag{2.5}$$

As a result, one obtains a symmetrical impedance matrix (2.3), whose symmetry is of advantage; however, one must accept certain deterioration of accuracy.

Owing to the symmetry, it is possible to apply a method of symmetrical components, developed by C.L. Fortescue in 1918, which is known from numerous references. Based on this method, a linear transformation from phase components to a set of symmetrical components, as for example for the sending-end voltage $\underline{U}_\text{A}$, is performed according to:

$$\begin{bmatrix} \underline{U}_\text{A0} \\ \underline{U}_\text{A1} \\ \underline{U}_\text{A2} \end{bmatrix} = \frac{1}{3} \begin{bmatrix} 1 & 1 & 1 \\ 1 & \underline{a} & \underline{a}^2 \\ 1 & \underline{a}^2 & \underline{a} \end{bmatrix} \cdot \begin{bmatrix} \underline{U}_\text{Aa} \\ \underline{U}_\text{Ab} \\ \underline{U}_\text{Ac} \end{bmatrix} \tag{2.6}$$

where:

$\underline{U}_\text{Aa}$, $\underline{U}_\text{Ab}$, $\underline{U}_\text{Ac}$ – voltage from phases: a, b, c,

$\underline{U}_\text{A0}$, $\underline{U}_\text{A1}$, $\underline{U}_\text{A2}$ – zero-, positive- and negative-sequence voltage,

$\underline{a} = 1\angle 120° = -0.5 + \text{j}0.5\sqrt{3}$ is a complex number with unit magnitude and 120° phase angle.

Note that multiplying any phasor by $\underline{a}$ results in rotating it by 120° counterclockwise.

For the transformation (2.6) the sequence of phases: (a, b, c) is assumed to be the base, in which phase (a) is considered as the first one in this sequence. Sometimes it is convenient to apply the transformation from phase components into the symmetrical components, assuming the other sequences of phases: (b, c, a) or (c, a, b), in which the phase b or phase c starts the sequence, respectively.

Transformation from symmetrical components into the phase components is defined as follows:

$$\begin{bmatrix} \underline{U}_{\mathrm{Aa}} \\ \underline{U}_{\mathrm{Ab}} \\ \underline{U}_{\mathrm{Ac}} \end{bmatrix} = \begin{bmatrix} 1 & 1 & 1 \\ 1 & \underline{a}^2 & \underline{a} \\ 1 & \underline{a} & \underline{a}^2 \end{bmatrix} \cdot \begin{bmatrix} \underline{U}_{\mathrm{A0}} \\ \underline{U}_{\mathrm{A1}} \\ \underline{U}_{\mathrm{A2}} \end{bmatrix} \tag{2.7}$$

Use of the symmetrical-components method to a three-phase network, which is represented by the symmetrical impedance matrix, such as (2.3), allows this network to be decoupled into three sequence networks, which are simpler to analyze. These sequence networks are called the zero-sequence, positive-sequence and negative-sequence networks. The sequence networks results can then be superposed to obtain three-phase network results by using (2.7).

In the sequence networks, the line is represented by its respective sequence impedances:

- positive- and negative-sequence impedance, which are identical:

$$\underline{Z}_{\mathrm{1L}} = \underline{Z}_{\mathrm{2L}} = \underline{Z}_{\mathrm{Ls}} - \underline{Z}_{\mathrm{Lm}} \tag{2.8}$$

- zero-sequence impedance:

$$\underline{Z}_{\mathrm{0L}} = \underline{Z}_{\mathrm{Ls}} + 2\underline{Z}_{\mathrm{Lm}} \tag{2.9}$$

Note that the positive- and negative-sequence impedances, as stated in (2.8), are equal. This is so for linear, symmetric impedances representing non-rotating power system items such as overhead lines and transformers.

Figure 2.15 presents models of a faulted single-circuit overhead line, together with the equivalent sources behind the line terminals, for the respective sequences. There are three circuits, which can be analyzed separately. These circuits can be composed into one resultant circuit by connecting them at points of unbalance and including the fault-path resistance R_{F}. At these points of unbalance, the respective sequence components of the total fault current ($\underline{I}_{\mathrm{F1}}$, $\underline{I}_{\mathrm{F2}}$, $\underline{I}_{\mathrm{F0}}$) flow into the sequence circuit, and flow out of the respective circuit. The particular sequence networks are connected in such a way as to satisfy the particular fault-type constraints.

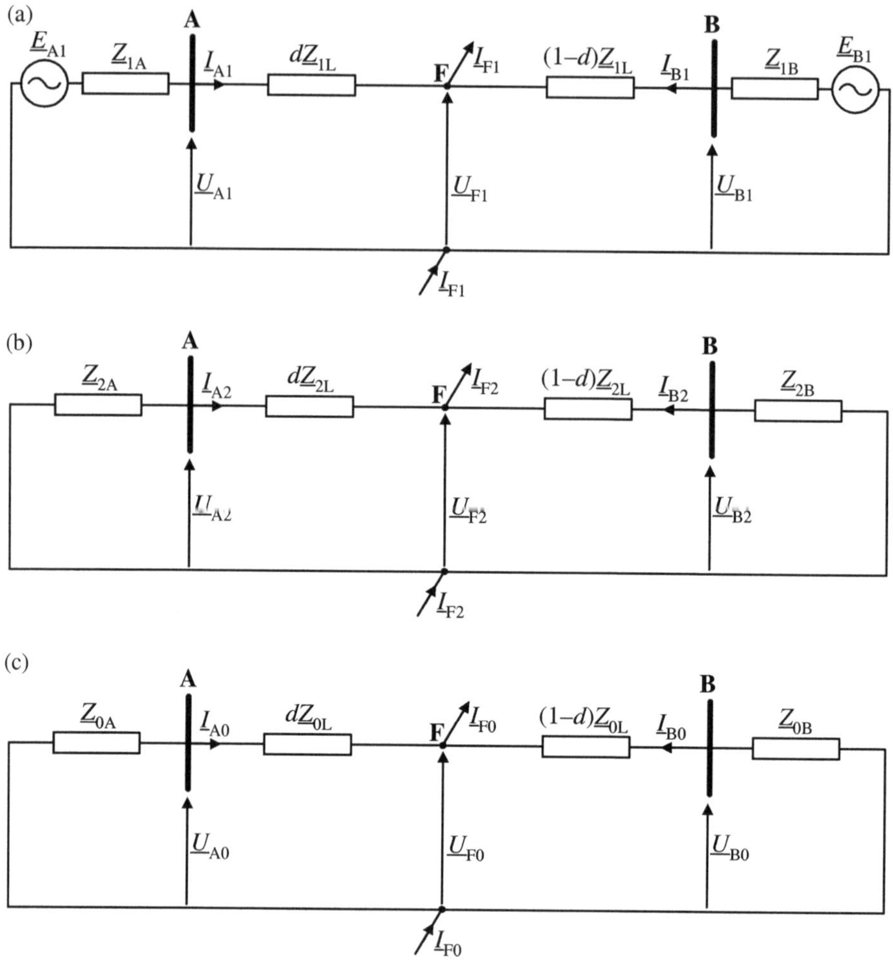

Fig. 2.15 Equivalent networks of single-circuit faulted line for: (a) positive-sequence, (b) negative-sequence, and (c) zero-sequence

Figure 2.16 shows the connection of the sequence networks for a single phase-to-ground fault: a–g fault. The sequence networks are connected in series and additionally the triple fault resistance ($3R_F$) is included. The series connection of Fig. 2.16 can also be applied for the remaining single phase-to-ground faults: b–g, c–g, however, this makes it necessary to use the following sequences of phases: (b, c, a) and (c, a, b), respectively.

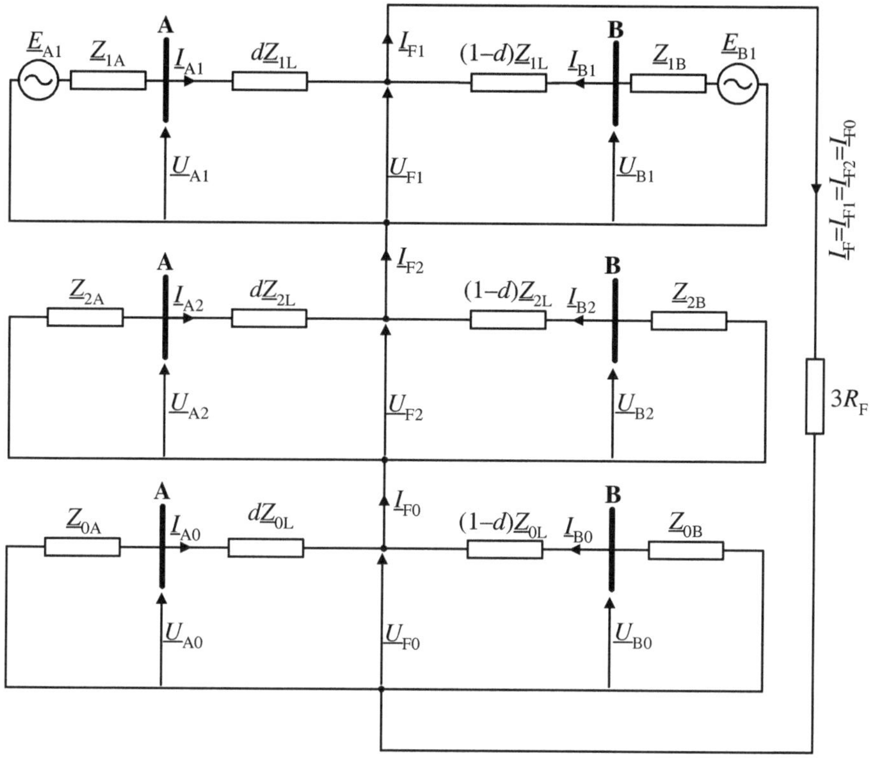

Fig. 2.16 Connection of sequence networks for single phase-to-ground fault (a–g fault) involving fault resistance R_F

In Fig. 2.17, equivalent circuit diagrams of a double-circuit line for the positive- and negative-sequence are shown.

In Fig. 2.18, equivalent circuits of a double-circuit line, with both circuits in operation [11, 109, 114, 122, 140, 286, 293, 345], for the zero-sequence are presented. As a result of mutual coupling of the line circuits, the current flowing in the faulted line AI–BI influences the voltage profile in the healthy parallel line AII–BII, and *vice versa*. In particular, in the faulted line (AI–BI) one can distinguish the following voltage drops (Fig. 2.18a):

- voltage drops resulting from the flow of the faulted-line current:

$$\underline{U}_0^{\mathrm{A}} = d\underline{Z}_{0\mathrm{LI}}\underline{I}_{\mathrm{AI0}} \tag{2.10}$$

$$\underline{U}_0^{\mathrm{B}} = (1-d)\underline{Z}_{0\mathrm{LI}}(\underline{I}_{\mathrm{AI0}} - \underline{I}_{\mathrm{F0}}) \tag{2.11}$$

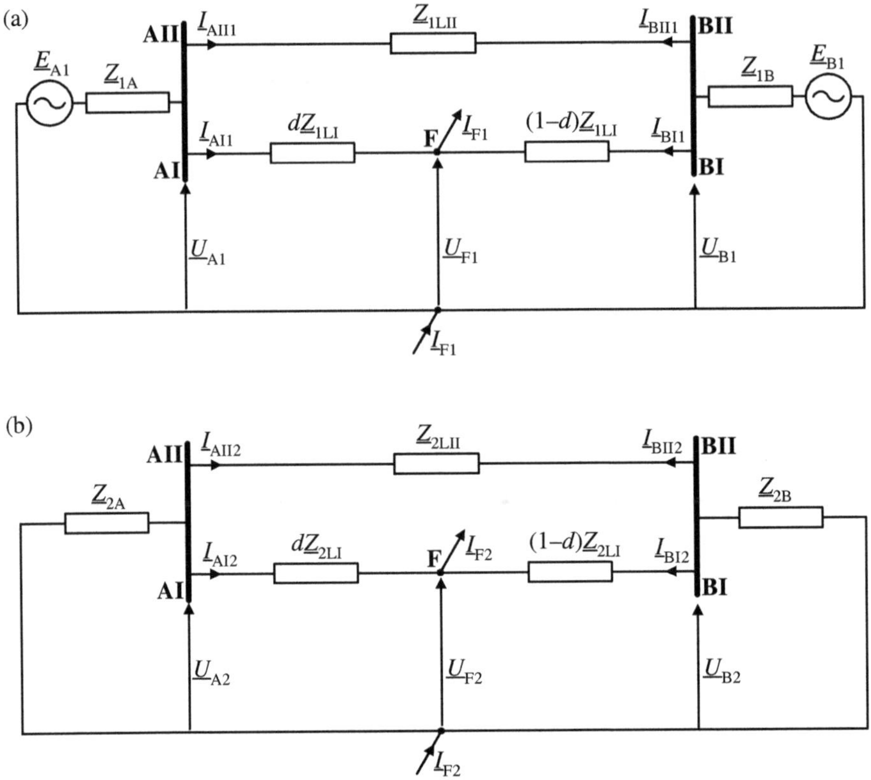

Fig. 2.17 Equivalent networks of double-circuit faulted line for: (a) positive-sequence, and (b) negative-sequence

- voltage drops resulting from the flow of the current in the healthy parallel line:

$$\underline{U}_0^{C} = d\underline{Z}_{0m}\underline{I}_{AII0} \tag{2.12}$$

$$\underline{U}_0^{D} = (1-d)\underline{Z}_{0m}\underline{I}_{AII0} \tag{2.13}$$

In the healthy line (AII–BII) there are the following voltage drops (Fig. 2.18a):

- voltage drops resulting from the flow of the healthy line current:

$$U_0^{E} = d\underline{Z}_{0LII}\underline{I}_{AII0} \tag{2.14}$$

$$\underline{U}_0^{F} = (1-d)\underline{Z}_{0LII}\underline{I}_{AII0} \tag{2.15}$$

- voltage drops resulting from the flow of the current from the faulted line:

$$\underline{U}_0^{\mathrm{G}} = d\underline{Z}_{0\mathrm{m}}\underline{I}_{\mathrm{AI0}} \tag{2.16}$$

$$\underline{U}_0^{\mathrm{H}} = (1-d)\underline{Z}_{0\mathrm{m}}(\underline{I}_{\mathrm{AI0}} - \underline{I}_{\mathrm{F0}}) \tag{2.17}$$

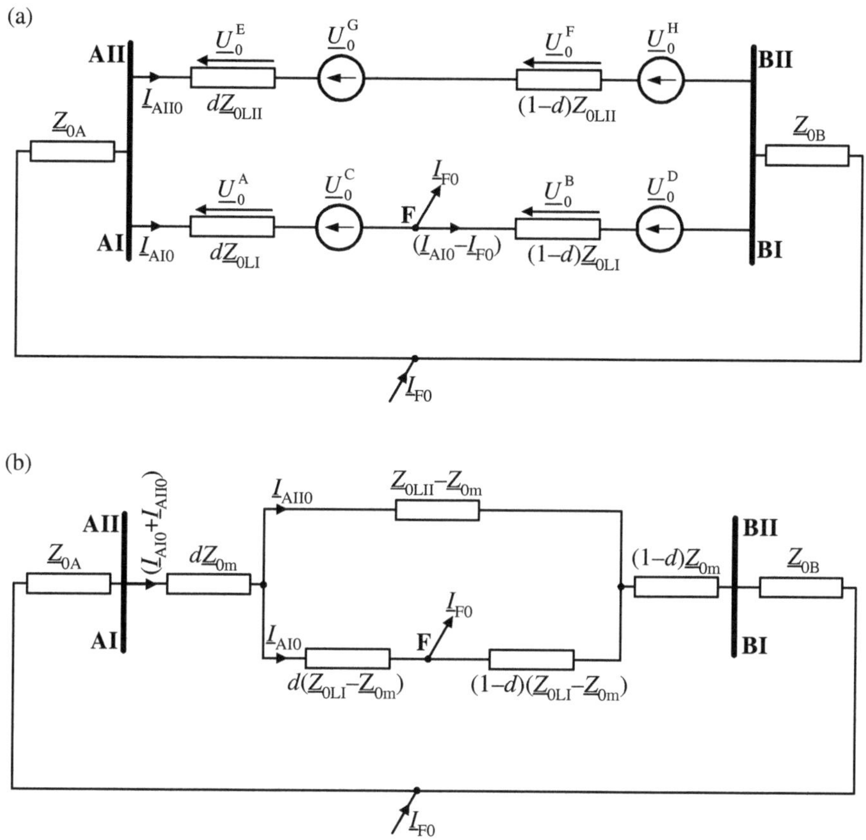

Fig. 2.18 Equivalent networks of double-circuit faulted line with both lines in operation for zero-sequence: (a) general circuit, and (b) alternative circuit

The circuit of Fig. 2.18a can be transformed to the form shown in Fig. 2.18b, which is more convenient for use. For this purpose the voltage drop between the bus AI and fault point F is determined, taking into account (2.10) and (2.12):

$$\underline{U}_0^{(\mathrm{AI-F})} = d\underline{Z}_{0\mathrm{LI}}\underline{I}_{\mathrm{AI0}} + d\underline{Z}_{0\mathrm{m}}\underline{I}_{\mathrm{AII0}} \tag{2.18}$$

Adding and subtracting the term $d\underline{Z}_{0m}\underline{I}_{AI0}$ to the right-hand side of (2.18) leads to the following alternative form of (2.18):

$$\underline{U}_0^{(AI-F)} = d\underline{Z}_{0m}(\underline{I}_{AI0} + \underline{I}_{AII0}) + d(\underline{Z}_{0LI} - \underline{Z}_{0m})\underline{I}_{AI0} \tag{2.19}$$

Analogously, after taking (2.11) and (2.13), one obtains for the voltage drop between the fault point F and the bus BI:

$$\underline{U}_0^{(F-BI)} = (1-d)(\underline{Z}_{0LI} - \underline{Z}_{0m})(\underline{I}_{AI0} - \underline{I}_{F0}) + (1-d)\underline{Z}_{0m}(\underline{I}_{AI0} + \underline{I}_{AII0} - \underline{I}_{F0}) \tag{2.20}$$

Similarly, for the healthy-line path (between buses AII, BII) one obtains:

$$\begin{aligned}\underline{U}_0^{(AII-BII)} &= d\underline{Z}_{0m}(\underline{I}_{AI0} + \underline{I}_{AII0}) + (\underline{Z}_{0LII} - \underline{Z}_{0m})\underline{I}_{AII0} \\ &\quad + (1-d)\underline{Z}_{0m}(\underline{I}_{AI0} + \underline{I}_{AII0} - \underline{I}_{F0})\end{aligned} \tag{2.21}$$

Taking into account (2.19)–(2.21), the circuit of Fig. 2.18b is obtained.

In Fig. 2.19, the mutual coupling effect is depicted for a double-circuit line with the healthy parallel line switched off and grounded at both ends [114].

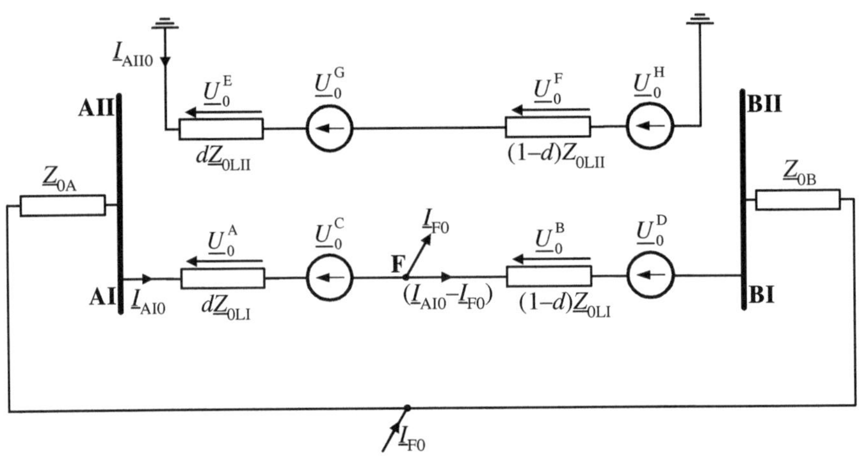

Fig. 2.19 Zero-sequence equivalent network for double-circuit line with faulted line in operation and parallel healthy line switched off and grounded at both ends

The particular voltage drops for both lines are expressed in the same way as for the case of both lines in operation (2.14)–(2.17). The usual unavailability of the zero-sequence current from the healthy parallel line $\underline{I}_{AII0}$ causes difficulty in reflecting the mutual coupling effect in the fault-location algorithms for this mode of operation. Therefore, it remains to estimate this current, based on other measurements available [114].

In the models presented so far (Figs. 2.12 and 2.13 and Figs. 2.15–2.19), only the series parameters of the line have been accounted for. These models can be used for short lines. For medium-length lines, typically ranging from 80 to 250 km, it is common to incorporate the shunt admittance to the line model [84]. Shunt conductance is usually neglected and only shunt capacitances of the line are considered for that. It is a common practice to lump the total shunt capacitance and insert half at each of the line sections. In this way the nominal π circuit is obtained [84].

Figure 2.20 shows a positive-sequence circuit of the faulted line, for which both sections (A–F and F–B) are represented using the nominal π circuits. The parameter $\underline{Y}_{1L}$ used in describing admittances of shunt branches denotes:

$$\underline{Y}_{1L} = \mathrm{j}\omega_1 C'_{1L}\ell \tag{2.22}$$

where:

C'_{1L} – line capacitance for the positive-sequence per unit length,
ℓ – line length,
ω_1 – angular fundamental frequency.

The equivalent circuit diagrams for the remaining sequences are obtained analogously.

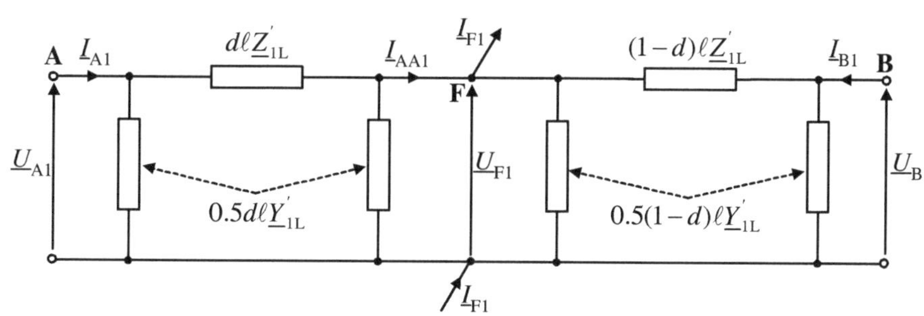

Fig. 2.20 Equivalent circuit diagram of the network for the positive-sequence with the use of nominal π circuit for faulted line

Figure 2.21a presents a model of the line section between the bus A and the fault point F (as applied in Fig. 2.20). Note that a sought distance to fault (d) is used for determining both the series impedance $d\underline{Z}_{1L}$ and shunt admittance $0.5d\underline{Y}_{1L}$. However, it is inconvenient for performing fault-location calculations. In order to make the calculation simpler, the iterative calculations are usually performed in such a way that the unknown fault distance is left in the current iteration (n) as related to the series impedance $d_{(n)}\underline{Z}_{1L}$ (Fig. 2.21b). On the other hand, the shunt admittance is determined using the fault distance from the previous iteration (n – 1): $0.5d_{(n-1)}\underline{Y}_{1L}$ at both ends of the faulted section (Fig. 2.21b). When starting the iterative calculations (iteration number: n = 1) one takes the fault distance obtained when neglecting the shunt admittance ($\underline{Y}_{1L} = 0$) as the fault distance from the previous iteration (iteration number: $n-1 = 0$).

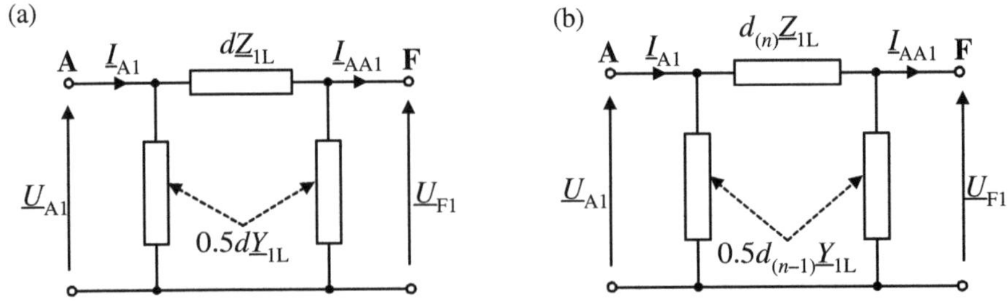

Fig. 2.21 Model of faulted-line section from the side A for the positive-sequence: (a) basic model, and (b) model used in simple iterative fault-location calculations

2.3.2 Distributed-parameter Models

For short and medium-length lines using the lumped model is usually sufficient. In order to improve fault-location accuracy, especially in the case of long-length lines, the distributed nature of overhead-line parameters has to be considered.

In the distributed-parameter line models the voltage and current along the line are functions of the distance x (point X) from the sending end (A) of the line and the time t (Fig. 2.22). The voltage $u(x, t)$ and current $i(x, t)$ are related with the parameters of the line ($R_L^{'}, L_L^{'}, C_L^{'}$ – resistance, inductance and capacitance of the line per unit length) by the so-called *telegrapher's equations* [19, 88, 157]:

$$\frac{\partial u(x,t)}{\partial x} + L_L^{'}\frac{\partial i(x,t)}{\partial t} = -R_L^{'} i(x,t) \tag{2.23}$$

$$C_L^{'}\frac{\partial u(x,t)}{\partial t} + \frac{\partial i(x,t)}{\partial x} = 0 \tag{2.24}$$

Note that in (2.24) the line conductance is neglected, which is a common practice.

Partial differential equations (2.23) and (2.24) can be solved using the method of characteristics developed by Collatz [37]. For this purpose the *modified telegrapher's equations* are formulated:

$$\frac{\partial v(x,t)}{\partial x} - \chi^2\frac{\partial i(x,t)}{\partial t} = -\eta\, i(x,t) \tag{2.25}$$

$$\frac{\partial v(x,t)}{\partial t} - \frac{\partial i(x,t)}{\partial x} = 0 \tag{2.26}$$

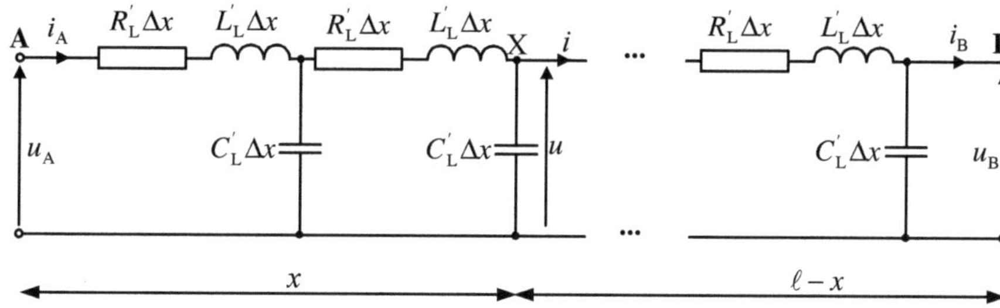

Fig. 2.22 Distributed-parameter model of long overhead line

where:

$v(x,t) = -C_{\mathrm{L}}^{'} u(x,t)$,

$\chi = \sqrt{L_{\mathrm{L}}^{'} C_{\mathrm{L}}^{'}}$,

$\eta = R_{\mathrm{L}}^{'} C_{\mathrm{L}}^{'}$.

The traveling-waves method is applied as the alternative to solving the partial differential equations (2.25) and (2.26). In this method, the voltage and current are considered as two components: the forward and backward traveling waves [157].

A distributed-parameter model of an overhead line can also be applied for phasors. In this case, the so-called equivalent π circuit [84] is utilized. In Fig. 2.23, two such circuits are used for representing both line sections A–F and F–B. The model of Fig. 2.23 is for the general case, i.e., for the *i*th symmetrical component, where: $i = 1$ – positive-sequence, $i = 2$ – negative-sequence, $i = 0$ – zero-sequence.

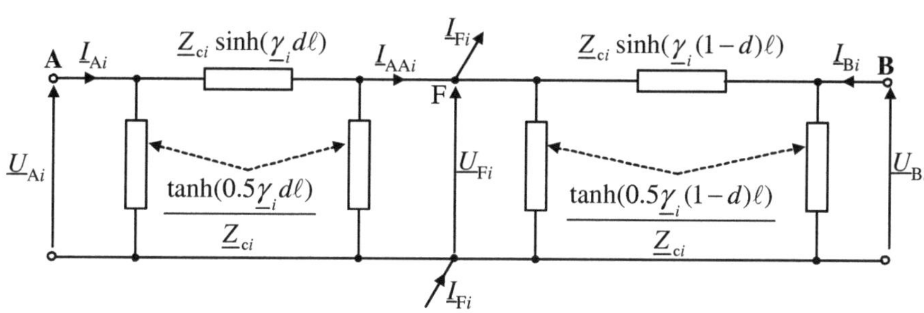

Fig. 2.23 Distributed-parameter model of faulted line for the *i*th symmetrical component

In Fig. 2.23, both the series and shunt parameters of the line are distributed parameters and are expressed using:

- surge impedance of the line for the *i*th sequence:

$$\underline{Z}_{ci} = \sqrt{\frac{\underline{Z}_{i\mathrm{L}}^{'}}{\underline{Y}_{i\mathrm{L}}^{'}}} \qquad (2.27)$$

- propagation constant of the line for the ith sequence:

$$\underline{\gamma}_i = \sqrt{\underline{Z}'_{i\text{L}}\underline{Y}'_{i\text{L}}} \tag{2.28}$$

Using the equivalent π circuit model, the voltage and current, for example, from the end A can be analytically transferred to the fault point F (Fig. 2.23) according to:

$$\underline{U}_{\text{F}i} = \cosh(\underline{\gamma}_i d\ell)\underline{U}_{\text{A}i} - \underline{Z}_{ci}\sinh(\underline{\gamma}_i d\ell)\underline{I}_{\text{A}i} \tag{2.29}$$

$$\underline{I}_{\text{AA}i} = -\frac{1}{\underline{Z}_{ci}}\sinh(\underline{\gamma}_i d\ell)\underline{U}_{\text{A}i} + \cosh(\underline{\gamma}_i d\ell)\underline{I}_{\text{A}i} \tag{2.30}$$

An alternative representation of the faulted line, based on the distributed-parameter line model is depicted in Fig. 2.24 [84]. Both the series impedances and shunt admittances are expressed as the lumped parameter multiplied by the respective correction factor:

- correction factors for series impedances:

$$\underline{A}_i^{\text{sh}} = \frac{\sinh(\underline{\gamma}_i d\ell)}{\underline{\gamma}_i d\ell} \tag{2.31}$$

$$\underline{B}_i^{\text{sh}} = \frac{\sinh(\underline{\gamma}_i (1-d)\ell)}{\underline{\gamma}_i (1-d)\ell} \tag{2.32}$$

- correction factors for shunt admittances:

$$\underline{A}_i^{\text{th}} = \frac{\tanh(0.5\underline{\gamma}_i d\ell)}{0.5\underline{\gamma}_i d\ell} \tag{2.33}$$

$$\underline{B}_i^{\text{th}} = \frac{\tanh(0.5\underline{\gamma}_i (1-d)\ell)}{0.5\underline{\gamma}_i (1-d)\ell} \tag{2.34}$$

Using the correction factors (2.31)–(2.34) allows us to recalculate the lumped parameters into the distributed ones.

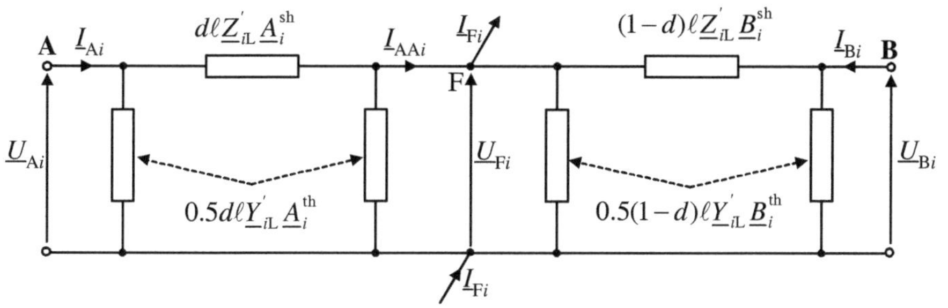

Fig. 2.24 Distributed-parameter model of faulted line for the *i*th symmetrical component, with use of the correction factors for representing series and shunt parameters

In Fig. 2.25, the distributed-parameter model of a faulted line for the *i*th symmetrical component for application to simplified iterative fault-location calculations is presented.

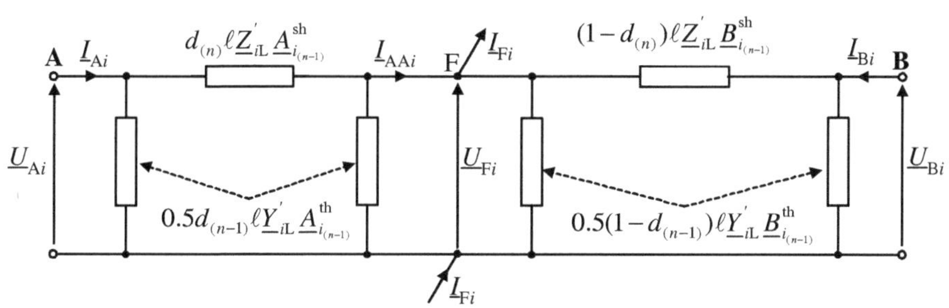

Fig. 2.25 Distributed-parameter model of faulted line for the *i*th symmetrical component for application to simplified iterative calculations of fault location

The model of Fig. 2.25 is simpler in comparison to the strict model of Fig. 2.24. The unknown fault distance in the model of Fig. 2.25, which is calculated in the current iteration (iteration number: n): $d_{(n)}$, is involved only to represent the series impedances. The shunt parameters are represented by the fault-distance value obtained in the previous iteration: $d_{(n-1)}$. Also, this value of fault distance is used for calculating all correction factors:

$$\underline{A}_{i_{(n-1)}}^{sh} = \frac{\sinh(\underline{\gamma}_i d_{(n-1)} \ell)}{\underline{\gamma}_i d_{(n-1)} \ell} \tag{2.35}$$

$$\underline{B}_{i_{(n-1)}}^{sh} = \frac{\sinh(\underline{\gamma}_i (1-d_{(n-1)}) \ell)}{\underline{\gamma}_i (1-d_{(n-1)}) \ell} \tag{2.36}$$

$$\underline{A}^{\text{th}}_{i_{(n-1)}} = \frac{\tanh(0.5\underline{\gamma}_i d_{(n-1)}\ell)}{0.5\underline{\gamma}_i d_{(n-1)}\ell} \tag{2.37}$$

$$\underline{B}^{\text{th}}_{i_{(n-1)}} = \frac{\tanh(0.5\underline{\gamma}_i (1-d_{(n-1)})\ell)}{0.5\underline{\gamma}_i (1-d_{(n-1)})\ell} \tag{2.38}$$

2.3.3 Modal Transformation

The symmetrical components approach was used in the preceding sections for presenting different line models. Yet, there is another technique based on modal transformation that considers the general case of an untransposed line [56].

Using the modal transformation [56, 135] the line-impedance matrix $\mathbf{Z}$ and admittance matrix $\mathbf{Y}$ are transformed into the matrices $\mathbf{Z}_{\text{mode}}$, $\mathbf{Y}_{\text{mode}}$:

$$\mathbf{Z}_{\text{mode}} = \mathbf{T}_{\text{u}}^{-1}\mathbf{Z}\mathbf{T}_{\text{i}} \tag{2.39}$$

$$\mathbf{Y}_{\text{mode}} = \mathbf{T}_{\text{i}}^{-1}\mathbf{Y}\mathbf{T}_{\text{u}} \tag{2.40}$$

where the superscript (–1) denotes the matrix inversion.

The transformation (2.39) and (2.40) is performed in such a way that the matrices $\mathbf{Z}_{\text{mode}}$, $\mathbf{Y}_{\text{mode}}$ are diagonal, which means that the three-phase coupled network becomes decoupled into three decoupled single-phase networks.

Three-phase voltage $\mathbf{U}$ and current $\mathbf{I}$ vectors are transformed into the modal vectors $\mathbf{U}_{\text{mode}}$, $\mathbf{I}_{\text{mode}}$. This is performed using the matrices $\mathbf{T}_{\text{i}}$, $\mathbf{T}_{\text{u}}$ (used in (2.39) and (2.40)):

$$\mathbf{U}_{\text{mode}} = \mathbf{T}_{\text{u}}^{-1}\mathbf{U} \tag{2.41}$$

$$\mathbf{I}_{\text{mode}} = \mathbf{T}_{\text{i}}^{-1}\mathbf{I} \tag{2.42}$$

For balanced (equally transposed) three-phase lines, both matrices $\mathbf{T}_{\text{i}}$, $\mathbf{T}_{\text{u}}$ are identical and are composed of the different real-value elements such as:

- Clarke transformation (also called the α–β transform) [25]:

$$\mathbf{T}_u = \mathbf{T}_i = \begin{bmatrix} 1 & 1 & 0 \\ 1 & -\frac{1}{2} & \frac{\sqrt{3}}{2} \\ 1 & -\frac{1}{2} & -\frac{\sqrt{3}}{2} \end{bmatrix} \tag{2.43}$$

$$\mathbf{T}_u^{-1} = \mathbf{T}_i^{-1} = \frac{1}{3}\begin{bmatrix} 1 & 1 & 1 \\ 2 & -1 & -1 \\ 0 & \sqrt{3} & -\sqrt{3} \end{bmatrix} \tag{2.44}$$

- Karrenbauer transformation:

$$\mathbf{T}_u = \mathbf{T}_i = \begin{bmatrix} 1 & 1 & 1 \\ 1 & -2 & 1 \\ 1 & 1 & -2 \end{bmatrix} \tag{2.45}$$

$$\mathbf{T}_u^{-1} = \mathbf{T}_i^{-1} = \frac{1}{3}\begin{bmatrix} 1 & 1 & 1 \\ 1 & -1 & 0 \\ 1 & 0 & -1 \end{bmatrix} \tag{2.46}$$

- Wedepohl transformation:

$$\mathbf{T}_u = \mathbf{T}_i = \begin{bmatrix} 1 & 1 & 1 \\ 1 & 0 & -2 \\ 1 & -1 & 1 \end{bmatrix} \tag{2.47}$$

$$\mathbf{T}_u^{-1} = \mathbf{T}_i^{-1} = \frac{1}{3}\begin{bmatrix} 1 & 1 & 1 \\ \frac{3}{2} & 0 & -\frac{3}{2} \\ \frac{1}{2} & -1 & \frac{1}{2} \end{bmatrix} \tag{2.48}$$

In the case of untransposed lines, there is also a possibility of determining the transformation matrices: $\mathbf{T}_u$, $\mathbf{T}_i$, which are not identical. They can be applied for transforming the coupled phase quantities to decoupled modal quantities based on eigenvalue/eigenvector theory [135].

2.4 Series-compensated Lines

Power (P) transfer capability of a traditional uncompensated transmission line (Fig. 2.1) is determined by the well-known formula (with line resistance and capacitance being neglected):

$$P = \frac{|\underline{U}_A| \cdot |\underline{U}_B|}{X_L} \sin(\delta) \tag{2.49}$$

where:

$\underline{U}_A$, $\underline{U}_B$ – phasors of voltage from the end A and B, respectively,
X_L – line reactance,
δ – electric angle between the terminal voltage phasors.

The maximum value of δ is limited by the stability constraints, and thus an increase in the power transfer capability can be obtained by reducing the line reactance. This can be done by adding series-compensating capacitors to counteract series inductance. As a result, the total reactance of the series compensated line (X_{total}) is equal to:

$$X_{total} = X_L - X_C \tag{2.50}$$

where X_C is the capacitor reactance.

The compensation degree is expressed by the following ratio:

$$k_{SC} = \frac{X_C}{X_L} 100\% \tag{2.51}$$

and usually falls within the range of 50 up to 90%.

The capacitor compensation in high-voltage transmission networks is performed by adding series capacitors of the fixed value or of the value controlled with the thyristor circuits.

Use of the series capacitors besides increasing the power-transfer capability brings about several advantages to power-system operation [350], such as:

- improving power-system stability;
- reduced transmission losses;
- enhanced voltage control; and
- flexible power-flow control.

The environmental concerns are also of importance here since instead of constructing a new line, the power-transfer capability of the existing line is increased. The cost of introducing the series capacitor compensation is much lower than that of constructing a new equivalent overhead power line [350].

Usually, only one three-phase capacitor bank is installed on a power transmission line. As far as a single line is concerned, the one-line circuit diagram of the series-compensated line is as presented in Fig. 2.26. Series capacitors (SCs) are installed on the line at a distance d_{SC} (p.u.) from the bus A. In order to protect SCs against overvoltages they are equipped with MOVs. The SC and its MOV are the main components of the compensating bank installed in each phase of the line. Therefore, for the sake of simplifying the series-compensated transmission networks presented, only these components are indicated in the schemes (Fig. 2.26, Fig. 2.27 and further figures showing configurations of series-compensated networks).

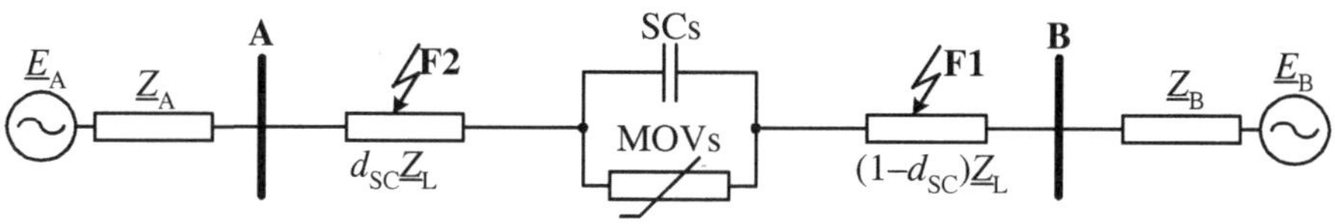

Fig. 2.26 Single transmission line compensated with SCs and MOVs installed at midpoint

Figure 2.27a presents a scheme of the compensating bank from one phase of a line, which contains a fixed series capacitor [159, 194, 268]. Besides the SC and MOV there is a protection of MOV against overheating. This thermal (overload) protection (OP) measures the current conducted by the MOV. If the energy absorbed by the MOV exceeds its pre-defined limit the MOV becomes shunted by firing the air gap. Figure 2.27b presents a compensating bank with thyristor controlled capacitor [32, 46, 149, 340]. MOVs are non-linear resistors commonly approximated by the standard exponential formula:

$$\frac{i_{\mathrm{MOV}}}{p} = \left(\frac{u_{\mathrm{x}}}{U_{\mathrm{REF}}}\right)^{q} \tag{2.52}$$

Figure 2.27c shows the voltage–current characteristic for the following parameters of the approximation (2.52): $q = 23$, $p = 1$ kA, $U_{REF} = 150$ kV.

Series capacitors equipped with MOVs, when set on a transmission line, create certain problems for its protective relays [61, 74, 75, 83, 91, 126, 146, 150, 159, 210, 227, 241, 251, 252, 267, 277, 282, 287] and fault locators [10, 64, 106, 111, 222, 228, 254, 264, 268, 272, 274, 276, 279, 284, 285, 340]. Under faults behind the SCs and MOVs (fault F1 in Fig. 2.19), a fault loop seen from the bus A becomes strongly non-linear, and as a consequence, the nature of transients as well as the steady-state situation are entirely different, compared with traditional uncompensated lines. In the case of faults in front of the SCs and MOVs (fault F2 in Fig. 2.26) the SCs and MOVs are outside the fault loop seen from the bus A; however, they influence the infeed of the fault from the remote substation B.

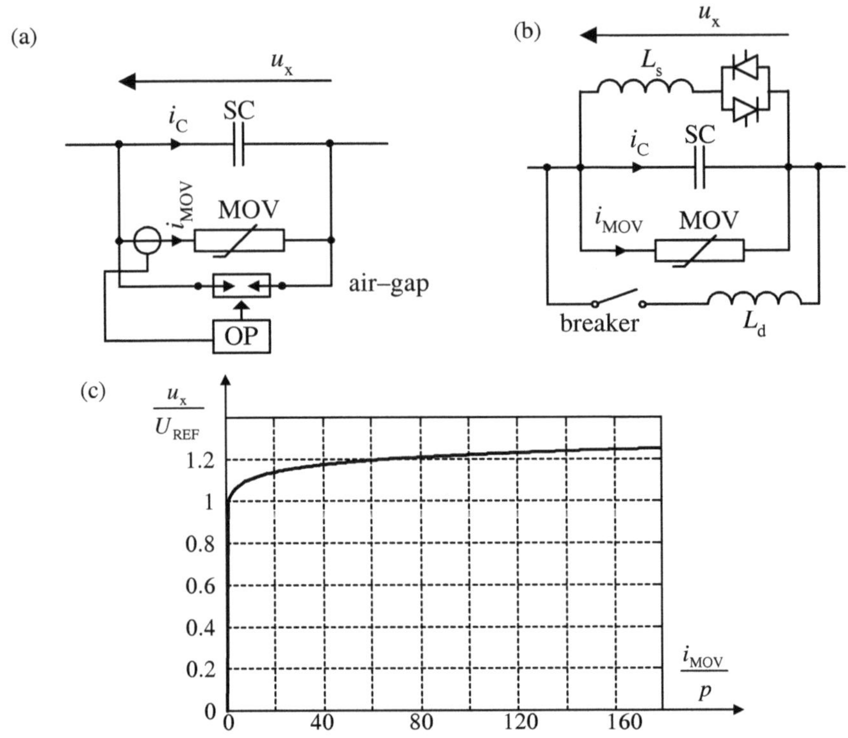

Fig. 2.27 Series capacitor bank: (a) scheme of bank with fixed capacitor, (b) scheme of bank with thyristor-controlled capacitor, and (c) typical voltage–current characteristic of MOV

Adequate representation of the SCs and MOVs has to be applied for both protective relays and fault location. The form of this representation depends on the type of protection and fault-location algorithms. If these algorithms are based on the phasor technique, then the SCs and MOVs from a particular phase can be represented with the fundamental-frequency equivalent [86, 106, 111, 272, 274, 276, 284, 285] in the form of resistance–capacitive reactance series branch with parameters dependent on an amplitude of the current (fundamental-frequency component) measured in the phase of interest. When considering protection and fault-location algorithms based on a differential equation approach, the SC and MOV from a particular phase is represented with use of the estimated instantaneous voltage drop across the compensating bank [10, 74, 75, 268, 277, 282].

Figure 2.28 shows operation of SCs and MOVs under the sample fault on a 400-kV, 300-km transmission line compensated at $k_{SC} = 80\%$. The parameters of the approximation (2.52) are as taken for plotting the voltage–current characteristic from Fig. 2.27c. A single phase-to-ground fault (a–g fault) with fault resistance of 10 Ω was applied just behind the SCs and MOVs. In Fig. 2.28a, the three-phase currents entering the SCs and MOVs are shown. The voltage drops across the SCs and MOVs are shown in Fig. 2.28b.

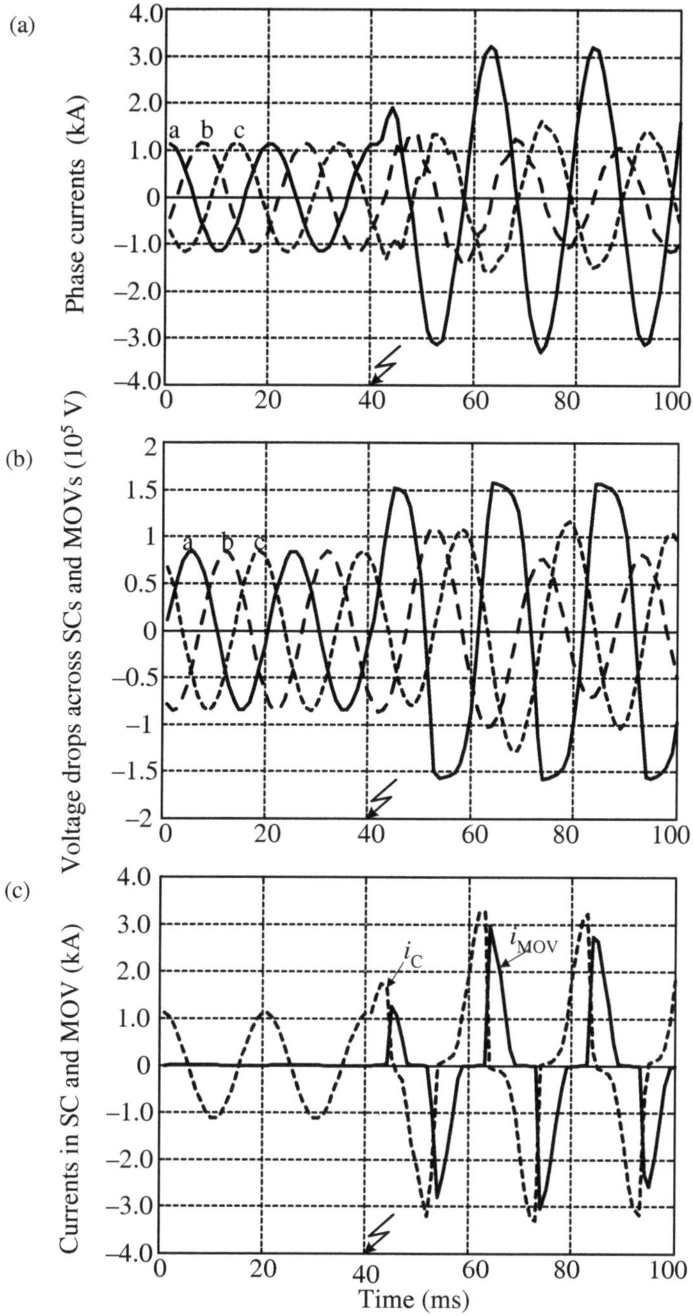

Fig. 2.28 Operation of SCS and MOVs under the sample a–g fault: (a) phase currents entering SCs and MOVs, (b) voltage drops across SCs and MOVs, and (c) currents flowing in SC (i_C) and in MOV (i_{MOV}) from the faulted phase

It can be observed that the voltage drop in the faulted phase (a) is limited to around ±150 kV. Such a limitation results from applying the MOVs with the reference voltage: V_{REF} = 150 kV. The waveforms of the voltage drop from the healthy phases (b, c) are distorted by sub-synchronous resonance oscillations. Such oscillations appear since MOVs from these phases operate in the linear range, conducting low current. The sub-synchronous resonance oscillations are also visible in currents entering the SCs and MOVs from the healthy phases (Fig. 2.28a).

Figure 2.28c shows division of the fault current from the faulted phase (a) into the parallel branches of the SC and its MOV. The SC and MOV conduct the fault current alternately, around for the quarter of the fundamental period. Such voltage limitation results from applying the MOVs with the reference voltage: V_{REF} = 150 kV. The waveforms of the voltage drop from the healthy phases (b, c) are distorted by sub-synchronous resonance oscillations. Such oscillations appear since MOVs from these phases operate in the linear range, conducting low current. The sub-synchronous resonance oscillations are also visible in currents entering the SCs and MOVs from the healthy phases (Fig. 2.28a). Figure 2.28c shows division of the fault current from the faulted phase (a) into the parallel branches of the SC and its MOV. The SC and MOV conduct the fault current alternately, around for a quarter of the fundamental period.

Figure 2.29 depicts series capacitors compensation of the transmission line using the compensation banks installed at both ends [272, 350].

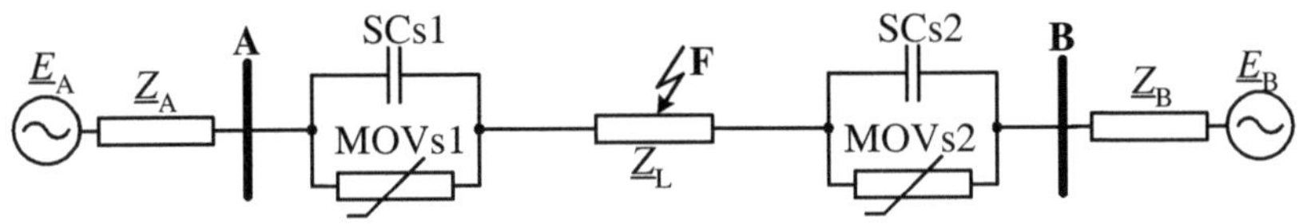

Fig. 2.29 Transmission line compensated with SCs and MOVs banks at both ends

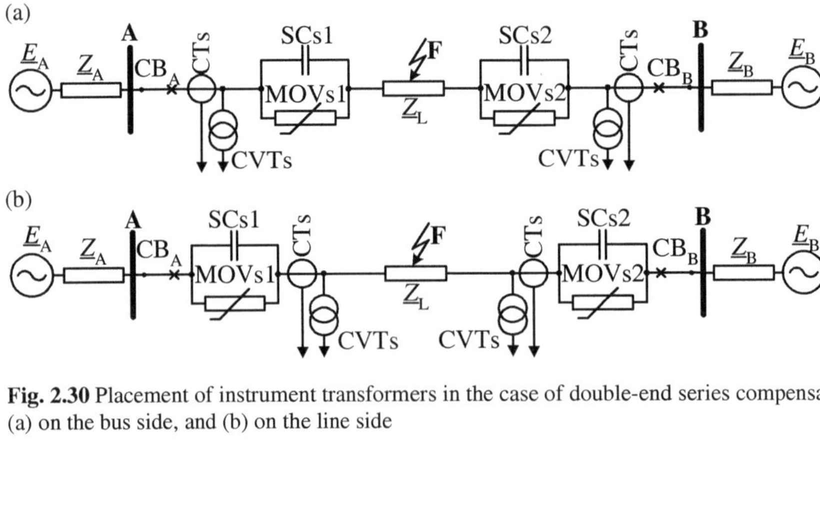

Fig. 2.30 Placement of instrument transformers in the case of double-end series compensation: (a) on the bus side, and (b) on the line side

In the case of such compensation, the placement of current and voltage instrument transformers (CTs – current transformers, CVTs – capacitive voltage transformers) at the line ends is important for considering fault location. The instrument transformers can be placed on the bus side (Fig. 2.30a) or on the line side (Fig. 2.30b) [350].

Similarly, double-circuit transmission lines, analogously to the single line, can be compensated using the capacitor compensating banks installed at the midpoint [279, 284, 285] (Fig. 2.31) or at the line ends (Fig. 2.32).

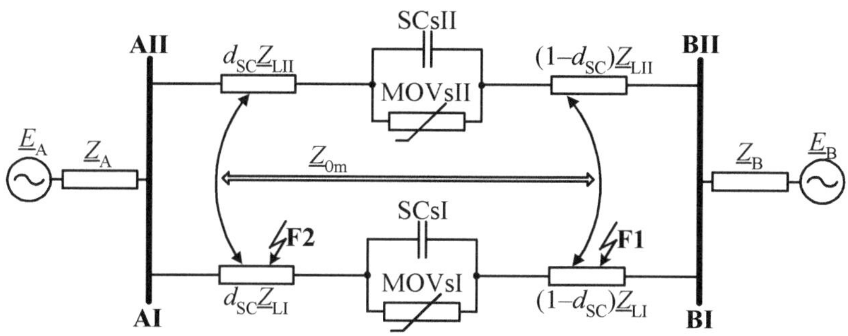

Fig. 2.31 Double-circuit transmission lines with capacitor compensating banks installed at the midpoint in both circuits

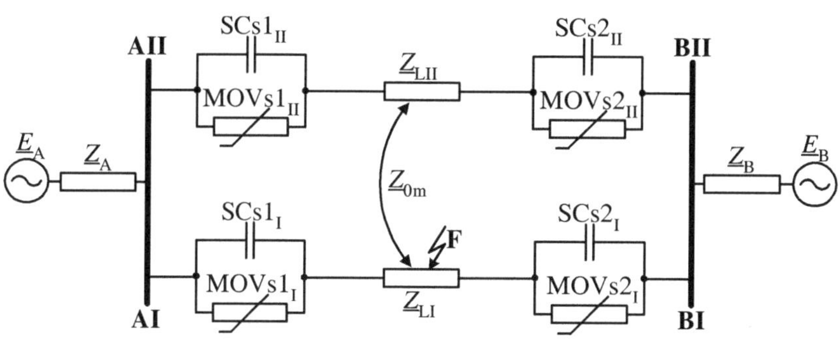

Fig. 2.32 Double-circuit transmission lines with capacitor compensating banks installed at two ends in both circuits

2.5 Distribution Networks

2.5.1 Basic Principles of Distribution Systems

Distribution systems, also called medium voltage (MV) systems deliver the electrical energy to final step-down transformers at voltages below 132 kV (this value may be different in particular systems). The MV networks provide supplies direct to large customers, but the vast majority of consumers are connected at low voltage via distribution transformers (MV/LV) [73, 87, 327]

A scheme of a typical distribution system is presented in Fig. 2.33. The MV networks are supplied from HV/MV-transforming substations. The electrical energy is then transported via a series of underground-cable and overhead-line circuits to the customers. In some cases a dispersed generation (DG) can be connected to the network. The particular line sections are of length from a few hundred meters to some dozens of kilometers. The distribution voltages in a specific service territory are likely similar because it is easier and more cost effective to stock spare parts when the system voltages are consistent. Although due to different loads and the network development, separate feeders are made up of sections with different technical data: varying cable cross-sections and overhead-line parameters.

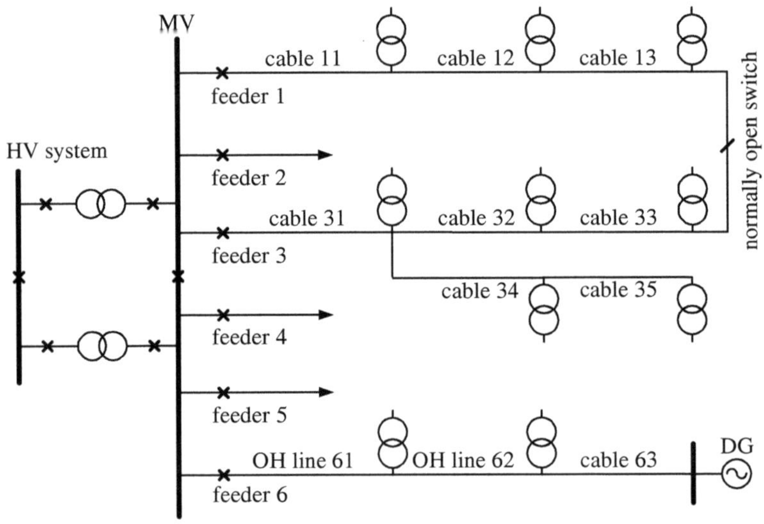

Fig. 2.33 Typical configurations of MV network

The configuration layout and complexity of distribution networks vary widely depending on the application. From that point of view one can specify: rural systems; suburban systems; urban systems, and industrial systems.

Apart from the application, typically the distribution system is radial in nature, including feeders and laterals. Some loads directly in MV level or in LV level are one- or two-phase loads. The majority of distribution circuits are three phase, although in some rural areas single-phase circuits may be used.

Generally, distribution networks are radial systems supplied from one step-down point. However, sometimes also closed-loop networks are used. Moreover, more and more frequently DG sources are connected to distribution networks. A similar effect is observed during a fault introducing motor loads. This is due to the fact that motors behave as generators just after the fault inception. Similarly, if there are motor load taps along a radial distribution line it is equivalent to a line connected to multiple energy sources [327].

2.5.2 Methods of Neutral Grounding

The fault current during a fault depends on the fault-loop impedance and the equivalent supplying voltage. In the case of a phase-to-phase fault the fault loop has small impedance and the fault current reaches a relatively high level. The total fault-loop impedance is basically determined by the longitudinal parameters of the line sections covered by the fault.

Phase-to-ground fault characteristics significantly depend on the type of the network neutral grounding (grounding). System grounding, or the intentional connection of a phase or neutral conductor to ground, is used for the purpose of controlling the voltage to ground (earth), within predictable limits. It also provides for a flow of current that will allow detection of an unwanted connection between system conductors and ground (a phase-to-ground fault). The question of how the network neutral shall be grounded is determined by particular regulations. Local utility practice also influences the choice of the neutral-grounding arrangement on the MV circuits. The general purpose of grounding is to protect life and property in the event of faults and transient phenomena resulting from switching operations. Another reason for adequate grounding of the network is to allow high sensitivity of the relays operation. Fault detection and location invariably relies on the presence of a considerable amount of fault current and this is met in the solidly grounded networks or when the grounding impedance is small – as far as a fault between one or more phases and ground is considered.

Regarding the method of the neutral grounding, the MV networks can be divided into the following categories [96, 97, 265, 327]:

- directly or solidly grounded;
- network grounded through impedance; and
- isolated network.

Adequate method can be assured by grounding the star point delta/wye step-down transformer (Fig. 2.34a). In the case when the MV side of the transformer has delta connection, a separate grounding zigzag connected transformer is applied as a, generally, more efficient solution (Fig. 2.34b).

In the solidly grounded network the only impedance between the neutral and ground is that represented by the grounding transformer impedance, grounding conductor and the ground plate itself (Fig. 2.34). Solid neutral grounding is cheaper as it requires no extra equipment but can lead to high ground-fault currents that may cause damage and high step or touch voltages. On the other hand, this solution establishes a low-impedance path for ground-fault currents and hence satisfactory operation of relay protection.

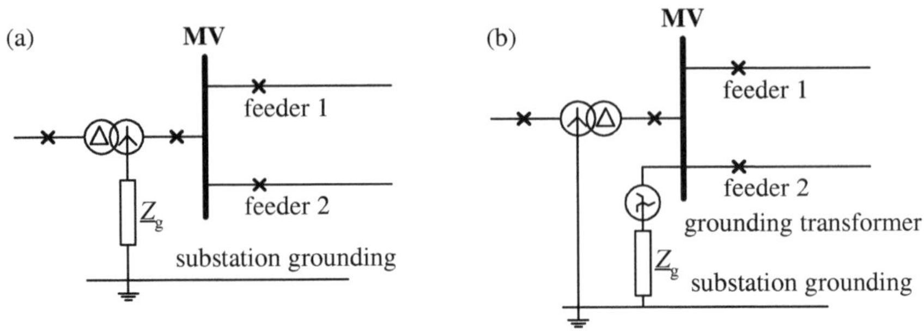

Fig. 2.34 Methods of MV network grounding: (a) with delta-wye step-down grounding transformer; and (b) with grounding zigzag connected transformer; Z_g – grounding impedance

If the grounding impedance $\underline{Z}_g$ is intentionally selected for reducing a phase-to-ground fault, the grounding becomes non-effective (Fig. 2.35). For high grounding impedance the network capacitance plays a significant role.

There are two types of impedance grounding [96, 278]:

- grounding through resistor; and
- grounding through arc-suppressing coil (Petersen coil).

Resistance grounding is broadly dividing into two categories: high-resistance (when the fault current is limited to about 10 A) and low-resistance grounding (with fault current of some hundreds of amps). Generally, the high grounding resistance decreases the selectivity of the protective relays because the fault detection and the fault-place estimation are less accurate.

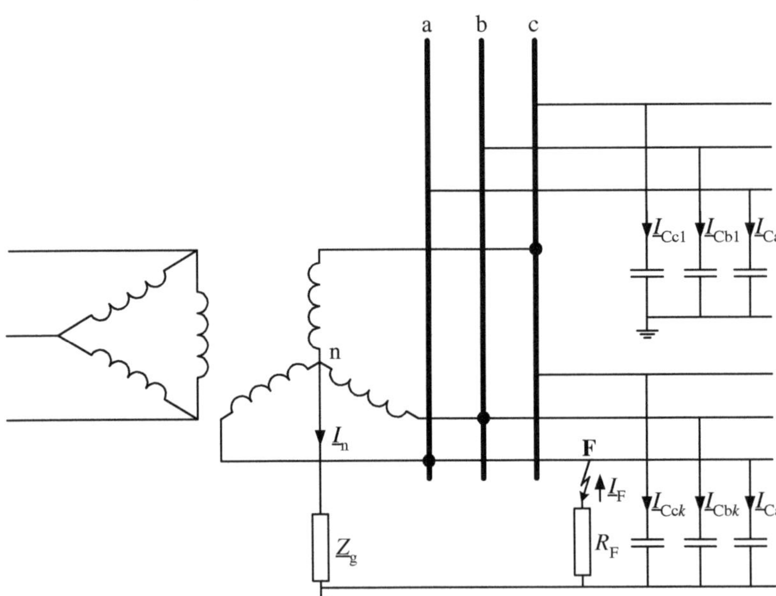

Fig. 2.35 Equivalent scheme of network grounding through impedance

Grounding the neutral through the Petersen coil is used for reducing the capacitive current in isolated networks. The adequate inductance in the neutral is selected for compensation of the capacitive current during a phase-to-ground fault; thus this method is also known as the ground-fault compensation method. The equivalent scheme for this arrangement is shown in Fig. 2.36. The coil reactance is adjustable in relatively coarse steps, to allow for changes in network zero-sequence capacitance resulting from the switching out of lines. If the coil inductance exactly matches the system capacitance, the fault current has resistive character and may be insufficient for selective relay operation. Frequently, the grounding equipment is completed with a resistor (resistor R_g in Fig. 2.36) that, with some delay, automatically paralleling the coil what permits fault current sufficient to correct relay operation.

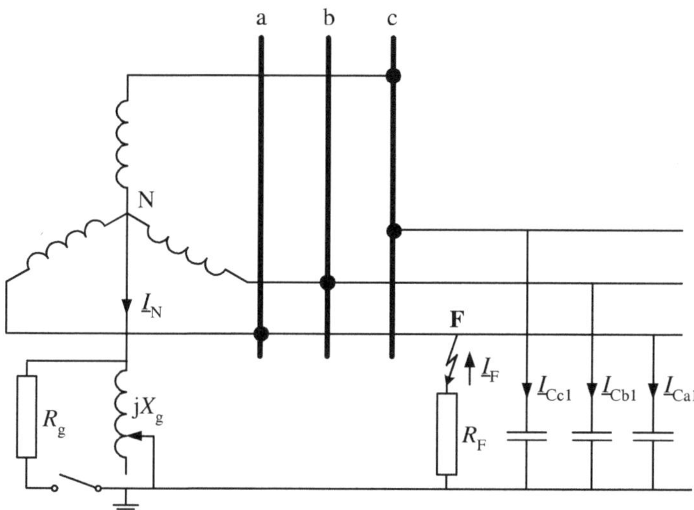

Fig. 2.36 Equivalent scheme of the network grounding through a Petersen coil

An isolated network is one in which there is no intentional connection between the circuit and ground (Fig. 2.37). The currents during single phase-to-ground faults are low and depend mostly on the phase to ground capacitances of the lines.

The voltage between faulted equipment and ground is small, which improves safety. However, transient and power-frequency overvoltages can be higher than those obtained for other type of neutral grounding. Moreover, a one-phase grounding results in the full line-to-line voltage appearing on the other two sound phases. This situation can frequently cause failures in motors and transformers, due to insulation breakdown. There are known difficulties in detecting and locating the first line-to-ground fault: the overcurrent principle is not suitable here because the capacitive current in a sound line may be even greater than that measured at the faulted line [95, 97].

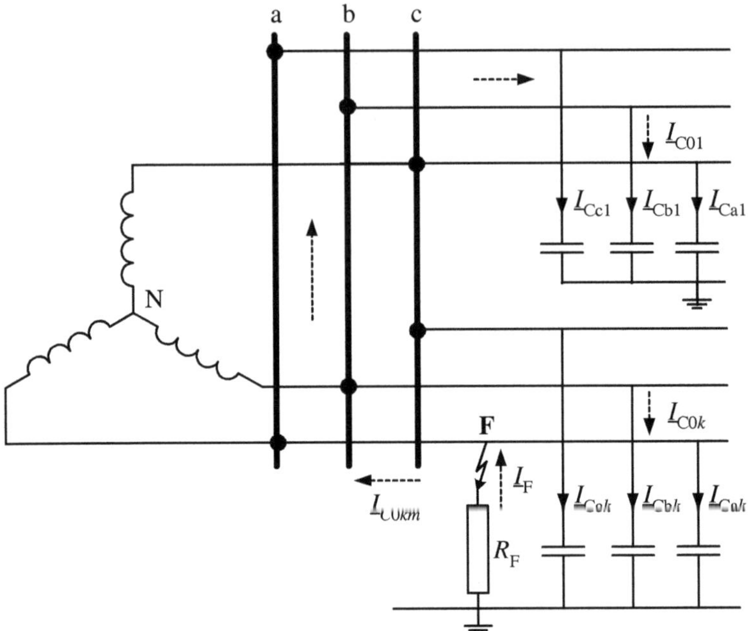

Fig. 2.37 Equivalent scheme of the isolated network; dotted lines show path for the fault current

2.5.3 Network Representation

Distribution networks are different from transmission and sub-transmission systems in the following characteristics:

- they operate at lower voltages than transmission lines;
- they are usually used radially;
- they usually have tapped loads along the line, not just at the terminals (where the measurements are available); and
- due to no line transposition and single-phase lateral and load, such networks usually should be considered as unbalanced networks.

Usually we consider a distribution system as a three-phase main line with lateral branches, some of which may be one- or two-phase circuits. Frequently, the system is wye connected with a ground at the supply transformer and often with multiple additional neutral grounds. The method of the neutral grounding has a considerable influence on the value and character of the phase-to-ground fault current. Practically, the impedance-based fault-location algorithms for phase-to-ground faults can be used only in effectively grounding network. On the other hand, the grounding method does not influence the accuracy for the other types of faults (as for example phase-to-phase).

The above-mentioned remarks are particularly important for cable networks. In branching points or laterals along the feeders, the cable sheaths are connected together and, generally, grounded through different resistances (Fig. 2.38a). Cable sheaths usually remain connected even if the respective sections are disconnected (Fig. 2.38b). This causes different equivalent schemes for positive/negative and zero-sequence of the network.

The circuit model described earlier did not include the effects of frequency dependence of the circuit parameters. Although this is unlikely to be a problem for the overhead line sections, it is possible that some cable configurations may exhibit a frequency dependence in the range 50–300 Hz, due to, for example, the non-linearity of steel armor.

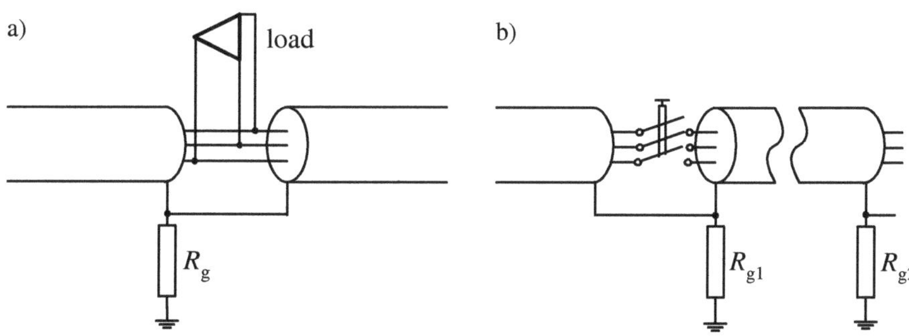

Fig. 2.38 Cable-connection principle: (a) load connection; and (b) connection with switches

This could readily be incorporated into the circuit model using similar techniques adopted for time-domain solutions of frequency-dependent cable-parameter modeling [96, 97].

3 Power-line Faults – Models and Analysis

Abstract This chapter presents the basics of power-line faults. The fault models are formulated using symmetrical-components and phase-coordinates approaches. These models will be utilized in the fault-location algorithms presented in this book. The analysis of arcing faults, including typical waveforms of current and voltage signals, obtained from the ATP–EMTP simulation, is presented. Fault-identification algorithms for fault detection, phase selection and directional discrimination are briefly discussed.

3.1 Introduction

There are various causes of faults occurring in power systems. Breakdown of the insulation can be caused by lightning strokes on overhead lines. As a result, the connection with ground via a ground wire is established. Also, such ground connection occurs when a tree or a man-made object provides the connecting path. Some faults are also caused by switching mistakes of the station personnel.

Power-system faults can be shunt, series or a combination of shunt and series faults. Under a shunt fault there is a flow of current between two or more phases, or between phase(s) and ground. A series fault is an abnormality at which the impedances of the three phases are not equal, usually caused by the interruption of one or two phases.

Most of the power-system faults occur in transmission networks, especially on overhead lines. Lines are the elements in the system that are in charge of transporting important bulks of energy from generator plants to load centers. Due to their inherent characteristic of being exposed to atmospheric conditions, transmission lines have the highest fault rate in the system. Therefore, a number of researchers have devoted great effort to the line protection and fault location.

There are various known fault statistics, which are related to different voltage levels, technical and weather conditions. All of them unambiguously indicate that around 75% of the total number of power-system faults occurs in transmission networks. This fact reveals the very high importance of fault analysis for transmission networks.

3.2 Fault on Power Lines

Faults on EHV/UHV overhead lines are in majority single-phase-to-ground arcing faults and are mostly temporary. Therefore, protective relays are provided with the automatic reclosing function [8, 53, 300]. This function allows the line to be reclosed and kept in operation after the fault has disappeared because the arc can self-extinguish. The circuit breakers can operate on a single phase (single pole) or on all three phases. For applying a proper autoreclosing option the fault type is required to be correctly recognized by the fault-location techniques [55, 171, 177].

The main characteristic of faults is related to the fault impedance involved, which can basically be considered as fault resistance. In this respect, the faults are categorized as bold and resistive ones, respectively. Usually, for inter-phase faults, fault resistances are small and in general do not exceed 0.5 Ω. They may, however, become much higher during ground faults, because tower footing resistance may be as high as 10 Ω [174]. If there is a flashover of an insulator, the connection of towers with ground wires makes the resulting fault resistance smaller. In practice, it seldom exceeds 3 Ω. For some ground faults the fault resistances may become much higher, which happens in cases of fallen trees, or if a broken conductor lies on the high-resistive soil [326].

3.2.1 Fault Types

For formulating fault-location algorithms mostly the basic linear models of faults, such as presented in Fig. 3.1, are considered. The fault resistance involved is denoted by R_F and the resistance connected to ground in the case of inter-phase faults involving ground (Figs. 3.1c and e) by R_g. Note that it is assumed that the fault resistance R_F for inter-phase faults (Figs. 3.1b–e) is identical in all phases. Such basic fault models are considered for both the symmetrical components approach and the phase co-ordinates approach as well.

Besides numerous fault-location techniques considering the basic linear fault models, there have also been developed some algorithms considering the electric-arc phenomenon [51, 54, 190, 191, 244]. Also, this phenomenon is widely used in digital simulations [169, 170, 217] aimed at evaluating the performance of the algorithms based on the linear fault models.

Broken-conductor or open-conductor failure in one phase (phase a) is shown in Fig. 3.2a. However, such a failure may also happen in two phases. Broken-conductor failure may also occur coupled with this phase-to-ground fault (Figs. 3.2b and c). For such combined faults, different sequences, as seen from the measuring point (M in Figs. 3.2b and c) can be considered. For a fault from Fig. 3.2b, an open-conductor failure is located outside the fault loop seen from the measuring point, while it is inside for the case of Fig. 3.2c. Such two combined faults (Figs. 3.2b and c) impose different conditions on fault location [110].

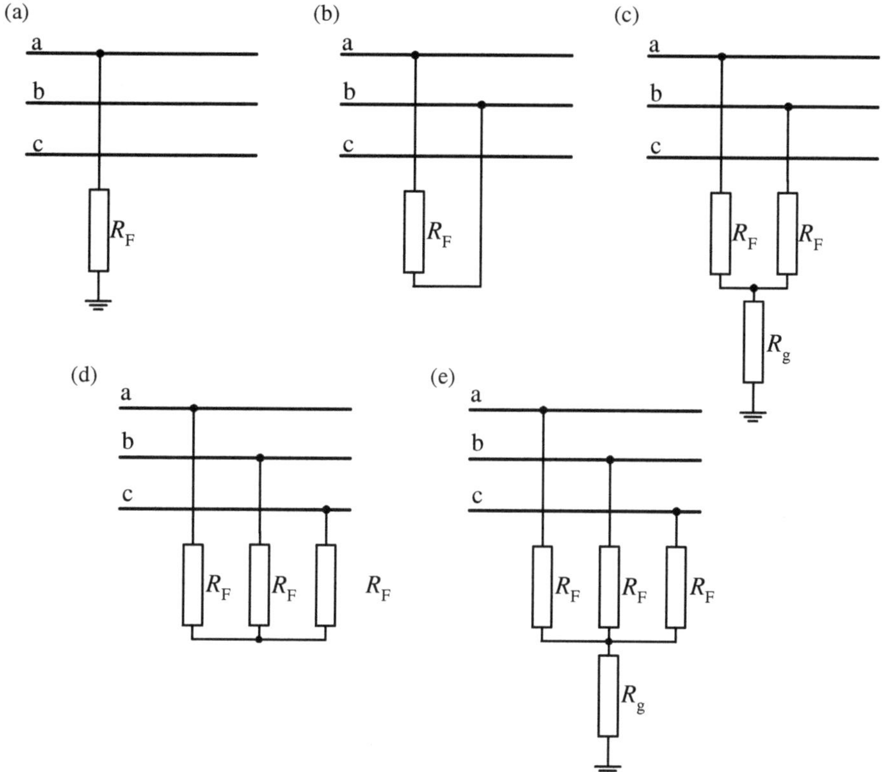

Fig. 3.1 Typical shunt faults: (a) phase-to-ground, (b) phase-to-phase, (c) phase-to-phase-to-ground, (d) three-phase, and (e) three-phase-to-ground

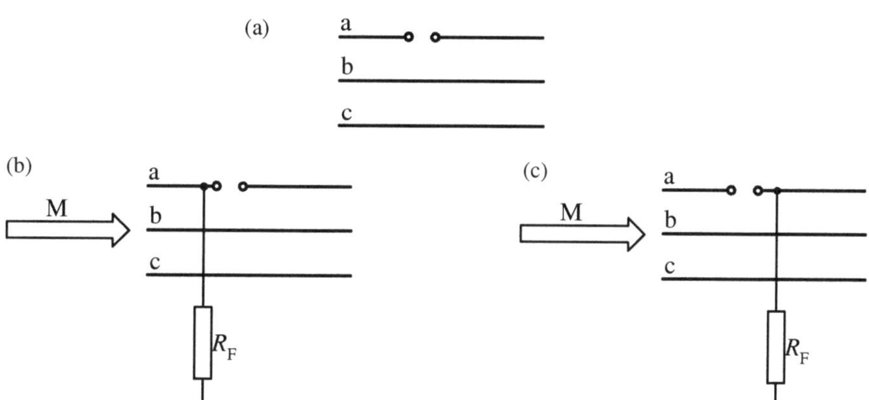

Fig. 3.2 Broken-conductor faults: (a) broken-conductor failure alone, (b) phase-to-ground fault with broken conductor, and (c) broken conductor with phase-to-ground fault

Fig. 3.3 Phase-to-ground fault combined with phase-to-phase fault

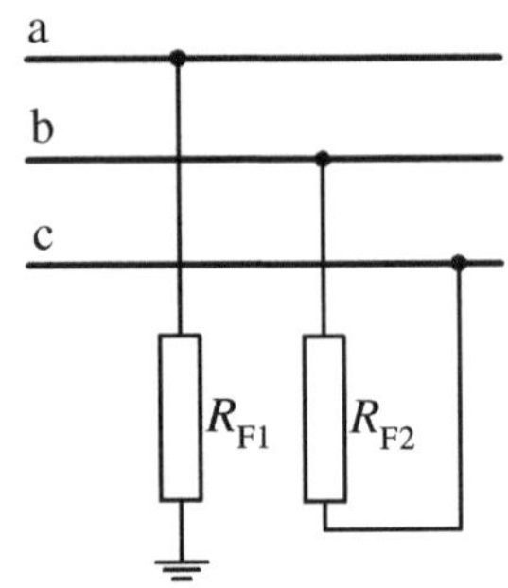

Sometimes more than one fault can occur simultaneously. For example, these may all be shunt faults, as shown in Fig. 3.3, where a phase-to-ground fault occurs in combination with a phase-to-phase fault for the remaining phases. In general, different fault resistances (R_{F1}, R_{F2}) can be involved in these faults.

Double faults [238] are considered as faults to ground, occurring simultaneously at two different locations in one or several circuits. In Fig. 3.4a, such a fault, also called a cross-country fault is shown as occurring on a double-circuit line.

Flashover faults on a double-circuit line (Fig. 3.4b) are usually caused by a lightning stroke to ground wire or tower, or due to a direct lightning stroke to a phase conductor [39].

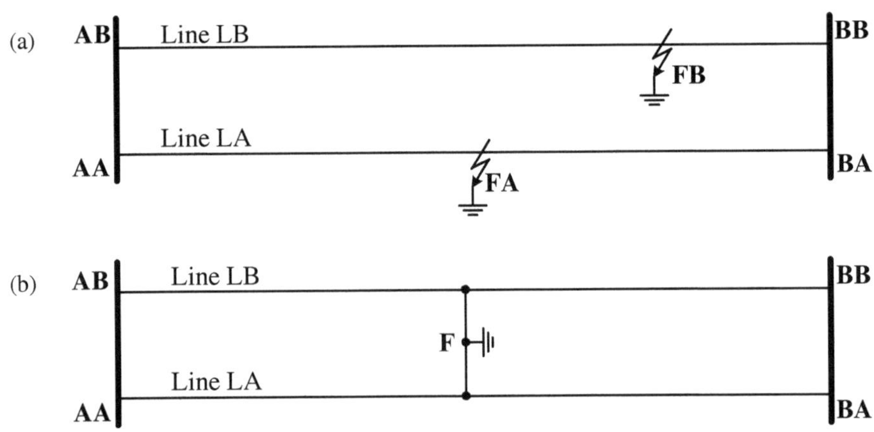

Fig. 3.4 Faults on double-circuit lines: (a) cross-country ground fault, and (b) flashover fault to ground

A fault involving two different nominal power-system voltages is called an inter-system fault. Such faults can occur on transmission lines hung on the same tower and rated at different voltages.

Besides the faults described, there are also multiple faults, as for example, faults to ground occurring simultaneously at more than two different locations in one or several circuits originated from a common source.

3.2.2 Fault Statistics

Numerous statistics data concerning power-system faults are available in the literature and Internet. They differ depending on which power-system element is considered and on the voltage level. In order to be acquainted briefly with the fault statistics for power transmission lines, in Tables 3.1 and 3.2 the representative data [109] is provided. In general, the number of faults decreases with an increase of the voltage level (Table 3.1). As regards the fault type, it is the single phase-to-ground faults that most frequently occur (Table 3.2).

Table 3.1 Number of faults on lines of different voltage levels (fault/year/100 km)

Source of information	**Voltage level**	
	(200–250) kV	**(300–500) kV**
Poland	3–5	1–3
CIGRE	0.4–10.4	0.4–4.68
IEEE	1.24	0.83
NORDEL (Denmark, Norway, Iceland, Finland, Sweden)*	1.0	0.3
Former Soviet Union	1.5	1

* NORDEL statistics is available at: www.nordel.org

Table 3.2 Statistics of different fault types occurring on lines of different voltage levels (fault/year/100 km)

Fault type	**Voltage level**	
	(200–250) kV	**(300–500) kV**
Single phase-to-ground faults	2.64	2.2
Phase-to-phase-to-ground faults	0.56	0.16
Faults involving more than one circuit of a line	0.11	0.06
Faults involving circuits of lines of different voltage levels	0.005	0.004

3.2.3 Models of Resistive Faults in Symmetrical Components

There is a wide group of fault-location algorithms that require determining the total fault current ($\underline{I}_F$), i.e., the current flowing through the fault-path resistance. These are mainly fault-location algorithms utilizing one-end measurement. Also, the majority of two-end or multi-end fault-location algorithms, but not utilizing complete three-phase voltage and current measurements, belong to this group. Use of complete measurements of three-phase voltage and current at both ends, or at all ends in the case of multi-terminal lines, allows determination of the total fault current to be avoided.

In general, since the total fault current ($\underline{I}_F$) is an immeasurable quantity, it can be calculated [115, 124] or estimated [62, 109, 121] by expressing it with the measured quantities, the network parameters and the unknown distance to fault. If the currents from all line terminals are available, then the total fault current can be calculated. However, since the required total fault current is calculated with use of the distant measurements, a special calculation method aimed at minimizing the influence of the line shunt capacitances charging has to be applied [124]. When the currents from all line terminals are not available, then the total fault current has to be estimated.

However, for both the above-mentioned cases (calculation or estimation of the total fault current) the total fault current can be expressed as the following weighted sum of its symmetrical components:

$$\underline{I}_F = \underline{a}_{F0}\underline{I}_{F0} + \underline{a}_{F1}\underline{I}_{F1} + \underline{a}_{F2}\underline{I}_{F2} \tag{3.1}$$

where:

$\underline{a}_{F0}$, $\underline{a}_{F1}$, $\underline{a}_{F2}$ – weighting coefficients (complex numbers), dependent on fault type and the assumed priority for using particular symmetrical components,

$\underline{I}_{F0}$, $\underline{I}_{F1}$, $\underline{I}_{F2}$ – zero-, positive- and negative-sequence components of total fault current, which are to be calculated or estimated.

Determination of the total fault current (3.1) is required for reflecting the voltage drop across the fault path ($\underline{U}_F$) in the fault loops considered in the fault-location algorithms:

$$\underline{U}_F = R_F \underline{I}_F \tag{3.2}$$

Depending on the fault type, the phase-to-ground or phase-to-phase fault loops are considered, as for the protective distance relays [39, 62, 109, 121].

It appears that there is some freedom in setting the weighting coefficients in (3.1). Example 3.1 illustrates this for a phase ‘a’ to ground fault.

Example 3.1 Determination of the weighting coefficients for a–g fault

At the place of the fault occurring in the phase ‘a’ there is a flow of the total fault current: $\underline{I}_F = \underline{I}_{Fa}$, while in the remaining phases $\underline{I}_{Fb} = 0$, $\underline{I}_{Fc} = 0$. Calculating the symmetrical components of the total fault current and taking into account the constraints for the fault considered one obtains:

$$\begin{bmatrix} \underline{I}_{F0} \\ \underline{I}_{F1} \\ \underline{I}_{F2} \end{bmatrix} = \frac{1}{3}\begin{bmatrix} 1 & 1 & 1 \\ 1 & \underline{a} & \underline{a}^2 \\ 1 & \underline{a}^2 & \underline{a} \end{bmatrix}\begin{bmatrix} \underline{I}_{Fa} \\ 0 \\ 0 \end{bmatrix} = \frac{1}{3}\begin{bmatrix} \underline{I}_{Fa} \\ \underline{I}_{Fa} \\ \underline{I}_{Fa} \end{bmatrix}$$

which results in:

$$\underline{I}_{F0} = \underline{I}_{F1} = \underline{I}_{F2} = \frac{1}{3}\underline{I}_{Fa}$$

It follows from the above that the total fault current ($\underline{I}_F = \underline{I}_{Fa}$) can be expressed in the following alternative ways, depending on which symmetrical component is preferred:

- $\underline{I}_F = \underline{I}_{F0} + \underline{I}_{F1} + \underline{I}_{F2}$,
- $\underline{I}_F = 3\underline{I}_{F1}$,
- $\underline{I}_F = 3\underline{I}_{F2}$,
- $\underline{I}_F = 3\underline{I}_{F0}$,
- $\underline{I}_F = 1.5\underline{I}_{F1} + 1.5\underline{I}_{F2}$, and others not listed here.

Analogously, determination of the total fault current can be considered for the other fault types.

Different priority with respect to using the particular symmetrical components [62, 109, 121] can be applied, as for example:

- use of the zero-sequence is generally avoided since the zero-sequence impedance for the overhead line is considered as an uncertain parameter;
- use of the negative- instead of positive-sequence is preferred, because the line shunt capacitance charging is more extensive for the positive-sequence;
- use of both positive- and negative-sequence, excluding the zero-sequence ($\underline{I}_F = 1.5\underline{I}_{F1} + 1.5\underline{I}_{F2}$ for the fault considered in the above example) is preferred, as it allows the calculations to be made simpler [62, 122].

Instead of using the positive-sequence components, which for three-phase balanced faults are present alone, the incremental (superimposed) positive-sequence components are recommended due to a reduced shunt capacitance effect [3–5, 225].

Tables 3.3–3.6 gather alternative sets of the weighting coefficients for different faults, depending on the assumed priority for using respective sequences. In Tables 3.3–3.5 the zero-sequence components are avoided ($\underline{a}_{F0}=0$) and different priority for using particular sequences is applied:

- Table 3.3 – the priority for the negative-sequence quantities;
- Table 3.4 – the priority for the positive-sequence quantities; and
- Table 3.5 – the positive- and negative-sequence components are uniformly utilized.

In turn, in Table 3.6 the positive-sequence components are avoided ($\underline{a}_{F1}=0$) for all faults, except three-phase balanced faults.

Table 3.3 Set of weighting coefficients from (3.1) with eliminating zero-sequence and giving priority to using negative- over positive-sequence

Fault type	Total fault current	$\underline{a}_{F1}$	$\underline{a}_{F2}$	$\underline{a}_{F0}$
a–g	$\underline{I}_{Fa}$	0	3	0
b–g	$\underline{I}_{Fb}$	0	$-1.5+\text{j}1.5\sqrt{3}$	0
c–g	$\underline{I}_{Fc}$	0	$-1.5-\text{j}1.5\sqrt{3}$	0
a–b	$\underline{I}_{Fa}-\underline{I}_{Fb}$	0	$1.5-\text{j}0.5\sqrt{3}$	0
b–c	$\underline{I}_{Fb}-\underline{I}_{Fc}$	0	$\text{j}\sqrt{3}$	0
c–a	$\underline{I}_{Fc}-\underline{I}_{Fa}$	0	$-1.5-\text{j}0.5\sqrt{3}$	0
a–b–g	$\underline{I}_{Fa}-\underline{I}_{Fb}$	$1.5+\text{j}0.5\sqrt{3}$	$1.5-\text{j}0.5\sqrt{3}$	0
b–c–g	$\underline{I}_{Fb}-\underline{I}_{Fc}$	$-\text{j}\sqrt{3}$	$\text{j}\sqrt{3}$	0
c–a–g	$\underline{I}_{Fc}-\underline{I}_{Fa}$	$1.5-\text{j}0.5\sqrt{3}$	$1.5+\text{j}0.5\sqrt{3}$	0
a–b–c (a–b–c–g)*	$\underline{I}_{Fa}-\underline{I}_{Fb}$	$1.5+\text{j}0.5\sqrt{3}$	$(1.5-\text{j}0.5\sqrt{3})$**	0
	* – inter-phase fault loop (a–b) is considered, however, the other fault loops (b–c), (c–a) can be taken as well, ** – this coefficient is different from zero; however, the negative-sequence is not present in signals.			

Table 3.4 Set of weighting coefficients from (3.1) with eliminating zero-sequence and giving priority to using positive- over negative-sequence

Fault type	Total fault current	$\underline{a}_{F1}$	$\underline{a}_{F2}$	$\underline{a}_{F0}$
a–g	$\underline{I}_{Fa}$	3	0	0
b–g	$\underline{I}_{Fb}$	$-1.5-\text{j}1.5\sqrt{3}$	0	0
c–g	$\underline{I}_{Fc}$	$-1.5+\text{j}1.5\sqrt{3}$	0	0
a–b	$\underline{I}_{Fa}-\underline{I}_{Fb}$	$1.5+\text{j}0.5\sqrt{3}$	0	0
b–c	$\underline{I}_{Fb}-\underline{I}_{Fc}$	$-\text{j}\sqrt{3}$	0	0
c–a	$\underline{I}_{Fc}-\underline{I}_{Fa}$	$-1.5+\text{j}0.5\sqrt{3}$	0	0
a–b–g	$\underline{I}_{Fa}-\underline{I}_{Fb}$	$1.5+\text{j}0.5\sqrt{3}$	$1.5-\text{j}0.5\sqrt{3}$	0
b–c–g	$\underline{I}_{Fb}-\underline{I}_{Fc}$	$-\text{j}\sqrt{3}$	$\text{j}\sqrt{3}$	0
c–a–g	$\underline{I}_{Fc}-\underline{I}_{Fa}$	$1.5-\text{j}0.5\sqrt{3}$	$1.5+\text{j}0.5\sqrt{3}$	0
a–b–c (a–b–c–g)*	$\underline{I}_{Fa}-\underline{I}_{Fb}$	$1.5+\text{j}0.5\sqrt{3}$	$(1.5-\text{j}0.5\sqrt{3})$**	0
	* and ** – remarks as in Table 3.3.			

Table 3.5 Set of weighting coefficients from (3.1) with eliminating zero-sequence and using both positive- and negative-sequence

Fault type	Total fault current	$\underline{a}_{F1}$	$\underline{a}_{F2}$	$\underline{a}_{F0}$
a–g	$\underline{I}_{Fa}$	1.5	1.5	0
b–g	$\underline{I}_{Fb}$	$-0.75-\mathrm{j}0.75\sqrt{3}$	$-0.75+\mathrm{j}0.75\sqrt{3}$	0
c–g	$\underline{I}_{Fc}$	$-0.75+\mathrm{j}0.75\sqrt{3}$	$-0.75-\mathrm{j}0.75\sqrt{3}$	0
a–b	$\underline{I}_{Fa}-\underline{I}_{Fb}$	$0.75+\mathrm{j}0.25\sqrt{3}$	$0.75-\mathrm{j}0.25\sqrt{3}$	0
b–c	$\underline{I}_{Fb}-\underline{I}_{Fc}$	$-\mathrm{j}0.5\sqrt{3}$	$\mathrm{j}0.5\sqrt{3}$	0
c–a	$\underline{I}_{Fc}-\underline{I}_{Fa}$	$-0.75+\mathrm{j}0.25\sqrt{3}$	$-0.75-\mathrm{j}0.25\sqrt{3}$	0
a–b–g	$\underline{I}_{Fa}-\underline{I}_{Fb}$	$1.5+\mathrm{j}0.5\sqrt{3}$	$1.5-\mathrm{j}0.5\sqrt{3}$	0
b–c–g	$\underline{I}_{Fb}-\underline{I}_{Fc}$	$-\mathrm{j}\sqrt{3}$	$\mathrm{j}\sqrt{3}$	0
c–a–g	$\underline{I}_{Fc}-\underline{I}_{Fa}$	$1.5-\mathrm{j}0.5\sqrt{3}$	$1.5+\mathrm{j}0.5\sqrt{3}$	0
a–b–c (a–b–c–g)*	$\underline{I}_{Fa}-\underline{I}_{Fb}$	$1.5+\mathrm{j}0.5\sqrt{3}$	$(1.5-\mathrm{j}0.5\sqrt{3})$**	0
	* and ** – remarks as in Table 3.3.			

Table 3.6 Set of weighting coefficients from (3.1) with possible elimination of using positive-sequence

Fault type	Total fault current	$\underline{a}_{F1}$	$\underline{a}_{F2}$	$\underline{a}_{F0}$
a–g	$\underline{I}_{Fa}$	0	3	0
b–g	$\underline{I}_{Fb}$	0	$-1.5+\mathrm{j}1.5\sqrt{3}$	0
c–g	$\underline{I}_{Fc}$	0	$-1.5-\mathrm{j}1.5\sqrt{3}$	0
a–b	$\underline{I}_{Fa}-\underline{I}_{Fb}$	0	$1.5-\mathrm{j}0.5\sqrt{3}$	0
b–c	$\underline{I}_{Fb}-\underline{I}_{Fc}$	0	$\mathrm{j}\sqrt{3}$	0
c–a	$\underline{I}_{Fc}-\underline{I}_{Fa}$	0	$-1.5-\mathrm{j}0.5\sqrt{3}$	0
a–b–g	$\underline{I}_{Fa}-\underline{I}_{Fb}$	0	$3-\mathrm{j}\sqrt{3}$	$\mathrm{j}\sqrt{3}$
b–c–g	$\underline{I}_{Fb}-\underline{I}_{Fc}$	0	$\mathrm{j}2\sqrt{3}$	$\mathrm{j}\sqrt{3}$
c–a–g	$\underline{I}_{Fc}-\underline{I}_{Fa}$	0	$3+\mathrm{j}\sqrt{3}$	$\mathrm{j}\sqrt{3}$
a–b–c (a–b–c–g)*	$\underline{I}_{Fa}-\underline{I}_{Fb}$	$1.5+\mathrm{j}0.5\sqrt{3}$	$(1.5-\mathrm{j}0.5\sqrt{3})$**	0
	* and ** – remarks as in Table 3.3.			

In some fault-location algorithms [52] the following relation between the zero-sequence component of the total fault current and the remaining components for faults involving ground is utilized:

$$\underline{I}_{\mathrm{F0}} = \underline{b}_{\mathrm{F1}}\underline{I}_{\mathrm{F1}} + \underline{b}_{\mathrm{F2}}\underline{I}_{\mathrm{F2}} \tag{3.3}$$

where $\underline{b}_{\mathrm{F1}}$, $\underline{b}_{\mathrm{F2}}$ – coefficients dependent on fault type (Table 3.7). They are derived taking into account the constraints of the particular fault (Example 3.2 for b–c–g fault). There are two alternative sets (SET I and SET II in Table 3.7).

Table 3.7 Coefficients used in relation (3.3)

Fault type	SET I		SET II	
	$\underline{b}_{\mathrm{F1}}$	$\underline{b}_{\mathrm{F2}}$	$\underline{b}_{\mathrm{F1}}$	$\underline{b}_{\mathrm{F2}}$
a–g	0	1	1	0
b–g	0	$-0.5+\mathrm{j}0.5\sqrt{3}$	$-0.5-\mathrm{j}0.5\sqrt{3}$	0
c–g	0	$-0.5-\mathrm{j}0.5\sqrt{3}$	$-0.5+\mathrm{j}0.5\sqrt{3}$	0
a–b–g	$0.5-\mathrm{j}0.5\sqrt{3}$	$0.5+\mathrm{j}0.5\sqrt{3}$	as in SET I	
b–c–g	–1	–1		
c–a–g	$0.5+\mathrm{j}0.5\sqrt{3}$	$0.5-\mathrm{j}0.5\sqrt{3}$		

Example 3.2 Determination of the coefficients involved in (3.3) for b–c–g fault
At the fault place in the healthy phase 'a' there is no current, $\underline{I}_{\mathrm{Fa}} = 0$. Taking this, the symmetrical components of the total fault current are as follows:

$$\begin{bmatrix} \underline{I}_{\mathrm{F0}} \\ \underline{I}_{\mathrm{F1}} \\ \underline{I}_{\mathrm{F2}} \end{bmatrix} = \frac{1}{3}\begin{bmatrix} 1 & 1 & 1 \\ 1 & \underline{a} & \underline{a}^2 \\ 1 & \underline{a}^2 & \underline{a} \end{bmatrix}\begin{bmatrix} 0 \\ \underline{I}_{\mathrm{Fb}} \\ \underline{I}_{\mathrm{Fc}} \end{bmatrix} = \frac{1}{3}\begin{bmatrix} \underline{I}_{\mathrm{Fb}} + \underline{I}_{\mathrm{Fc}} \\ \underline{a}\underline{I}_{\mathrm{Fb}} + \underline{a}^2\underline{I}_{\mathrm{Fc}} \\ \underline{a}^2\underline{I}_{\mathrm{Fb}} + \underline{a}\underline{I}_{\mathrm{Fc}} \end{bmatrix}$$

The sum of positive- and negative-sequence currents equals:

$$\underline{I}_{\mathrm{F1}} + \underline{I}_{\mathrm{F2}} = \frac{1}{3}((\underline{a} + \underline{a}^2)\underline{I}_{\mathrm{Fb}} + (\underline{a}^2 + \underline{a})\underline{I}_{\mathrm{Fc}})$$

Taking into account the identity $1 + \underline{a} + \underline{a}^2 = 0$, one obtains:

$$\underline{I}_{\mathrm{F1}} + \underline{I}_{\mathrm{F2}} = \frac{1}{3}(-\underline{I}_{\mathrm{Fb}} - \underline{I}_{\mathrm{Fc}})$$

Finally, one obtains:

$$\underline{I}_{F0} = -\underline{I}_{F1} - \underline{I}_{F2}$$

The coefficients for the b–c–g fault considered are thus: $\underline{b}_{F1} = -1$, $b_{F2} = -1$ (Table 3.7).

3.2.4 Models of Resistive Faults in Phase Coordinates

If a line considered is untransposed or if there are devices switched into the line that during faults introduce additional asymmetry, a description of the faulted network can be performed using the *phase-coordinates approach* [135, 136, 216, 274]. Series capacitors equipped with metal-oxide varistors are such devices [86, 111, 274] introducing the phase asymmetry.

A generalized fault model [109, 272, 284, 285] is presented in Fig. 3.5. It allows different faults to be represented by assuming for the resistors R:

- $R = R_F$, if the particular connection exists due to the fault (R_F denotes the fault resistance); and
- normally open switch, if there is no such connection.

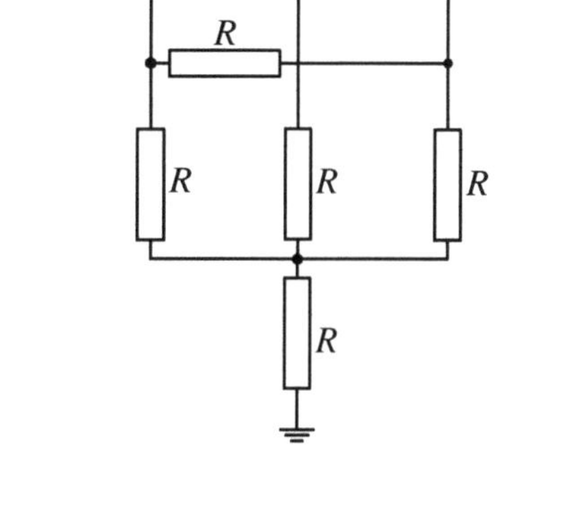

Fig. 3.5 Generalized fault model

Using the phase coordinates approach, a fault can be described with the following matrix formula:

$$\mathbf{I}_{\mathrm{F}} = \frac{1}{R_{\mathrm{F}}} \mathbf{K}_{\mathrm{F}} \mathbf{U}_{\mathrm{F}} \tag{3.4}$$

where:

$\mathbf{I}_{\mathrm{F}} = \begin{bmatrix} \underline{I}_{\mathrm{Fa}} \\ \underline{I}_{\mathrm{Fb}} \\ \underline{I}_{\mathrm{Fc}} \end{bmatrix}, \mathbf{U}_{\mathrm{F}} = \begin{bmatrix} \underline{U}_{\mathrm{Fa}} \\ \underline{U}_{\mathrm{Fb}} \\ \underline{U}_{\mathrm{Fc}} \end{bmatrix}$ – column matrices of total fault current and voltage drop at fault,

$\mathbf{K}_{\mathrm{F}} = \begin{bmatrix} k_{\mathrm{aa}} & k_{\mathrm{ab}} & k_{\mathrm{ac}} \\ k_{\mathrm{ba}} & k_{\mathrm{bb}} & k_{\mathrm{bc}} \\ k_{\mathrm{ca}} & k_{\mathrm{cb}} & k_{\mathrm{cc}} \end{bmatrix}$ – fault matrix, the elements of which are dependent on fault type, R_{F} – fault resistance (Fig. 3.5).

Fault matrix $\mathbf{K}_{\mathrm{F}}$ for different fault types is built in the following two-step procedure:

- Step I. Calculate the diagonal and off-diagonal elements of the auxiliary matrix ($\mathbf{K}_{\mathrm{F}}$):

$$k_{ij} = \begin{cases} -1, & \text{if phases } i, j \text{ are involved in the fault} \\ 0, & \text{otherwise} \end{cases} \qquad i, j = \mathrm{a, b, c} \tag{3.5}$$

Note that the diagonal elements of the auxiliary matrix, which is to be recalculated in Step II, are given in parentheses (...).

- Step II. Substitute the result of summing of absolute values in the respective column for each diagonal element of the auxiliary matrix obtained in Step I:

$$k_{ii} = \sum_{j=\mathrm{a}}^{j=\mathrm{c}} |\, k_{ij} \,| \qquad i = \mathrm{a, b, c} \tag{3.6}$$

Use of this two-step procedure is explained in detail in Examples 3.3 and 3.4.

Example 3.3 Determination of the fault matrix $\mathbf{K}_{\mathrm{F}}$ for a–b–g fault
Phases 'a' and 'b' are involved in this fault and thus the following settings in the auxiliary matrix ($\mathbf{K}_{\mathrm{F}}$) are made:

- $k_{aa} = -1$, $k_{bb} = -1$ (since there are connections of phases 'a' and 'b' to ground),
- $k_{ab} = k_{ba} = -1$ (since there is an interconnection between phases 'a' and 'b'),

while the remaining elements are set to zero.

As a result, one obtains the auxiliary matrix in the form:

$$(\mathbf{K}_F) = \begin{bmatrix} (-1) & -1 & 0 \\ -1 & (-1) & 0 \\ 0 & 0 & (0) \end{bmatrix}$$

The sums of absolute values of the elements in the respective columns are substituted for the related diagonal elements of the auxiliary matrix:

- 1st column: $|(-1)| + |-1| + 0 = 2 = k_{aa}$;
- 2nd column: $|-1| + |(-1)| + 0 = 2 = k_{bb}$;
- 3rd column: $0 + 0 + (0) = 0 = k_{cc}$.

Making these substitutions, one obtains the final form of the fault matrix for an a–b–g fault, which is as in the third column and the row 'a–b–g' of Table 3.8:

$$\mathbf{K}_F = \begin{bmatrix} 2 & -1 & 0 \\ -1 & 2 & 0 \\ 0 & 0 & 0 \end{bmatrix}$$

Example 3.4 Determination of the fault matrix $\mathbf{K_F}$ for an a–g fault
For an a–g fault only phase 'a' is involved in a fault and thus one obtains according to the Step I procedure:

$$(\mathbf{K}_F) = \begin{bmatrix} (-1) & 0 & 0 \\ 0 & (0) & 0 \\ 0 & 0 & (0) \end{bmatrix}$$

Applying then the procedure of the Step II the following final form of the fault matrix for an a–g fault (Table 3.8: third column and 'a–g' row) is obtained:

$$\mathbf{K}_F = \begin{bmatrix} 1 & 0 & 0 \\ 0 & 0 & 0 \\ 0 & 0 & 0 \end{bmatrix}$$

Table 3.8 Steps I and II of determining fault matrix $\mathbf{K}_F$ for different faults

Fault type	STEP I (3.5)	STEP II (3.6)
a–g	$(\mathbf{K}_F)=\begin{bmatrix}(-1) & 0 & 0\\ 0 & (0) & 0\\ 0 & 0 & (0)\end{bmatrix}$	$\mathbf{K}_F=\begin{bmatrix}1 & 0 & 0\\ 0 & 0 & 0\\ 0 & 0 & 0\end{bmatrix}$
b–g	$(\mathbf{K}_F)=\begin{bmatrix}(0) & 0 & 0\\ 0 & (-1) & 0\\ 0 & 0 & (0)\end{bmatrix}$	$\mathbf{K}_F=\begin{bmatrix}0 & 0 & 0\\ 0 & 1 & 0\\ 0 & 0 & 0\end{bmatrix}$
c–g	$(\mathbf{K}_F)=\begin{bmatrix}(0) & 0 & 0\\ 0 & (0) & 0\\ 0 & 0 & (-1)\end{bmatrix}$	$\mathbf{K}_F=\begin{bmatrix}0 & 0 & 0\\ 0 & 0 & 0\\ 0 & 0 & 1\end{bmatrix}$
a–b	$(\mathbf{K}_F)=\begin{bmatrix}(0) & -1 & 0\\ -1 & (0) & 0\\ 0 & 0 & (0)\end{bmatrix}$	$\mathbf{K}_F=\begin{bmatrix}1 & -1 & 0\\ -1 & 1 & 0\\ 0 & 0 & 0\end{bmatrix}$
b–c	$(\mathbf{K}_F)=\begin{bmatrix}(0) & 0 & 0\\ 0 & (0) & -1\\ 0 & -1 & (0)\end{bmatrix}$	$\mathbf{K}_F=\begin{bmatrix}0 & 0 & 0\\ 0 & 1 & -1\\ 0 & -1 & 1\end{bmatrix}$
c–a	$(\mathbf{K}_F)=\begin{bmatrix}(0) & 0 & -1\\ 0 & (0) & 0\\ -1 & 0 & (0)\end{bmatrix}$	$\mathbf{K}_F=\begin{bmatrix}1 & 0 & -1\\ 0 & 0 & 0\\ -1 & 0 & 1\end{bmatrix}$
a–b–g	$(\mathbf{K}_F)=\begin{bmatrix}(-1) & -1 & 0\\ -1 & (-1) & 0\\ 0 & 0 & (0)\end{bmatrix}$	$\mathbf{K}_F=\begin{bmatrix}2 & -1 & 0\\ -1 & 2 & 0\\ 0 & 0 & 0\end{bmatrix}$
b–c–g	$(\mathbf{K}_F)=\begin{bmatrix}(0) & 0 & 0\\ 0 & (-1) & -1\\ 0 & -1 & (-1)\end{bmatrix}$	$\mathbf{K}_F=\begin{bmatrix}0 & 0 & 0\\ 0 & 2 & -1\\ 0 & -1 & 2\end{bmatrix}$
c–a–g	$(\mathbf{K}_F)=\begin{bmatrix}(-1) & 0 & -1\\ 0 & (0) & 0\\ -1 & 0 & (-1)\end{bmatrix}$	$\mathbf{K}_F=\begin{bmatrix}2 & 0 & -1\\ 0 & 0 & 0\\ -1 & 0 & 2\end{bmatrix}$
a–b–c	$(\mathbf{K}_F)=\begin{bmatrix}(0) & -1 & -1\\ -1 & (0) & -1\\ -1 & -1 & (0)\end{bmatrix}$	$\mathbf{K}_F=\begin{bmatrix}2 & -1 & -1\\ -1 & 2 & -1\\ -1 & -1 & 2\end{bmatrix}$
a–b–c–g	$(\mathbf{K}_F)=\begin{bmatrix}(-1) & -1 & -1\\ -1 & (-1) & -1\\ -1 & -1 & (-1)\end{bmatrix}$	$\mathbf{K}_F=\begin{bmatrix}3 & -1 & -1\\ -1 & 3 & -1\\ -1 & -1 & 3\end{bmatrix}$

3.2.5 Arcing Faults

According to the fault current state, the fault arc [134, 170, 300] is classified as:

- primary arc; and
- secondary arc.

The *primary arc* occurs during flashover of the line insulator string, caused by lightning stroke or other reasons. The *secondary arc* follows the primary one when the faulted phase circuit breaker trips, as is sustained by mutual coupling between the healthy and faulted phases.

The primary arc appears after fault inception and lasts until single-phase tripping of the faulted phase. It shows generally a deterministic behavior as observed in the field and laboratory arc tests [314]. After isolating the fault (by single-phase tripping) there is a secondary arc, which is sustained by the capacitive and inductive coupling to the sound phases. The secondary arc usually self-extinguishes. The secondary arc has extremely random characteristics affected by the external conditions around the arc channel.

A vast majority of fault-location algorithms process current and voltage signals from the fault interval (starting from the fault inception until the circuit-breaker operation) and in some cases of the pre-fault interval (just before the fault inception). For these algorithms the primary arc is of interest. However, it mainly concerns simulations performed for evaluating the fault-location algorithms under study. This is so since the vast majority of fault-location algorithms apply a linear model of the fault path for their formulation. Only a few fault-location algorithms take into account the primary-arc model.

In [304], it is shown that by measuring voltage on the line side of a circuit breaker, the location of a permanent fault can be calculated using the transient caused by the fault-clearing operation of the circuit breaker. Due to the fault-clearing operation of the circuit breaker, a surge is initiated and travels between the opened circuit breaker and the fault, if the latter is still present. The distance to fault is determined by measuring the propagation time of the surge from the opened circuit breaker to the fault. In this relation, modeling the secondary arc is important.

Dynamic Model of Arc

The dynamic voltage–current characteristics of the electric arc have features of hysteresis. Extensive studies in [134, 170, 300] have shown that the dynamic volt-ampere characteristics of the electric arc can be exactly simulated by the empirical differential equation:

$$\frac{\mathrm{d}g_k}{\mathrm{d}t} = \frac{1}{T_k}(G_k - g_k) \tag{3.7}$$

where:
the subscript k indicates the kind of arc as follows: k = p for primary arc, k = s for secondary arc,
g_k – dynamic arc conductance,
G_k – stationary arc conductance,
T_k – time constant.

The stationary arc conductance G_k can be physically interpreted as the arc conductance value when the arc current is maintained for a sufficiently long time under constant external conditions. So, G_k is the static characteristic of the arc, which can be evaluated from:

$$G_k = \frac{|i|}{(u_{0k} + R|i|)l_k} \tag{3.8}$$

where:
i – instantaneous arc current,
u_{0k} – arc voltage drop per unit length along the main arc column,
R – characteristic arc resistance per unit length,
l_k – arc length.

For the primary arc $u_{0\mathrm{p}}$ is constant and equal to about 15 V/cm for the range of current 1.3–24.0 kA [58] and l_p may be assumed constant and somehow wider than the length of the line insulator string. The value of the constant-voltage parameter of the secondary arc $v_{0\mathrm{s}}$ is evaluated empirically on the basis of numerous investigation results in the range of low values of current, collected in [58]. For the range of peak currents I_s, from approximately (1–55) A it can be roughly defined as $u_{0\mathrm{s}} = 75 I_\mathrm{s}^{-0.4}$ V/cm [300].

The arc length of the secondary current l_s changes with time, and for relatively low wind velocities (up to 1 m/s), it can be approximated as $l_\mathrm{s} = 10 l_\mathrm{p} t_\mathrm{r}$ for $t_\mathrm{r} > 0.1\,\mathrm{s}$ but when the secondary arc re-ignition time $t_\mathrm{r} \le 0.1\,\mathrm{s}$: $l_\mathrm{s} = l_\mathrm{p}$.

The secondary-arc re-ignition voltage (in V/cm) can be calculated using the empirical formula [300]:

$$u_\mathrm{r} = \frac{5 + 1620 T_\mathrm{e}}{(2.15 + I_\mathrm{s})(t_\mathrm{r} - T_\mathrm{e})} \tag{3.9}$$

where:
T_e – secondary arc extinguishing time (when $t_\mathrm{r} \le T_\mathrm{e}$, $u_\mathrm{r} = 0$),
I_s – peak value of current on the volt-ampere arc characteristic.

Time constants are determined as follows [300]:

$$T_k = \frac{\alpha_k I_k}{l_k} \tag{3.10}$$

where α_k – empirical coefficients.

The empirical coefficients α_k can be obtained by fitting (3.7) with (3.8) and (3.10) to match the experimental dynamic volt-ampere characteristics of the heavy- and low-current arcs, accordingly.

The model (3.7) allows the arc conductance $g(t)$ to be determined, from which the arc resistance $r_{\text{arc}}(t) = 1/g(t)$ is calculated.

Using the ATP-EMTP program [56] for arc fault simulation, the arc can be reflected with the non-linear resistor – defined in the ELECTRICAL NETWORK unit, while the arc model – in the MODELS. Interaction of these units of the ATP-EMTP program is shown in Fig. 3.6. The arc current as the input quantity is measured on-line and the non-linear differential equation (3.7) is being solved. As a result, the arc resistance is determined and transferred for fixing the resistance of the resistor modeling the arc. Typical waveforms of arc voltage and currents are shown in Fig. 3.7. Considering these signals one can observe a hysteresis feature of the arc, which is shown in Fig. 3.8a for a single fundamental-frequency cycle. In Fig. 3.8b a change of arc resistance with the passing of time is presented. The resistance of the arc was determined from the voltage and current signals presented in Fig. 3.7.

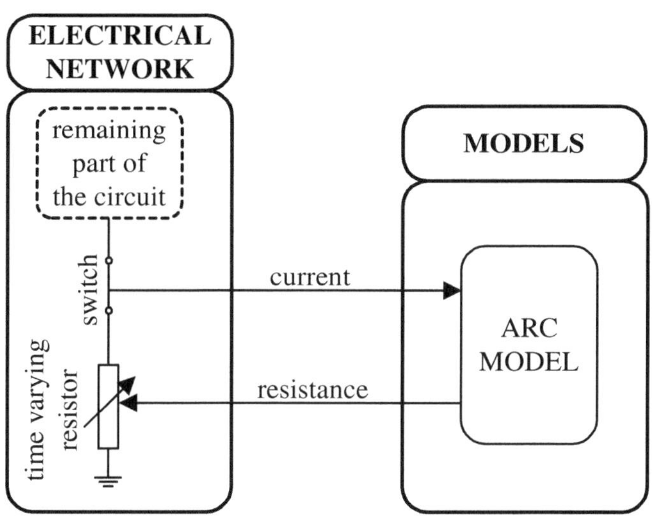

Fig. 3.6 Modeling of primary arc with ATP-EMTP – interaction between the program units

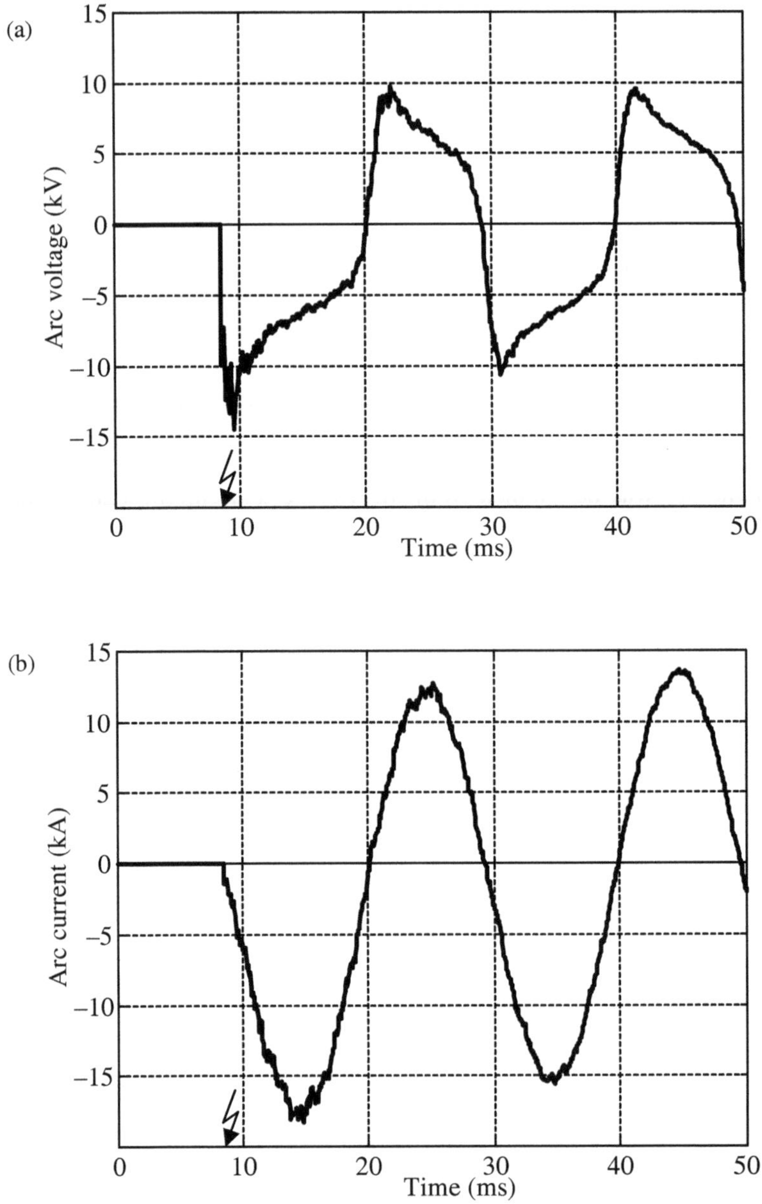

Fig. 3.7 Modeling of primary arc with ATP-EMTP: (a) arc voltage, and (b) arc current

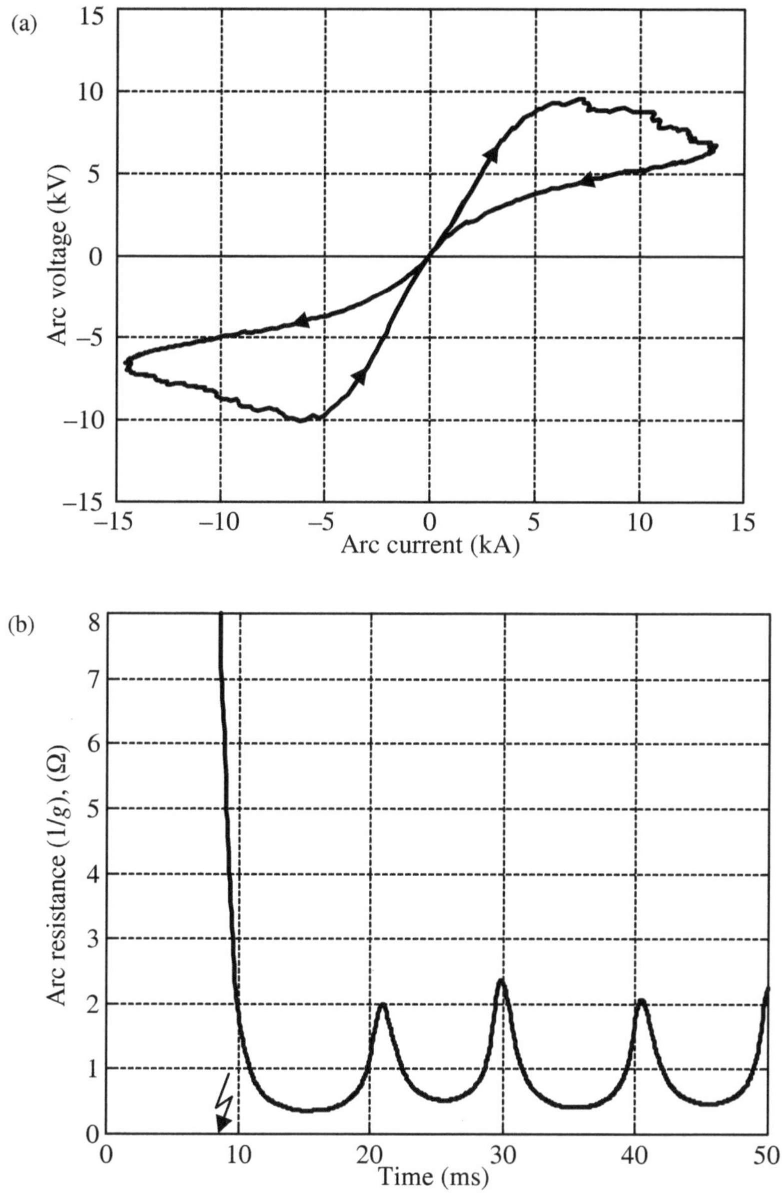

Fig. 3.8 Modeling of primary arc with ATP-EMTP: (a) arc voltage *versus* arc current (visualized for a single cycle only), and (b) arc resistance

Static Model of Primary Arc

For many applications a simpler static model of the primary arc (Figs. 3.9 and 3.10) is utilized [51, 53, 67]. The voltage drop across an arc is determined as:

$$u(t) = U_{\mathrm{a}}\,\mathrm{signum}[i(t)] + \xi(t)\,, \tag{3.11}$$

where:

$U_{\mathrm{a}} = U_{\mathrm{p}} l_{\mathrm{p}}$ – magnitude of rectangular wave (U_{p}, l_{p} – as in (3.8)),

$\xi(t)$ – Gaussian noise with zero average value.

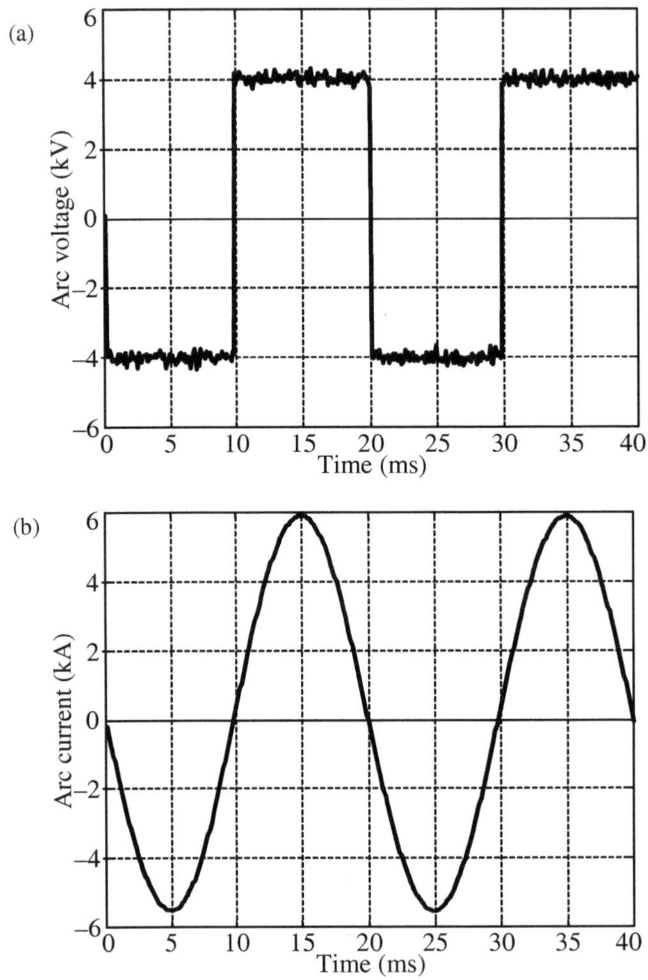

Fig. 3.9 Static model of primary arc: (a) arc voltage, and (b) arc current

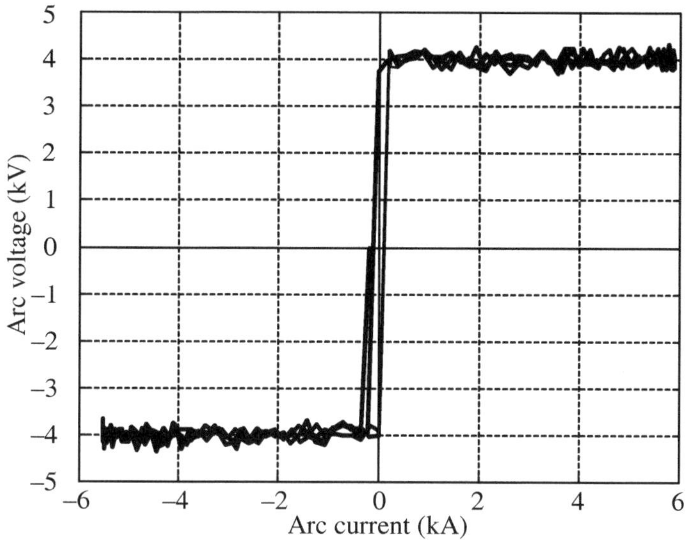

Fig. 3.10 Static model of primary arc: arc voltage *versus* arc current

Use of the static model of primary arc for time-domain solution of fault-distance estimation and arcing-fault detection on overhead lines was introduced in [51, 53].

3.3 Fault-identification Algorithms

Fault identification is the first task in the fault-location algorithm. It includes fault detection, type of fault and direction determination that are needed for correct fault-loop determination. Generally, the used algorithms are similar to that applied for protective relaying purposes, however, as realized off-line, may adopt more sophisticated principles for error reducing. Frequently, that part of the common procedure is realized by the protective relay that delivers to the fault locator required information on the identified fault and the recorded data.

3.3.1 Fault Detection

A number of approaches to fault detection are proposed in the literature. The abnormal conditions (not necessarily faults) are detected by watching the phase impedances and/or phase-current amplitudes and/or phase-voltage amplitudes and/or zero-sequence current amplitude. Depending on a particular application, different activation criteria are combined in a different way. To speed up the fault detection, one may also apply derivatives of the relevant signals. It is quite easy to introduce adaptivity to such approaches. Knowing the breaker positions, monitoring the average load of a line, etc., the thresholds may be self-adjusted to improve the sensitivity and reliability of fault detection [148].

The general structure of the fault-detection algorithm is presented in Fig. 3.11. The used criteria are based on a statistical approach. Two different statistical measures of the voltage or current signal are determined in parallel filters F1 and F2 and some criterion value is formulated to assess the fault inception. Below, we present more useable realizations.

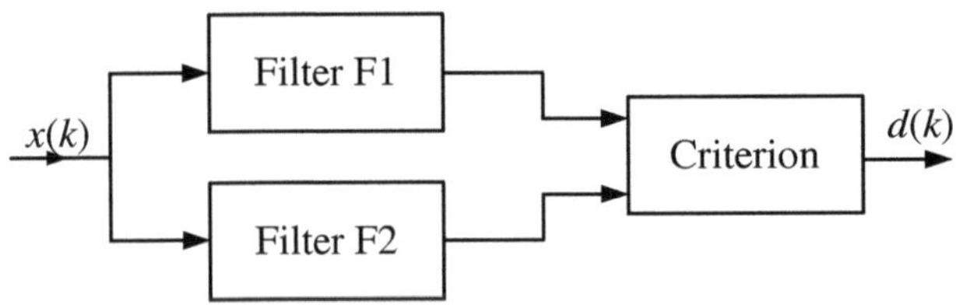

Fig. 3.11 General structure of fault-detection algorithm

Much easier methods refer directly to samples of current and/or voltage waveforms (Fig. 3.12). Disregarding a particular solution, two approaches are considered [77]:

- A *sample-by-sample* method computing numerically the first derivative of a watched signal. If this derivative overruns a pre-set value, an auxiliary counter starts to count up. This counter is incremented by the absolute value of the derivative. When it reaches another pre-set threshold, a fault is confirmed. Certainly, the first threshold must be set above the maximum value of the scaled derivative under normal conditions.
- A *cycle-by-cycle* algorithm compares a present sample with the sample one cycle back. The threshold for such a difference may be set much lower than in the sample-by-sample method. An auxiliary counter may be used to initiate when the absolute value of the defined difference overruns its threshold [77]. The fault is detected when the counter increased by the successive differences overreaches the second threshold.

Another method relies on the absolute difference between the MEAN and MEDIAN filters outputs [77]:

$$e_L(k) = |\,\text{median}(x(k), L) - \text{mean}(x(k), L)\,| \tag{3.12}$$

where:
x – considered relaying signal (voltage or current),
k – present sampling instant,
L – length of the data window.

The length of the data window, L, may be either fixed or adaptable. On-line data window adjustment can be applied for improving the fault detection.

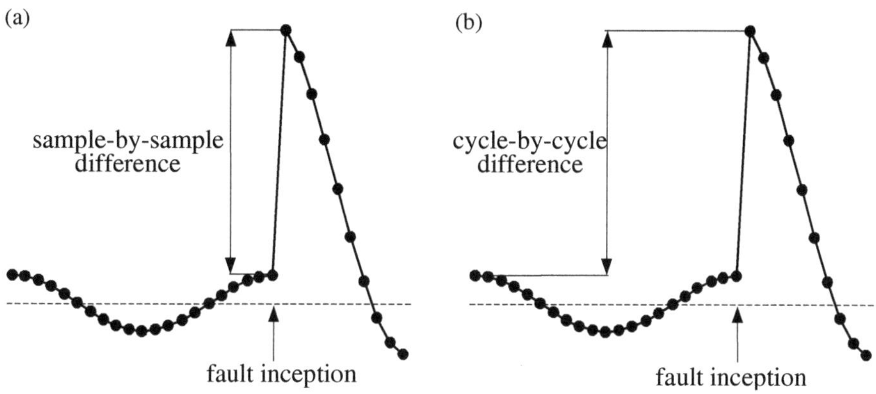

Fig. 3.12 Fault detection – illustration of differentiation methods: (a) sample-by-sample, and (b) cycle-by-cycle

There is a group of methods specially intended for the high-impedance fault detection. One of the known methods is based on harmonic spectrum analysis [176]. The algorithm detects high-impedance faults on an overhead line based upon the harmonic spectrum of the relaying signals. The algorithm utilizes the NEPM and NERM measures.

Normalized even-order power measure (NEPM) is computed once a cycle according to:

$$\mathrm{NEPM} = \sum_{k=6}^{32} \frac{I_{k(i)}^2 / I_{1(i)}^2}{I_{k(i-1)}^2 / I_{1(i-1)}^2} \tag{3.13}$$

where:
I_1, I_k – fundamental frequency and kth harmonic amplitudes, i – number of a cycle.

The other measure, proposed to avoid fault detection under some normal switching events, relates the NEPM from (3.13) to the normalized odd-order power measure (NOPM) – defined analogously to (3.13), but for odd harmonics from the range 7–33:

$$\mathrm{NERM} = \frac{\mathrm{NEPM}}{\mathrm{NOPM}} \tag{3.14}$$

In order to gain sensitivity to ground faults with very low fault currents the most advanced measure (even-order incremental variance measure – EIVM) has been proposed:

$$\mathrm{EIVM}(n) = \sum_{k=0}^{n} (\mathrm{NERM}(n-i) - \mathrm{NERM}(n-i-1))^2 \tag{3.15}$$

where n is a number of the considered cycles.

If the defined measures (3.13)–(3.15) overreach their pre-defined threshold a fault is being confirmed.

The observed signals in the case of series-compensated lines are affected by the series capacitors and their MOVs transients. During faults MOVs switch between linear and non-linear modes generating a variety of harmonics in the signals. Under high-resistance faults sub-synchronous oscillations may occur also. The performed study of series-compensated line transients proved that the measures (3.13)–(3.15) can be successfully adopted for fault detection in such lines [146]. Moreover, such fault detectors can be designed for detecting a certain class of faults. Namely, it is of vital importance to provide the decision whether or not the series capacitor (-s) is involved in a fault loop.

When considering a fault-detection issue it is worth listing traveling-wave approaches too. Correct classification of all types of faults from the information available at the local bus appears difficult, therefore a communication channel is required. However, significant improvement has been made by utilizing some extra information present at the local bus and related to the subsequent wave reflections [20, 40].

There are also several applications of artificial neural networks (ANNs) to fault detection. The most typical is aimed towards arcing fault detection [171]. Such algorithms are usually combined with other fault-identification purposes [239]. Some of them are considered in Chap. 9.

3.3.2 Phase Selection

Again, changing of different electrical values during fault are utilized for type of fault determination. The simplest method assumes that a phase impedance (resulting from a phase voltage and current) for a phase involved in a fault is below a certain level – correlated with the maximum load. In addition, the zero-sequence quantities (current and/or voltage) are used as indicators of faults with the ground. Instead of processing voltage and current samples directly, superimposed current and voltage samples can be used for fault classification too [12, 17].

Another family of methods uses relationships between the symmetrical components of the fault current and/or voltage. For speeding up of detection and to obtain more reliable decision, information about phase angle is used [148]. Two fault signatures: negative-sequence vs. positive-sequence and negative-sequence vs. zero-sequence current and/or voltage are used simultaneously. The criterion is based on checking the relations between adequate post-fault angles, which change significantly faster just after fault than do magnitudes. Let us present this method in more detail.

Generally, the approach based on using symmetrical components introduces the following three criteria spaces [148]:

- negative-sequence *versus* positive-sequence relation;
- negative-sequence *versus* zero-sequence relation; and
- significant increase of the positive-sequence quantity with absence of the negative- and zero-sequence components.

Using these criteria spaces it is possible to identify different types of fault. Detailed criteria can be defined with respect to different features of the considered sequence components. The second condition was added because it is well known that the presence of large negative-sequence components in the relay input signals reveals the occurrence of a fault in the supervised system. This is so for all faults but excluding the case of the balanced three-phase fault, for which only the positive-sequence current is present, however, with the increased value after a fault occurrence.

Unfortunately, during faults there are transients in the measured currents and thus the symmetrical components are determined with certain errors. Moreover, accuracy and speed of operation are in opposition, so the developer should decide on some kind of compromise.

Symmetrical components are defined for phasors of a three-phase system. There are known correct methods that are based on the full-cycle Fourier algorithms [193, 237], state observers [257] or Kalman filters [82]. In fast procedures the current phasors should be calculated with use of an adequately simple method. Such a method is proposed below.

Let the phasor of phase-a current (similarly for phase-b and -c currents) is defined as:

$$\underline{I}_{\mathrm{a}}(k) = I_{\mathrm{aR}}(k) + \mathrm{j}\, I_{\mathrm{aI}}(k) \tag{3.16}$$

Both orthogonal components in (3.16) can be obtained from two consecutive samples:

$$I_{\mathrm{aR}}(k) = \frac{i_{\mathrm{a}}(k) + i_{\mathrm{a}}(k-1)}{2\cos(\omega_1 T_{\mathrm{s}}/2)} \tag{3.17}$$

$$I_{\mathrm{aI}}(k) = \frac{-i_{\mathrm{a}}(k) + i_{\mathrm{a}}(k-1)}{2\sin(\omega_1 T_{\mathrm{s}}/2)} \tag{3.18}$$

where: $i_{\mathrm{a}}(k)$, $i_{\mathrm{a}}(k-1)$ – kth and (k–1)th samples of the input current (from the phase a); ω_1 – angular frequency of the fundamental component, T_{s} – sampling period.

The orthogonal components (3.17) and (3.18) are sinusoidal in the steady state. Unfortunately, during faults the currents contain high-frequency noise and thus the phasor estimates (3.16) are considerably deformed, which should be taken into considerations in the following steps. Having the phasor estimates (3.16) for a three-phase system it is possible to determine the symmetrical components of currents. Applying the known transformation [249] one obtains

$$\begin{aligned}
&\underline{I}_0\begin{cases}3I_{0\mathrm{R}}=I_{\mathrm{aR}}+I_{\mathrm{bR}}+I_{\mathrm{cR}}\\ 3I_{0\mathrm{I}}=I_{\mathrm{aI}}+I_{\mathrm{bI}}+I_{\mathrm{cI}}\end{cases}\\
&\underline{I}_1\begin{cases}3I_{1\mathrm{R}}=I_{\mathrm{aR}}-\frac{1}{2}I_{\mathrm{bR}}-\frac{\sqrt{3}}{2}I_{\mathrm{bI}}-\frac{1}{2}I_{\mathrm{cR}}+\frac{\sqrt{3}}{2}I_{\mathrm{cI}}\\ 3I_{1\mathrm{I}}=I_{\mathrm{aI}}-\frac{1}{2}I_{\mathrm{bI}}+\frac{\sqrt{3}}{2}I_{\mathrm{bR}}-\frac{1}{2}I_{\mathrm{cI}}-\frac{\sqrt{3}}{2}I_{\mathrm{cR}}\end{cases}\\
&\underline{I}_2\begin{cases}3I_{2\mathrm{R}}=I_{\mathrm{aR}}-\frac{1}{2}I_{\mathrm{bR}}+\frac{\sqrt{3}}{2}I_{\mathrm{bI}}-\frac{1}{2}I_{\mathrm{cR}}-\frac{\sqrt{3}}{2}I_{\mathrm{cI}}\\ 3I_{2\mathrm{I}}=I_{\mathrm{aI}}-\frac{1}{2}I_{\mathrm{bI}}-\frac{\sqrt{3}}{2}I_{\mathrm{bR}}-\frac{1}{2}I_{\mathrm{cI}}+\frac{\sqrt{3}}{2}I_{\mathrm{cR}}\end{cases}
\end{aligned} \tag{3.19}$$

where the particular phasors of phase currents from (3.19) are determined as in (3.17) and (3.18).

The complex symmetrical components calculated as in (3.19) can then be represented in an exponential form:

$$\underline{I}_m(k)=I_{m\mathrm{R}}(k)+\mathrm{j}I_{m\mathrm{I}}(k)=I_m\,\mathrm{e}^{\mathrm{j}(ak+\gamma_m)} \tag{3.20}$$

for m =0, 1, 2 – components.

Each type of fault is characterized by the appearance of the specific set of symmetrical component (or components) of currents [148]. An abnormal situation is identified if the adequate symmetrical component exceeds the pre-defined threshold. It is important that this threshold setting should be relatively high in order to avoid false operation during changes in normal load conditions of the system.

In the proposed algorithm the criteria values are defined in a form of the phase shift angles between the respective symmetrical components phasors, i.e., the angles determined for the ratios of the symmetrical components phasors:

$$\alpha=\mathrm{angle}\left(\frac{\underline{I}_2}{\underline{I}_1}\right)=\gamma_2-\gamma_1 \tag{3.21}$$

$$\beta=\mathrm{angle}\left(\frac{\underline{I}_2}{\underline{I}_0}\right)=\gamma_2-\gamma_0 \tag{3.22}$$

where the angle γ_m is defined for m =0, 1, 2 – components as in (3.20).

Using in (3.21) the incremental positive-sequence current ($\Delta \underline{I}_1$) instead of the positive-sequence current ($\underline{I}_1$) is advantageous and assures better accuracy.

The positions of the vectors representing the ratios of the respective symmetrical components phasors for different fault types are depicted in Fig. 3.13. It is shown in Fig. 3.13a that six vectors representing the ratio of the negative- and positive-sequence phasors (3.20) define nine different types of a fault: a–g, b–g, c–g, a–b, a–c, b–c, a–b–g, a–c–g and b–c–g. For distinguishing between phase-to-phase and phase-to-phase-to-ground faults it is required to consider also the zero-sequence current (Fig. 3.13b).

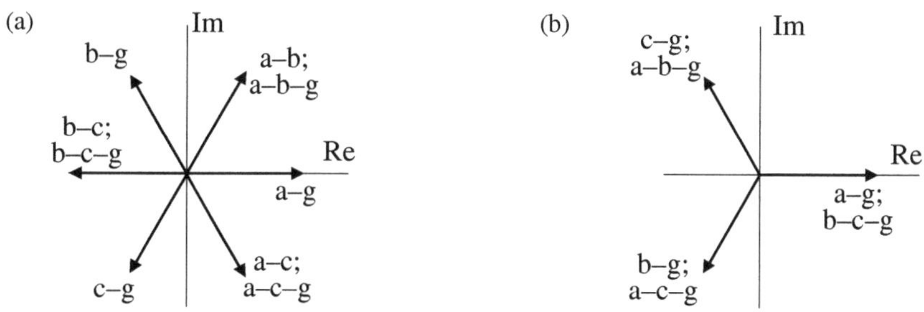

Fig. 3.13 Position of the vectors representing: (a) ratio of negative- and positive-sequence phasors, and (b) ratio of negative- and zero-sequence phasors

It should be pointed out that in a normal (no-fault) state, the zero- and the negative-sequence currents have negligible values and, therefore, the angles γ_0 and γ_2 could not be determined. To avoid undesirable errors, the criteria angles (3.21) and (3.22) are calculated only if i_1 and i_0 have non-zero values.

The structure of the algorithm [253] for determining the proposed criteria quantities (phase shift angles: α, β– defined in (3.21) and (3.22)) is shown in Fig. 3.14.

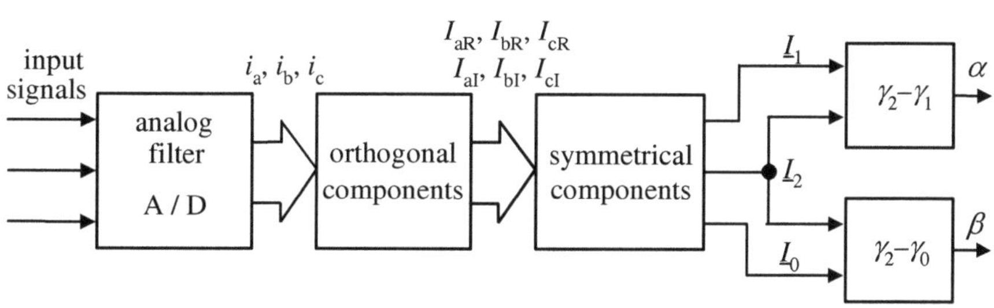

Fig. 3.14 Structure of the algorithm for determination of criteria quantities α and β

The final decision may be improved by using fuzzy logic reasoning (see Chap. 9).

3.3.3 Directional Detection

A directional element becomes essential for a complete distance relay when a fault to be cleared occurs very close to a relaying point. In the fault locators such a tool may be important when both lines (forward and backward) have different characteristics, e.g., one of them is series-compensated. The traditional approach to mitigate the problem of the voltage fall in distance relays is to apply an additional polarization voltage. It may be done by [287]:

- cross-polarization of a voltage signal; and
- memory polarization of a voltage signal.

In the fault locators information on direction to fault is immediately used for distance to fault estimation.

The traditional concept is based on computing the increments in positive-sequence voltage and current (by subtracting the pre-fault positive sequence quantities from the post-fault ones) [209]. The pre-fault values are recommended to be understood as delayed by 5 cycles. The impedance resulting from the increments of positive-sequence current and voltage is calculated as follows:

$$\Delta \underline{Z}_1 = \frac{\underline{U}_1 - \underline{U}_1^{\mathrm{pre}}}{\underline{I}_1 - \underline{I}_1^{\mathrm{pre}}} \tag{3.23}$$

where $\underline{U}_1^{\mathrm{pre}}$, $\underline{U}_1$ are positive-sequence voltages for pre- and fault periods, and analogously for currents.

Relation (3.23) enables very reliable direction detection. For a forward fault such an impedance is a local source impedance (as to the value and with the negative sign to account for the fact that the current is flowing from a source to a forward fault location). While for a reverse fault, this impedance equals the line impedance plus the remote source impedance. Certainly, for a series-compensated line, the line impedance must be understood as the line inductive impedance itself plus the capacitive impedance of the series *RC* equivalent of the parallel arrangement of a series capacitor and its MOV.

Figure 3.15 presents the performance of the positive-sequence directional element under a sample forward fault close to the substation (Fig. 3.15a) and reverse fault (Fig. 3.15b) in a 400-kV system. In the first case (forward fault) the impedance (3.23) establishing the steady-state position in the third quadrant of the complex plane Its value is equal to the local system source impedance multiplied by (–1). For the backward fault the impedance resulting from the increments of positive-sequence voltage and current locates in the 1st quadrant and settles at the value equal to the line impedance plus the remote source impedance.

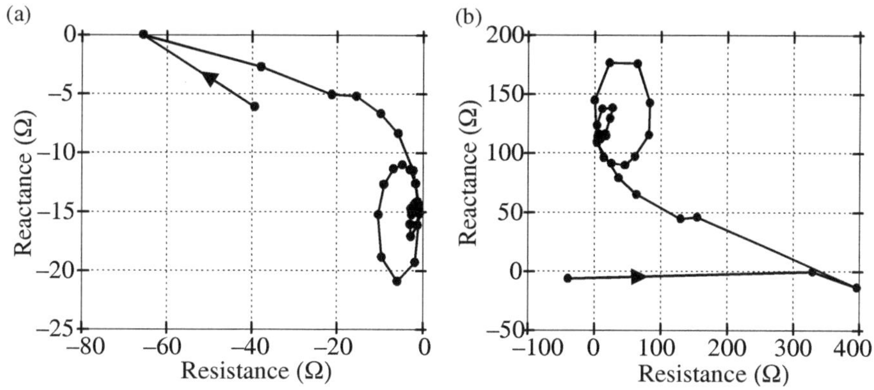

Fig. 3.15 Principle of directional detection: (a) forward fault, and (b) backward fault

Full-cycle Fourier filtration was used when determining the sequence components for both the tests of Fig. 3.15. The final criterion can be formulated with respect to the angle of the incremental impedance:

$$\theta_d < \text{angle}\left(\Delta \underline{Z}_1\right) < \theta_u \tag{3.24}$$

where quantities θ_d , θ_u define lower and upper thresholds.

The results of the tests performed show that the algorithm enables reliable and fast fault-direction detection.

4 Signal Processing for Fault Location

Abstract This chapter presents the basic signal-processing methods applied in the fault location for measurement of the electrical quantities and formulation the estimation procedures. This methods can be divided into three categories depending on signals representation and analytical tools used: frequency domain, time domain and time-frequency representation methods. All the three methods are discussed in this chapter from the point of view of their application to the fault-location algorithms.

4.1 Introduction

The estimate of distance to fault in a line is a function of network model parameters and measurements. This chapter gives a general review of the methods used for measuring the adequate quantities for fault location. Representation of the measured variables depends on the assumed network model. From that point of view one can distinguish the following methods for description of the state of an analyzed network:

- Specification of the relation between currents and voltages by means of time-domain functions. This representation is closely related to the classical circuit theory where transients phenomena are described in a form of a set of algebraic (copying Kirchhoff's laws) and differential equations (imitating a circuit-elements models). In the discrete-time domain differential equations are substituted by adequate difference relations. Such an approach is suitable for appropriate representation of a circuit dynamics [30, 50].
- The circuit description under assumption that the observed process is stable. In an AC system that is equivalent to the foundation that all the process variables (voltages and currents) are sinusoidal with constant magnitude and angle velocity ω. In such a condition a network may be entirely defined by use of voltage and current phasors and impedance (admittance) for the network parameters representation that is originated from the Fourier-transform (FT). Unlike in previous approaches, principally the time is not directly represented in phasors: all process variables are stationary. Such a treatment is also called frequency-domain circuit analysis [50, 84].

- Representation of an observed process as a dynamic one (as in the first mentioned approach) but with utilization of some of the tools and results obtained from the frequency-domain analysis. Such an approach arose from a desire to apply frequency-domain tools to analysis of a non-stationary process. This method is called the multi-resolution analysis and is connected with application of tools like the short time Fourier transform (STFT), the wavelet transform (WT) and others [167, 246, 249].

Parameters of the considered network (fault-loop parameters in the fault-location procedures) are determined by using the network model and measurements. As the network model is considered one can distinguish lumped- and distributed-parameters models. In this book we consider digital measurements, which is connected with conversion of continuous-time voltages and currents to their discrete representation at the input to the measurement unit.

Since the vast majority of digital protective relays and other controlling devices (including fault locators) is still based on phasors, the issue of phasor estimation is probably the single most investigated and documented subject in research on digital relaying and fault location [250]. Irrespective of the chosen circuit representation the voltage and current phasor estimation is inherently connected with a signal processing applied in fault location. Therefore, this chapter is mainly concentrated on the phasor-estimation problem.

4.2 Time-*Versus* Frequency-domain Analysis

The fault locator acquires information on the supervised process (connected with a fault on the line) through analog-to-digital converters (ADC) in a form of recorded voltages and currents samples. Naturally, the sampling proceeds in time converting this independent variable from continuous to discrete form. Observed voltage or current waveforms that vary over time (time-domain function) may be considered as examples of the more general concept of the signal. Consequently, the voltage or current measured at the output of instrument transformers represents the continuous-time signal, while the recorded samples correspond with the discrete-time signal.

A continuous-time circuit may be analyzed by using the classical circuit theory where dynamical relations are represented by differential equations. For example, for a lumped-parameters circuit as in Fig. 4.1a one can write as follows:

$$u(t) = Ri(t) + L\frac{\mathrm{d}i(t)}{\mathrm{d}t} + \frac{1}{C}\int_0^t i(\tau)\mathrm{d}\tau \tag{4.1}$$

which can be represented in more practical (from the calculation point of view) state-variables form:

$$\begin{aligned}\frac{\mathrm{d}u_{\mathrm{C}}(t)}{\mathrm{d}t} &= \frac{1}{C}i(t)\\ \frac{\mathrm{d}i(t)}{\mathrm{d}t} &= \frac{1}{L}u(t)-\frac{R}{L}i(t)-\frac{1}{L}u_{\mathrm{C}}(t)\end{aligned} \tag{4.2}$$

where: $u_{\mathrm{C}}(t)$, $i(t)$ – state variables [50, 84].

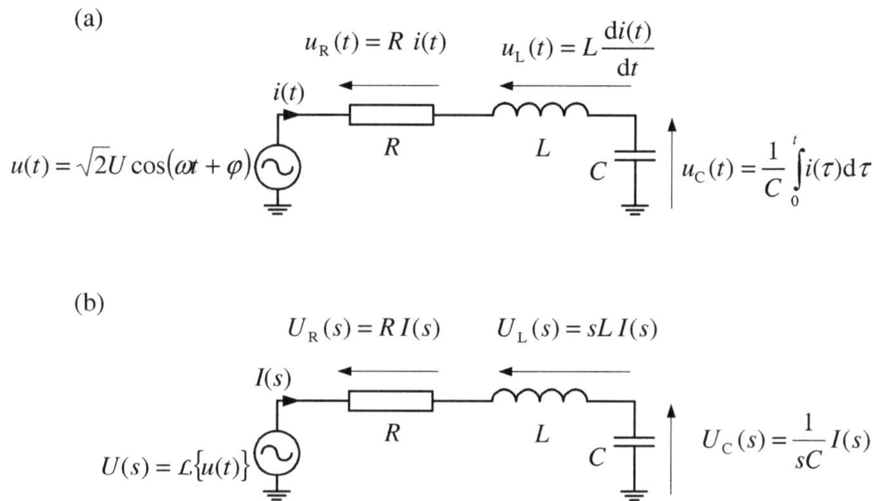

Fig. 4.1 Continuous-time circuit analysis: (a) in time domain, and (b) in Laplace domain (frequency domain)

The above can be represented in the more general matrix form:

$$\frac{\mathrm{d}}{\mathrm{d}t}\mathbf{x}(t)=\mathbf{A}\mathbf{x}(t)+\mathbf{B}\mathbf{u}(t) \tag{4.3}$$

where:

$$\mathbf{x}(t)=\begin{bmatrix}u_{\mathrm{C}}(t)\\ i(t)\end{bmatrix},\ \mathbf{A}=\begin{bmatrix}\frac{1}{C} & 0\\ -\frac{1}{L} & -\frac{R}{L}\end{bmatrix},\ \mathbf{B}=\begin{bmatrix}0 & 0\\ 0 & \frac{1}{L}\end{bmatrix},\ \mathbf{u}(t)=\begin{bmatrix}0\\ u(t)\end{bmatrix}.$$

Solution of (4.2) for given network parameters, and defined initial conditions: $u_{\mathrm{C}}(t_0)$, $i(t_0)$ delivers adequate waveforms for any profiles of excitation $u(t)$. Then, the state vector $\mathbf{x}(t)$ may be determined by integration of (4.3) [50]:

$$\mathbf{x}(t)=\mathrm{e}^{\mathbf{A}(t-t_0)}\mathbf{x}(t_0)+\int_0^t \mathrm{e}^{\mathbf{A}(t-\tau)}\mathbf{B}\mathbf{u}(\tau)\mathrm{d}\tau \tag{4.4}$$

Substituting the first equation in (4.2) into the second one yields the second-order differential equation:

$$u(t) = LC\frac{\mathrm{d}^2 u_\mathrm{C}(t)}{\mathrm{d}t^2} + RC\frac{\mathrm{d}u_\mathrm{C}(t)}{\mathrm{d}t} + u_\mathrm{C}(t) \tag{4.5}$$

Relation (4.1) can be easily transformed into the frequency domain. Assuming, for simplicity, zero initial conditions: t_0=0, $u_\mathrm{C}(0)$=0, $i(0)$=0 and applying Laplace transformation one obtains:

$$U(s) = R\cdot I(s) + sL\cdot I(s) + \frac{1}{sC}I(s) = Z(s)I(s) \tag{4.6}$$

where: $Z(s) = \frac{U(s)}{I(s)} = R + sL + \frac{1}{sC}$ is the Laplace impedance of the considered circuit, s is the complex variable: $s = \sigma + \mathrm{j}\omega$ in which, physically, σ plays the role of the signal attenuation (damping) and ω represents an angular velocity.

Similarly, the differential equation (4.4) takes the following form in the Laplace domain:

$$U(s) = s^2 LCU_\mathrm{C}(s) + sRCU_\mathrm{C}(s) + U_\mathrm{C}(s) = G(s)U_\mathrm{C}(s) \tag{4.7}$$

where: $G(s) = \frac{U_\mathrm{C}(s)}{U(s)} = \frac{1}{s^2 LC + sRC + 1}$ is called the Laplace transfer function. Note that the impedance in (4.6) is also some kind of transfer function.

The equivalent scheme of the considered circuit represented in the Laplace domain is presented in Fig. 4.1b. Functions $U(s)$, $I(s)$ in (4.6) are Laplace transforms of signals $u(t)$ and $i(t)$, respectively, obtained according to the well-known integral transformation:

$$U(s) = \mathcal{L}\{u(t)\} = \int_0^\infty u(t)\mathrm{e}^{-st}\mathrm{d}t \tag{4.8}$$

and similarly for current $i(t)$.

The transformation (4.8) is called unilateral or one-sided, in contrast to a bilateral or two-sided Laplace transform where the signal may be non-zero in both positive and negative time. The inverse Laplace transform defines the method for recovering the original time-domain function from its Laplace-domain image:

$$u(t) = \mathcal{L}^{-1}\{U(s)\} = \int_{\sigma-\mathrm{j}\infty}^{\sigma+\mathrm{j}\infty} U(s)\mathrm{e}^{st}\mathrm{d}s \tag{4.9}$$

Relations (4.8) and (4.9) should be considered as definitions and are rarely directly used; different calculation techniques are applied instead [30, 50].

A Laplace transform (e.g., $U(s)$) may be considered as a function in the frequency domain. If it is assumed that the original time-domain function $u(t)$ has zero initial conditions then the real part of the variable s may be taken equal to zero ($s = \mathrm{j}\omega$) and the Laplace transform is adequate to the Fourier transform (if the last exists): $\mathcal{L}\{u(t)\} = \mathcal{F}\{u(t)\}$, which is equivalent to: $U(s)\big|_{s=\mathrm{j}\omega} = U(\mathrm{j}\omega)$.

Replacing the circuit equations in the form of differential equations as in (4.1) by adequate Laplace-domain equations – as in (4.6) – considerably simplifies an analysis of the circuit dynamics (differential equations are reduced to algebraic equations). Therefore, the Laplace transform plays an important role in analysis of such systems. It is the cornerstone of classical circuits, systems and control theory.

A direct relation between time and frequency domains is defined by the Fourier transform, which, as was shown, is immediately connected with the Laplace transform. The classical Fourier transform is defined by the following pair of integrals:

$$U(\mathrm{j}\omega) = \mathcal{F}\{u(t)\} = \int_{-\infty}^{\infty} u(t)\mathrm{e}^{-\mathrm{j}\omega t}\mathrm{d}t \tag{4.10}$$

$$u(t) = \mathcal{F}^{-1}\{U(\mathrm{j}\omega)\} = \frac{1}{2\pi}\int_{-\infty}^{\infty} U(\mathrm{j}\omega)\mathrm{e}^{\mathrm{j}\omega t}\mathrm{d}\omega \tag{4.11}$$

The function $U(\mathrm{j}\omega)$ may be interpreted as the spectral representation of $u(t)$. Because it is complex-valued the Fourier-domain representation has both a magnitude and phase spectrum:

$$U(\mathrm{j}\omega) = \left|U(\mathrm{j}\omega)\right|\mathrm{e}^{\mathrm{j}\psi(\omega)} \tag{4.12}$$

where:

$$\left|U(\mathrm{j}\omega)\right| = \sqrt{U_{\mathrm{R}}^2(\omega) + U_{\mathrm{I}}^2(\omega)}\,, \tag{4.13}$$

$$\psi(\omega) = \tan^{-1}\frac{U_{\mathrm{I}}(\omega)}{U_{\mathrm{R}}(\omega)}\,. \tag{4.14}$$

A transform pair $u(t) \leftrightarrow U(\mathrm{j}\omega)$ represents a one-to-one mapping as long as a function $u(t)$ satisfies conditions, which guarantee the Fourier integral convergence. The Laplace transform can handle a more general class of functions. Both transforms can be considered as equivalent only for functions that vanish for $t<0$ [30]. Therefore, the Fourier transform has little significance in analysis of a system dynamics.[1] Yet, the clear interpretation of the Fourier transform and its properties are particularly useful in a signal analysis and filters design.

In the digital measurement systems considered here the continuous-time functions represented by voltage and current waveforms are transformed into discrete-time functions with adequately chosen sampling frequency f_s. Sampling process changes continuous time t into discrete time $t_k = kT_s$, k – sample number, $T_s = 1/f_s$ – sampling period, and:

$$u(t) \xrightarrow{\text{sampling}} u(t_k) = u(kT_s) \rightarrow u(k) \tag{4.15}$$

There is a known relation between the continuous and discrete-time domain of the signal. It is defined by the sampling theorem (known also as the Shannon sampling theorem or the Nyquist sampling theorem). The sampling theorem indicates that a continuous-time signal can be unambiguously represented in a discrete form only if it does not contain frequency components above one-half of the sampling rate [230]. Under such conditions the samples are a unique representation of the signal. The proper sampling (with respect to the sampling theorem) enables the exact reconstruction of the analog signal from the samples.

The Laplace transform applied for discrete-time signals and functions leads to a Z-transform, where for convenience, the operator s is adequately replaced by an operator z [30, 50]:

$$\begin{gathered} \mathcal{L}\{u(t_k)\} \leftrightarrow \mathcal{Z}\{u(k)\} \\ U(s) \leftrightarrow U(z) \\ z = \mathrm{e}^{sT_s} \end{gathered} \tag{4.16}$$

The last relation gives an easy way for frequency representation of the Z-transforms:

$$U(\mathrm{e}^{\mathrm{j}\omega T_s}) = U(z)\big|_{z=\mathrm{e}^{\mathrm{j}\omega T_s}} \tag{4.17}$$

Note that here we use $U(\mathrm{e}^{\mathrm{j}\omega T_s})$ instead of $U(\mathrm{j}\omega)$ to make it explicit that this is a function of $\mathrm{e}^{\mathrm{j}\omega T_s}$.

Similarly as in (4.8), the Z-transform is defined as:

$$U(z) = \mathcal{Z}\{u(k)\} = \sum_{k=-\infty}^{k=\infty} u(k) z^{-k} \tag{4.18}$$

---

[1] In some literature the generalized Fourier transform is defined in a similar way as the bilateral Laplace transform and both concepts become the same [30].

In digital signal processing the index k is counted from 0 that reduces (4.18) to a so-called unilateral Z-transform.

The Z-transform is a powerful tool for analysis of discrete-time systems. Similarly to the Laplace transform, it can be used to solve linear constant-coefficient difference equations. In other words, instead of solving these equations directly, we transform them into a set of algebraic equations first, and then solve in this transformed domain. Therefore, we can consider the Z-transform as a discrete-time counterpart of the Laplace transform.

It is a well-known consequence of the continuous-time signal sampling in the frequency domain: discrete-time signals have a periodic frequency spectrum and the period depends on sampling frequency [230]. For illustration of this principle let us consider an exponential continuous-time signal of the form (Fig. 4.2a):

$$x(t)=\begin{cases}\mathrm{e}^{-t/\tau} & \text{for } t\geq 0,\\ 0 & \text{for } t<0,\end{cases}\qquad \tau=0.02\ \text{s}$$

Applying the Fourier transform we obtain:

$$X(\mathrm{j}\omega)=\int_0^{+\infty}\mathrm{e}^{-t/\tau}\mathrm{e}^{-\mathrm{j}\omega t}\mathrm{d}t=\int_0^{+\infty}\mathrm{e}^{-(1/\tau+\mathrm{j}\omega)t}\mathrm{d}t=-\frac{1}{1/\tau+\mathrm{j}\omega}\mathrm{e}^{-(1/\tau+\mathrm{j}\omega)t}\Big|_0^{\infty}=\frac{1}{1/\tau+\mathrm{j}\omega}$$

(in that case we can use transformation (4.8) with $s=\mathrm{j}\omega$).

The real and imaginary parts of the spectrum are represented as the following real-valued functions (Fig. 4.2b):

$$X_{\mathrm{R}}(\omega)=\frac{1/\tau}{(1/\tau)^2-\omega^2},\quad X_{\mathrm{I}}(\omega)=\frac{-\omega}{(1/\tau)^2-\omega^2}$$

Let us now consider the same signal but sampled every $T_{\mathrm{s}}=0.001$ s (sampling frequency $f_{\mathrm{s}}=1000$ Hz). The discrete-time representation of such a signal has the form (Fig. 4.3a):

$$x(k)=\begin{cases}\mathrm{e}^{-T_{\mathrm{s}}k/\tau} & \text{for } k\geq 0,\\ 0 & \text{for } k<0,\end{cases}\qquad \tau=0.02\ \text{s},\ T_{\mathrm{s}}=0.001\ \text{s}$$

Applying the Z-transform (unilateral) one obtains:

$$X(z)=\sum_{k=0}^{+\infty}\mathrm{e}^{-T_{\mathrm{s}}k/\tau}z^{-k}=\sum_{k=0}^{+\infty}\left(\mathrm{e}^{T_{\mathrm{s}}/\tau}z\right)^{-k}=\frac{1}{1-\mathrm{e}^{-T_{\mathrm{s}}/\tau}z^{-1}}=\frac{z}{z-\mathrm{e}^{-T_{\mathrm{s}}/\tau}}$$

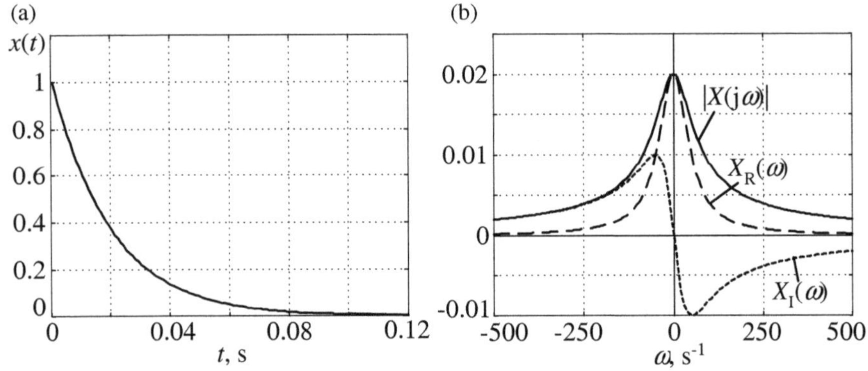

Fig. 4.2 Continuous-time signal (a) and its spectrum (b)

Applying the Z-transform (unilateral) one obtains:

$$X(z) = \sum_{k=0}^{+\infty} e^{-T_s k/\tau} z^{-k} = \sum_{k=0}^{+\infty} \left(e^{T_s/\tau} z\right)^{-k} = \frac{1}{1 - e^{-T_s/\tau} z^{-1}} = \frac{z}{z - e^{-T_s/\tau}}$$

The similar relation for the frequency domain leads to:

$$X(e^{j\omega T_s}) = \sum_{k=0}^{+\infty} e^{-T_s k/\tau} e^{-j\omega T_s k} = \sum_{k=0}^{+\infty} e^{-(1/\tau + j\omega)T_s k} = \frac{1}{1 - e^{-(1/\tau + j\omega)T_s}} = \frac{e^{j\omega T_s}}{e^{j\omega T_s} - e^{-T_s/\tau}}$$

Using trigonometric representation of the exponential function:

$$e^{j\omega T_s} = \cos\omega T_s + j\sin\omega T_s$$

we can obtain both components of the complex function:

$X(e^{j\omega T_s}) = X_R(e^{j\omega T_s}) + jX_I(e^{j\omega T_s})$, with:

$$X_R(e^{j\omega T_s}) = \frac{1 - e^{-T_s/\tau}\cos\omega T_s}{1 + e^{-T_s/\tau}\left(e^{-T_s/\tau} - 2\cos\omega T_s\right)}, \quad X_I(e^{j\omega T_s}) = \frac{-e^{-T_s/\tau}\sin\omega T_s}{1 - e^{-T_s/\tau}\left(e^{-T_s/\tau} - 2\cos\omega T_s\right)}$$

It is evident that the above functions are periodic with an interval of 2π (Fig. 4.4b). Taking into account the symmetry of the considered function it can be seen that all information on the original discrete-time signal is included in its spectrum with the frequency interval $(0, \pi/T_s)$ (the frequency $\omega = \pi/T_s$ is known as the Nyquist (Shannon) frequency).

A very useful property of Z-transform is connected with transformation of the shifted-in-time signal:

$$X(z)z^{-m} = Z\{x(k-m)\} \tag{4.19}$$

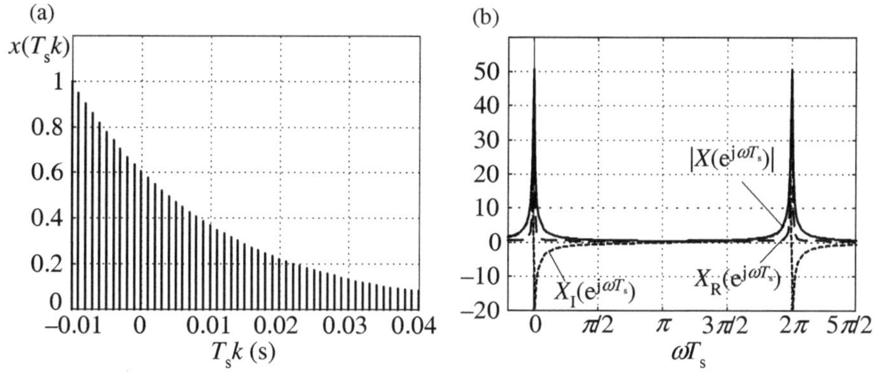

Fig. 4.3 Discrete-time signal (a) and its spectrum (b)

This property significantly facilitates analysis of difference equations.

Returning to our example with the simple circuit (Fig. 4.1) let us define the equivalent circuit for the discrete-time domain. Transition of continuous to discrete time involves transformation of differentiation and integration formulas in a circuit equation into its numerical equivalents. That is not an unambiguous operation because one can assume different numerical procedure, of integration or differentiation. Let us consider the trapezoidal formula for numerical integration.

Numerical formula for voltage determination on a capacitor can be derived from numerical integration of the first equation in (4.2):

$$u_{\mathrm{C}}(t_k) = u_{\mathrm{C}}(t_{k-1}) + \frac{1}{C}\int_{t_{k-1}}^{t_k} i(\tau)\mathrm{d}\tau \approx u_{\mathrm{C}}(t_{k-1}) + \frac{T_{\mathrm{s}}}{2C}\big(i(t_{k-1}) + i(t_k)\big) \tag{4.20}$$

from which:

$$u_{\mathrm{C}}(k) = r_{\mathrm{C}} i(k) + e_{\mathrm{C}}(k-1) \tag{4.21}$$

where: $r_{\mathrm{C}} = \dfrac{T_{\mathrm{s}}}{2C}$, $e_{\mathrm{C}}(k-1) = r_{\mathrm{C}} i(k-1) + u_{\mathrm{C}}(k-1)$, $t_k = kT_{\mathrm{s}}$.

Similarly, for inductance one can write:

$$i_{\mathrm{L}}(t_k) = i_{\mathrm{L}}(t_{k-1}) + \frac{1}{L}\int_{t_{k-1}}^{t_k} u(\tau)\mathrm{d}\tau \approx i_{\mathrm{L}}(t_{k-1}) + \frac{T_{\mathrm{s}}}{2L}\big(u(t_{k-1}) + u(t_k)\big) \tag{4.22}$$

from which:

$$u_{\mathrm{L}}(k) = r_{\mathrm{L}} i(k) + e_{\mathrm{L}}(k-1) \tag{4.23}$$

where: $r_{\mathrm{L}} = \dfrac{2L}{T_{\mathrm{s}}}$, $e_{\mathrm{L}}(k-1) = -r_{\mathrm{L}} i(k-1) - u_{\mathrm{L}}(k-1)$.

The above relations can be considered as digital models of a capacitor (4.21) and an inductance (4.23), associated with the trapezoidal rule. The equivalent discrete-time domain of the considered circuit is presented in Fig. 4.4a.

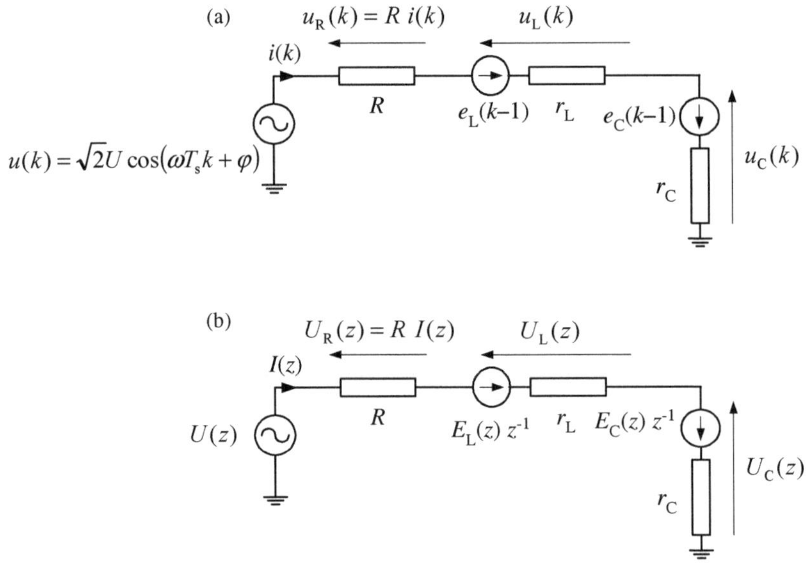

Fig. 4.4 Discrete-time circuit analysis: in the time domain (a) and in the Z-domain (b)

Transforming individual signals in Fig. 4.4a to Z-domain gives the equivalent circuit as in Fig. 4.4b. Note that:

$$E_C(z)z^{-1} = Z\{e_C(k-1)\} = r_C I(z)z^{-1} + U_C(z)z^{-1}$$

and similarly for $E_L(z)z^{-1}$.

The circuit equation (4.6) in the Z-domain takes the form:

$$U(z) = (R + r_L + r_C)I(z) + E_C(z)z^{-1} + E_L(z)z^{-1}$$

The time-domain version of this equation is as follows:

$$u(k) = (R + r_L + r_C)i(k) + e_C(k-1) + e_L(k-1)$$

It can be seen from this brief analysis that the Z-transform may be easily used for designing numerical algorithms: formulas represented as Z-transforms can be translated in an easy manner to the discrete-time-domain relations (algorithms). On the other hand, there is a close connection between the Laplace transform and the Z-transform. It may be deduced from the above-applied trapezoidal rule for numerical integration. Let us compare the Laplace transform of the function integration (omitting an initial condition):

$$y(t) = \int_0^t x(\tau)\mathrm{d}\tau \leftrightarrow Y(s) = \mathcal{L}\left\{\int_0^t x(\tau)\mathrm{d}\tau\right\} = \frac{1}{s}X(s) \tag{4.24}$$

with the similar operation in the discrete-time domain with application of the trapezoidal method for numerical integration:

$$y(k) = y(k-1) + \frac{T_\mathrm{s}}{2}\left(x(k) + x(k-1)\right) \leftrightarrow Y(z) = Y(z)z^{-1} + \frac{T_\mathrm{s}}{2}(1+z^{-1})X(z) \tag{4.25}$$

If we demand for both the Laplace and z-domain that for equal excitations: $X(s) \leftrightarrow X(z)$ also the outputs need to be the same: $Y(s) \leftrightarrow Y(z)$ then it results from the above that both expanses are connected with the relation:

$$s = \frac{2}{T_\mathrm{s}}\frac{1-z^{-1}}{1+z^{-1}} \tag{4.26}$$

known as the bilinear transformation [30, 230]. Using this transformation the continuous-time (analog) filter (or any input–output relation) is converted to a z-domain formula (digital filter) by replacing s in the Laplace transform according to (4.26) and substituting all s-transforms to Z-transforms: $X(s) \rightarrow X(z)$. Next, the discrete-time-domain algorithm may be deduced from the given Z-transform, similarly as in (4.24) and (4.25). It should be noted that application of the transformation (4.26) entailing the frequency warping. That is a consequence of the non-linear relation between frequencies of the analog system ω_a and its digital ω_d image. Substituting in (4.26), adequately: $s = \mathrm{j}\omega$ and $z = \mathrm{e}^{\mathrm{j}\omega T_\mathrm{s}}$ yields (with respect to the magnitude):

$$\omega_\mathrm{a} = \alpha \tan(\omega_\mathrm{d} T_\mathrm{s}/2) \tag{4.27}$$

where the coefficient α may be selected to match specified frequency conditions [30, 249]. For example, if we demand: $\omega_\mathrm{a} = \omega_\mathrm{d} = \omega_1$ then

$$s = \alpha \frac{1-z^{-1}}{1+z^{-1}} \tag{4.28}$$

where: $\alpha = \dfrac{\omega_1}{\tan(\omega_1 T_\mathrm{s}/2)}$.

Similarly as the Laplace transfer function (4.7), we can define the z-domain transfer function as a relation between Z-transforms in the output $Y(z)$ and input $X(z)$. Such elements play a very important role in digital signal processing the transfer function of which can be represented in the following form:

$$G(z) = \frac{Y(z)}{X(z)} = \frac{\sum_{m=0}^{M} b_m z^{-m}}{1 + \sum_{l=1}^{L} a_l z^{-l}} \tag{4.29}$$

They are known as digital filters. If the denominator in (4.29) does not depend on the variable z (L=0) then we have a finite impulse response (FIR) filter. The output signal $y(k)$ is obtained according to:

$$y(k) = \sum_{m=0}^{M} b_m x(k-m) \tag{4.30}$$

where b_m is known as the filter impulse response.

FIR filters may be easily designed with linear phase characteristic and therefore they are frequently applied as estimators of signal parameters [137, 237, 249].

The algorithms designed for power-system automation and control purposes should meet two contradictory demands:

- results of measurement (estimation) should be available as soon as possible; and
- measurement accuracy should satisfy expectations related to the sensitivity and selectivity of the final decision made by the controlling device.

In the case of fault-location procedures the requirements for the ratio of velocity action are somewhat less but, on the other hand, expectations regarding the accuracy increase. The important circumstances that need to be taken into consideration when designing the measurement algorithms are the fact that the observed system was corrupted during a fault. This causes a sudden dynamical process that changes the measurement condition and that is exactly the most interesting moment from the control devices point of view. But in transient conditions the observed process (represented by a signal) has too many unknown parameters that need to be identified for describing it. The usual way is to simplify the considered process model leaving only those characteristics that are important for the given task. In the next section, the development of this approach is described that has led to the methods used contemporarily.

4.3 Phasor-based Algorithms

Consider a linear time-invariant system (circuit) containing a periodical source. Such a system can be described by a non-homogeneous differential equation – as in (4.5). Response of the system to a given forcing, in the absence of any transients (after adequate long time), takes the form of a steady-state response. For periodical excitation the response is tracking the input signal with the same frequency. For mathematical convenience in finding the output, we let the input signal $x(t)$ be of the form:[2]

$$x(t) = X\mathrm{e}^{\mathrm{j}(\omega_x t+\psi)} = X\cos(\omega_x t+\psi) + \mathrm{j}X\sin(\omega_x t+\psi) = X\mathrm{e}^{\mathrm{j}\psi}\mathrm{e}^{\mathrm{j}\omega_x t} = \underline{X}\mathrm{e}^{\mathrm{j}\omega_x t} \tag{4.31}$$

where $\underline{X} = X\mathrm{e}^{\mathrm{j}\psi}$ is called the phasor of the input signal,[3] ω_x is its angular velocity.

It can be shown that putting the excitation (4.31) into the circuit differential equation we obtain:

$$y(t) = |G(\mathrm{j}\omega_x)|X\mathrm{e}^{\mathrm{j}(\omega_x t+\psi+\arg\{H(\mathrm{j}\omega_x)\})} = |G(\mathrm{j}\omega_x)|\underline{X}\mathrm{e}^{\mathrm{j}(\omega_x t+\arg\{H(\mathrm{j}\omega_x)\})} = \underline{Y}\mathrm{e}^{\mathrm{j}\omega_x t} \tag{4.32}$$

where $G(\mathrm{j}\omega_x)$ is the transfer function as in (4.7) and

$$\underline{Y} = G(\mathrm{j}\omega_x)\underline{X} \tag{4.33}$$

is the output phasor.

Returning to our example from Fig. 4.1a and assuming a voltage of the form:

$$u(t) = U\cos(\omega_1 t+\psi)$$

we can easily determine the steady-state current flowing through the circuit. First, we need the share of the voltage in the complex form:

$$u(t) = U\cos(\omega_1 t+\psi) + \mathrm{j}U\cos(\omega_1 t+\psi) = \underline{U}\mathrm{e}^{\mathrm{j}\omega_1 t}\,,\ \underline{U} = U\mathrm{e}^{\mathrm{j}\psi} = U\angle\psi$$

Next, from the Laplace-domain circuit impedance (4.6) we can determine the Fourier-domain impedance (Fig. 4.5):

$$Z(\mathrm{j}\omega_1) = \left(R + sL + \frac{1}{sC}\right)_{s=\mathrm{j}\omega_1} = R + \mathrm{j}\left(\omega_1 L - \frac{1}{\omega_1 C}\right) = |Z|\mathrm{e}^{\mathrm{j}\varphi}\,,\ \varphi = \tan^{-1}\frac{\omega_1^2 LC - 1}{\omega_1 RC}$$

---

[2] In the general case the forcing signal may be damped (decayed).

[3] We will not underline the phasor symbol if it is clearly signified from the context.

Note that we are using only one point in the frequency area: $\omega = \omega_1$ (which is defined by the input signal). Finally:

$$\underline{I} = \frac{\underline{U}}{Z(\mathrm{j}\omega_1)} = \frac{U}{|Z|}\mathrm{e}^{\mathrm{j}(\psi-\varphi)}$$

and the real instantaneous current:

$$i(t) = \frac{U}{|Z|}\cos(\omega_1 t + \psi - \varphi).$$

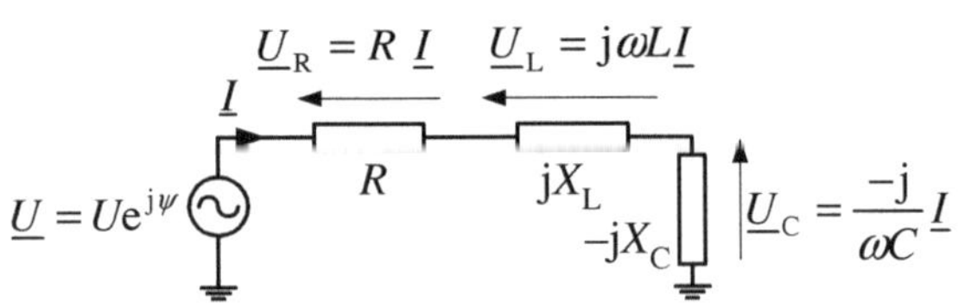

Fig. 4.5 Steady-state circuit analysis with complex phasors application

In the considered linear system the real and imaginary parts of the output correspond to the real and imaginary parts of the complex forcing signal. Practically, we want the response to the real part (the cosine) so we simply take the real part of the output as the result.

The application of the phasor concept to the solution of AC circuit steady-state problems was proposed by Ch. Steinmetz[4] and was first presented to an astonished audience at the International Electrical Congress in 1893 [302]. The main simplification of the method consists in transformation of the set of differential equations that described a linear electric circuit into a set of complex algebraic equations. The input–output relations are reduced to multiplication of the input phasor by the transfer function (which may have the form of an impedance, an admittance or another parameters) represented by a complex number. In fact, the transformation between the time domain and the frequency domain is applied here (regarding a signal and a network parameter) but thanks to an introduced simplification the concept delivers effective tool for AC circuits analysis and measurements.

Besides the network calculations the phasor transformation also plays an important role in the measurements and the circuit-parameters identification. In a computer-based control system in both the aforesaid applications we need to estimate the voltage and current phasors from sampled instantaneous waveforms. Note that the phasor concept meets the pre-supposed demand concerning the simplification of the signal model: transients are neglected and only one harmonic

[4] Charles Proteus Steinmetz (1865–1923), born Carl August Rudolph Steinmetz in Breslau, Germany (now Wroclaw, Poland).

component is considered. The next section presents the voltage- and current-phasors estimation algorithms.

4.3.1 Phasor Measurement

Since the vast majority of computer-based power-system control devices is still based on phasors, the issue of phasor estimation is probably the single most investigated and documented subject in research on power-system relaying and control. The digital phasor estimators are designed to meet several contradictory requirements, namely: to be fast, immune to frequency excursions, and insensitive to signal pollution including harmonics, sub-harmonics, high-frequency oscillations, noise and decaying DC components [137, 237, 249].

Basic fault-location algorithms are derived on an assumption that the post-fault state can be considered as a steady-state one. Consequently, in an AC system fault-loop voltage and current can be represented in the form of sinusoidal waveforms. These waveforms may be expressed in the frequency domain as phasors. We can see that the phasor in (4.30) has the same form as the exponential representation of the signal spectrum (4.12) and both representations will be the same for $\omega = \omega_1$.

Below, we present the basic methods for determination of voltage and current phasors from instantaneous sampled waveforms (signals). They are closely related to the general Fourier transform so we start from a presentation of a family of the Fourier methods suitable for digital signal processing.

We have seen in the previous section that the detailed rule of the Fourier transform depends on the type of considered signals. From the practical point of view the following types of signals are important [230, 298]:

- A continuous-time signal $x(t)$ for which the general Fourier transform (4.10), (4.11) is applied (if the signal fulfils the Fourier integral convergence).
- A continuous-time signal $x(t)$ is periodic with a period T; such a signal can be represented as the complex Fourier series and its coefficients can be interpreted as the spectral representation of $x(t)$.
- A discrete-time signal $x^{\#}(k)$ is obtained from sampling a signal $x(t)$ with sampling period T_s. The pair of the Fourier transform is as follows:

$$X(\mathrm{e}^{\mathrm{j}\omega T_s}) = X_{\mathrm{R}}(\mathrm{e}^{\mathrm{j}\omega T_s}) + \mathrm{j}X_{\mathrm{I}}(\mathrm{e}^{\mathrm{j}\omega T_s}) = \sum_{k=-\infty}^{+\infty} x^{\#}(k)\mathrm{e}^{-\mathrm{j}\omega T_s k} \tag{4.34}$$

$$x^{\#}(k) = \frac{1}{2\pi} \int_{-\pi/T_s}^{\pi/T_s} X(\mathrm{e}^{\mathrm{j}\omega T_s})\mathrm{e}^{\mathrm{j}\omega T_s k}\mathrm{d}\omega \tag{4.35}$$

It is called the discrete-time Fourier transform (DTFT). Note that the above transformation is directly connected with the Z-transform with $z = \mathrm{e}^{\mathrm{j}\omega T_s}$ (4.18).

- A discrete-time signal $x(k)$ is obtained as a finite duration of length M of the infinite-time signal $x^{\#}(k)$, such that:

$$x(k) = \begin{cases} x^{\#}(k), \ 0 \le k \le M-1 \\ 0, \ \text{otherwise} \end{cases} \tag{4.36}$$

The pair of the Fourier transform is as follows:

$$X(n) = \sum_{k=0}^{M-1} x(k) \mathrm{e}^{-\mathrm{j}2\pi nk/M} \tag{4.37}$$

$$x(k) = \frac{1}{M} \sum_{n=0}^{M-1} X(n) \mathrm{e}^{\mathrm{j}2\pi nk/M} \ , k\text{=0, 1, .., } M-1 \tag{4.38}$$

and is known as the discrete Fourier transform (DFT).

Practically important is the last case defining the DFT representation. It is the basic method used in digital signal processing. In practical application a discrete signal is limited to M samples (even if it is longer) assuming zeros outside this interval. Results of the direct transformation (4.37) are samples of the Fourier series coefficients $X(n)$ that define the discrete periodic spectrum of series $x(k)$. It can be seen that $X(n)=X(n+rM)$, r – an integer number, but we are interested in values of $X(n)$ only in the interval $0 \le n \le M-1$ because they have important physical interpretation: considering the signal $x(k)$ has the period on N, then $X(n)= X_R(n)+\mathrm{j}\, X_I(n)$ can be interpreted as the complex magnitude of the nth harmonic, n=0, 1, .., $M-1$. Similarly, the signal $x(k)$ is either a finite-length sequence of length M, or it is a periodic sequence with period M.

The above is illustrated in Fig. 4.6. The frame in Fig. 4.6a indicates the consecutive signal samples (data window) that are selected for transformation. The original continuous-time signal $x(t)$ is sampled with a period of T_s=1.25 ms. The interval taken for DFT has a length of M=8 that is equivalent to the period of $T_0=MT_s$=10 ms, and, adequately:

$$\omega_0 = \frac{\omega_s}{M} = \frac{2\pi}{MT_s} = 2\pi f_0 \tag{4.39}$$

Transformation (4.37) may then be represented as:

$$X_R(n) + \mathrm{j}X_I(n) = \sum_{k=0}^{M-1} x(k)\cos\left(k\omega_0 nT_s\right) - \mathrm{j}\sum_{k=0}^{M-1} x(k)\sin\left(k\omega_0 nT_s\right) \tag{4.40}$$

Another representation is given by the spectrum magnitude:

$$|X(n)| = \sqrt{X_R^2(n) + jX_I^2(n)} \tag{4.41}$$

and the phase characteristic:

$$\theta(n) = \arg\{X(n)\} = \tan^{-1}\left(X_I(n)/X_R(n)\right) \tag{4.42}$$

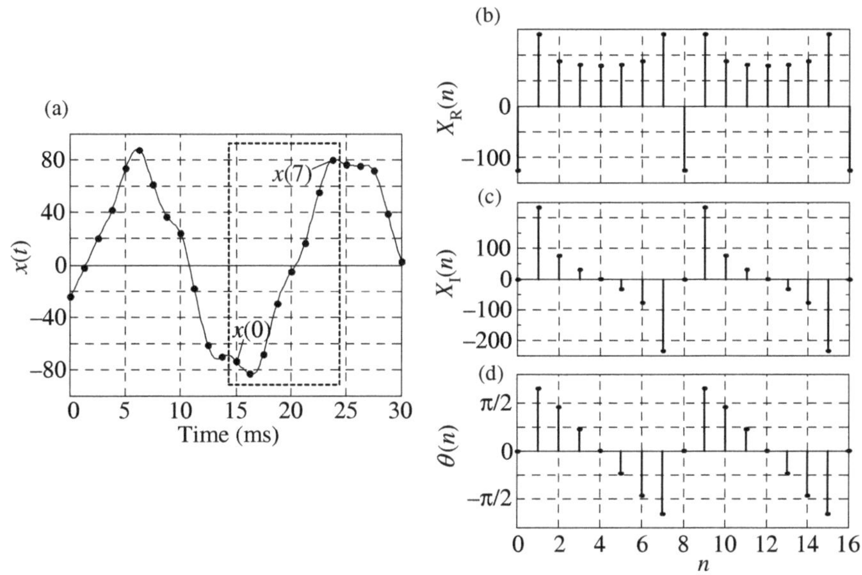

Fig. 4.6 DFT application to the signal-parameters estimation: (a) sampled signal, (b), (c) the spectrum characteristics, and (d) the phase characteristic

It can be seen from Figs 4.6b–d that the signal spectrum characteristics are periodic and all the information is included in the first M coefficients.

The coefficients $X(n)$ may also be interpreted as samples of the continuous-frequency domain of the DTFT at M points uniformly spaced around the unit circle in the Z-plane, i.e., at the points $\omega_n = (2\pi n/M)$, $n = 0, 1, \ldots, M-1$. This will be further illustrated.

Fourier Algorithm

The DFT method may be directly used for voltage and current phasors estimation in AC networks. Let f_1 be the system nominal frequency, and N – number of samples per signal period, so: $T_s = T_1/N = 1/(f_1 N)$. Moreover, let as assume that the transformation period is equal to the signal period: $M=N$. Therefore, components of the complex coefficients in (4.40) may be determined as follows:

$$X_{\mathrm{R}}(n)=p_{n\mathrm{R}}\sum_{i=0}^{N-1}x(i)\cos\big(n\nu(i+0.5)\big)$$
$$X_{\mathrm{I}}(n)=-p_{n\mathrm{I}}\sum_{i=0}^{N-1}x(i)\sin\big(n\nu(i+0.5)\big)\quad,\ n=0,1,..,\frac{N}{2}\tag{4.43}$$

where: $\nu=2\pi/N$ – angle between samples, $p_{n\mathrm{R}}=p_{n\mathrm{I}}$ – scaling coefficients.

In determining the scaling coefficients we should consider the phasor definition. In (4.31) we have assumed that the phasor magnitude is equal to the signal magnitude. That is the generally accepted assumption in the signal processing. In such a case we obtain:

$$p_{n\mathrm{R}}=p_{n\mathrm{I}}=2n/N\tag{4.44}$$

In the electrical power engineering practice it is convenient to operate the RMS phasors. This involves division of (4.44) by $\sqrt{2}$;

$$p_{n\mathrm{R}}=p_{n\mathrm{I}}=\sqrt{2}n/N\tag{4.45}$$

The transformation (4.43) represents the digital correlation of the signal $x(t)$ with functions cosine and sine, which are known as the correlation functions:

$$h_{n\mathrm{R}}(i)=\cos\big(n\nu(i+0.5)\big)$$
$$h_{n\mathrm{I}}(i)=\sin\big(n\nu(i+0.5)\big)\tag{4.46}$$

Increasing the phase in the above functions by $n\nu/2$ is not obligatory but thanks to that, when N is even (which is a typical solution), the correlation functions are symmetrically placed in the data window. Such a solution simplifies adjusting the phasors in the DFT outputs of the complex system.

Returning to the example in Fig. 4.6 we can see that for analysis of a dynamic process we have to apply the DFT algorithm for successive sets of the signal samples on the lapse of time. In other words, the frame defined a set of samples in Fig. 4.6a is moving with time-step index k. Consequently, the signal spectrum becomes a function of discrete time. According to this we need to modify the algorithm (4.43). Assuming that we are interested in estimation of the first-harmonic phasor (n=1) we obtain:

$$X_{\mathrm{R}}(k)=\frac{2}{N}\sum_{i=0}^{N-1}x(k-i)h_{\mathrm{R}}(i)$$
$$X_{\mathrm{I}}(k)=-\frac{2}{N}\sum_{i=0}^{N-1}x(k-i)h_{\mathrm{I}}(i)\quad k=0,1,...\tag{4.47}$$

where: $h_{\mathrm{R}}(i)=h_{1\mathrm{R}}(i)=\cos\big(\nu(N-i-0.5)\big)$, $h_{\mathrm{I}}(i)=h_{1\mathrm{I}}(i)=\sin\big(\nu(N-i-0.5)\big)$, $\nu=2\pi/N$.

Note that the phasor components are a function of time determined by the index k (a harmonic number n was omitted). The scheme of the signal processing according to (4.47) for N=16 is presented in Fig. 4.7. It is assumed that calculation starts when the first signal sample (k=0) turns up into the data window. In the example shown the calculations at each step are carried out with full window data, so at the start the initial window should be cleared: $x(k)=0$ for $k<0$. The first and the last (k=24) positions of the data window are indicated in the figure. The correlation functions (cosine and sine) are shifted with the data frame.

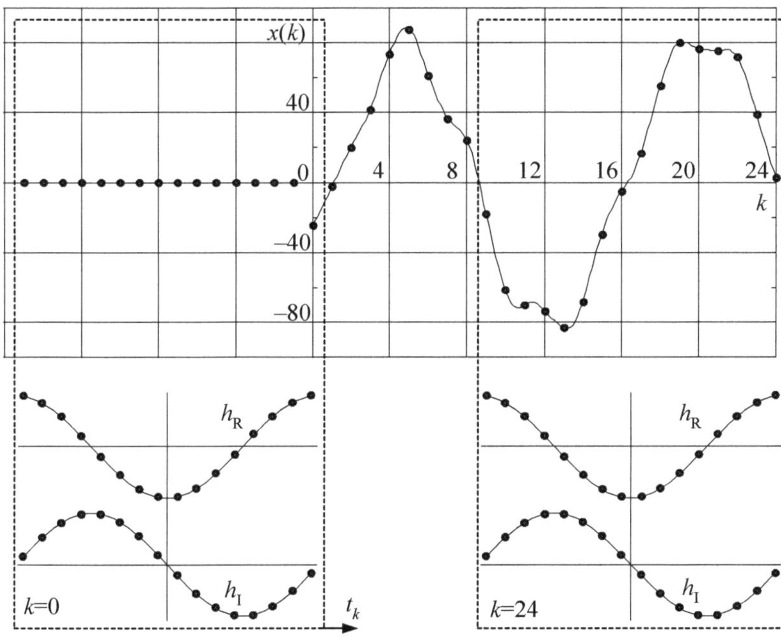

Fig. 4.7 Scheme of the full-period Fourier algorithm according to (4.46)

The algorithm (4.47) is known as the full-period Fourier method for a signal phasor orthogonal (real and imaginary) components estimation. It may be organized in different ways. In particular, the correlation functions may represent the infinite waveforms (cosine and sine in the considered case) and the data frame of length M moving along it (Fig. 4.8).

In such a case the full-period phasor estimator takes the following form:

$$\begin{aligned} X_{\mathrm{R}}(k) &= \frac{2}{N}\sum_{i=0}^{N-1} x(k-i)h_{\mathrm{R}}(k,i) \\ X_{\mathrm{I}}(k) &= -\frac{2}{N}\sum_{i=0}^{N-1} x(k-i)h_{\mathrm{I}}(k,i) \end{aligned} \qquad k = 0,1,\ldots \tag{4.48}$$

where: $h_{\mathrm{R}}(k,i) = \cos\big(\nu(k-i-0.5)\big)$, $h_{\mathrm{I}}(k,i) = \sin\big(\nu(k-i-0.5)\big)$.

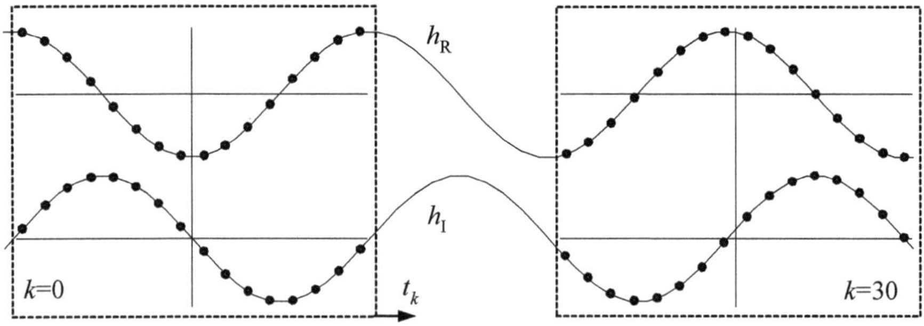

Fig. 4.8 Scheme of the full-period Fourier algorithm with variable correlation functions

The condition for orthogonality of the phasor components: $X_R(k)$, $X_I(k)$ requires that the correlation functions would be orthogonal in the data window. The latter is defined by:

$$\sum_{i=0}^{M-1} h_R(k,i)\cdot h_I(k,i)=0 \text{ for any } k \tag{4.49}$$

Substituting in (4.49) the cosine and sine functions from (4.48) we can check that the orthogonality condition is satisfied if the data window length is a multiple of half the period of the correlation function: $M=mN/2$, m – natural value.

It would be interesting to compare both the presented methods defined by (4.47) and (4.48). Some results are shown in Fig. 4.9. The input signal is the same as in Fig. 4.7. Application of the algorithm (4.47) gives results as in Fig. 4.9a (the phasor components) and Fig. 4.9b – changing of the phasor phase, while Figs 4.9c and d present results obtained from the algorithm (4.48). Note that in both cases the data window is filling up to k= 15 (the algorithm transient period). The phasor components of the first algorithm form a sinusoidal waveform with the same frequency as the correlation functions. At the steady state the phase increases at each step by the angle ν. The results of the algorithm (4.48) form the constant components (the observed small variations result from the signal transients). It is not surprising that both algorithms give the same magnitude of the estimated phasors.

Analyzing the input–output relation in the estimator (4.47) (see Fig. 4.7 and Fig. 4.9a) we can note that this algorithm is similar to the pair of FIR filters if their impulse be chosen as the cosine/sine waveforms of one period length. Indeed, a FIR filter, which realizes convolution of an input signal with an impulse function will be here defined as:

$$\begin{aligned} X_R(k) &= \frac{2}{N}\sum_{i=0}^{N-1} x(k-i)h_R(i) \\ X_I(k) &= \frac{2}{N}\sum_{i=0}^{N-1} x(k-i)h_I(i) \end{aligned} \qquad k=0,1,\ldots \tag{4.50}$$

where: $h_R(i)=\cos(\nu(i+0.5))$, $h_I(i)=\sin(\nu(i+0.5))$.

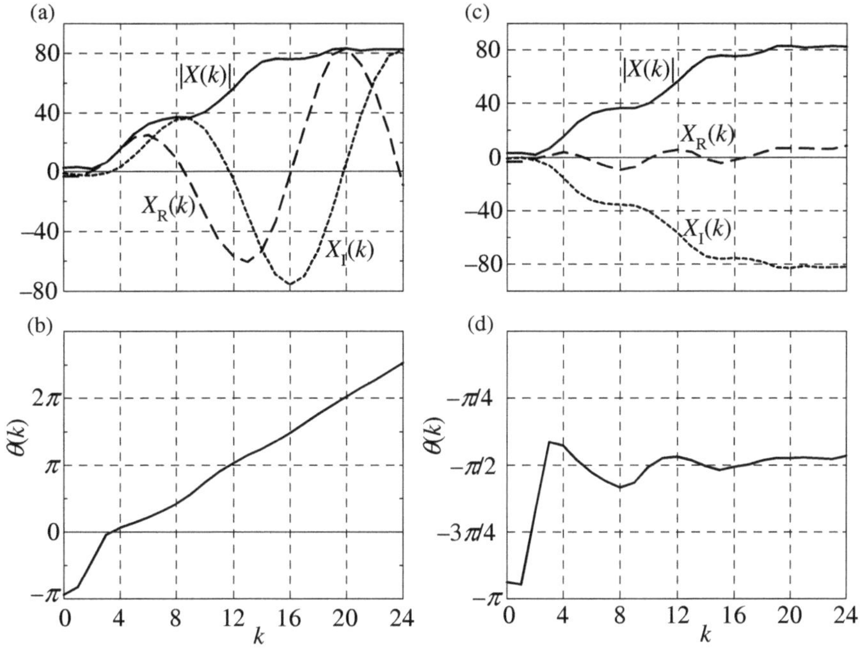

Fig. 4.9 Time (a), (c) and phase (b), (d) response of different Fourier algorithms

One can see the small difference between (4.47) and (4.50). We use the same symbols for the output signals as in (4.47) although in the filters (4.50) the physical meaning of the outputs is the same as the input (they are signals) that is in opposition to the interpretation of the correlation process as in (4.47). That is, however, of no importance from a practical point of view. We should also remember that both procedures give results that as complex numbers are mutually conjugated: $\underline{X}(k)_{\text{convolution}}=\underline{X}(k)^{*}_{\text{correlation}}$. That is why in the second equation in (4.50) there is no minus sign.

Thanks to symmetry of the cosine/sine correlation functions in the Fourier algorithm it can be written in a more simple recursive form. Writing (4.48) for k–1 and k samples, after some rearrangements we obtain:

$$
\begin{aligned}
X_{\mathrm{R}}(k) &= X_{\mathrm{R}}(k-1)+\frac{2}{N}\cos(\nu k)\big(x(k)-x(k-N)\big) \\
X_{\mathrm{I}}(k) &= X_{\mathrm{I}}(k-1)-\frac{2}{N}\sin(\nu k)\big(x(k)-x(k-N)\big)
\end{aligned}
\tag{4.51}
$$

The recursive form of the full-period Fourier algorithm (4.51) reduces calculations to only two multiplications at each time step. For starting at k=1 we should take an initial values for $X_R(0)$, $X_I(0)$ and $x(k)$, k=1–N, 2–N, .., 0.

The similar forms can also be defined for other types of the algorithms. We should remember, however, that the recursive form (4.51) is hazardous when the algorithm is continuously realized in a floating-point processor due to possible errors accumulating in successive steps.

Stable output for using the presented full-period Fourier algorithms is available only after one period from the beginning of the processing. Such a long delay may not be acceptable if the operation action is faster or if an after-fault stage is shorter than one period. In such a case we can use the half-period Fourier algorithm. Similarly to (4.47) we can write:

$$\begin{aligned} X_{\mathrm{R}}(k) &= \frac{4}{N}\sum_{i=0}^{\frac{N}{2}-1} x(k-i)h_{\mathrm{R}}(i) \\ X_{\mathrm{I}}(k) &= \frac{4}{N}\sum_{i=0}^{\frac{N}{2}-1} x(k-i)h_{\mathrm{I}}(i) \end{aligned} \qquad k = 0,1,\ldots \tag{4.52}$$

where: $h_{\mathrm{R}}(i) = -\cos\left(\nu(N/4-i-0.5)\right)$, $h_{\mathrm{I}}(i) = \sin\left(\nu(N/4-i-0.5)\right)$.

It is also possible to formulate the half-cycle algorithm with variable correlation functions as in (4.48).

The half-cycle methods give the stable output with a delay of only half a cycle. However, for this speeding up with time we pay the price of accuracy. The estimation accuracy may be analyzed by using the estimator magnitude response as a function of frequency. Let us see how we can analyze the frequency characteristics of the considered algorithms.

For evaluation of the frequency characteristics of the relations (4.47) we can apply the Z-transform:

$$\begin{aligned} z\{X_{\mathrm{R}}(k)\} &= \frac{2}{N} z\{x(k)\}\sum_{i=0}^{N-1}\cos\left(\nu(N-i-0.5)\right)z^{-1} \\ z\{X_{\mathrm{I}}(k)\} &= -\frac{2}{N} z\{x(k)\}\sum_{i=0}^{N-1}\sin\left(\nu(N-i-0.5)\right)z^{-i} \end{aligned} \tag{4.53}$$

Samples of cosine/sine functions are simple coefficients of the algorithms. Taking the output–input relation we obtain the Z-transform transfer functions:

$$\begin{aligned} H_{\mathrm{R}}(z) &= \frac{z\{X_{\mathrm{R}}(k)\}}{z\{x(k)\}} = \frac{2}{N}\sum_{i=0}^{N-1}\cos\left(\nu(N-i-0.5)\right)z^{-i} \\ H_{\mathrm{I}}(z) &= \frac{z\{X_{\mathrm{I}}(k)\}}{z\{x(k)\}} = -\frac{2}{N}\sum_{i=0}^{N-1}\sin\left(\nu(N-i-0.5)\right)z^{-i} \end{aligned} \tag{4.54}$$

The frequency form is easily obtained by substitution (4.17):

$$
\begin{aligned}
H_{\mathrm{R}}(\mathrm{e}^{\mathrm{j}\omega T_s}) &= \frac{2}{N}\sum_{i=0}^{N-1}\cos\big(\nu(N-i-0.5)\big)\mathrm{e}^{-\mathrm{j}\omega T_s i} \\
H_{\mathrm{I}}(\mathrm{e}^{\mathrm{j}\omega T_s}) &= -\frac{2}{N}\sum_{i=0}^{N-1}\sin\big(\nu(N-i-0.5)\big)\mathrm{e}^{-\mathrm{j}\omega T_s i}
\end{aligned}
\tag{4.55}
$$

It is more convenient in the given application to consider the relative frequency with reference to the fundamental frequency (the first harmonic): $\omega \leftrightarrow \omega/\omega_1$. After simple rearrangements we obtain:

$$
\begin{aligned}
H_{\mathrm{R}}(\mathrm{e}^{\mathrm{j}\omega}) &= H_{\mathrm{R}}(\omega)\mathrm{e}^{-\mathrm{j}\left(\frac{N-1}{2}\omega\nu\right)} \\
H_{\mathrm{I}}(\mathrm{e}^{\mathrm{j}\omega}) &= H_{\mathrm{I}}(\omega)\mathrm{e}^{-\mathrm{j}\left(\frac{N-1}{2}\omega\nu-\frac{\pi}{2}\right)}
\end{aligned}
\tag{4.56}
$$

where:

$H_{\mathrm{R}}(\omega) = p(\omega)\cos(\nu/2)\sin(\omega\nu/2)$, $H_{\mathrm{I}}(\omega) = -p(\omega)\sin(\nu/2)\cos(\omega\nu/2)$,

$$p(\omega) = \frac{2\sin(\pi\omega)}{N\sin\big((\omega+1)\nu/2\big)\sin\big((\omega-1)\nu/2\big)}.$$

We can see that $|H_{\mathrm{R}}(\omega)|$ and $|H_{\mathrm{I}}(\omega)|$ represent the magnitude response of the considered full-period Fourier algorithms. It can be seen from (4.56) that the phase response for both estimators linearly depend on frequency and are shifted from each other by $\pi/2$. Figure 4.10 presents this characteristics for N=20 (which is equivalent to the sampling frequency f_s=1000 Hz for the 50-Hz system). The algorithm rejects DC component and all harmonics with the exception of the first one for which the gain is equal to 1 for both components. That is the important feature of this method.

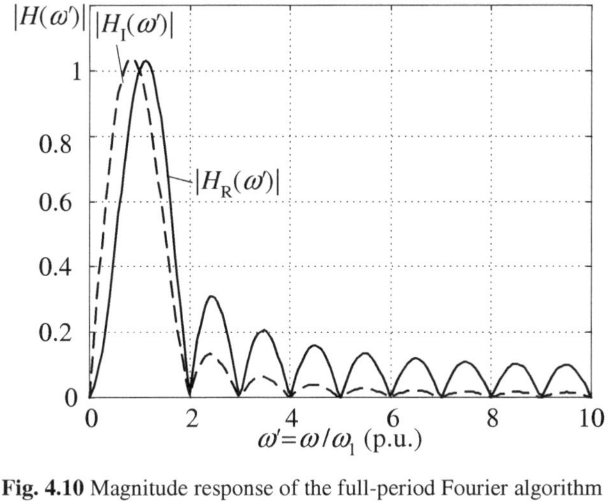

Fig. 4.10 Magnitude response of the full-period Fourier algorithm

In a similar way we can analyze the other considered algorithm. We can see that all three presented versions of the full-cycle Fourier method ((4.47), (4.48), and (4.50)) have the same magnitude response and only the phase responses are slightly different.

The magnitude response of the half-cycle Fourier response is presented in Fig. 4.11. We see the consequence of the data window shortening: DC component is not rejected by the real component filter and only odd harmonics are eliminated.

From the provided analysis we can see that the phasors for both orthogonal components are calculated from a real-valued signal by using a pair of filters (no matter the details) the phase responses of which are shifted by $\pi/2$: the phase of the imaginary component filter precedes the other one. If this relation is met for any frequency then we have the phasor estimator with the linear phase characteristic. On the other hand, the magnitude responses of the considered filters, generally, are different. Let us show the consequence of such a difference.

Consider that the phasor is calculated according to (4.47) from a harmonic input signal:

$$x(k) = X\cos(\omega v k + \beta) \tag{4.57}$$

The obtained phasor is as follows:

$$\underline{X}(k) = X_{\mathrm{R}}(k) + \mathrm{j}X_{\mathrm{I}}(k) = H_{\mathrm{R}}(\mathrm{e}^{\mathrm{j}\omega})x(k) + \mathrm{j}H_{\mathrm{I}}(\mathrm{e}^{\mathrm{j}\omega})x(k) = H(\mathrm{e}^{\mathrm{j}\omega})x(k) \tag{4.58}$$

where:

$$H(\mathrm{e}^{\mathrm{j}\omega}) = H_{\mathrm{R}}(\mathrm{e}^{\mathrm{j}\omega}) + \mathrm{j}H_{\mathrm{I}}(\mathrm{e}^{\mathrm{j}\omega}) = \left|H(\mathrm{e}^{\mathrm{j}\omega})\right|\mathrm{e}^{\mathrm{j}\theta(\omega)} \tag{4.59}$$

can be considered as the phasor estimator 'transfer function'.

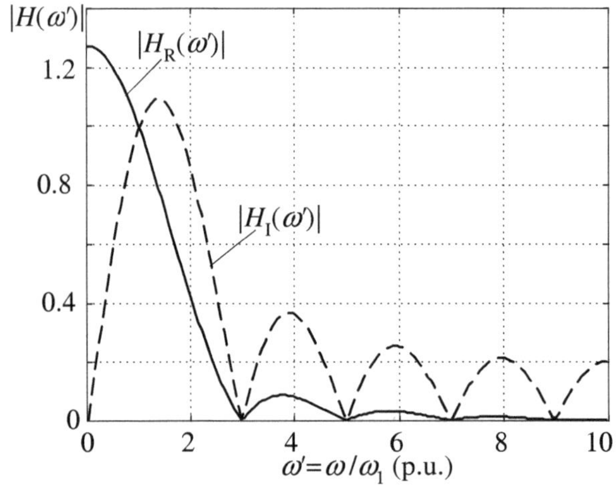

Fig. 4.11 Magnitude response of the half-cycle Fourier algorithm

Assume that both component filters were designed to estimate the phasor of frequency ω_0 (the gain for that frequency is equal to one – Fig. 4.10). Then we have:

$$H(\mathrm{e}^{\mathrm{j}\omega_0}) = (1+\mathrm{j})\mathrm{e}^{\mathrm{j}\theta_{\mathrm{R}}(\omega_0)} / \sqrt{2}\,,\ \text{where}\ \theta_{\mathrm{R}}(\omega_0) = \arg\{H_{\mathrm{R}}(\mathrm{e}^{\mathrm{j}\omega_0})\}$$

The question is how the $H(\mathrm{e}^{\mathrm{j}\omega})$ is looking for any frequency ω. For easier analysis we represent the input signal (4.57) as a sum of two complex functions:

$$x(k) = X\cos(\omega vk + \beta) = \frac{X}{2}\left(\mathrm{e}^{\mathrm{j}\omega vk} + \mathrm{e}^{-\mathrm{j}\omega vk}\right) \tag{4.60}$$

Now the output phasor of frequency ω can be represented as follows [249]:

$$\underline{X}(k) = X_{\mathrm{R}}(k) + \mathrm{j}X_{\mathrm{I}}(k) = \frac{X}{2}H(\mathrm{e}^{\mathrm{j}\omega})\mathrm{e}^{\mathrm{j}\omega vk} + \frac{X}{2}H(\mathrm{e}^{-\mathrm{j}\omega})\mathrm{e}^{-\mathrm{j}\omega vk} \tag{4.61}$$

Note that $H(\mathrm{e}^{\mathrm{j}\omega})$ represents the complex filter and, generally, it is not conjugate symmetric:

$$H(\mathrm{e}^{\mathrm{j}\omega}) \neq H^{*}(\mathrm{e}^{-\mathrm{j}\omega}) \tag{4.62}$$

Therefore, we should carefully rearrange (4.61):

$$\underline{X}(k) = X\left(H_{\mathrm{A}}(\mathrm{e}^{\mathrm{j}\omega})\cos(\omega vk + \beta) + \mathrm{j}H_{\mathrm{B}}(\mathrm{e}^{\mathrm{j}\omega})\sin(\omega vk + \beta)\right), \tag{4.63}$$

where:

$$H_{\mathrm{A}}(\mathrm{e}^{\mathrm{j}\omega}) = \frac{H(\mathrm{e}^{\mathrm{j}\omega}) + H(\mathrm{e}^{-\mathrm{j}\omega})}{2}\,,\quad H_{\mathrm{B}}(\mathrm{e}^{\mathrm{j}\omega}) = \frac{H(\mathrm{e}^{\mathrm{j}\omega}) - H(\mathrm{e}^{-\mathrm{j}\omega})}{2}\,. \tag{4.64}$$

Selecting in (4.63) the real and imaginary parts we obtain [249]:

$$\underline{X}(k) = X\mathrm{e}^{\mathrm{j}\theta_{\mathrm{A}}(\omega)}\left[A(\omega)\cos\left(\omega vk + \beta + \theta_{\mathrm{B}}(\omega)\right) + \mathrm{j}B(\omega)\sin\left(\omega vk + \beta + \theta_{\mathrm{B}}(\omega)\right)\right] \tag{4.65}$$

where:

$$\begin{aligned} A(\omega) &= \frac{|H(\mathrm{e}^{\mathrm{j}\omega})| + |H(\mathrm{e}^{-\mathrm{j}\omega})|}{2} \\ B(\omega) &= \frac{\left||H(\mathrm{e}^{\mathrm{j}\omega})| - |H(\mathrm{e}^{\mathrm{j}\omega})|\right|}{2} \end{aligned} \tag{4.66}$$

$$\theta_A(\omega) = \frac{\theta(\omega) + \theta(-\omega)}{2}$$
$$\theta_B(\omega) = \frac{\theta(\omega) - \theta(-\omega)}{2} \tag{4.67}$$

$\theta(\omega)$ – as in (4.59) and: $\theta(-\omega) = \arg\{H(\mathrm{e}^{-\mathrm{j}\omega})\}$: $H(\mathrm{e}^{-\mathrm{j}\omega}) = |H(\mathrm{e}^{-\mathrm{j}\omega})|\mathrm{e}^{\mathrm{j}\theta(-\omega)}$.

Equation (4.65) represents an ellipse, the longer axis of which precedes the real axis of the complex plane by an angle $\theta_B(\omega)$ – Fig. 4.12. The value of $A(\omega)$ is equal to half the length of a longer axis, and $B(\omega)$ – the shorter one. Position and shape of the ellipse depend on parameters of the filter and frequency of the input signal. It can be verified that for non-recursive filters and correlation algorithms $\theta(\omega) = \theta(-\omega)$ and $\theta_A(\omega) = \theta(\omega)$, $\theta_B(\omega) = 0$. Furthermore, for the chosen designed frequency $\omega = \omega_0$ is: $H(-\mathrm{j}\omega_0) = 0$ and $A(\omega_0) = B(\omega_0)$.

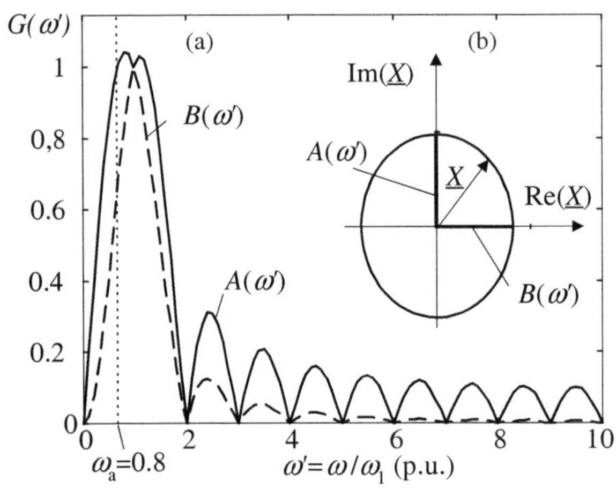

Fig. 4.12 Magnitude response of the full-period Fourier algorithm (a) with the envelope for frequency ω=0.8 (b)

The magnitude and phase response of the estimator (4.47) can be directly obtained from (4.65)

$$|\underline{X}(k,\omega)| = \sqrt{X_R^2(k) + X_I^2(k)} = XG(k,\omega) \tag{4.68}$$

where:

$$G(\omega) = \sqrt{A^2(\omega)\cos^2\xi + B^2(\omega)\sin^2\xi} = \frac{1}{\sqrt{2}}\sqrt{|H(\mathrm{j}\omega)|^2 + |H(-\mathrm{j}\omega)|^2 + 2|H(\mathrm{j}\omega)||H(-\mathrm{j}\omega)|\cos(2\xi)} \tag{4.69}$$

$$\arg(\underline{X}(k)) = \tan^{-1}\left(\frac{X_{\mathrm{I}}(k)}{X_{\mathrm{R}}(k)}\right) = \theta_{\mathrm{A}}(\omega) + \tan^{-1}\left(\frac{B(\omega)}{A(\omega)}\tan\xi\right) \tag{4.70}$$

and: $\xi = \xi(k,\omega) = \omega v k + \beta + \theta_{\mathrm{B}}(\omega)$.

We can seen from (4.68) that, in general, the phasor estimator is represented by a non-linear time-variant algorithm. For a harmonic input signal of frequency ω the estimate of the magnitude changes in time with double frequency. A time-invariant response is obtained only for separate frequencies, including frequency ω_0 for which the estimator was designed. In the right side of Fig. 4.12 is shown the elliptic envelop for frequency of $\omega = 0.8\omega_1$ (with reduced size).

The output phasor $\underline{X}(k)$ slides on the ellipsis depending on the actual signal phase, which changes with time. Its magnitude can vary between the higher ($A(\omega)$) and lower ($B(\omega)$) boundaries of the estimator frequency characteristic.

Least Squares Method Application

The general problem in designing the signal-parameters estimator is to choose an appropriate model for the signal representation and to define a procedure for the model-parameters estimation. In the methods proposed above the model was formulated in the form of an adequately defined correlation function (DFT) or impulse function of the FIR filter. In the following part of this section we present the tool for estimators that uses the least squares method (LSM).

In the digital signal processing the LSM technique applies to problems where data are given from measurements with associated uncertainties that influence the fit made to them. The LSM criterion specifies that the sum of squared differences between the postulated signal model and its estimates are minimal.[5] For a two-state model (4.61) successive measurements can be arranged in the form [137, 249]:

$$\begin{array}{rcccc} x(0) = & X_{\mathrm{R}} & +0 & +\varepsilon(0) \\ x(1) = & X_{\mathrm{R}}\cos\nu & -X_{\mathrm{I}}\sin\nu & +\varepsilon(1) \\ \vdots & \vdots & \vdots & \vdots \\ x(M-1) = & X_{\mathrm{R}}\cos\big((M-1)\nu\big) & -X_{\mathrm{I}}\sin\big((M-1)\nu\big) & +\varepsilon(M-1) \end{array} \tag{4.71}$$

or in matrix notation:

$$\mathbf{x} = \mathbf{H}\mathbf{X} + \boldsymbol{\varepsilon} \tag{4.72}$$

where:

[5] This model is also known as the linear regression model [240]

$\mathbf{x} = [x(0)\ x(1) \ldots x(M-1)]^{\mathrm{T}}$ – vector of measurements,

$\mathbf{H} = [\mathbf{h}(0)\ \ \mathbf{h}(1) \ldots \mathbf{h}(M-1)]^{\mathrm{T}}$ – signal model matrix:

$$\mathbf{h}(i) = [\cos(\nu i)\ \ -\sin(\nu i)] = [h_{\mathrm{R}}(i)\ \ h_{\mathrm{I}}(i)],$$

$\boldsymbol{\varepsilon} = [\varepsilon(0)\ \ \varepsilon(1) \ldots \varepsilon(M-1)]^{\mathrm{T}}$ – vector of measurement errors,

$\mathbf{X} = [X_{\mathrm{R}}\ \ X_{\mathrm{I}}]^{\mathrm{T}}$ – vector of estimated parameters.

In the signal model considered the vector $\mathbf{X}$ of the estimated parameters consists of two searching values. In the more general case vector $\mathbf{X}$ can be denoted as:

$$\mathbf{X} = [X_1\ \ X_2 \cdots X_L]^{\mathrm{T}} \tag{4.73}$$

where L – number of estimated values (number of states in the signal model).

Estimated vector parameters $\hat{\mathbf{X}} = \mathbf{X}$ determines the model of measured variables

$$\hat{\mathbf{x}} = \mathbf{H}\hat{\mathbf{X}} \tag{4.74}$$

The aim of LSM estimation is to minimize an error vector

$$\mathbf{e} = \mathbf{x} - \hat{\mathbf{x}} = \{e_i\}, i = 1,.., M \tag{4.75}$$

in a sense of defined criterion. The objective function to be minimized in a least squares is chosen as [249]:

$$S(\hat{x}) = \mathbf{e}^{\mathrm{T}}\mathbf{G}\mathbf{e} = \sum_{l=1}^{M}\sum_{j=1}^{M} g_{lj} e_j e_l \tag{4.76}$$

where $\mathbf{G}=\{g_{lj}\}$ is a square weighting matrix that can be chosen for minimizing (4.76).

Minimizing of (4.76) with respect to $\hat{\mathbf{x}}$ is accomplished for

$$\frac{\partial S(\hat{\mathbf{X}})}{\partial \hat{\mathbf{X}}} = \frac{\partial}{\partial \hat{\mathbf{X}}}\left[(\mathbf{x} - \mathbf{H}\hat{\mathbf{X}})^T \mathbf{G}(\mathbf{x} - \mathbf{H}\hat{\mathbf{X}})\right] = 2\mathbf{H}^{\mathrm{T}}\mathbf{G}\mathbf{H}\hat{\mathbf{X}} - 2\mathbf{H}^{\mathrm{T}}\mathbf{G}\mathbf{x} = 0 \tag{4.77}$$

that gives the so-called normal equation

$$\mathbf{H}^{\mathrm{T}}\mathbf{G}\mathbf{H}\hat{\mathbf{X}} = \mathbf{H}^{\mathrm{T}}\mathbf{G}\mathbf{x}\,. \tag{4.78}$$

From (4.78) it is easy to derive the vector of searching estimates

$$\hat{\mathbf{X}} = \mathbf{P}\mathbf{H}^{\mathrm{T}}\mathbf{G}\mathbf{x} \tag{4.79}$$

where

$$\mathbf{P} = (\mathbf{H}^{\mathrm{T}}\mathbf{G}\mathbf{H})^{-1} \tag{4.80}$$

The weighting matrix **G** that minimizes the function $S(\hat{\mathbf{X}})$ is defined as

$$\mathbf{G} = \mathbf{R}^{-1} \tag{4.81}$$

where $\mathbf{R} = \mathrm{E}(\varepsilon\varepsilon^{\mathrm{T}})$ – correlation matrix of measurements.
If the expected value of measurement errors $\mathrm{E}(\boldsymbol{\varepsilon})=0$ and weighting matrix is chosen as in (4.81) then the least squares estimator (4.79) is optimal [338].

In a deterministic process when no information about measurement noise is available the weighting matrix should be assumed as $\mathbf{G} = \mathbf{I}$ and algorithm (4.79) takes the classical form

$$\hat{\mathbf{X}} = \mathbf{H}^{\#}\mathbf{x} \tag{4.82}$$

where

$$\mathbf{H}^{\#} = \mathbf{P}\mathbf{H}^{\mathrm{T}}, \quad \mathbf{P} = (\mathbf{H}^{\mathrm{T}}\mathbf{H})^{-1} \tag{4.83}$$

We can note that a quadratic fit utilized in the LSM method has close connection with the pseudo-inverse of measurement model matrix $\mathbf{H}^{\#}$. The measurement model is overdetermined when it consists of $M>L$ number of measurements (4.71). The pseudo-inverse (4.83) exists whenever matrix $(\mathbf{H}^{\mathrm{T}}\mathbf{H})$ has an inverse. The operation described by $\hat{\mathbf{X}}$ is referred to as linear estimation [237, 240]. It should be pointed out that when one speaks of a linear model in LSM estimation the term linear means linear in parameters **X** and not in the independent variable *k* (discrete time step).

Mathematical formulation of the LSM algorithm also delivers information on estimation errors. Diagonal elements of matrix **P** are variances of the obtained results. This can be used for comparison of different algorithms. Direct application of the general LSM equation (4.79) leads to the non-recursive LSM algorithm. We can see that elements of the estimated vector $\hat{\mathbf{X}}$ result from multiplication of the adequate row in the matrix $\mathbf{P}\mathbf{H}^{\mathrm{T}}\mathbf{G}$ by vector **x**. This can also be described in the form

$$\hat{\mathbf{X}} = \mathbf{P}_{\mathrm{G}}\mathbf{H}_{\mathrm{G}}\mathbf{x} \tag{4.84}$$

where

$$\mathbf{P}_{\mathrm{G}} = (\mathbf{H}_{\mathrm{G}}\mathbf{H})^{-1}, \quad \mathbf{H}_{\mathrm{G}} = \mathbf{H}^{\mathrm{T}}\mathbf{G} \tag{4.85}$$

and elements of estimates $\hat{\mathbf{X}}$ are defined as

$$\hat{X}_i = \sum_{j=1}^{M} h_{ij} x_j \text{ , } i=1,2,..L \tag{4.86}$$

where:

$$h_{ij} = \sum_{l=1}^{L} p_{il} h_{jl}^{g} \text{ ,} \tag{4.87}$$

p_{il} – elements of matrix $\mathbf{P}_{\mathrm{G}}$

$$\mathbf{P}_{\mathrm{G}} = \{p_{il}\} = (\mathbf{H}_{\mathrm{G}}^{T}\mathbf{H})^{-1} = \{q_{il}\}^{-1} \tag{4.88}$$

h_{jl}^{g} – elements of matrix $\mathbf{H}_{\mathrm{G}}$

$$\mathbf{H}_{G} = \mathbf{H}^{T}\mathbf{G} = \{h_{jl}^{g}\}, \tag{4.89}$$

$$h_{jl}^{g} = \sum_{i=1}^{M} h_{ij} g_{il} \text{ and } q_{il} = \sum_{i=1}^{M} h_{ji}^{g} h_{il}$$

From (4.86) one can note that in practical consideration (as for example, in protective algorithms) only separate components of the searching vector $\hat{\mathbf{X}}$ have to be estimated. Therefore, for the *L*-state signal model only $K \leq L$ variables are to be estimated. Expression (4.84) performs a correlation algorithm of the input signal *y* with rows of matrix $\mathbf{P}_{\mathrm{G}}\mathbf{H}_{\mathrm{G}}$. If consecutive results are obtained from the last *M* samples of the input signal (*M* – constant or expanded window length), the algorithm is called non-recursive [137].

The non-recursive LSM identification has similar features to FIR filters (e.g., linear phase of the transfer function, finite duration response) and, moreover, criterion used provides a very effective method for synthesis of algorithms with defined properties.

Characteristics of the algorithm used depend on the applied window length *M*, the signal model (matrix **H**) and weighting matrix **G**. The last one depends on the measurement condition (4.80) and together with matrix **H** defines matrix $\mathbf{H}_{\mathrm{G}}$ (4.87). However, matrix $\mathbf{H}_{\mathrm{G}}$ can also be established by appropriate changing of the measurement model due to a desirable modification of the algorithm. Hence, it is reasonable to distinguish a signal model used in matrix **H** and a measurement model applied in matrix $\mathbf{H}_{\mathrm{G}}$. That is further explained in the example of the 3-state signal model.

Matrix **P** ($\mathbf{P}_{\mathrm{G}}$) is defined as the inverse of $\{q_{il}\}$ and, therefore, the analytical expression for coefficients h_{ij} in (4.87) is, in general, complicated and can be formulated only for simple cases. To illustrate, consider the design of more usable algorithms.

Considered previously, the Fourier algorithm for estimation of the fundamental frequency phasor is based on the sinusoidal-waveform signal model as in (4.71). Let us, for generalization, examine the deterministic signal model shifted by angle β with respect to the center of the measurement window (Fig. 4.13):

$$\mathbf{h}(j) = [h_{\mathrm{R}}(j) \;\; h_{\mathrm{I}}(j)] = [\cos(\nu j + \beta - \nu M + \nu/2) \;\; -\sin(\nu j + \beta - \nu M + \nu/2)] \quad (4.90)$$

$\mathbf{H} = [\mathbf{h}(0)\mathbf{h}(1)\ldots\mathbf{h}(M-1)]^{\mathrm{T}}$ and **G=I** (no weighting factors are considered).

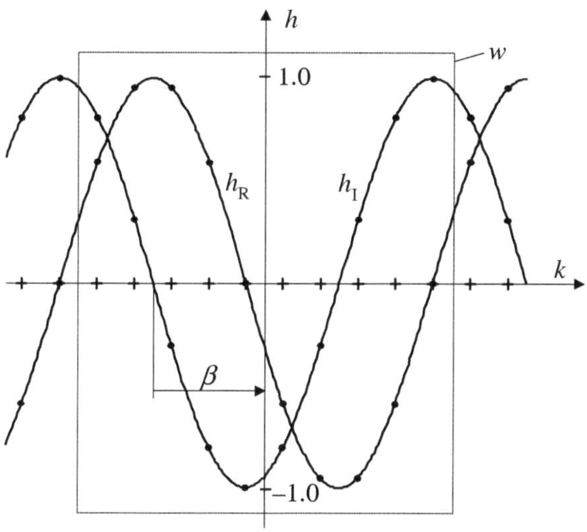

Fig. 4.13 Two-states signal model

The estimated vector has two unknowns:

$$\hat{\mathbf{X}} = \begin{bmatrix} \hat{X}_{\mathrm{R}} \\ \hat{X}_{\mathrm{I}} \end{bmatrix} \quad (4.91)$$

Matrix **P** takes the form

$$\mathbf{P} = (\mathbf{H}^{\mathrm{T}}\mathbf{H})^{-1} = \begin{bmatrix} p_{11} & p_{12} \\ p_{21} & p_{22} \end{bmatrix} = \begin{bmatrix} q_{11} & q_{12} \\ q_{21} & q_{22} \end{bmatrix}^{-1} \quad (4.92)$$

Substituting (4.90) into (4.92) yields [249]:

$$q_{11} = \sum_{i=0}^{M-1} \cos^2(\nu i + \beta - \nu M + \nu/2) = \frac{M}{2} + \frac{\sin(\nu M)\cos(2\beta - \nu M)}{2\sin\nu}$$

$$q_{12} = q_{21} = -\sum_{i=0}^{M-1} \sin(\nu i + \beta - \nu M + \nu/2)\cos(\nu j + \beta - \nu M + \nu/2) \\ = -\frac{\sin(\nu M)\sin(2\beta - \nu M)}{2\sin\nu} \tag{4.93}$$

$$q_{22} = \sum_{i=0}^{M-1} \sin^2(\nu i + \beta - \nu M + \nu/2) = \frac{M}{2} - \frac{\sin(\nu M)\cos(2\beta - \nu M)}{2\sin\nu}$$

It is easy to verify that for a data window equal to a multiple of the half-period of the signal model:

$$M = l\frac{N}{2}, \; l - \text{integer} \tag{4.94}$$

matrix **P** has the value

$$\mathbf{P} = \begin{bmatrix} p_{\mathrm{R}} & 0 \\ 0 & p_{\mathrm{I}} \end{bmatrix} = \frac{2}{M}\begin{bmatrix} 1 & 0 \\ 0 & 1 \end{bmatrix} \tag{4.95}$$

For one data window (without time consideration), estimated values are determined similarly as in (4.40). In the consecutive time steps that formula is applied for the stored last M samples of the signal $x(k)$. Moreover, if functions h_{R} and h_{I} of length $M = N$ are moving with the data window one obtains well-known full-period correlation procedure of the form

$$\begin{aligned} \hat{X}_{\mathrm{R}}(k) &= \frac{2}{N}\sum_{j=0}^{N-1} h_{\mathrm{R}}(j)x(k+j-N+1) \\ \hat{X}_{\mathrm{I}}(k) &= \frac{2}{N}\sum_{j=0}^{N-1} h_{\mathrm{I}}(j)x(k+j-N+1) \end{aligned} \qquad k = 0,1,\cdots \tag{4.96}$$

where for $\beta = 0$:

$$\begin{aligned} h_{\mathrm{R}}(j) &= \cos(\nu j + \nu/2) \\ h_{\mathrm{I}}(j) &= -\sin(\nu j + \nu/2) \end{aligned} \qquad j = 0 \cdots N-1 \tag{4.97}$$

For starting the algorithm (4.96) (k=0) the initial input data window should be known $\left(x(-N+1)\ldots x(0)\right)$.

In fact, the algorithm (4.96) is identical with the full-cycle Fourier algorithm with moving window (4.47). The difference is only in the description of summation order. We can check that the above algorithm can also be described in the form with a stationary correlation function (4.48).

The obtained relations are identical with the results from the DFT approach. For example, substituting in (4.95): $M = N/2$ we obtain the half-period Fourier algorithm (4.52). Moreover, if we demand any window length *M*, the resulting algorithm has the form:

$$\begin{bmatrix} \hat{X}_R(k) \\ \hat{X}_I(k) \end{bmatrix} = \begin{bmatrix} p_{11}(k) & p_{12}(k) \\ p_{21}(k) & p_{22}(k) \end{bmatrix} \begin{bmatrix} X'_R(k) \\ X'_I(k) \end{bmatrix}, \quad k = 0,1,\cdots \tag{4.98}$$

where:

$$\begin{aligned} X'_R(k) &= \sum_{j=0}^{M-1} h_R(k+j-M+1)x(k+j-M+1) \\ X'_I(k) &= \sum_{j=0}^{M-1} h_I(k+j-M+1)x(k+j-M+1) \end{aligned} \tag{4.99}$$

$h_R(j)$, $h_I(j)$ – as in (4.97); coefficients $p_{ij}(k)$ in (4.98) are calculated as in (4.91) after substitution: $\beta = \nu k$.

Another example is presented in the next section where we apply the idea of LSM to reject of DC from the input signal.

4.3.2 DC Component Consideration

While voltage signals are practically not contaminated with decaying DC components, currents can contain exponentially decaying DC offsets with unpredictable initial magnitude and unknown time constant. This is so due to both the fault resistance and the fault incipience angle being random factors. The decaying DC offset in a fault current can result in nearly twenty per cent overshoot in the estimated magnitude depending on the used algorithm [307]. This has to be taken into account when setting fast digital relays. In fault location a DC component influences the distance to fault estimation error, especially when the fault record is relatively short.

Generally, the phasor estimators can be easily designed to remove a pre-defined DC component [237, 258, 306]. The problem arises, however, when the time constant of the DC offset is not known. Several approaches have been put forward so far to minimize the influence of a decaying DC component on phasor estimation. The most practical solution merges a digital mimic filter with any robust phasor estimator such as the full-period Fourier algorithm. The digital mimic filter reproduces numerically an analog mimic circuit commonly applied in traditional line protection [137, 237]. The original mimic circuit is a differentiating element, and therefore, its digital replica is a high-pass digital filter [18, 237]. This, in turn, results in a worsened frequency characteristic of the resulting phasor estimator. The strong asset of a digital mimic filter is its significant robustness to the actual (random) value of the time constant.

The other approach is to include an exponentially decaying DC component into the signal model and apply certain curve-fitting techniques to design a phasor estimator. Usually, the DC offset is dealt with as a noise, represented by either deterministic or probabilistic characteristics, rather than a parameter to be fitted [18, 237]. Several specific estimators originate from this general approach:

- Linear estimators when applied to suppress a DC component deal with the non-linear exponential function by using the first few terms of its Taylor series [258].
- The least squares method (LSM) yields non-recursive or recursive estimators [108, 137, 262]. The latter can be modified by introducing resetting or forgetting factors for better absorption of non-harmonic frequencies including the decaying DC component.
- Application of a state-variable model brings forth estimators in the form of a state observer [257] or a Kalman filter [78]. The requirement for rarely available *a priori* noise characteristics is, however, the key disadvantage of the latter measuring technique.

Yet another, numerically simpler, approach to a decaying DC component is to apply a cosine filter for estimating both the direct (cosine filter) and quadrature (cosine filter delayed by one-fourth of a cycle) orthogonal components [307]. Additional delay slowing down the estimator's response is, though, the main disadvantage of this approach.

The aforementioned methods, although they improve the measurement, do not reject a DC component perfectly. Below, we present yet another algorithm that fully rejects a DC component from the input signal regardless of its time constant and the initial magnitude.

For further considerations let us assume the following signal model:

$$x(t) = X_0 \exp(-t/\tau) + \sum_{m=1}^{K} X_m \cos(m\omega_1 t - \varphi_m) \tag{4.100}$$

where:
X_0, τ – magnitude and time constant of the decaying DC component, respectively; K – number of considered harmonics; $\omega_1 = 2\pi / T_1$ radial system frequency; T_1 – period of the fundamental frequency; φ_m, Y_m – initial phase angle and magnitude of the *m*th harmonic.

To estimate the phasor out of M consecutive samples of the signal (4.100) we may use the three-state LSM algorithm with the signal model matrix $\mathbf{H}(k)$ composed from the following rows [250]:

$$\mathbf{h}(k-j)=[h_{\mathrm{R}}(k-j),\ h_{\mathrm{I}}(k-j),\ h_0(k-j)],\ j=0..M-1 \tag{4.101}$$

$$h_{\mathrm{R}}(k-j)=\cos(\nu(k-j)),\ h_{\mathrm{I}}(k-j)=-\sin(\nu(k-j))$$
$$h_0(k-j)=\exp(b(k-j))$$

where: $\nu=\dfrac{2\pi}{N}$, $b=-\dfrac{T_1}{N\tau}$, N – number of samples per cycle, k – sample index.

Consider the full-period ($M = N$) standard LSM algorithm based on the signal model matrix (4.101) with the time constant τ known. The vector of estimates is updated at every sampling instant using the following fundamental equation:

$$\hat{\mathbf{X}}(k)=\mathbf{P}(k)\mathbf{H}^T(k)\mathbf{x}(k) \tag{4.102}$$

where the covariance matrix $\mathbf{P}=\left(\mathbf{H}^{\mathrm{T}}(k)\mathbf{H}(k)\right)^{-1}$ and:
$\mathbf{x}(k)=[x(k-N+1),\ x(k-N+2),\ ..,\ x(k)]^{\mathrm{T}}$, $\hat{\mathbf{X}}(k)=[X_{\mathrm{R}}(k),\ X_{\mathrm{I}}(k),\ X_0(k)]^{\mathrm{T}}$,
while the superscript T stands for matrix transposition.
Practically, only two variables $X_{\mathrm{R}}(k)$ and $X_{\mathrm{I}}(k)$ are to be determined by (4.101) as they form the estimated phasor.

The algorithm (4.102) rejects a DC component perfectly as long as the input signal matches the assumed signal model (4.1). The standard LSM algorithm is, however, very sensitive to the time constant τ, and, therefore, cannot be used directly in digital relaying [237].

Due to the presence of the exponentially decaying DC component in the signal model (4.102), the matrix $\mathbf{P}(k)$ in (4.102) is a full matrix with off-diagonal elements depending on the exponential function. As a consequence, the filters of the resulting non-recursive estimator have non-regular coefficients and, unlike the full-period Fourier algorithm, they cannot be represented in the recursive form. The recursive algorithm, though, can be designed for certain special LSM estimators as shown below.

Let us consider the weighting LSM estimator with the measurement model matrix defined as in (4.85) with the rows of $\mathbf{H}_{\mathrm{G}}(k)$ given as:

$$\mathbf{h}_{\mathrm{G}}(k-j)=[h_{\mathrm{c}}(k-j),\ h_{\mathrm{s}}(k-j),\ h_0(k-j)],\ j=0..N-1 \tag{4.103}$$

where: $h_{\mathrm{c}}(k-j)=\cos(\nu(k-j))-d_{\mathrm{c}}(k)$, $h_{\mathrm{s}}(k-j)=-\sin(\nu(k-j))-d_{\mathrm{s}}(k)$.

It is worth emphasizing that the matrix $\mathbf{H}_{\mathrm{G}}(k)$ is obtained according to (4.103) by adequate modification of the original signal model matrix $\mathbf{H}(k)$ (4.101). For this operation the weighting matrix $\mathbf{G}(k)$ need not be known. Instead, the adequate values of the coefficients $d_c(k)$ and $d_s(k)$ need to be determined.

For the measurement model (4.85), the estimator is determined as in (4.88) with the following covariance matrix:

$$\mathbf{P}_G(k) = \left(\mathbf{H}_G^T(k)\mathbf{H}(k)\right)^{-1} \tag{4.104}$$

The functions $d_c(k)$ and $d_s(k)$ in (4.103) are fixed in the signal data window. When designing the estimator (4.102), it is justified to state the following two requirements for these two functions:

- the covariance matrix $\mathbf{P}_G(k)$ of the final algorithm should be diagonal; and
- the functions $d_c(k)$ and $d_s(k)$ should be easy to determine from the DC component of the input signal.

The first requirement is aimed at achieving the simplest form of the final measuring algorithm. Actually, it is enough to impose this requirement only on the first two rows of the matrix $\mathbf{P}_G(k)$ that are related to the estimated orthogonal components. It is worth noting, however, that diagonalization of the first two rows in a 3×3 matrix leads to diagonalization of the same rows in the inverse matrix. Eventually, the first requirement can be written as follows:

$$q_{ij} = 0 \text{ for } i \neq j\text{ , } i = 1{,}2\text{ , } j = 1{,}2{,}3 \tag{4.105}$$

where $\{q_{ij}\} = \mathbf{H}_G^T(k)\mathbf{H}(k)$ – see (4.91).

From that condition we can obtain the following relations defining functions $d_c(k)$ and $d_s(k)$ [250]:

$$d_c(k) = \frac{\sum_{j=0}^{N-1} \cos\left(\nu(k-j)\right)}{\exp(bk)\sum_{j=0}^{N-1} \exp(-bj)} \tag{4.106}$$

$$d_s(k) = \frac{-\sum_{j=0}^{N-1} \sin\left(\nu(k-j)\right)}{\exp(bk)\sum_{j=0}^{N-1} \exp(-bj)} \tag{4.107}$$

which, after simplification, yield

$$d_{\mathrm{c}}(k) = D\cos(\nu k + \vartheta)$$
$$d_{\mathrm{s}}(k) = -D\sin(\nu k + \vartheta) \tag{4.108}$$

where: $\vartheta = \tan^{-1}\left(\dfrac{\sin\nu}{\cos\nu - \exp(b)}\right)$, $D = \dfrac{1-\exp(b)}{\sqrt{(1-\exp(b))^2 + \sin^2\nu}}$.

Note that the diagonal elements in the first two rows of the matrix $\mathbf{P}_{\mathrm{G}}(k)$ are constant: $1/q_{11} = 1/q_{22} = 2/N$ (similarly to the full-period Fourier algorithm).

From (4.108) it is obvious that the obtained functions $d_{\mathrm{c}}(k)$ and $d_{\mathrm{s}}(k)$ are directly related to the time constant of the DC component. This is in accord with the second postulated condition for these two functions.

Analogously to (4.96) we derive the algorithm for phasor estimation from (4.102) and (4.103) as follows:

$$X_{\mathrm{c}}(k) = \frac{2}{N}\sum_{j=0}^{N-1} h_{\mathrm{c}}(k-j)x(k-j) = X_{\mathrm{R}}(k) - \delta_{\mathrm{c}}(k) \tag{4.109}$$

$$X_{\mathrm{s}}(k) = \frac{2}{N}\sum_{j=0}^{N-1} h_{\mathrm{s}}(k-j)x(k-j) = X_{\mathrm{I}}(k) + \delta_{\mathrm{s}}(k) \tag{4.110}$$

where:

$$X_{\mathrm{R}}(k) = \frac{2}{N}\sum_{j=0}^{N-1} h_{\mathrm{R}}(k-j)x(k-j) \tag{4.111}$$

$$X_{\mathrm{I}}(k) = \frac{2}{N}\sum_{j=0}^{N-1} h_{\mathrm{I}}(k-j)x(k-j) \tag{4.112}$$

$$\delta_{\mathrm{c}}(k) = d_{\mathrm{c}}(k)X_{\mathrm{a}}(k)$$
$$\delta_{\mathrm{s}}(k) = d_{\mathrm{s}}(k)X_{\mathrm{a}}(k) \tag{4.113}$$

$$X_{\mathrm{a}}(k) = \frac{2}{N}\sum_{j=0}^{N-1} x(k-j) \tag{4.114}$$

It can be seen from (4.109) and (4.110) that the auxiliary (original) orthogonal components $X_{\mathrm{R}}(k)$, $X_{\mathrm{I}}(k)$ are estimated using the regular full-period Fourier algorithm. Next, the new (corrected) values of the orthogonal components $X_{\mathrm{c}}(k)$, and $X_{\mathrm{s}}(k)$ are obtained by applying an appropriate correction (4.113). The correcting functions $\delta_{\mathrm{c}}(k)$ and $\delta_{\mathrm{s}}(k)$ are proportional to the average DC component in the data window X_a (4.114) and the functions $d_c(k)$ and $d_s(k)$ (4.113). This algorithm can be also rewritten in recursive form [249].

The above algorithm still requires the time constant of the decaying DC component τ to be known *a priori*. However, the correlation function is now separated into two operations enabling easy modifications of this estimator. In particular, this form of the estimator allows for extracting the value of τ in the course of the phasor estimation.

Let us further simplify the functions $d_c(k)$ and $d_s(k)$. From (4.108) one obtains:

$$\begin{aligned} d_c(k) &= p_c h_R(k) + p_s h_I(k) \\ d_s(k) &= p_s h_R(k) - p_c h_I(k) \end{aligned} \tag{4.115}$$

where: $p_c = D\cos\vartheta, \quad p_s = D\sin\vartheta$.

The coefficients p_c and p_s depend on the constant ν (related to the sampling frequency) and the parameter $r = \exp(b)$ (related to the time constant τ).

The parameter r can be estimated from the measurements – the component $X_a(k)$ defined by (4.114) stands for a sum of N elements of the geometric progression with the multiplier of $r = \exp(b)$. Therefore:

$$r = r(k) = \exp(b) = \frac{X_a(k)}{X_a(k-1)} \tag{4.116}$$

The estimator (4.116) is the central element of the presented algorithm since it allows for accurate and direct estimation of an unknown time constant of the decaying DC component.

Physically, the time constant τ does not change during a given transient. The estimator (4.116) can produce, however, unrealistic values at the first few samples after the fault inception, and therefore, requires some forced stabilization. This can be done by checking the following simple condition:

$$r_{min} \le r \le r_{max} \tag{4.117}$$

and truncating the estimate if the limits get overrun. The limits are naturally chosen as:

$$r_{min} = \exp\frac{-T_1}{N\tau_{min}}, \quad r_{max} = 1 \tag{4.118}$$

where τ_{min} is the minimum value of the time constant recommended to be set at around 4 ms.

Further damping of the unwanted oscillations of the estimated factor r can be done by averaging it over a given number of samples L, i.e. by applying the equation:

$$r_{\text{filtered}}(k) = \frac{1}{L}\sum_{m=0}^{L-1} r_{\text{original}}(k-m) \tag{4.119}$$

As our simulations show, good results are obtained for $L = 2..4$. It is worth noting that, practically, the filter (4.119) does not delay the phasor estimation. This is so because (4.119) is applied to the signal parameter that physically does not change in time.

In order to simplify the application equations for the new algorithm, let us express the coefficients p_c and p_s as functions of ν and r. Substituting D (4.108) into (4.115) and taking into account ϑ (4.108) and (4.116) yields:

$$p_c = \frac{(1-r)(\cos\nu - r)}{(1-r)^2 + \sin^2\nu}, \quad p_s = \frac{(1-r)\sin\nu}{(1-r)^2 + \sin^2\nu} \tag{4.120}$$

Figure 4.14 presents waveforms of p_c and p_s as a functions of r for $N = 20$ samples per period. Some features of the algorithm are illustrated in the following example.

Example 4.1 Figure 4.15 presents the current signal that was recorded in that 50-Hz network with the sampling frequency of f_s=1000 Hz. Prepare the procedure for estimation the current phasor removing the DC component according to the presented algorithm.

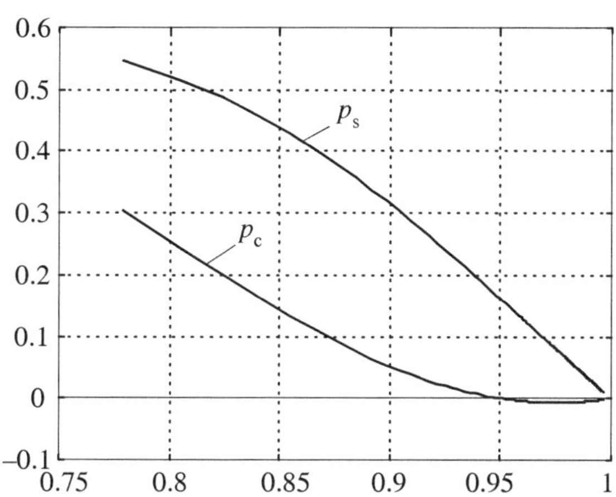

Fig. 4.14 Coefficients p_c and p_s as functions of r for $N = 20$

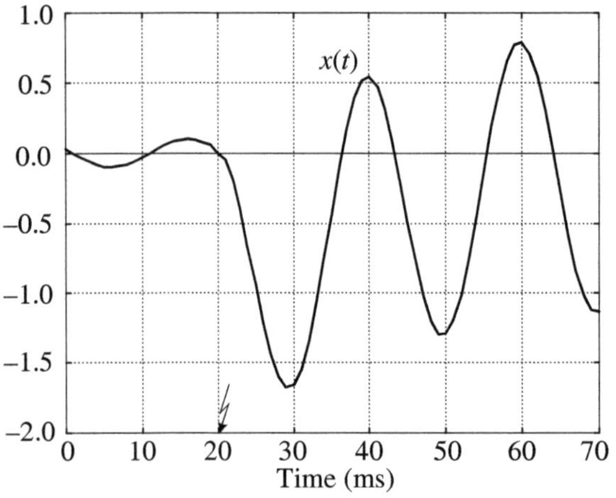

Fig. 4.15 Recorded signal

Following the above description of the algorithm we can prepare the MATLAB procedure as follows (function *dc_lsm.m*).

```
function [Xc,Xs,XR,XI,del_c,del_s]=dc_lsm(N,f1,x_k);
% inputs:     N - number of samples per cycle
%             f1 - system frequency
%             x_k - recorded input signal
%outputs:     Xc, Xs - components of corrected phasor
%             XR, XI - components of the phasor (Fourier alg.)
%             del_c, del_s - introduced corrected signals

 a=2*pi/N;
t1=1/f1;                    % system frequency period
 tamx=0.3;                  % max. decaying time const.
 tamn=0.004;                % min. decaying time const.

 rsmx=exp(-t1/(N*tamx));
 rsmn=exp(-t1/(N*tamn));

 for i=1:N,                 % correlation functions
        hR(i)=2*(cos((i-1)*a))/N;
        hI(i)=-2*(sin((i-1)*a))/N;
 end;

 ns=max(size(x_k));            % number of input data
 j=1;
% main time loop
 for i=1:ns,
```

```
      dx=x_k(i);
      if i>N,
       dx=dx-x_k(i-N);      % difference of signal
      end;
      if i>1,
        XR(i)=XR(i-1)+hR(j)*dx;  % full-period Fourier
        XI(i)=XI(i-1)+hI(j)*dx;  % recursive algorithm
        Xa(i)=Xa(i-1)+dx;        % dc component
        if abs(Xa(i-1))>0,
         rs(i)=Xa(i)/Xa(i-1);
         if (rs(i)>1)|(rs(i)<0),
           rs(i)=1;
          elseif rs(i)<rsmn,
           rs(i)=rsmn;
         end,
         else,
          rs(i)=1;
        end;
       else,
        XR(i)=hR(i)*dx;          % first step
        XI(i)=hI(i)*dx;
        Xa(i)=x_k(i);
        rs(i)=1;
      end;

      p=(1-rs(i))/(1+rs(i)*(rs(i)-2*cos(a)));
      pc=p*(cos(a)-rs(i));
      ps=p*sin(a);
      del_c(i)=(pc*hR(j)+ps*hI(j))*Xa(i);  % corrections
      del_s(i)=(ps*hR(j)-pc*hI(j))*Xa(i);
      Xc(i)=XR(i)-del_c(i);
      Xs(i)=XI(i)+del_s(i);

      j=j+1;
      if j>N,
       j=1;
      end;
  end;   %i

return;
```

The results of the phasor estimation are presented in Fig. 4.16. We can see that the correction of the DC component is fully effective after one period from the fault point.

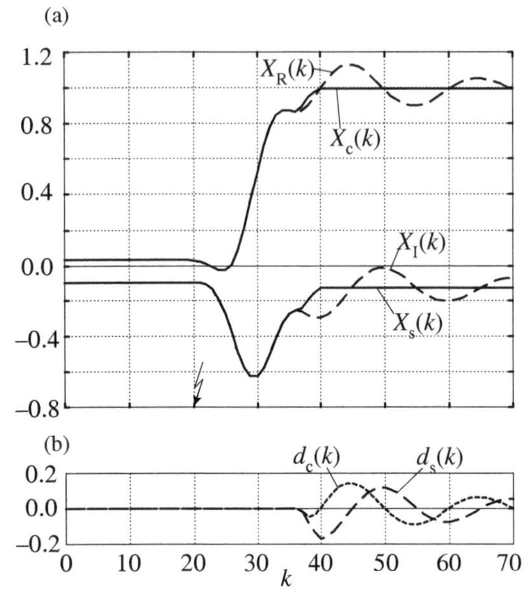

Fig. 4.16 Results of estimation

The real-time computations can be summarized as follows:

1. Calculate the orthogonal components $X_R(k)$, $X_I(k)$ by (4.111), (4.112), i.e., using the regular full-period Fourier algorithm. Calculate $X_a(k)$ by (4.114). (In the presented MATLAB function we are using the recursive version of the algorithm.)
2. If $X_a(k-1) < \xi$ (ξ – minimum value of the DC component to be considered in the correction), take $X_R(k)$, $X_I(k)$ as the final results, and skip steps 3 and 4.
3. Determine the functions $d_c(k)$ and $d_s(k)$ by the following sequence of computations:
   - estimate the parameter r by (4.116) and (4.117);
   - calculate the values of p_c and p_s by (4.120) or from memory-stored functions (Fig.4.14, for example);
   - determine $d_c(k)$ and $d_s(k)$ according to (4.115).
4. Calculate the correcting values $\delta_c(k)$ and $\delta_s(k)$ as well as the final values of the orthogonal components (4.109) and (4.110).

The check-point set in step 2 ensures that no division by zero will take place. The condition also by-passes the correction when the DC component is small and the correction is actually useless.

The steady-state performance of the algorithm is certainly the same as for the standard full-period Fourier filter (Fig. 4.10). It is possible to synthesize the half-cycle version of this algorithm, however, it is more sensitive to higher harmonics in the input current [250].

4.3.3 Synchrophasors

A synchrophasor (synchronized phasor) is a relatively new term that is related to the technique of phasor measurements that occur at 'the same time' in many places inside the power system. Therefore, synchrophasors are the time-stamped phasors. The synchronization process and the phasors measurement are realized in devices generally called phasor measurement units (PMU) that may be autonomous or integrated in other control devices, such as a relay, a recorder etc., [236]. The big progress in the synchrophasors technique application in power systems was made after an easy available time-synchronizing signal appeared. In the modern application the common time source (so-called universal time clock - UTC) is delivered by the satellite global positioning system (GPS)[6] in the form of a radio clock that guarantees a time resolution less than 1 μs that allows for phase-angle measurements with uncertainties of less than 0.02° [16]. In the standard application the synchrophasors measure voltages and currents at diverse locations on a power grid, and can output accurately time-stamped voltage and current phasors. Because these phasors are truly synchronized, comparison of a few quantities is possible in real time.

For utilizing the idea of synchrophasors it is required to respect some assumed standard agreements. First, it is assumed that the considered continuous-time signal model is referred to the cosinusoidal function:

$$x(t) = X\cos(\omega t + \psi) = \mathrm{Re}\left(\sqrt{2}\,\underline{X}\mathrm{e}^{\mathrm{j}\omega t}\right) \tag{4.121}$$

where the phasor magnitude is rescaled to the RMS value:

$$\underline{X} = \frac{X}{\sqrt{2}}\mathrm{e}^{\mathrm{j}\psi} = \frac{X}{\sqrt{2}}\angle\psi = X_{\mathrm{R}} + \mathrm{j}X_{\mathrm{I}} \tag{4.122}$$

The idea of the synchrophasors consists in representation of the measured phasor by its RMS: $X/\sqrt{2}$ and an angle for each reporting request. The rates of requested reports are also standardized to the maximum rate of 25 per second for a 50-Hz system and 30 rates per second for a 60-Hz system [107]. The value of an angle reported is limited to $\pm\pi$ as is shown in Fig. 4.17 [242].

---

[6] At present there are few such satellite systems available.

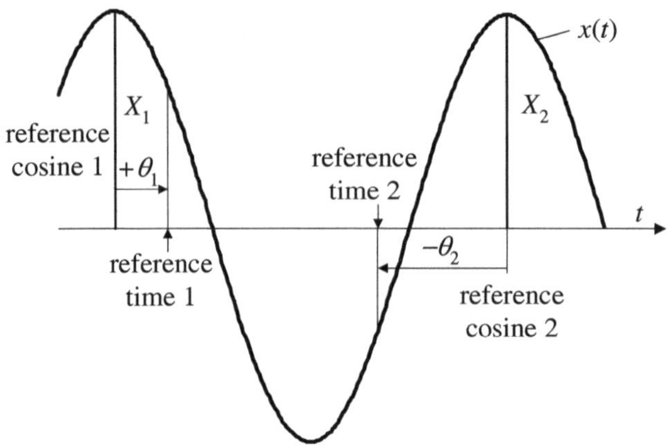

Fig. 4.17 Definition of synchrophasors

The phasor can be estimated by using a full-cycle Fourier algorithm. For example, the estimator (4.48) can be represented in the following form:

$$\underline{X}(k) = \frac{\sqrt{2}}{N} \sum_{i=0}^{N-1} x(k-i-0.5)\mathrm{e}^{-\mathrm{j}\nu(k-i-0.5)} \tag{4.123}$$

The result has index k but in fact due to the filter group delay it is referenced to the middle of the window that, for even N, points to a time index: k–(N–1)/2 that should be taken as a reference time.

For fulfillment of the high-accuracy requirement of synchrophasors the signal-processing procedures should be prepared very carefully taking into account the following details [16, 151, 242]:

- estimation of a phasor angle and a magnitude is influenced by the system frequency which may differ from the nominal value;
- the measurements need to be stable and accurate even during dynamical changes due to sub-synchronous oscillation power swings and other low-frequency transients;
- the measurement chain, including instrument transformers, transducers, analog filters and A/C devices are not ideal and adequate non-linear and dynamical corrections are needed;
- fast communication channels are needed to accommodate the data-transfer rates required for state updates that would reflect transients.

Some of the above-mentioned problems are not so far fully resolved. For example, it is not clear what phasor definition should be implemented during dynamic conditions [242, 347].

From the point of view of the applications considered here the synchrophasors technology delivers very powerful tools that are especially important for improving the fault-location accuracy in lines with application of both-side measurements. The methodology and examples are provided in Chap. 7.

4.4 Time-domain Analysis

The signal and network parameters estimation algorithms presented so far are based on the phasor concept. They were derived on an assumption that the post-fault state can be considered as a steady-state one. Consequently, fault-loop voltage and current are represented in a form of sinusoidal waveforms. However, there are some processes in which the important information about the system parameters is contained in decaying components with different frequency, including DC component. If such transients last a relatively long period then the phasor-based signal models may not be adequate to be the equivalent of the process. In that situation we can try to use directly the instantaneous-based state of the system representation.

4.4.1 Differential-equation-based Approach

In the distance-relaying and the fault-location algorithms based on the considered approach the fault-loop model is described as a time-domain process. The model parameters are involved with instantaneous current and voltage waveforms measured at the substation known also as the system-model-based approach.

For illustration, let us consider the *RL* circuit with lumped parameters (Fig. 4.18) that is represented by the following differential equation (the fault resistance R_F is neglected):

$$u(t) = Ri(t) + L\frac{\mathrm{d}i(t)}{\mathrm{d}t} \tag{4.124}$$

For representation of the above equation in a discrete-time domain a known numerical procedure for resolving the following differential equation can be used:

$$\frac{\mathrm{d}i(t)}{\mathrm{d}t} = -\frac{R}{L}i(t) + \frac{1}{L}u(t) = f\big(i(t), u(t)\big) \tag{4.125}$$

Applying the trapezoidal rule one obtains [34]:

$$i(k) = i(k-1) + \frac{T}{2}\Big(f\big(i(k-1), u(k-1)\big) + f\big(i(k), u(k)\big)\Big) \tag{4.126}$$

where it is convenient to use the integration step T equal to the sampling period of the input waveforms: $T{=}T_s$.

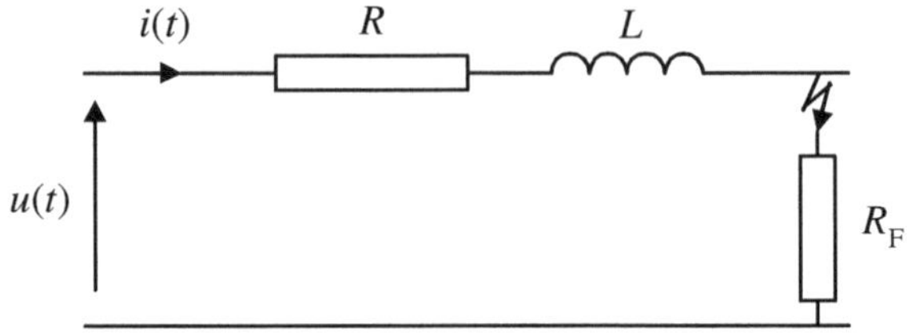

Fig. 4.18 Fault-loop model

The above immediately leads to the discrete-time version of (4.124):

$$u(k)+u(k-1)=R\big(i(k)+i(k-1)\big)+L\frac{2}{T_s}\big(i(k)-i(k-1)\big) \tag{4.127}$$

It can be seen that we should calculate two unknowns in (4.127): R and L, whereas we have only one equation. However, we can easily formulate an additional equation for a different time instant. Generally, the pair of such equations can be written in the form [256]:

$$\begin{aligned} b_1 &= Ra_1 + Ld_1 \\ b_2 &= Ra_2 + Ld_2 \end{aligned} \tag{4.128}$$

from which the sought values are derived by constituting the following impedance algorithm:

$$\begin{aligned} R &= \frac{d_1 b_2 - d_2 b_1}{a_2 d_1 - a_1 d_2} \\ X &= \frac{a_2 b_1 - a_1 b_2}{a_2 d_1 - a_1 d_2}\omega_1 \end{aligned} \tag{4.129}$$

Comparing (4.128) with (4.127) it can be noticed that coefficients a_1, a_2, b_1, b_2 estimate the value of voltage and current, respectively, with reference to the time instant t_k–$T_s/2$, e.g.: $b_1/2=b_1(t_k-T_s/2)=(u(t_k)+u(t_{k-1}))/2$, while the coefficients d_1, d_2 represent estimates of the current derivatives with reference to the same time instant, e.g.: $d_1=d_1(t_k-T_s/2)=(i(t_k)-i(t_{k-1}))/T_s$. It is important to assure the same reference time for all elements of the estimators (4.129) because this guarantees that there is no unwanted phase shifting between them. In other cases additional phase compensation is need.

In an AC system it is reasonable to assure that in the steady state the aforesaid estimates be equal to the adequate sinusoidal values. For the considered trapezoidal rule applied in (4.127) one can write the following relations:

$$p_a\big(\sin(\nu k)+\sin(\nu k-\nu)\big)=\sin(\nu k-\nu/2) \quad \text{– for approximation} \tag{4.130a}$$

$$p_d\big(\sin(\nu k)-\sin(\nu k-\nu)\big)=\omega_1\cos(\nu k-\nu/2) \quad \text{– for differentiation} \tag{4.130b}$$

from which it follows: $p_a=\dfrac{1}{2\cos(\nu/2)}$, $p_d=\dfrac{\omega_1}{2\sin(\nu/2)}$, where: $\nu=\omega_1 T_s$.

Rewriting (4.127) with respect to (4.130a and b) one obtains:

$$u(k)+u(k-1)=R\big(i(k)+i(k-1)\big)+L\alpha\big(i(k)-i(k-1)\big) \tag{4.131}$$

where: $\alpha=\dfrac{p_d}{p_a}=\dfrac{\omega_1}{\tan(\omega_1 T_s/2)}$.

It is no surprise that we have obtained the same result as in (4.127).

Following this description it is possible to use different methods for calculation of the coefficients in the estimators (4.129). Choosing the 2nd-order Gear method and introducing adequate correction to match the proper relation for sinusoidal signals, one obtains the following coefficients in (4.128) for four consecutive samples [256]:

$$\begin{aligned} a_1 &= i(k), \quad & a_2 &= i(k-1)\\ b_1 &= u(k), \quad & b_2 &= u(k-1)\end{aligned} \tag{4.132a}$$

$$\begin{aligned} d_1 &= \alpha\big(3i(k)-4i(k-1)+i(k-2)\big)\\ d_2 &= \alpha\big(3i(k-1)-4i(k-2)+i(k-3)\big)\end{aligned} \tag{4.132b}$$

where: $\alpha=\dfrac{\omega_1}{2\sqrt{\big(1-\cos(\omega_1 T_s)\big)^4+\big(2\sin(\omega_1 T_s)-\sin(2\omega_1 T_s)/2\big)^2}}$.

One can note that the 3-point differentiation rule according to the 2nd-order Gear method is related to the last (actual) point. For the nominal frequency the output signal is shifted by almost 90° (for N=20 samples per period the shifting angle is equal to 89.58°).

Because of transients in post-fault current and voltage waveforms the estimators give unstable results. As a remedy one can apply adequate pre-filtering. An interesting algorithm can be obtained when for pre-filtering one applies two short-window orthogonal filters. For example, using half-cycle Fourier filters (4.52) leads to the following coefficients in (4.128):

$$\begin{aligned} a_1 &= \frac{1}{2}\big(I_\mathrm{I}(k)+I_\mathrm{I}(k-1)\big), \quad & a_2 &= \frac{1}{2}\big(I_\mathrm{R}(k)+I_\mathrm{R}(k-1)\big)\\ b_1 &= \frac{1}{2}\big(U_\mathrm{I}(k)+U_\mathrm{I}(k-1)\big), \quad & b_2 &= \frac{1}{2}\big(U_\mathrm{R}(k)+U_\mathrm{R}(k-1)\big)\\ d_1 &= \frac{1}{T_s}\big(I_\mathrm{I}(k)-I_\mathrm{I}(k-1)\big), \quad & d_2 &= \frac{1}{T_s}\big(I_\mathrm{R}(k)-I_\mathrm{R}(k-1)\big)\end{aligned} \tag{4.133}$$

where indices R and I indicate adequate current and voltage orthogonal components obtained according to (4.52).

Another approach consists in utilizing more than two equations as in (4.128) formulated for different points. This leads to the LSM algorithm. For example, the measurement model for estimation of parameters in (4.124) with use of the trapezoidal scheme of integration over M consecutive points has the shape:

$$\begin{bmatrix} i(k-M+1)+i(k-M) & \alpha\big(i(k-M+1)-i(k-M)\big) \\ \vdots & \vdots \\ i(k-1)+i(k-2) & \alpha\big(i(k-1)-i(k-2)\big) \\ i(k)+i(k-1) & \alpha\big(i(k)-i(k-1)\big) \end{bmatrix} \begin{bmatrix} R \\ L \end{bmatrix} = \begin{bmatrix} u(k-M+1)+u(k-M) \\ \vdots \\ u(k-1)+u(k-2) \\ u(k)+u(k-1) \end{bmatrix} \tag{4.134}$$

where: $M \geq 2$, α is as in (4.131), which can be written in the matrix form:

$$\mathbf{HX} = \mathbf{x}$$

where the unknown vector: $\mathbf{X} = [X_1 \quad X_2]^T = [R \quad L]^T$ is determined from (4.80):

$$\mathbf{X} = \left(\mathbf{H}^T\mathbf{H}\right)^{-1}\mathbf{H}^T\mathbf{x}$$

For selection of the value of M we should take into consideration that the LSM algorithm approximates the considered process assuming that the parameters are constant in the data window. A small value of M makes the higher error of the parameters estimation, while an increase of M results in the measurement delay. A similar effect results when the differential equation is used for pre-filtered current and voltage signals. In this case some short-window smoothing filter is used at the input of the procedure.

4.4.2 Laplace-transform Application

Similar algorithms can be obtained by description of the considered fault-loop in a Laplace transform. Consider (4.120) in the Laplace domain (neglecting initial conditions):

$$U(s) = RI(s) + sLI(s) \tag{4.135}$$

and, next in the Z-domain, using the bilinear transformation (4.28):

$$\left(1+z^{-1}\right)U(z) = R\left(1+z^{-1}\right)I(z) + L\alpha\left(1-z^{-1}\right)I(z) \tag{4.136}$$

which directly leads to (4.131).

The above process can be described in a more general approach that consists of the following steps:

- formulate the circuit equations in the Laplace domain;
- transform it into the Z-domain;
- write the discrete-time equations; and
- formulate the unknown parameters estimator.

It can be seen that details of the time-domain algorithm as in (4.136) depend on the used transformation between three considered domains, which briefly can be shown as a transition between the following stages: $X(s) \rightarrow X(z) \rightarrow x(k)$. Different algorithms can be obtained by selection of an adequate procedure for digital approximation of the integration method. Representation of these methods in a Laplace transform makes easier analysis of its frequency features. Repeating the procedure (4.24)–(4.26) one can derive:

$$s = \alpha\left(1 - z^{-1}\right) \text{ with: } \alpha = \frac{\omega_1 / 2}{\sin\left(\omega_1 T_s / 2\right)} \text{ – for the Euler method} \tag{4.137a}$$

$$s = \alpha\left(3 - 4z^{-1} + z^{-2}\right) \text{ with: } \alpha \text{ as in (4.132) – for the 2nd-order Gear method} \tag{4.137b}$$

Frequency characteristics $D(\omega)$ of these transformations with respect to the continuous-time domain can be obtained by substituting: $z = e^{j\omega T_s}$ and $s = j\omega$ adequately for the Z-domain and Laplace-domain of the above relations. The resulting magnitude and phase responses for considered methods are presented in Fig. 4.19. One can see that the phase response for the trapezoidal method agrees with that expected for continuous-time differentiation. However, the response amplifies in the high-frequency band to infinity. Both Gear and Euler integration methods have deformed frequency responses. In the case of the Euler rule the resulting differentiation is calculated with a half-sampling period delay that gives the angle equal to half the angle between samples for the designed frequency (i.e., for the fundamental component in Fig. 4.19).

All these details should be carefully considered for choosing an adequate formula for numerical integration. Note that a relation between parameters R and L estimated from sampled current and voltage waveforms depends on both the considered characteristics: magnitude and phase.

This approach can also be applied for a complex network. In the case of a multi-phase line it is convenient to decouple the circuit model by using, for example, a $0\alpha\beta$ transformation [9]. This also enables correction of the shunt capacitive current.

It should be noted that the algorithms originating from the system-model approach usually apply more information regarding the system transients. In particular, a decaying DC component is included into the model as in (4.124) and this component does not introduce the addition error in the estimates.

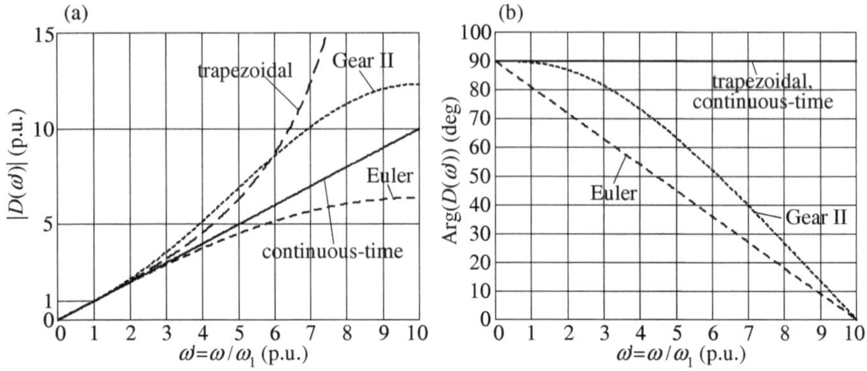

Fig. 4.19 Frequency characteristic of the considered numerical differentiation methods: magnitude response (a) and phase response (b)

This procedure can be easily applied even for more complicated equations as in (4.124), e.g., when integration and higher-order operators s are also used. Some examples are presented in Chap. 6.

4.5 Time-frequency Analysis

Standard procedures for measurement of electrical signals that represent voltage, current and other physical values are based on the assumption that the observed process is stable. We have shown in previous sections that such an assumption leads to application of a well-defined Fourier transform. One of the important and usable consequences of such an approach is easy interpretation of the measurements in AC system: voltage and current phasors may be directly used for calculation of other convenient electrical parameters, such as impedance or power. There are known, however, limitations of such a proposition: the method is not suitable for transient period analysis and results in relatively big errors, especially when we need to obtain a fast result.

In this section we present the wavelet transform (WT), which is well adjusted to non-stationary signals and, moreover, has good time resolution. We further show (Chap. 9) that this transformation may be used for fault location.

4.5.1 Short-time Fourier Transform

The aim of the signal analysis in measurements and control devices is to extract usable information through adequate transformation. As we have seen (Sect. 4.2) the general Fourier transform has a known practical limitation: the accurate information about the signal spectrum (in the frequency domain) needs the information about the signal waveform in an unlimited time period, so we can say that it is perfectly local in frequency, but is global (unlimited) in time. We should also assure that the signal is stationary. Any change observed in the signal will be averaged in its spectrum and the information about its time inception (when it appears) will be lost. That feature is characterized as the signal resolution in time and frequency domains. For the Fourier transform a temporal characteristic of the signal, such as the start and end of a finite signal and the instant of appearance of a singularity in a transient signal, are not preserved in the spectrum.

The Fourier spectrum analysis is global in time and is practically not suitable to analyze non-stationary and fast-changing transient signals. However, any signal that we are interested in is of finite extent. Practically used time-restricted Fourier transformations (as the DFT) also have limitations because transients in the data window are averaged in the transformation process. The intention to recover the usable information on the signal spectrum (magnitude of specified component) has limited value because we lose information on the time of occurrence.

The above-mentioned problem is the issue of time-frequency analysis that is based on the assumption that non-stationary signals (with varying parameters in time) require a tool that transforms a signal into a joint time-frequency domain [30, 246]. One such tool is known as the short time Fourier transform (STFT). It is based on the assumption that the non-stationary signal $x(t)$ can be divided into segments limited by a window $w(t)$ inside which it can be considered as the stationary one. STFT is defined as the Fourier transform of the function: $x(t)w(t-\tau)$:

$$X_w(\mathrm{j}\omega,\tau) = \int_{-\infty}^{+\infty} x(t)w(t-\tau)\mathrm{e}^{-\mathrm{j}\omega t}\mathrm{d}t \tag{4.138}$$

where: $w(t) = 0$ for $t < 0$ and $t > t_w$.

As the window function $w(t)$ is shifted in time: $w(t-\tau)$, the above transform describes evolution of the signal spectrum with time. Generally, the window function $w(t)$ can assume different forms but mostly it has rectangular shape.

The window function can also be connected with the complex exponential function in (4.138) giving so-called windowed basis functions:

$$\psi_{\omega,\tau}(t) = w(t-\tau)\mathrm{e}^{-\mathrm{j}\omega t} \tag{4.139}$$

with two parameters: the shifting coefficient τ and the frequency ω. The basis functions represent cosine/sine waveforms of limited length with an envelope formed by the window function. These set of functions can be used for approximation of the signal spectrum according to (4.138).

The discrete form of (4.138) is as follows:

$$X_w(n,m) = \sum_{k=0}^{M-1} x(k)w(k-m)e^{-j\nu nk} \tag{4.140}$$

where ν is an angle between samples, M – number of samples in the window $w(k)$.

Fig. 4.20 Recorded voltage waveform

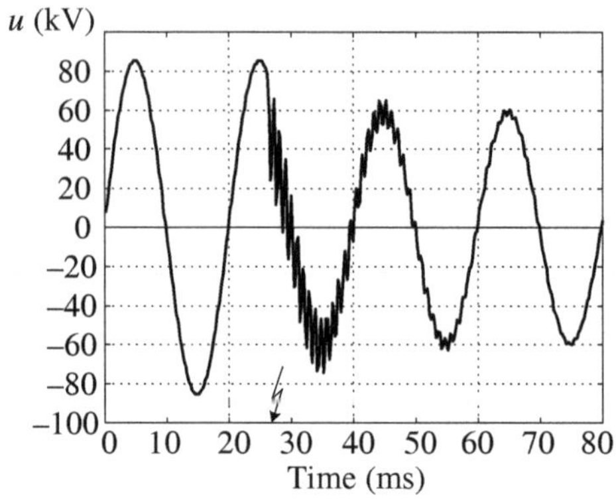

We can note that *n* characterizes a frequency: it is a harmonic number with respect to the window length *M* considered as the fundamental frequency, while *m* indicates the time (number of window with a processing sequence).

We can see that STFT allows the generation of a time-frequency description of a signal. It can be seen, however, that choosing a short window will result in a transform presenting good resolution in time. On the other hand, a short window results in a reduced number of samples used in transformation (4.139) and, therefore, the number of different frequencies represented in the resulting spectrum will be reduced. The following example illustrates the STFT feature.

Example 4.2 Figure 4.20 presents the voltage waveform recorded during a fault at a 110-kV line with sampling frequency f_s=4.0 kHz. The system frequency is: f_1=50 Hz. Determine the short-time Fourier transform of the recorded waveform for full- and half-cycle rectangle windows. In the considered case the transform (4.139) is the same as DFT with the number of samples in the full-period window: M=100. The time resolution is defined by the window length: T=20 ms. Discrete time is determined by parameter m=1, 2, … adequately to the window position. The results of transformation for some specified time are presented in Fig. 4.21. The frequency resolution is also connected with window length: Δf=1/T=50 Hz. The transient period after fault seen in Fig. 4.20 can be observed in higher harmonics in the signal spectrum evolved in time (Fig. 4.21).

Performing the STFT with a half-cycle window (M=50, T=10 ms) we obtain more precise determination of a time event but this inevitably will decrease the frequency resolution: Δf=1/T=100 Hz. That is illustrated in Fig. 4.22 for comparison with both: full- and half-cycle window lengths. The shade intensity indicates the value of signal magnitude for a given frequency.

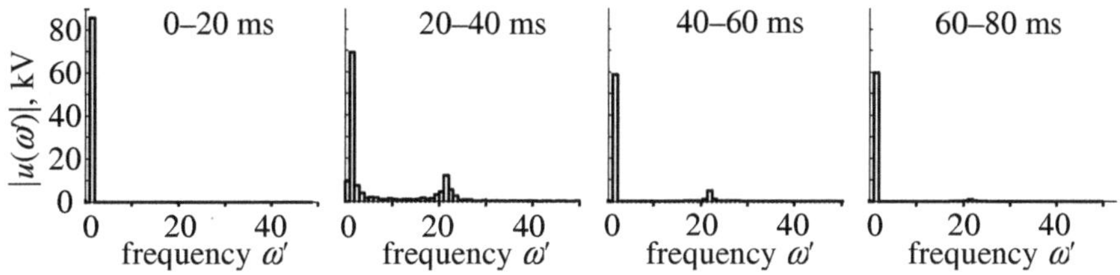

Fig. 4.21 Histograms of full-cycle STFT for different window placement

It can be shown that the product of the frequency resolution Δf and the time resolution T is constant. It is dependent upon the shape and length of the window function. Both these factors remain constant throughout an analysis. This relation is connected with the known Heisenberg's uncertainty principle [101].

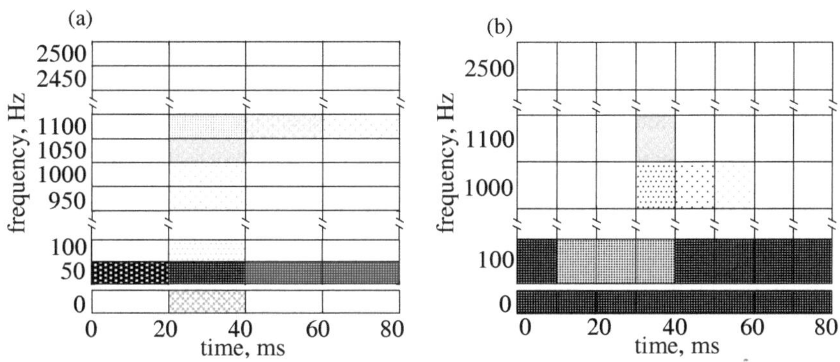

Fig. 4.22 STFT results for signal as in Fig. 4.20 for full- a) and half-period b) window

We can see that in the STFT method the frequency resolution is uniform at all frequencies. This means that the accuracy is poor for low frequencies and becomes increasingly better at high frequencies. The same window length is used for low and high spectrum analysis. The STFT, therefore, does not provide uniform accuracy for all frequencies and the computational resources are somehow poorly distributed.

4.5.2 Wavelet Transform

The wavelet transform concept is very similar to the above-presented STFT method. The main difference is with new formulation of the basis functions that are known as wavelets and are defined as:

$$\psi_{p,\tau}(t) = \frac{1}{\sqrt{p}}\psi^*\left(\frac{t-\tau}{p}\right) \tag{4.141}$$

where * means complex conjugate; p is a scaling factor and τ is the shifting.

The wavelet is an oscillating waveform that persists for only one or a few cycles and, generally, is a complex-valued function. A given wavelet has a location τ (position) and a scale p (duration), and an oscillation (frequency). Wavelets are used as a kernel function in an integral transform, in much the same way that complex exponentials are used in the Fourier analysis:

$$CWx(p,\tau)=\frac{1}{\sqrt{p}}\int_{-\infty}^{+\infty}x(t)\psi^{*}\left(\frac{t-\tau}{p}\right)\mathrm{d}t \tag{4.142}$$

where the value $1/\sqrt{p}$ represents a normalizing factor.

The relation (4.142) represents a so-called *continuous wavelet transform* – CWT, because p and τ are continuous parameters.

Wavelets with different p, τ parameters create a wavelet family with a basic (mother) wavelet $\psi_{10}(t)$ [30]. In contrast to the Fourier transform, WT is scaled with respect to time and frequency by the scale parameter p. If for wavelet $\psi(t)$ the time and frequency resolution is determined as (Δt, $\Delta\omega$), then for wavelet $\psi_p(t)=\psi(t/p)/\sqrt{p}$ the resolution be determined by the rectangle of ($p\Delta t$, $\Delta\omega/p$).

The original signal $x(t)$ may be obtained from its wavelet representation by use of the inversion wavelet transform:

$$x(t)=\frac{1}{C_{\psi}}\int_{-\infty}^{+\infty}\int_{-\infty}^{+\infty}CWx(p,\tau)\psi\left(\frac{t-\tau}{p}\right)\frac{\mathrm{d}p\mathrm{d}\tau}{p^{2}} \tag{4.143}$$

where: C_{ψ} is a normalizing constant.

An unambiguous pair of wavelet transforms defined by (4.142) and (4.143) needs adequate features of the basis functions $\psi_{p\tau}(t)$. Generally, such functions should be oscillating, limited in time and damped at both ends of the window [101]. Elements of the wavelet family have similar form and differ with scale p and shifting factor τ. As a result of transformation (4.142) we obtain the set of functions that characterize degree of similarities between the signal and particular wavelets. We can say that the signal $x(t)$ is spread into the set of wavelets. The resulting functions $CWx(\tau,p)$ characterize adequate signal features.

For a given scale p transform (4.142) represents a digital filter (realized as correlation) the impulse function of which is defined by an adequate wavelet. It is a band-pass filter with medium frequency defined by the scale coefficient p. We can see from (4.142) that a high value of p (p>>1) is related to a slow oscillating function $\psi_{p\tau}(t)$ with a long data window, while small values of scale p ($0<p<1$) create a fast oscillating wavelet with short data window that reflect high-frequency signal features. Shifting factor τ enables precise location of the event in time. The relation between the scaling factor p and the frequency shows the following example, which presents features of a Morlet wavelet. The Morlet wavelet represents an oscillating harmonic function modulated by the Gaussian function [316]:

$$\psi_{p\tau}(x) = \mathrm{e}^{-x^2/2}\mathrm{e}^{\mathrm{j}\omega_0 x}, \text{ where } x = \frac{t-\tau}{p} \tag{4.144}$$

The wavelet transform is here very similar to the Fourier transform but the basis function is suppressed by a bell-shaped envelope. The Fourier transform of the Morlet wavelet is as follows:

$$F\{\psi_{p\tau}(x)\} = \Psi(\mathrm{j}\omega) = \mathrm{e}^{-(p\omega-\omega_0)^2/2}\mathrm{e}^{-\mathrm{j}\omega\tau} \tag{4.145}$$

The Morlet wavelets and their spectrum for $p = 1$, 2 and 0.5 ($\tau = 0$) are presented in Fig. 4.23. The fundamental frequency was assumed: $\omega_0 = 4$. We can see that the wavelet frequency is inversely proportional to the scale factor p.

In contrast to the STFT, where the same window length is used, in WT the window length is related to scale p to assure the same frequency resolution [246]:

$$\frac{\Delta\omega}{\omega} = c \tag{4.146}$$

where: c is a constant, $\Delta\omega$ defines the signal resolution for frequency ω.

The condition (4.146) leads to a logarithmic frequency scale in the time-frequency plane. We can see that in WT the frequency is not the central parameter in a signal analysis; it was replaced by the scaling factor p. If the mother wavelet has frequency f_0, then the scaled wavelet takes frequency $f = f_0 / p$. In the wavelet analysis it is, therefore, more reasonable to say 'scale-time characteristic' (scalogram) instead of 'time-frequency characteristic' (spectrogram) used in STFT [246].

It should be stressed that in the considered CWT the analyzed signal may have a continuous- or discrete-time form but nevertheless the transformation parameters: a scale p and a shifting τ can take any real values: they are continuous-time functions. However, when both these functions are discrete time (after sampling of continuous-time functions), the adequate transform is called the *discrete-time wavelet transform* (DWT). It is convenient to assume that the scale p changes with degree of 2, i.e.: $p_0 = 1 = 2^0$ (mother wavelet), $p_1 = 2 = 2^1$ and, generally: $p_j = 2^j$.

Moreover, if an elementary shifting τ_0 is divided into n periods then changing of the shifting factor results in the following change of shifting values: $\tau_n = 2^j n \tau_0 = 2^j n$ (τ_0 – an elementary shifting). Therefore, wavelet (4.141) for DWT may be represented as:

$$\psi_{jn}(t) = 2^{-j/2}\psi\left(2^{-j}t - n\right) \tag{4.147}$$

where: $\psi(t) = \psi_{00}(t)$ is a basic (mother) wavelet.

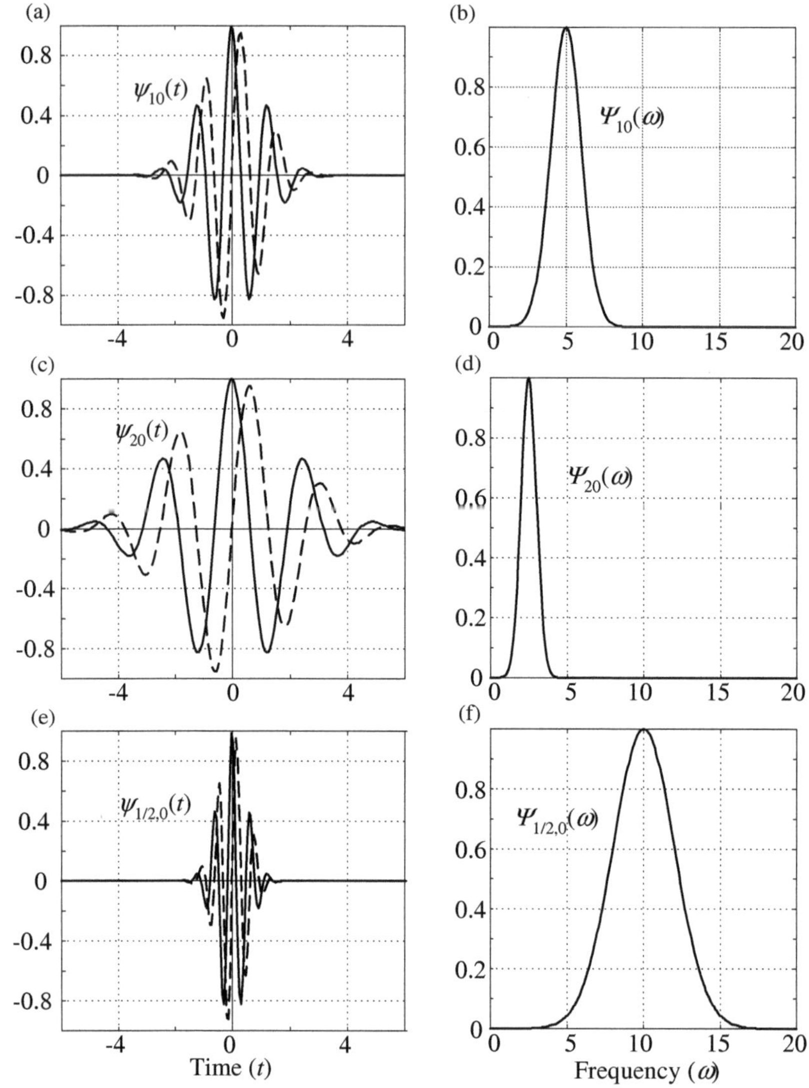

Fig. 4.23 Morlet wavelets and their spectrum for different scales: p=1 (a), (b), p=2 (c), d), p=0.5 (e) and (f); dashed line – imaginary part of wavelet function

We can see that the considered rate of scaling leads to the frequency changing in a grade of 2. This is so-called dyadic sampling: neighboring scales differ by one octave [246]. Transform (4.142) with wavelets as in (4.147) takes the form:

$$CWx(p_j,\tau_n) = DWx_{jn} = \int_{-\infty}^{+\infty} x(t)\psi_{jn}^{*}(t)\mathrm{d}t \tag{4.148}$$

The original signal may be reconstructed without loss of information (if the chosen mother wavelet fulfils adequate conditions [101]) by weighting summation the obtained coefficients DWx_{jn} :

$$x(t) = \sum_{j}\sum_{n} DWx_{jn}\psi_{jn}(t) \tag{4.149}$$

This is the inverse discrete wavelet transform (IDWT). We may note that even in a discrete transform, the analyzed function and used wavelets are continuous-time functions. That is because the form of the input signal does not depend on the used scale or shifting factor. In a computer application the integration in (4.148) may be represented in the form of an adequate discrete sum limited to non-zero values of the wavelet. This leads to the following form of (4.148):

$$DWx_{jn} = \sum_{k=0}^{M-1} x(k)\psi_{jn}^{*}(k) \tag{4.150}$$

where: $\psi_{jn}(k) = 2^{-j/2}\psi\left(2^{-j}k - n\right)$ and $kT = t_k$, T – sampling period of the signal $x(t)$, M – number of signal samples limited to non-zero values of $\psi_{jn}(k)$.

The DWT differs from CWT with clear steps in the time-frequency plane that is considered as multi-resolution wavelet analysis [101]. The purpose of the multi-resolution signal analysis is decomposing the signal in multiple frequency bands, in order to process the signal in multiple frequency bands differently and independently.

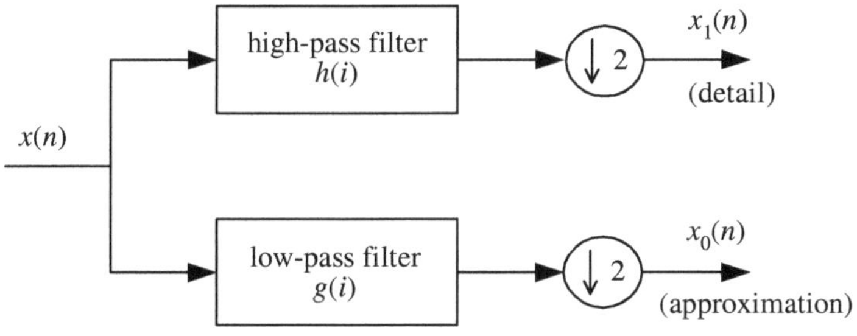

Fig. 4.24 Structure of one-level DWT algorithm

The advantage of DWT lies in the possibility to synthesize effective numerical algorithms for determination of wavelet expansion. At each stage of the wavelet analysis the signal may be divided into two components the spectra of which are placed in a half-section of a given level n of the frequency resolution $\Delta\omega_n$. These components may be obtained by low or high band-pass filtering and signal sampling with the grade of 2 (↓2) (taking only even samples – Fig. 4.24). Signal

$x_0(n)$ on the output of a low-pass filter is a smoothing (approximating) version of the input signal, while the $x_1(n)$ on the output of the high-pass filter contains only details that completed approximation. Both these signals result from the following filtering of a signal $x(n)$:

$$x_1(n) = \sum_k h(2n-k)x(k)$$
$$x_0(n) = \sum_k g(2n-k)x(k) \tag{4.151}$$

where: $h(2n-k)$ is the impulse function of high-pass and $g(2n-k)$ that of a low-pass filter.

The above pair of FIR filters are referred to as mirror filters. Their impulse functions are both connected with [246]:

$$h(M-1-k) = (-1)^k g(k) \tag{4.152}$$

where: $M = 2L$ defines the filter window length, L – wavelet level.

The reconstruction process (synthesis) of the input signal may be determined by the following procedure (Fig. 4.25):

$$\hat{x}(k) = \sum_l h'(l-2k)x_1(l) + \sum_l g'(l-2k)x_0(l) \tag{4.153}$$

which means that series of $x_1(n)$ and $x_0(n)$ should be completed by zeros inserting between given samples (operation $\uparrow 2$) and convolve with functions $h'(l-2k)$ and $g'(l-2k)$, respectively. These function may be obtained from the original impulse functions as follows:

$$\begin{matrix} g'(k) = (-1)^{k+1} h(k) \\ h'(k) = (-1)^k g(k) \end{matrix}, \quad k = 0,..M-1 \tag{4.154}$$

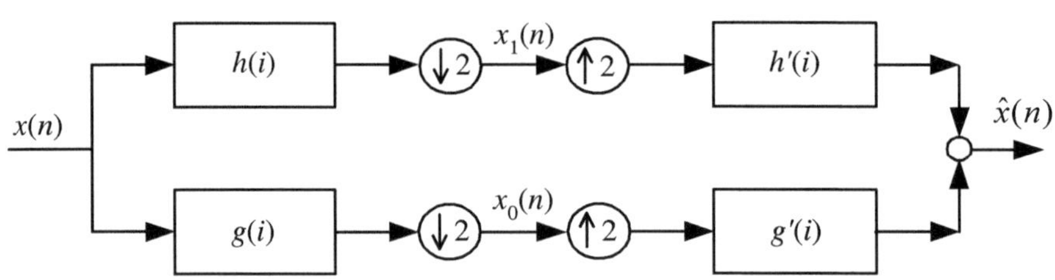

Fig. 4.25 Scheme of the one-level direct and inverse DWT

Examples of the impulse function applied in analysis and synthesis, for a 6-order Symmlet wavelet, are presented in Fig. 4.26 adequately.

The filtering process (4.151) can be repeated iteratively forming in the successive iterations a new level of resolution (signal scaling), where variable x_1^j (detail) is a coefficient of wavelet expansion of the input signal:

$$x_1^j(n) = DWx_{jn} \tag{4.155}$$

Multi-resolution signal decomposition is performed according to the scheme in Fig. 4.25. In the following stages of the algorithm only a low-pass signal (obtained from a low-pass filter) is further processed.

All successive details (signals x_1^j) forming wavelet decomposition of the input signal. Sampling is replaced by zero adding (Fig. 4.27). In practical application the user can apply suitable software for designing adequate impulse functions $g(k)$, $h(k)$ of mirror filters [167, 246].

The wavelet transform technique is a novel, powerful and fashionable tool in the field of signal processing. It has a few major advantages and is very suitable for analysis of non-stationary signals, short-term signals as well as transients, which are commonly present in power systems.

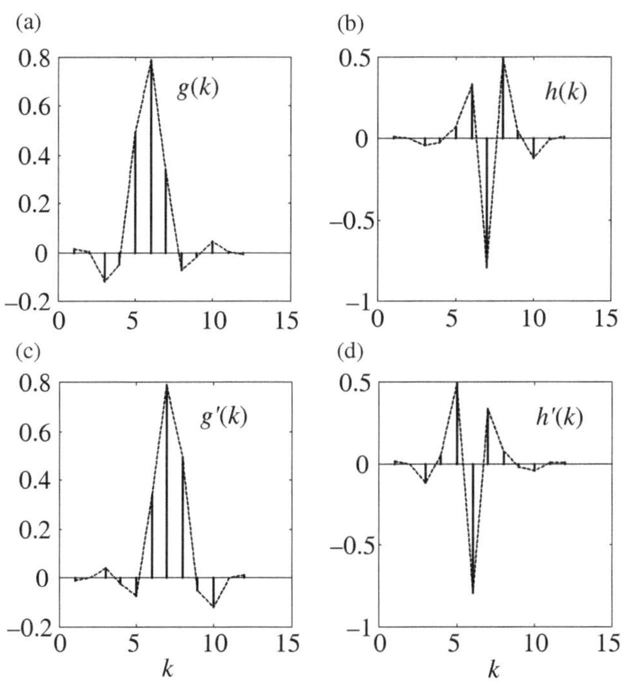

Fig. 4.26 Impulse functions applied for analysis (a), (b) and synthesis (c), (d) with Symmlet 6 wavelet

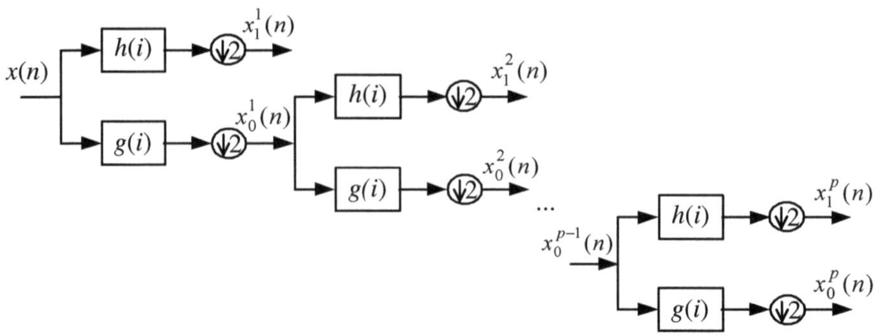

Fig. 4.27 Structure of the algorithm for signal wavelet decomposition

Recently, many researchers have put their efforts to use wavelet transform in application in power engineering. Possible use of wavelet transform in power engineering applications may be, but are not limited to:

- transient signal analysis and identification,
- non-stationary voltage distortion analysis,
- power-system protection – fault detection, identification, location, etc.;
- power-quality analysis and monitoring (integrated characterization of voltage disturbances, e.g., transients, harmonics, sags);
- power-system transient data compression and storage;
- power-system analysis in general; and
- analyzing the behavior of devices, apparatus and systems.

The presented consideration is only a brief introduction into wavelet transformation issues. Generally, wavelet analysis is becoming a common tool for analyzing localized transients of the observed process represented by signals and shows the advantage of measuring transients. Some application examples are presented in Chap. 9.

5 Measurement Chains of Fault Locators

Abstract This chapter is focused on the measurement chains of fault locators. Transient performance of capacitive voltage transformers and their dynamic compensation are considered. The basics for current transformers including the effects of the possible saturation of current transformers are discussed. Frequency characteristics of both capacitive voltage transformers and current transformers are presented. Design of analog low-pass anti-aliasing filters is addressed.

5.1 Introduction

Information on a fault, which is an abnormal power-system state, is brought to a fault locator by its input signals. Under power-system faults the current and voltage signals undergo changes with respect to both steady and transient states. These signals with the abrupt steady-state level change, and additionally being contaminated with the transient components, are delivered to fault-locator inputs via the measuring chains.

The functional structure of voltage and current measurement chains of a fault locator is shown in Fig. 5.1. Three-phase primary voltages (u_p) and currents (i_p) from a power system are transformed with use of instrument voltage and current transformers to the reduced secondary level (u_s, i_s). The secondary signals of these transformers are rated at around 100 V (voltage) and 1 A or 5 A (current). Then, matching transformers provide adequate level of the signals to electronic devices. Prior to an analog-to-digital (A/D) conversion, analog low-pass anti-aliasing filters are used for both voltage and current signals. Explanation on why such filters have to be applied is given in Sect. 5.4. At the outputs of A/D converters the sampled signals: $u_{2(n)}$, $i_{2(n)}$ are obtained. These signals are processed in digital fault-location algorithms.

Since in the majority of real applications, traditional electromagnetic voltage and current instrument transformers are utilized, the attention in this book is given to them. Moreover, there is a common opinion that a long time is still to pass before new unconventional instrument transformers [172, 173, 175, 220] become pre-dominant in transforming signals from a power system to protection, monitoring, control and measuring devices.

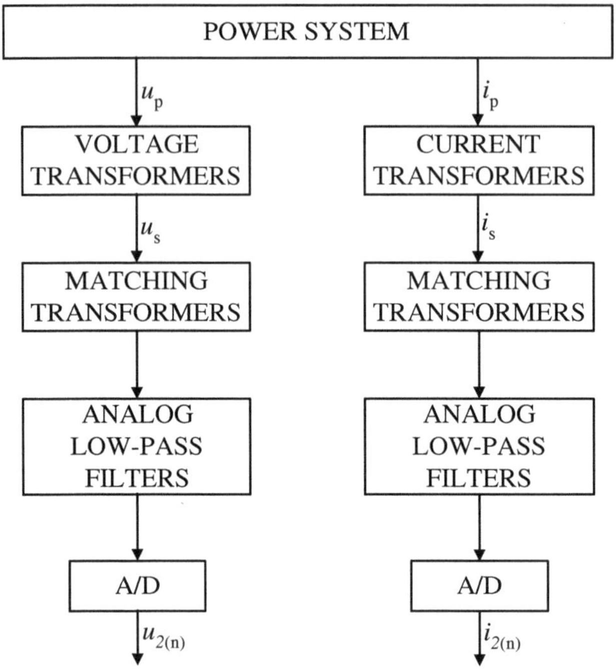

Fig. 5.1 Structure of voltage and current measurement chains

Due to certain construction limitations, both the instrument voltage (VTs) and current (CTs) transformers exhibit undesired dynamic behavior under short circuits in the power system. As a result, malfunctioning or substantial delay in the tripping of a protective relay may take place [72, 99, 100, 112, 154, 271, 317, 334]. Undesired steady-state and dynamic behavior of instrument transformers influence a fault-location process as well [5, 123, 232, 245, 271, 332].

Great efforts have been taken towards compensating the protective current and voltage instrument transformers for their transient errors. The aim of such compensation is to obtain a reasonably accurate replica of the primary current and voltage. The other possibility is based on minimizing the influence of transient errors of instrument transformers on the operation of both relaying and fault-location algorithms.

5.2 Voltage Transformers

At the transmission voltage level the instrument-level voltage signals for protective, monitoring and measuring devices are provided by means of capacitive voltage transformers (CVTs). A CVT provides a cost-effective way of obtaining a secondary voltage for HV and EHV systems [154, 161, 318]. Its functional scheme is depicted in Fig. 5.2. In this scheme, besides the primary (u_p) and secondary (u_s) voltages one can also distinguish the intermediate voltage (u_i), which is usually at the level of around 20 kV.

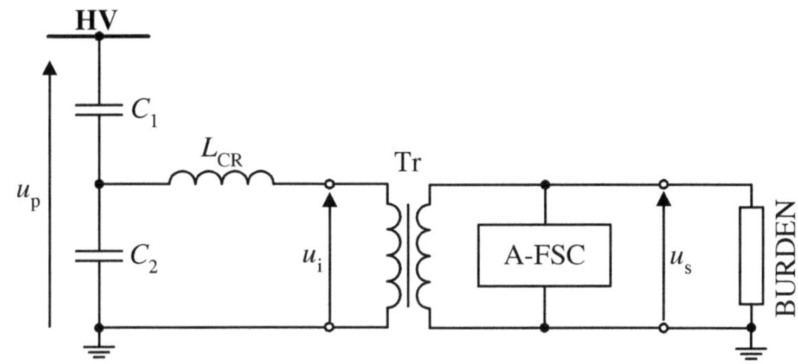

Fig. 5.2 Schematic diagram of CVT: C_1, C_2 – stack capacitors; L_{CR} – compensating reactor; Tr – inductive step-down transformer; A-FSC – anti-ferroresonance suppressing circuit; BURDEN – CVT burden imposed by connected protective and other devices

5.2.1 Transient Performance

The dynamics of a CVT is determined by two factors [112, 154]:

- non-linear oscillations under saturation of a magnetic core of a CVT step-down inductive voltage transformer; and
- discharging of a CVT internal energy during short circuits on the associated transmission line.

Non-linear oscillations can appear when the operating point of the magnetizing characteristic of the step-down transformer is shifted to the saturation region. CVTs are therefore equipped with special anti-ferroresonance circuit (Fig. 5.3) for avoiding stabilization of the sub-harmonics and assuring adequately fast damping of the oscillations [161, 333]. Examples of waveforms of CVT secondary voltage under interruption of the short-circuit of the secondary terminals are shown in Figs. 5.4 and 5.5. For a CVT unequipped with an anti-ferroresonance circuit (Fig. 5.4) considerable contamination with the third sub-harmonic component is observed. Equipping a CVT with properly designed anti-ferroresonance circuit allows effective damping of non-linear oscillations (Fig. 5.5). Sustained sub-harmonic oscillations do not appear and adequately high speed of damping of oscillations is assured by use of a correctly designed anti-ferroresonance circuit.

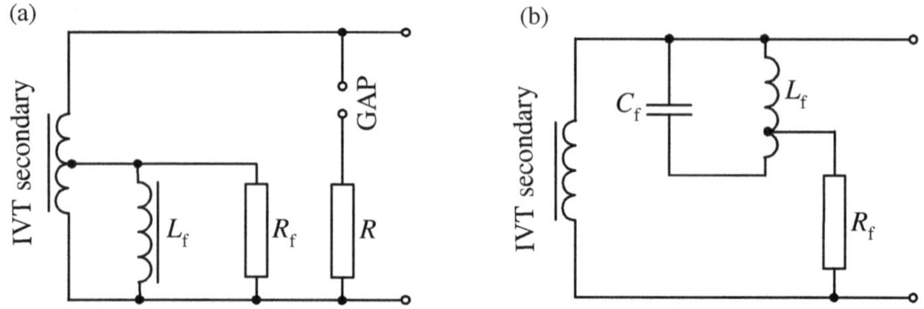

Fig. 5.3 Examples of anti-ferroresonance suppressing circuits: (a) passive, and (b) active

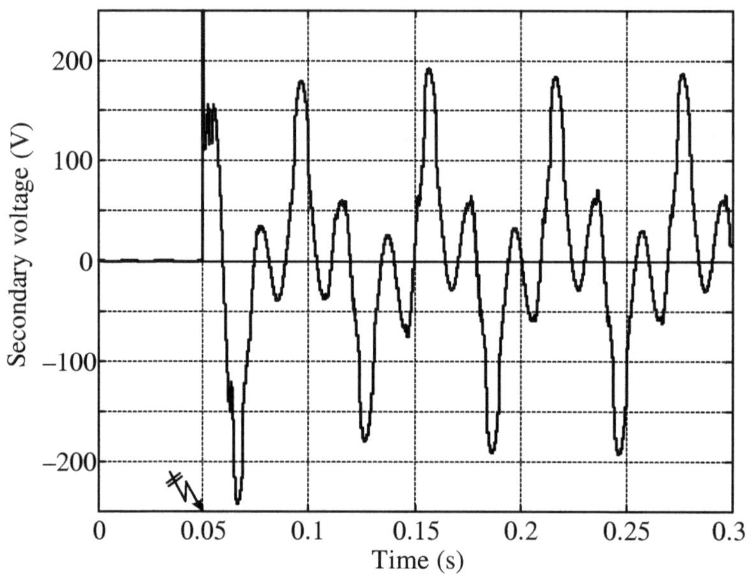

Fig. 5.4 Example of the waveform of a CVT secondary voltage for CVT unequipped with anti-ferroresonance suppressing circuit under interruption of short circuit of secondary terminals

Anti-ferroresonance circuits, however, affect the transients of the second kind. Discharging the CVT internal energy (accumulated in the stack capacitors and the compensating reactor of a CVT during the pre-fault state) – to the level that is determined by the reduced fault primary voltage – results in considerable distortion of the secondary wave [99, 112, 154, 161, 165, 266, 269].

The higher the reduction of the primary voltage, the more extensive transients induced by the CVT itself occur (Figs. 5.6 and 5.7). In particular, faults at zero crossing of the primary voltage result in substantial transient errors that, in turn, affect the operation of supplied protective relays and fault locators. CVT transients may occur during changes in system-operating states either due to normal switching operations or due to the occurrence of faults.

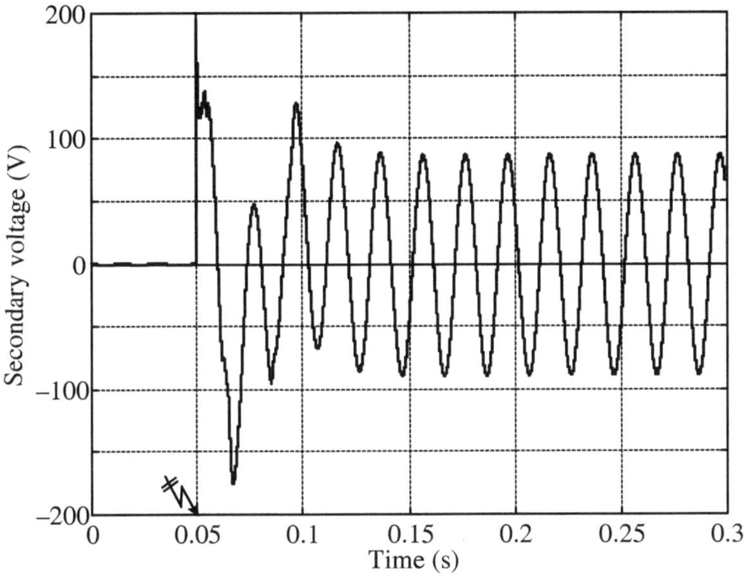

Fig. 5.5 Example of the waveform of a CVT secondary voltage for CVT equipped with anti-ferroresonance circuit effectively suppressing ferroresonance under interruption of short circuit of secondary terminals

Among different CVT parameters, the stack capacitances influence considerably the CVT-generated transients. Two types of CVT are distinguished in reference [154]:

- 'high-C CVT' – the sum of stack capacitances below some 100 nF; and
- 'extra high-C CVT' – the sum of stack capacitances above some 100 nF.

Typical CVT-generated transients for these CVT types are shown in Figs. 5.6 and 5.7. A detailed analysis of these transients is not given in this book but can be found in [154].

In Fig. 5.6 the case for which the primary voltage drops from the rated level to zero (for a solid fault at the primary terminals), under zero crossing, is presented. The CVT secondary voltage also tends to zero, however, after dissipating the CVT accumulated internal energy. The shape of the transient voltage depends highly on the stack capacitance value. For the high-C CVT, the transient voltage assumes higher values, but decays faster. In turn, for the extra high-C CVT, the transient voltage assumes lower values and decays much slower.

The case for which the primary voltage drops from the rated level to zero (a solid fault at the primary terminals) from the voltage peak is presented in Fig. 5.7. Decay of the transient voltage is now somehow faster than in the case presented in Fig. 5.6.

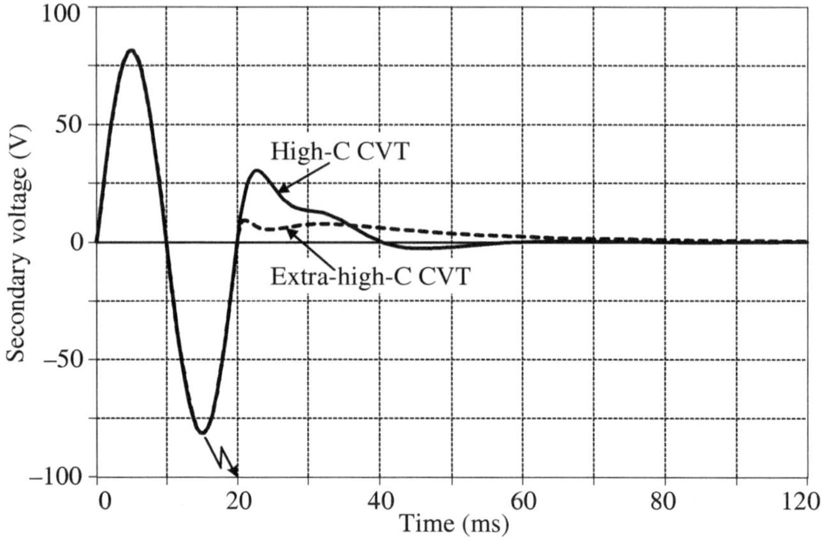

Fig. 5.6 Sample transients for high- and extra-high-C CVTs when primary voltage drops to zero under zero crossing

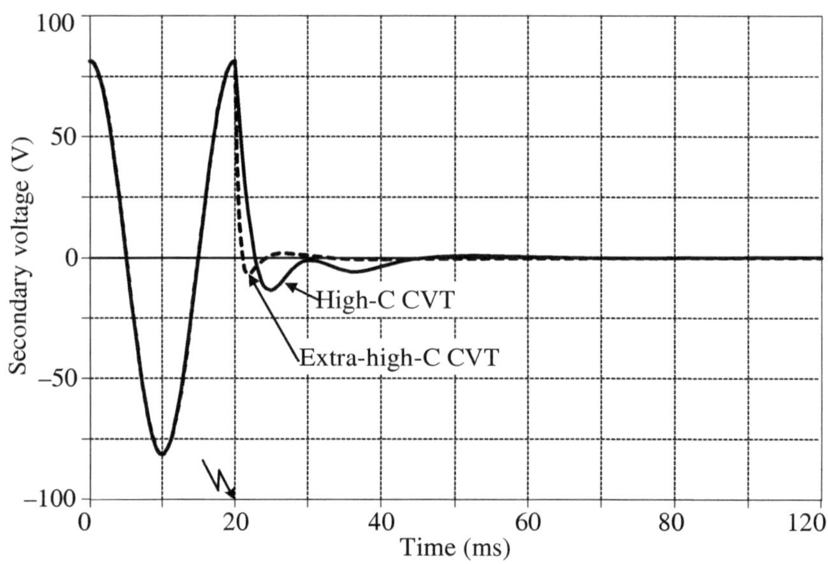

Fig. 5.7 Sample transients for high- and extra-high-C CVTs when primary voltage drops to zero from the voltage peak

5.2.2 Dynamic Compensation

In [112], it was proposed to reject the CVT-induced transients from the secondary voltage signal with use of the digital compensation algorithm based on inversion of the CVT simplified transfer function. In Fig. 5.8, a general CVT equivalent circuit diagram [161] is shown. After simplifying this model one obtains the model as shown in Fig. 5.9.

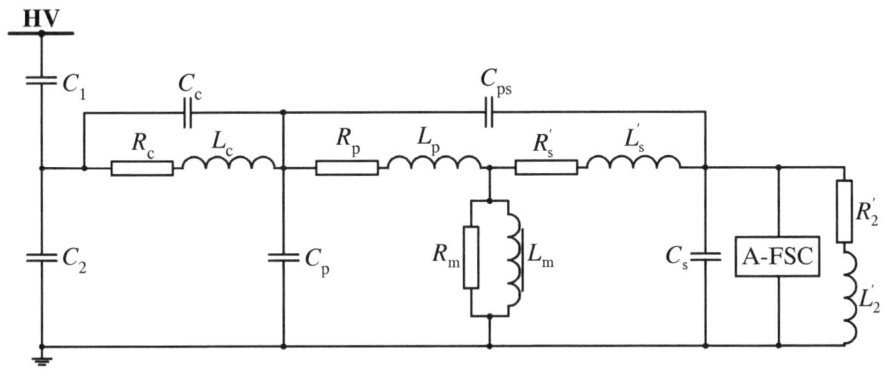

Fig. 5.8 General equivalent circuit diagram of CVT

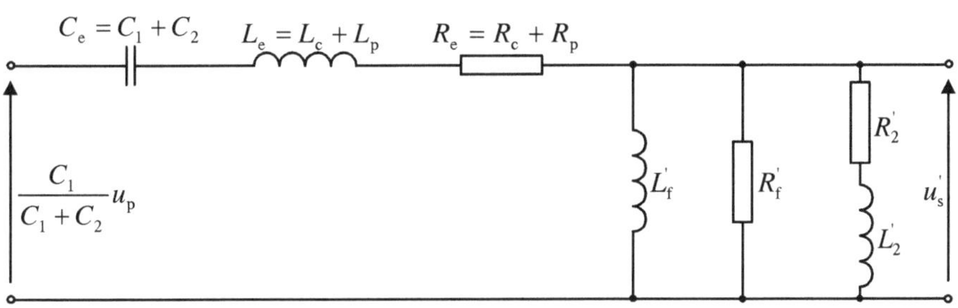

Fig. 5.9 Simplified circuit diagram of CVT equipped with passive anti-ferroresonance circuit

The following simplifications, facilitating the design of the compensating algorithm, have been made for obtaining the simplified circuit diagram of Fig. 5.9:

- saturation of the step-down inductive transformer is neglected since short circuits result rather in reduction of the primary voltage, which moves the operating point of the magnetic core down from the rated position;
- ideal transformation of the step-down transformer is assumed, which means that the parameters L_p, R_p, L_m, R_{Fe}, L_s, R_s and all stray capacitances in the equivalent circuit diagram from the circuit diagram of Fig. 5.8 are neglected;
- all the remaining parameters of the step-down inductive transformer are related to its primary side; and
- the *Thevenin theorem* is applied to the primary voltage and the capacitor stack.

Under these assumptions, the model from Fig. 5.8 reduces to the simple circuit displayed in Fig. 5.9, for which the transfer function takes the form:

$$G_{\mathrm{CVT}}(s)=\frac{A_3s^3+A_2s^2}{B_4s^4+B_3s^3+B_2s^2+B_1s+B_0} \tag{5.1}$$

where A_3, A_2, B_4, B_3, B_2, B_1, B_0 – coefficients duly expressed by the parameters of the equivalent circuit diagram from Fig. 5.9.

In order to get exact reproduction of the primary voltage, the ideal compensator at the CVT output has to be applied in such a way that:

$$G_{\mathrm{COMP\_ideal}}(s)=\frac{1}{G_{\mathrm{CVT}}(s)} \tag{5.2}$$

Due to a double zero at the origin of the transfer function (5.1) its direct inversion appears troublesome. Therefore, the following modified transfer function of the compensator has been proposed [112]:

$$G_{\mathrm{COMP}}(s)=\frac{(B_4s^4+B_3s^3+B_2s^2+B_1s+B_0)}{(A_3s+A_2)(A_4s^3+A_5s^2+A_6s+A_7)} \tag{5.3}$$

where A_4, A_5, A_6, A_7 – coefficients to be selected.

Applying the compensator of the transfer function (5.3) allows all the poles of the CVT transfer function (5.1) to be compensated, while only the single zero ($s=-A_2/A_3$) is cancelled. So, the double zero at the origin is left and some extra three poles are added. As a result, the transfer function between the primary voltage and the secondary compensated voltage being the result of the compensation is as follows:

$$G_{\mathrm{CVT}}(\mathrm{s})\cdot G_{\mathrm{COMP}}(s)=\frac{s^2}{A_4s^3+A_5s^2+A_6s+A_7} \tag{5.4}$$

Selection of the coefficients A_4, A_5, A_6, and A_7 may be done in a number of ways with the objective to obtain the desired dynamics of the compensated CVT [112]. Different numeric procedures can be applied for obtaining a discrete form of the compensator (5.3). The following trapezoidal rule (4.28) was applied:

$$s\Rightarrow\frac{\omega_1}{\tan(0.5\omega_1T_s)}\cdot\frac{(1-z^{-1})}{(1+z^{-1})} \tag{5.5}$$

where:
ω_1 – fundamental radian frequency,
T_s – sampling period,
z^{-1} – operator representing a time delay of a single sampling period.

The advantage of using (5.5) is that its usage gives the gain and the phase displacement at the fundamental frequency exactly the same as under continuous differentiation. After applying (5.5) to (5.3) and transforming to the time domain, the following digital compensator COMP (Fig. 5.10) in the form of a recursive filter is obtained:

$$u_{2cc(n)} = \sum_{i=0}^{i=4} \frac{N_i}{M_0} u_{2(n-i)} - \sum_{i=1}^{i=4} \frac{M_i}{M_0} u_{2cc(n-i)} \tag{5.6}$$

where:
n – current sampling instant,
u_2 – uncompensated secondary voltage (as supplied by an A/D converter),
u_{2cc} – compensated secondary voltage (the output from the compensator (5.6)).

The filter (5.6) constitutes the simplest compensator (COMP) for a CVT. This compensator may be even more optimized. The improved compensator $\text{COMP}_{\text{impr.}}$ (Fig. 5.10) is a cascade of the original compensator COMP, given by (5.6), and a short window non-recursive digital filter (F3) added to its output. The self-explanatory assumptions for the filter (F3) are summarized as follows:

- zero gain at half of the sampling frequency;
- unity gain and zero phase displacement at the fundamental frequency; and
- possibly short data window.

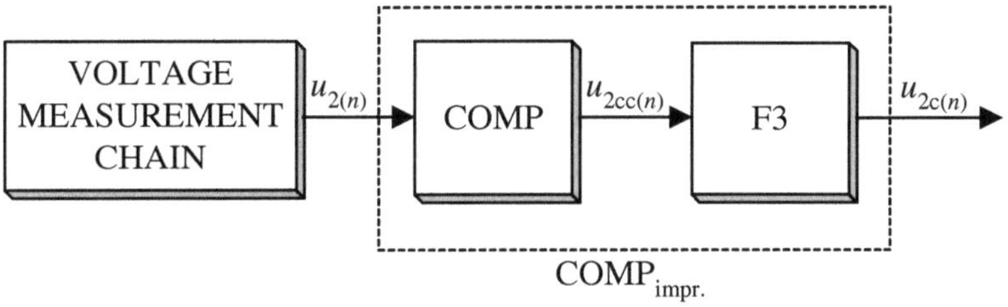

Fig. 5.10 Generic scheme of digital series compensation of CVT (COMP – original compensator, F3 – low-pass three-sample filter, $\text{COMP}_{\text{impr.}}$ – compensator with improved frequency response)

Use of a three-sample (the shortest possible window) low-pass filter F3 has been recommended in [112]. The output from the improved compensator (u_{2c}) is thus computed as:

$$u_{2c(n)} = A(u_{2cc(n)} + Bu_{2cc(n-1)} + Cu_{2cc(n-2)}) \tag{5.7}$$

where:

$$B = \frac{2\cos(\omega_1 T_s)}{1 + 2\cos(\omega_1 T_s)},$$

$$C = \frac{-1}{1 + 2\cos(\omega_1 T_s)},$$

$$A = \frac{1}{1 + B\cos(\omega_1 T_s) + C\cos(2\omega_1 T_s)}.$$

The cascade of the original compensator (5.6) and the low-pass filter (5.7) gives the improved compensation algorithm ($\mathrm{COMP}_{\mathrm{impr.}}$) of the following recursive form:

$$u_{2c(n)} = \sum_{i=0}^{i=6} P_i u_{2(n-i)} - \sum_{i=1}^{i=4} Q_i u_{2c(n-i)} \tag{5.8}$$

where P_i, Q_i – resultant coefficients of the improved compensator, dependent on the parameters of the CVT simplified transfer function and the sampling period width.

Both the original (5.6) and the improved (5.8) compensators, as the recursive digital filters require a kind of a starting procedure. To initiate the filters (5.6) or (5.8) one needs the last four samples of the compensated voltage. For this purpose the pure uncompensated secondary voltage may be used. The initiation is done once just after fault detection with the use of the frozen pre-fault data. However, to its advantage, the algorithm may be launched with the zero initial conditions, but necessarily at the maximum of the voltage wave.

Figures 5.11 through 5.14 present examples of performance of the improved compensator (5.8) for the simulated CVT transients appearing under a considerable decrease of the primary voltage during the transmission line fault. The applied compensation effectively removes the CVT-generated transients (Fig. 5.11). As a result, improved performance of the calculation of voltage magnitude (Fig. 5.12) and impedance components (resistance: Fig. 5.13, reactance: Fig. 5.14) is achieved.

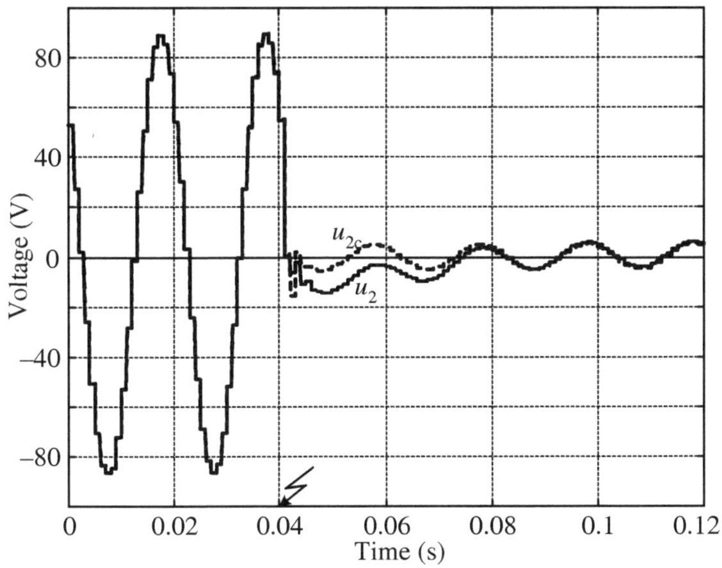

Fig. 5.11 Examples of voltage waveforms (presented in a staircase form): u_2 – CVT secondary voltage, u_{2c} – compensated secondary voltage

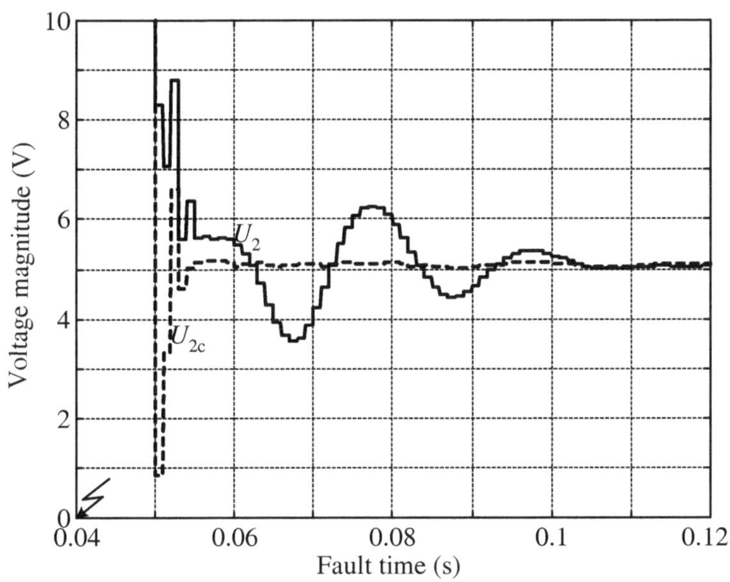

Fig. 5.12 Full-cycle Fourier calculation of voltage magnitude: U_2 – magnitude of CVT secondary voltage, U_{2c} – magnitude of compensated voltage

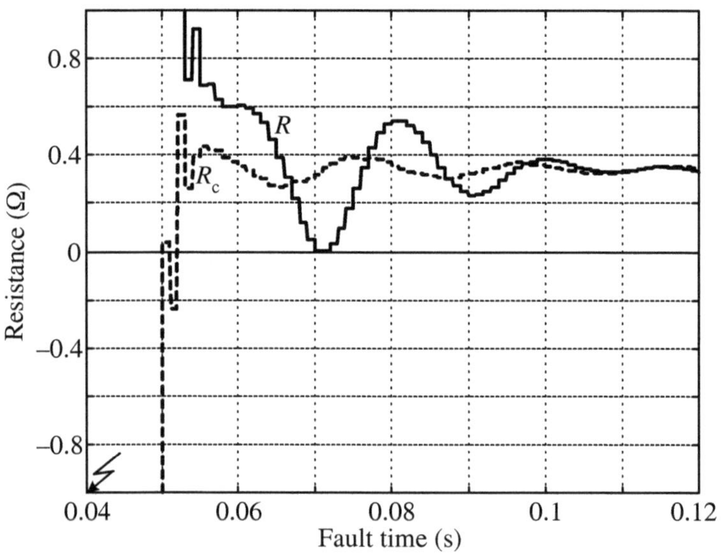

Fig. 5.13 Full-cycle Fourier calculation of the fault-loop resistance using secondary current and CVT secondary voltage (resistance R) or compensated voltage (resistance R_c)

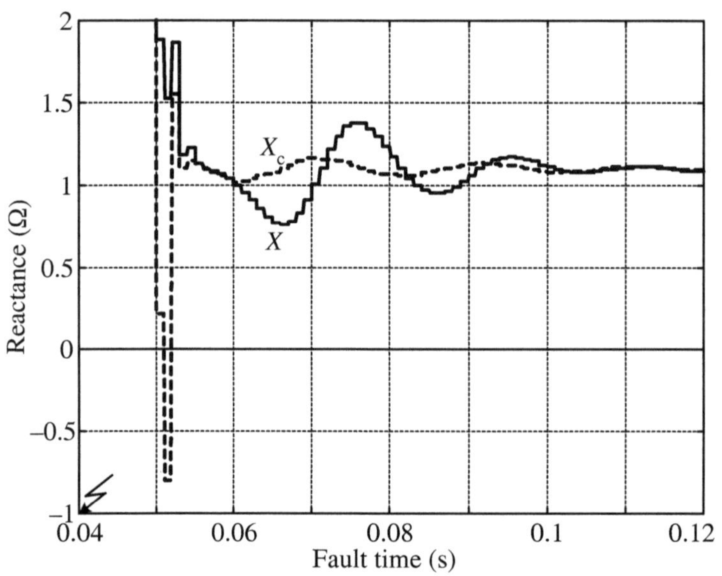

Fig. 5.14 Full-cycle Fourier calculation of the fault-loop reactance using secondary current and CVT secondary voltage (reactance X) or compensated voltage (reactance X_c)

An application of the compensation to fault location is depicted in Fig. 5.15. The compensation of the CVT secondary voltage results in decreased oscillations of the distance to fault (d_c). Averaging the fault distance exhibiting lower oscillations results in a more accurate final result. This is especially important when the fault quantities are recorded from a relatively short fault interval.

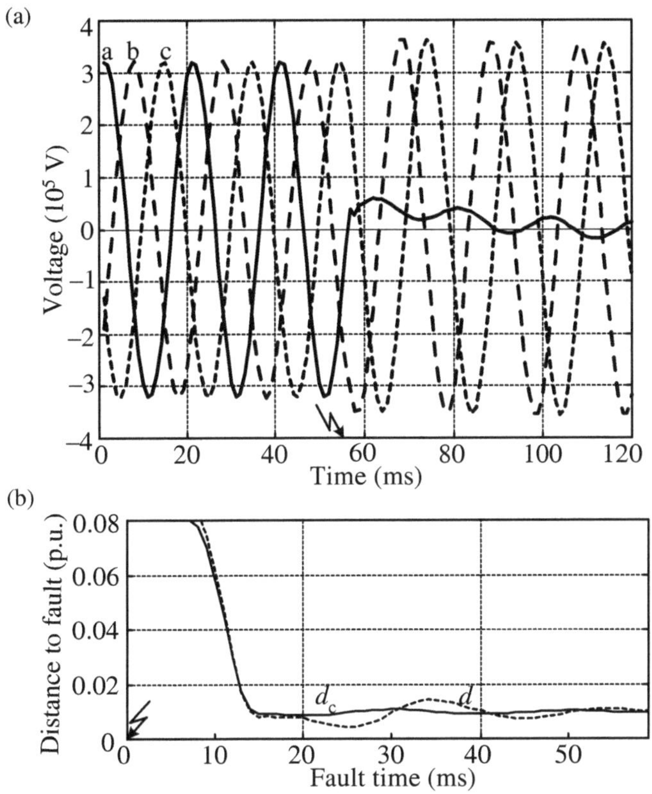

Fig. 5.15 Application of CVT compensation to fault location: (a) three-phase voltage under single-phase to ground fault, and (b) calculated distance to fault: d – under no CVT dynamic compensation, d_c – with CVT compensation

5.2.3 Frequency Response

The frequency characteristics of the measuring devices are of concern mainly when traveling-wave or high-frequency methods are considered.

The frequency response of a capacitive voltage transformer can be measured or calculated by including the stray capacitances into the equivalent circuit diagram. In Fig. 5.16 a typical frequency characteristic of a capacitive voltage transformer [161] is presented. Only within some range of frequencies around the fundamental frequency is the gain close to unity (0 dB).

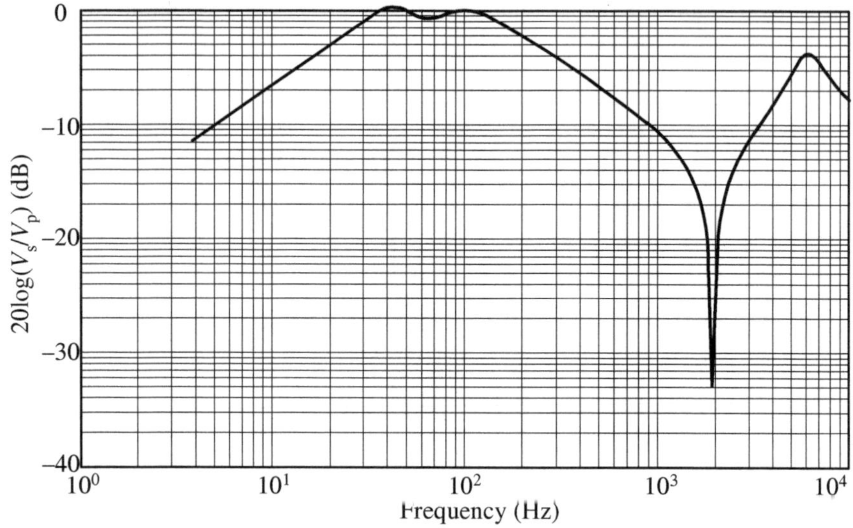

Fig. 5.16 Frequency response of capacitive voltage transformer

In contrast to capacitive voltage transformers, current transformers have a frequency response (Fig. 5.25) with almost equal magnitude in a wider range.

5.3 Current Transformers

5.3.1 Basics of Current Transformers

Instrument current transformers (CTs) transform power-system currents to the secondary level rated typically at 1 A or 5 A. The CT secondary current is substantially proportional to the primary current under normal conditions of operation, and differs in phase from it by an angle that is approximately zero for an appropriate direction of the connections. The steady-state error of a CT is classified into two: the current or ratio error, and the phase error. Both steady-state and transient performance of CTs is covered by the IEC Standard, as well as by national standards.

Figure 5.17 depicts a CT circuit model [160, 318], which includes:

i'_p, i_s – primary (recalculated to secondary side) and secondary currents,

i_e, i_r, i_m – exciting current and its active and reactive components,

R'_p, L'_p – primary winding resistance and leakage inductance recalculated to the secondary side,

R_s, L_s – secondary winding resistance and leakage inductance,
R_m, L_m – iron loss equivalent resistance, magnetizing non-linear inductance,
R_2, L_2 – load resistance and inductance.

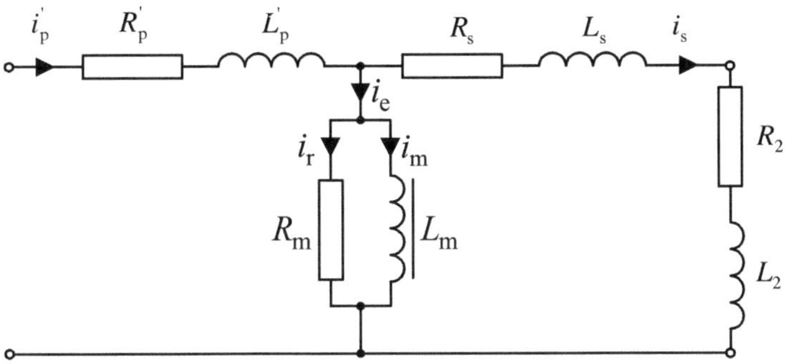

Fig. 5.17 Generic CT circuit model

CTs are designed to operate under load conditions, i.e., on the lower part of the linear region of the *U–I* characteristic of the magnetizing branch. The knee point of the magnetizing characteristic divides it into the linear and the non-linear regions. The *knee-point voltage* is understood as the point on the magnetizing curve where an increase of 10% in the flux density (voltage) causes an increase of 50% in the magnetizing force (current). For high fault primary currents without the DC component, the operating point remains in the linear range without exceeding the knee point of the characteristic. However, if the fault conditions are such that the DC component is present in the CT primary current, then a considerable increase of a flux in the CT magnetic core can take place. As a consequence of such an increase of a flux, the CT magnetic core gets saturated. Also, CTs can retain the remanent flux that may be left on the core after the fault is cleared [72, 141, 142, 349]. The remanent flux can either oppose or aid the build-up of the CT core flux, depending on the remanent flux polarity.

When a CT gets saturated, its secondary signal becomes distorted and a high transformation error appears. An example of waveforms of primary and secondary currents under CT transient saturation is shown in Fig. 5.18. It can be observed that there are intervals in which the recalculated secondary current differs considerably from the primary current.

Besides transient saturation CTs may suffer permanent saturation, under which there is no linear CT transformation at all.

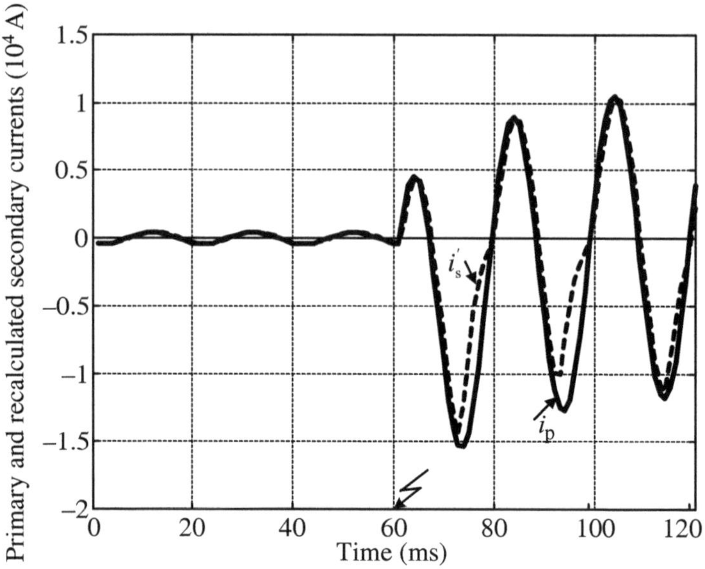

Fig. 5.18 CT saturation – primary and secondary (recalculated to the primary side) waveforms of currents

5.3.2 Fault Location Under CT Saturation

Many studies related to the analysis of the steady-steady and transient behavior of CTs have been reported so far in numerous references. The focus of attention is the problem of how distorted secondary currents due to CT saturation can cause malfunction or operating delays of protection relays and how to prevent saturation or to design protective algorithms insensitive to the effects of saturation [26, 334]. An issue of fault location in relation to CT saturation has been considered as well [332].

The remedies for assuring adequately high accuracy under CT saturation can be categorized as follows:

- use of hardware means for preventing CT saturation [22];
- use of voltage signals alone [23, 24, 184, 233, 341];
- use of voltage signals and current signals, but excluding currents from saturated CTs [123, 232, 271];
- minimizing fault-location errors caused by CT saturation and application of digital algorithms for reconstructing the CT primary current [142, 145, 182, 245]; and
- allowing currents from saturated CTs to be used but only from intervals of linear transformation (when there is no saturation) [152].

All the remedies listed, except an intentional use of voltage signals alone, require identifying the CT saturation. In general, the CT saturation identification (detection) is understood as recognizing instants when the saturation starts and when it ends. For this purpose saturation detectors are utilized. In general, we distinguish two families of methods for saturation detection:

- hardware-oriented methods based on superimposing an extra low power, and high-frequency signal to the secondary circuit of a CT and monitoring the core inductance using the superimposed signal (the value of the inductance indicates whether or not and to what degree the supervised CT is saturated); and
- waveform-oriented methods based on analyzing only the waveform of the secondary current of a CT.

Hardware-oriented Methods

McLaren *et al.* [208] proposed a scheme in which a 15-kHz signal is superimposed to the secondary current of a CT. The apparent impedance for this signal depends directly on the incremental inductance of the CT core, and consequently, on the degree of saturation. Therefore, the amplitude of the high-frequency current driven by an external voltage source acts as the saturation detector.

A similar approach was presented by Sanderson *et al.* in [290]. The authors use a 10-kHz externally driven signal to monitor the value of the equivalent magnetizing inductance of a CT. Keeping the 10-kHz driving force constant, the authors use a 10-kHz current as the indicator of saturation.

The obvious disadvantage of this family of methods is the need for connecting an extra circuit between the main CTs and the relay (fault locator). However, as measuring directly an equivalent magnetizing inductance of a CT, these methods appear as very robust.

Waveform-oriented Methods

The methods from this group process directly the waveform of the CT secondary current in order to distinguish between the linear and saturated operation of a CT. They are investigated as numerical procedures to be run exclusively on digital relays and fault locators. They call for comparatively high sampling frequencies in order to assure adequately short delays in detecting instants when a saturation starts and when it stops. Let us present two further examples of saturation detectors of this kind proposed in [152] and the modal saturation detector [103]. Another saturation detector that is based on processing the secondary current with use of an algorithm that evaluates the first, second and third difference functions is presented in [143, 144].

Saturation-detection Algorithm Based on Second Derivative

The algorithm from [152] combines the secondary current with its appropriately rescaled second-order derivative and is based upon the fact that during linear operation of a CT these two signals mutually compensate. Based on this principle, the following auxiliary error signal e_1 is defined:

$$e_{1(n)} = \frac{x_{(n)} - p \cdot x^{(2)}{}_{(n)}}{X_{(n)}} \tag{5.9}$$

where $x^{(2)}$ is a second-order derivative of the secondary current (i), computed numerically using the following five-point formula:

$$x^{(2)}{}_{(n)} = \frac{1}{12T_s}\left(-i_{(n)} + 16i_{(n-1)} - 30i_{(n-2)} + 16i_{(n-3)} - i_{(n-4)}\right) \tag{5.10}$$

where T_s stands for the sampling period. Since (5.10) gives the second-order derivative but for the time marker (n–2), the signal x in (5.9) should be, thus, delayed by two samples as well:

$$x_{(n)} = i_{(n-2)} \tag{5.11}$$

In (5.9) X stands for the amplitude of signal x, while the scaling factor p assumes the value:

$$p = \frac{1}{(2\pi f_1)^2} \tag{5.12}$$

where f_1 is a fundamental frequency.

The error signal e_1 (5.9) displays clear peaks when the supervised CT both gets in and out of saturation. Therefore, the flag SAT (SAT=1: saturation; SAT=0: no saturation) is handled as follows:

$$\left.\begin{array}{l}(e_{1(n)} > \mathrm{A}_1)\ \&\ (\mathrm{SAT}_{(n-1)} = 0) \rightarrow \mathrm{SAT}_{(n)} := 1 \\ (e_{1(n)} > \mathrm{B}_1)\ \&\ (\mathrm{SAT}_{(n-1)} = 1) \rightarrow \mathrm{SAT}_{(n)} := 0\end{array}\right\} \tag{5.13}$$

where A_1 and B_1 are the thresholds optimized for a given CT.

Figure 5.19 illustrates operation of the algorithm. As is seen, the beginnings of the saturated periods are easier to detect. They reflect in higher values of the error signal (e_1). This is the reason for using two different thresholds in (5.13). The method provides very good saturation recognition, but requires at least a 3-kHz sampling frequency (due to the need of computing the second-order derivative

(5.10). It is also worth emphasizing that due to the delay introduced by the derivative computations, the saturation periods are detected with a small delay.

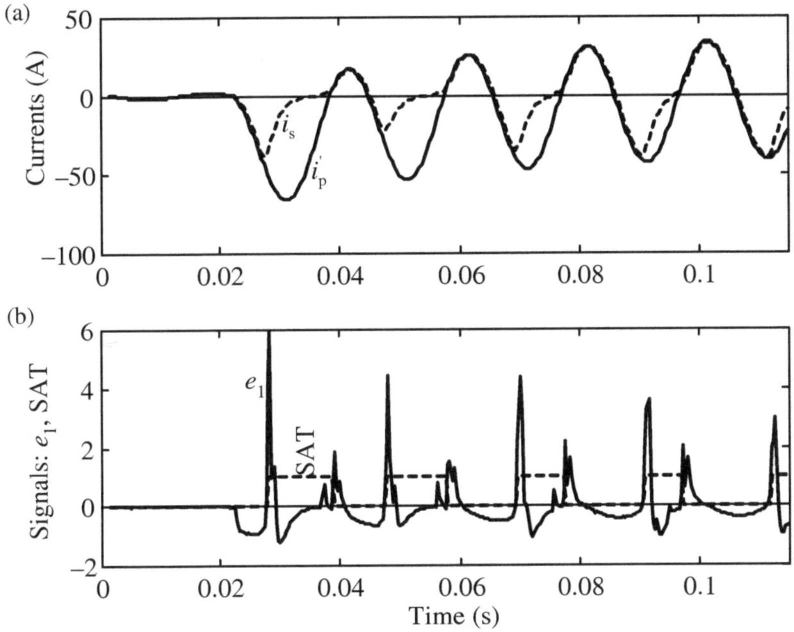

Fig. 5.19 Example performance of a saturation-detection algorithm based on a second-order derivative of current

Algorithm Based on Mean and Median Filters

The algorithm from [152] combines the mean and the median digital filters applied to the waveform of the secondary current, and is based on the ability of such a pair of filters to detect changes in the shape of a signal. The auxiliary error signal for this principle is defined as follows:

$$e_{2(n)} = \frac{\text{mean}(i)_{(n)} - \text{median}(i)_{(n)}}{I_{(n)}} \tag{5.14}$$

where I is the amplitude of the secondary current, while 'mean' and 'median' are mean and median functions [202], determined over a selected number of samples, say W.

The error signal e_2 (5.14) displays clear peaks when the supervised CT both enters and leaves its saturation periods. The performed simulation studies [152] indicated that the optimum value of the length of the data window (W) is around one twelfth of a cycle.

Algorithm Based on Modal Transformation

Another interesting saturation detector (Fig. 5.20), presented by Hosemann *et al.* [103], transforms modally the three-phase current vector ($\mathbf{i}_{abc}$) into the $0\alpha\beta$ components (the Clarke transformation (2.44)):

$$\mathbf{i}_{0\alpha\beta} = \frac{1}{3}\begin{bmatrix} 1 & 1 & 1 \\ 2 & -1 & -1 \\ 0 & \sqrt{3} & \sqrt{3} \end{bmatrix} \mathbf{i}_{abc} \tag{5.15}$$

The α and β components are further processed with use of a pair of orthogonal filters (F_d, F_q) for obtaining two pairs of orthogonal components: ($i_{\alpha d}$, $i_{\alpha q}$) and ($i_{\beta d}$, $i_{\beta q}$). Then, these components are combined into the real and imaginary part of a blocked backward-rotating space vector:

$$i_d = i_{\alpha d} - i_{\beta q} \tag{5.16}$$

$$i_q = i_{\alpha q} + i_{\beta d} \tag{5.17}$$

In [103] it is stated that if none of the CTs from a three-phase bank is under saturation, then the trajectory of the blocked backward-rotating space current (i_d, i_q) forms a circle. If the trajectory departs from the ideal circle shape, this means that at least one of CTs gets saturated (Fig. 5.21b). This observation may be formalized as follows:

$$\left.\begin{array}{l} i_d^2 + i_q^2 = \text{const.} \rightarrow \text{SAT} = 0 \\ i_d^2 + i_q^2 = \text{var.} \rightarrow \text{SAT} = 1 \end{array}\right\} \tag{5.18}$$

The saturation flag SAT is activated (SAT=1) if the trajectory departs from the circle shape for a given number of consecutive samples.

The method from [103] has two distinctive features compared with the previous algorithms [152]. First, it does not seek the periods of linear transformation during transient CT saturation. Instead, it yields only a global indication for CT saturation. Second, it uses one detector for both single-phase and three-phase currents. If the three-phase detector is used, one saves certain computation time, but misses information in which phase there is a CT saturation. Taking into account these features of the modal saturation detector [103], it can be applied as an auxiliary procedure only to two-end fault-location algorithms, which assure complete immunity to CT saturation appearing only at one line end [123]. Such a fault-location algorithm requires indicating from which line end three phase-current, as grossly contaminated due to CT saturation, has to be rejected.

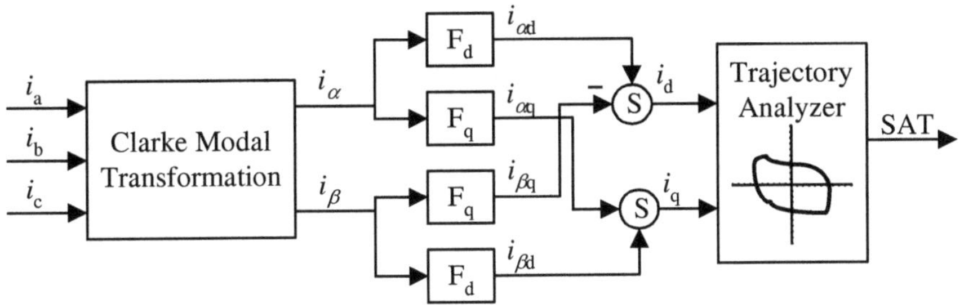

Fig. 5.20 Modal saturation detector

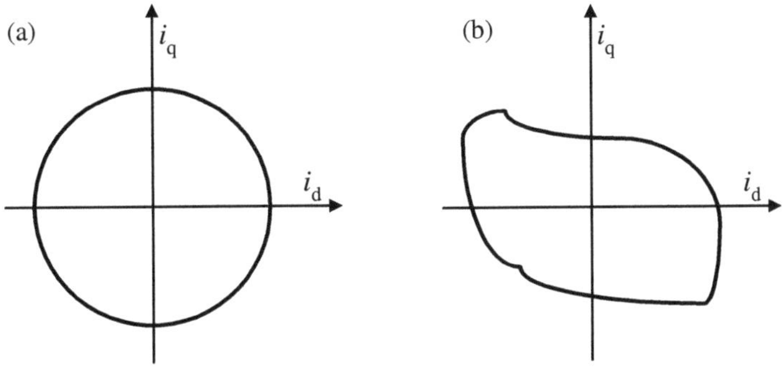

Fig. 5.21 Trajectories of the blocked backward-rotating space current in cases: (a) no CT saturation, and (b) CT saturation in at least one phase

Adaptive Measuring Technique

There is a chance that CT saturation, if it occurs, is a transient saturation only. In such a situation, in every cycle of the fundamental frequency, the CTs show periods of linear transformation. If a saturation detector is capable of indicating such periods, information brought by the samples from the linear operation of CTs may be used for determining criteria quantities, such as: amplitude of current, resistance, reactance, more accurately than by use of the standard Fourier technique [152].

However, the periods of saturation-free CT transformation may be quite short. Therefore, the use of variable data window techniques was recommended in [152]. In a variable data window algorithm, the data window extends after activation of the measurement and when it reaches its pre-defined full length (N samples) it stops extending and starts to slide. This technique, which combines both detection of saturated/linear operation of a CT and the variable data window measuring algorithm, is further named the adaptive method (Figs. 5.22 and 5.23).

As a result of applying the variable data window measuring algorithm one earns ultra-high-speed response of a measuring algorithm due to removing pre-fault samples from the data window. Deterioration of the filtration quality of such an algorithm is the price to pay for this gain [331, 336].

The presented adaptive method [152] follows the orthogonal-component based approach. The direct (i_d) and quadrate (i_q) orthogonal components of the secondary current (i) are filtered out as follows [331, 336]:

$$i_{d(k)} = \sum_{n=0}^{L_{(k)}-1} a(n, L_{(k)}) i_{(k-n)} \tag{5.19}$$

$$i_{q(k)} = \sum_{n=0}^{L_{(k)}-1} b(n, L_{(k)}) i_{(k-n)} \tag{5.20}$$

were:

- $L_{(k)}$ – window length, which increases until reaching its rated length (usually a cycle or half a cycle);
- data window coefficients are assumed to be located on the cos/sin functions:

  $$a(n, L_{(k)}) = k_d \left(\cos\left(\frac{\pi}{N}(L_{(k)} - 2n - 1) \right) - A(L_{(k)}) \right),$$

  $$b(n, L_{(k)}) = k_q \sin\left(\frac{\pi}{N}(L_{(k)} - 2n - 1) \right);$$
- $A(L_{(k)})$ – such shift of the 'cos' function that the sum of the coefficients $a(n, L_{(k)})$ within the current window is equal to zero (this guarantees perfect rejection of the constant component); and
- k_d, k_q – scaling coefficients, applied for getting unity gain at the fundamental frequency.

If the CT is not saturated, the variable data window algorithm is executed and gives the estimate of the criteria quantity based first on initial window length $L_{initial}$ (in [152] it was applied: $L_{initial}$=3 samples), next, on increased number of samples, until the window length reaches its full length (L=N). If so, the window starts to slide instead of extending. When saturation is detected (SAT=1), the amplitude is not measured, but the previous amplitude estimate is frozen (Fig. 5.22). When the flag SAT is deleted (the CT out of saturation), the algorithm re-activates and starts from L=$L_{initial}$ again.

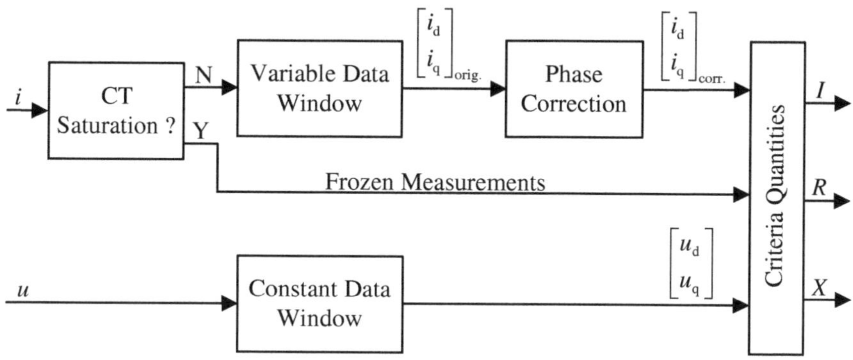

Fig. 5.22 Basic flow chart of the adaptive method for measuring criteria quantities (current magnitude, resistance, reactance)

The impedance (R, X) is measured based upon orthogonal components of both voltage (u) and current (i) signals. Voltage can be processed using a constant width data window (for example the full-cycle Fourier), while current signal is split into its orthogonal components using the adaptive method (5.19) and (5.20). Since the voltage data window slides, thus its midpoint moves by one sample. On the other hand, the data window for a current extends, thus its midpoint moves by half a sample. Therefore, an appropriate phase-compensating procedure must be applied. The group delay assumes:

$$\Delta\varphi = \frac{\pi}{N}(N - L_{(k)}) \tag{5.21}$$

Therefore, the corrected orthogonal components of the current are:

$$\begin{bmatrix} i_{\mathrm{d}(k)} \\ i_{\mathrm{q}(k)} \end{bmatrix}_{\mathrm{corr.}} = \begin{bmatrix} \cos(\Delta\varphi_{(k)}) & \sin(\Delta\varphi_{(k)}) \\ -\sin(\Delta\varphi_{(k)}) & \cos(\Delta\varphi_{(k)}) \end{bmatrix} \begin{bmatrix} i_{\mathrm{d}(k)} \\ i_{\mathrm{q}(k)} \end{bmatrix}_{\mathrm{orig.}} \tag{5.22}$$

The phase-correcting algorithm (5.22) is non-stationary and for short windows the correction is significant, while when the data window for the current reaches its full length (L=N) the correction vanishes.

Figure 5.23 presents an example performance of the adaptive algorithm for measuring current magnitude. In comparison to the full-cycle Fourier algorithm the adaptive algorithm provides much better response, both faster time response and better steady-state accuracy. Similarly, measurement of resistance and reactance (Fig. 5.24) appears superior with use of the adaptive method. Features of the presented adaptive method indicate the possibility of directly applying it to impedance-based fault-location algorithms [109, 352]. Successful application of this method to overcurrent relays is reported in [26], and to distance relays in [334].

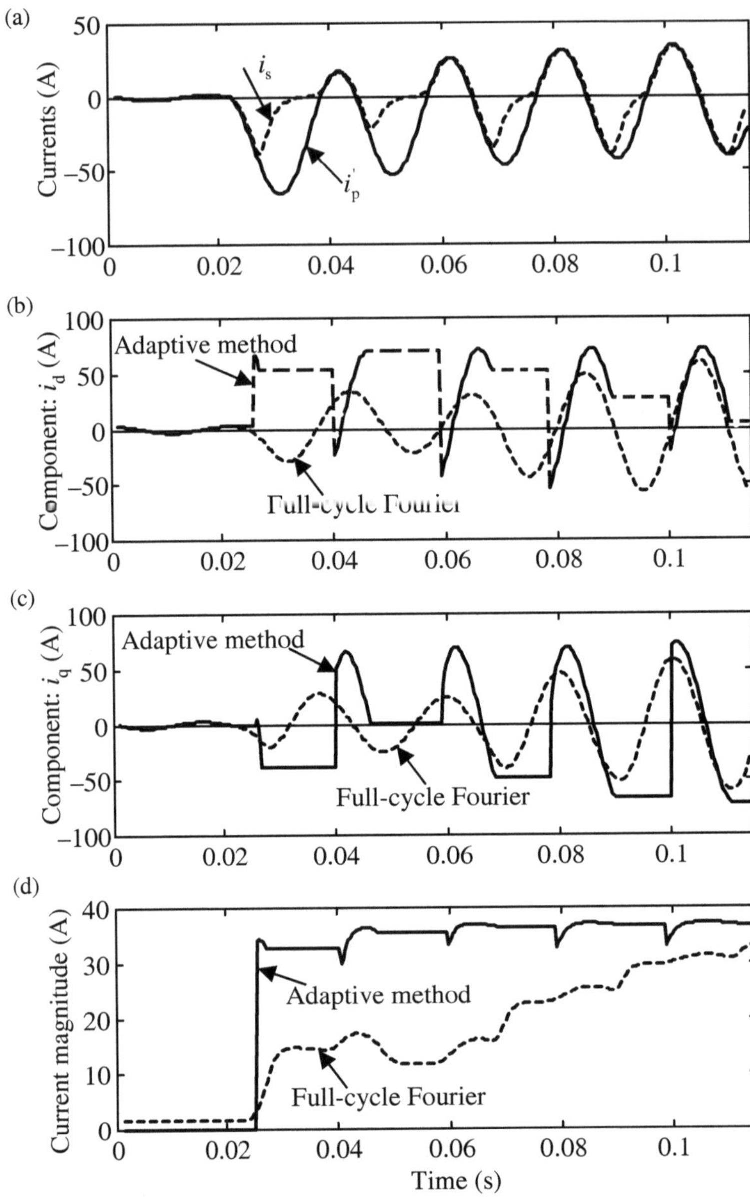

Fig. 5.23 Example performance of adaptive algorithm for measuring current magnitude, in comparison to full-cycle Fourier algorithm: (a) primary- and secondary-current waveforms, (b) orthogonal component i_d, (c) orthogonal component i_q, and (d) current magnitude

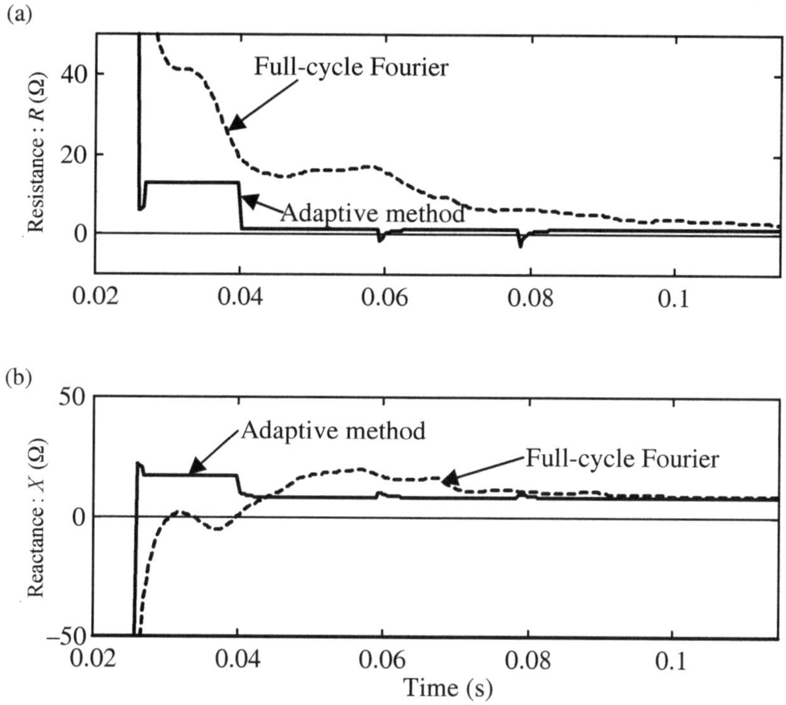

Fig. 5.24 Example performance of adaptive algorithm for measuring resistance and reactance, in comparison to full-cycle Fourier algorithm

Hardware Methods for Compensating Distortion in Secondary Current

Besides digital algorithms, analog (hardware) methods for compensating the distortion in the secondary current have also been developed.

In [22], an analog circuit is connected to the secondary terminals of a CT and used to generate a DC component equal and opposite to that seen in the primary one. Then, by injecting the generated DC component into the secondary winding component prevents saturation of a CT.

A digital algorithm for compensating the secondary current is put forward in [145]. Then, an advanced compensating algorithm of the distorted secondary current immune to the remanent flux is proposed in [141,142]. These algorithms estimate the secondary current corresponding to the CT ratio under CT saturation using the flux–current curve. In addition, it is stated that this approach allows for successful compensation of the secondary current even when a smaller CT than the rated size is used, resulting in secondary currents being more severely distorted. Moreover, it is shown [141, 142] that the proposed compensating algorithm can be implemented in real time into a digital-signal-processor hardware as part of the main protective relaying algorithm.

5.3.3 Frequency Response

Typical frequency response of a CT with respect to its magnitude is presented in Fig. 5.25.

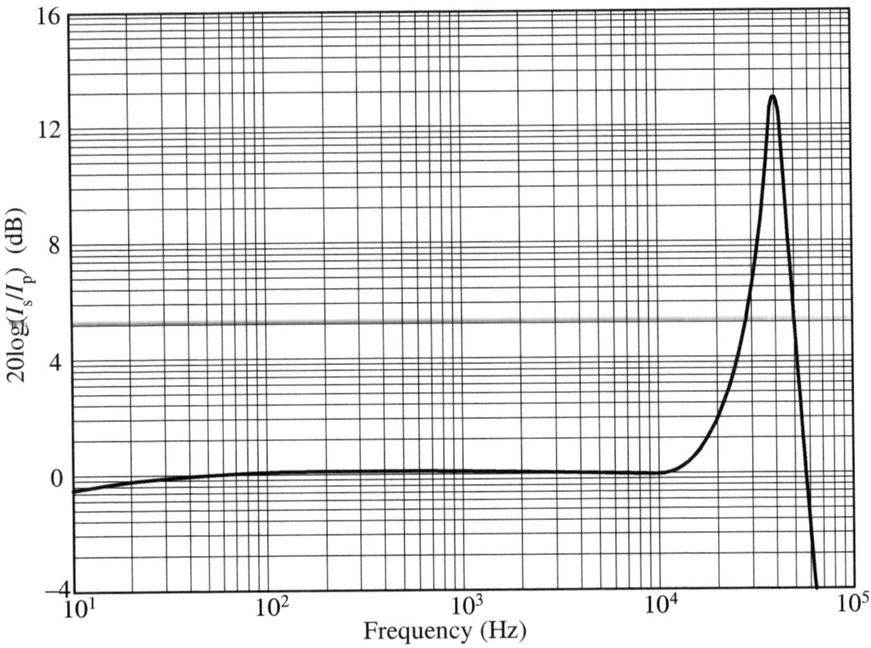

Fig. 5.25 Frequency response of current transformer

There is a wide range of frequency in the characteristic of Fig. 5.25 in which the gain is close to unity (0 dB). Thus, high fidelity of transformation of the components is observed for this range. The difference between the actual gain and the unity can be compensated for by applying the ratio correction factors tabulated as the function of frequency. Such ratio-correction factors can be obtained from measurements.

5.4 Analog Anti-aliasing Filters

The sampling of analog signals is performed with use of A/D converters. Digital information contained in the set of samples obtained differs from that provided by analog signals [230, 298]. The digital frequency is equal to the analog frequency if the frequency of the sampled analog signal is smaller than half of the sampling frequency ($0.5f_s$). This threshold value is commonly called the *Nyquist frequency*. Sampling the sine analog wave of the frequency higher than this threshold value results in obtaining a set of samples that represent the sine wave of the frequency different from that at the input of the A/D converter. Figure 5.26 shows the digital frequency *versus* the analog frequency.

Suppose that the calculations are based on fundamental frequency (f_1) components of the processed signals. It is seen (Fig. 5.26) that the sine analog waves of different frequencies: f_1, $f_s - f_1$, $f_s + f_1$, $2f_s - f_1$, $2f_s + f_1$, ..., after sampling give sets of samples representing the sine wave of the fundamental frequency (f_1). So, the sampled sinusoids assume the frequency that is not their own. This phenomenon of sinusoids changing frequency during sampling is called *aliasing*. The term 'aliasing' is originated here from comparing the effect of the frequency change to the crime on an identity (an alias), which is understood here as the frequency of an analog sine wave. In addition to the frequency-change effect, the aliasing also changes the phase of the signal by π for the respective ranges of frequencies of the analog signal, as shown in Fig. 5.27.

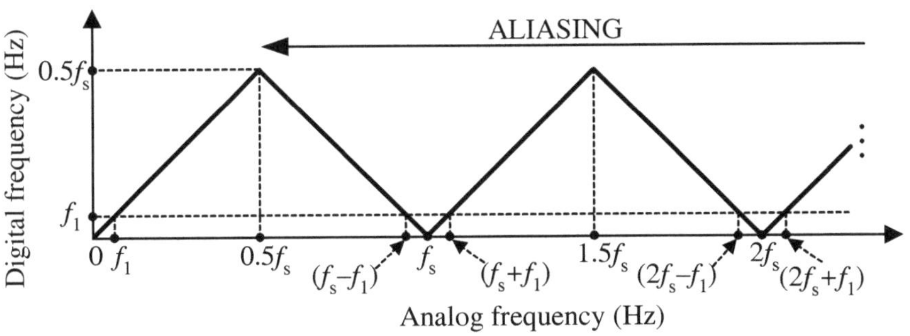

Fig. 5.26 Illustrating the fact that digital data is not uniquely related to a particular analog signal with respect to frequency

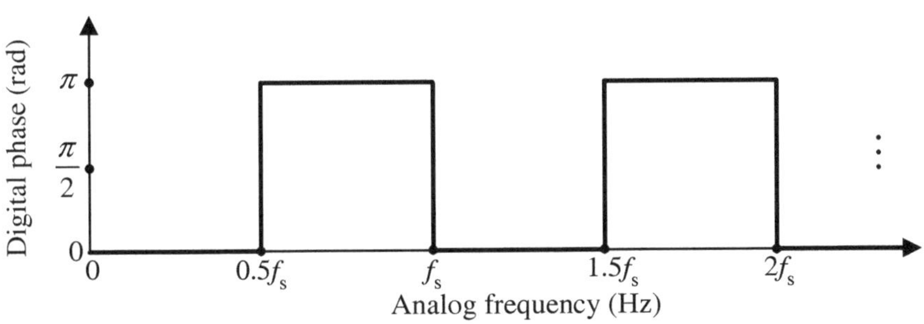

Fig. 5.27 Digital phase *versus* analog frequency

Figure 5.28 shows an example of sampling two analog sine signals. The sampling frequency is f_s = 1000 Hz. The first signal (S_1) is of frequency f_1 = 50 Hz (thus, below the Nyquist frequency) while the second signal (S_2) has the frequency $f_2 = f_s - f_1$ = 950 Hz (thus, above the Nyquist frequency). The aliasing is present in the case of sampling the signal S_2. As a result, both sampled signals give the sets of samples that represent the sinusoids of the fundamental frequency, shifted by an angle π. The set of samples for the signal S_1 allows its analog form to be reconstructed, while information contained in analog signal S_2 is lost completely. Sampling the signal of frequency 950 Hz at 1000 Hz sampling frequency creates a new frequency of 50 Hz for the digital signal. In this case, sampling destroys information encoded in the frequency domain of the analog signal S_2.

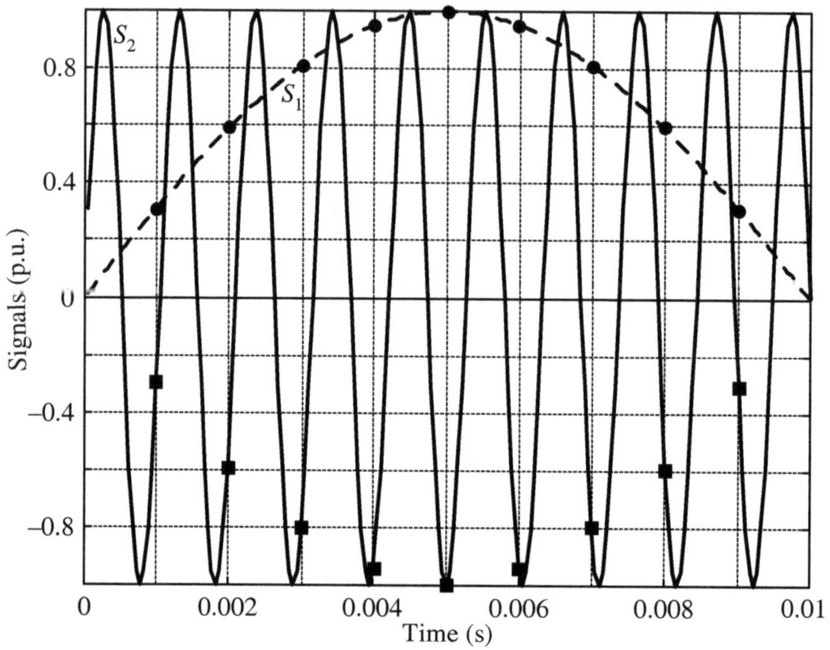

Fig. 5.28 Example of sampling two analog sine signals with sampling frequency f_s = 1000 Hz: sampled sine signals: S_1: signal of frequency f_1 = 50 Hz (samples denoted by circles), S_2: signal of frequency $f_2 = f_s - f_1$ = 950 Hz (samples denoted by squares)

In order not to lose information contained in analog signals proper sampling frequency of analog-to-digital (A/D) converters has to be applied. Claude E. Shannon in 1949 in his famous *sampling theorem* [230, 298] proved that if the signal contains no frequencies above f_{mx}, then the continuous time signal can be reconstructed from a periodically sampled sequence, provided that the sampling frequency f_s satisfies the condition:

$$f_s > 2f_{mx} \tag{5.23}$$

The sampling theorem indicates that a continuous signal can be properly sampled, *only if it does not contain frequency components above one-half of the sampling frequency* (*Nyquist frequency*). The other possibility calls for removing the frequencies higher than Nyquist frequency from the analog signal. This can be obtained by applying, prior to sampling, an analog low-pass filter, which is referred to as an anti-aliasing filter.

Figure 5.29 presents a simple *RC* four-port network of low-pass transfer function, which can be applied as the anti-aliasing filter. Assuming for the circuit burden impedance $Z_B \to \infty$, the transfer function of the circuit from Fig. 5.29 is obtained as:

$$G(s) = \frac{1}{(RC)^2 s^2 + 3RCs + 1} \tag{5.24}$$

The anti-aliasing cut-off frequency (f_c) of the low-pass analog filter is defined as:

$$|G_{AF}(j\omega_c)| = \frac{\sqrt{2}}{2} \tag{5.25}$$

where $\omega_c = 2\pi f_c$.

From (5.24) with respect to (5.25) one obtains:

$$\omega_c^{\ 4}(RC)^4 + 7\omega_c^{\ 2}(RC)^2 - 1 = 0 \tag{5.26}$$

After solving (5.26) and taking the solution for which: $(RC)^2 > 0$, the following formula for the circuit parameters is obtained:

$$RC = \frac{\sqrt{2\sqrt{53} - 14}}{4\pi f_c} \tag{5.27}$$

The time constant (*RC*) can be calculated from (5.27) after assuming the required cut-off frequency, which is usually set in the range: one-third up to one-half of the sampling frequency:

$$\frac{1}{3} f_s < f_c < \frac{1}{2} f_s \tag{5.28}$$

In the general case, the value of the cut-off frequency depends also on the filter order.

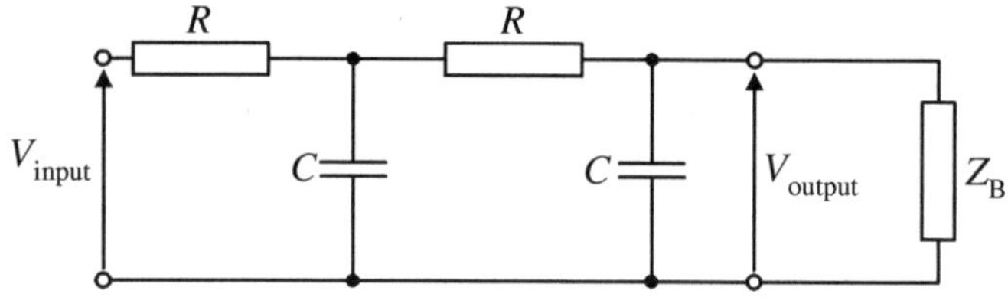

Fig. 5.29 *RC* four-port network of low-pass transfer function

Figures 5.30 and 5.31 show common building blocks: *Sallen–Key circuits* with operational amplifiers for analog anti-aliasing filters and their transfer functions [298].

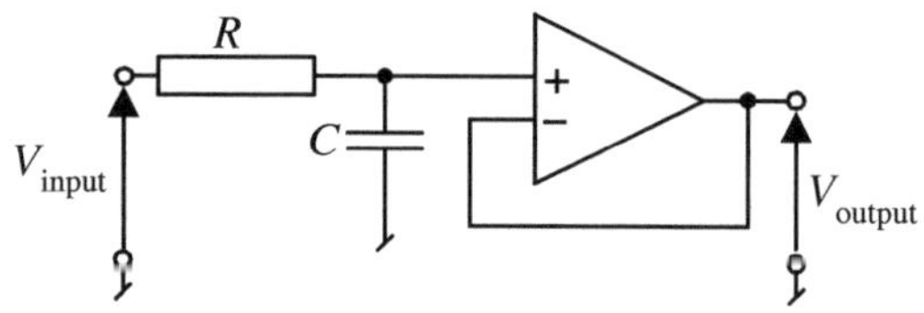

$$G(s) = \frac{1}{RCs+1}$$

Fig. 5.30 First-order Sallen–Key low-pass circuit

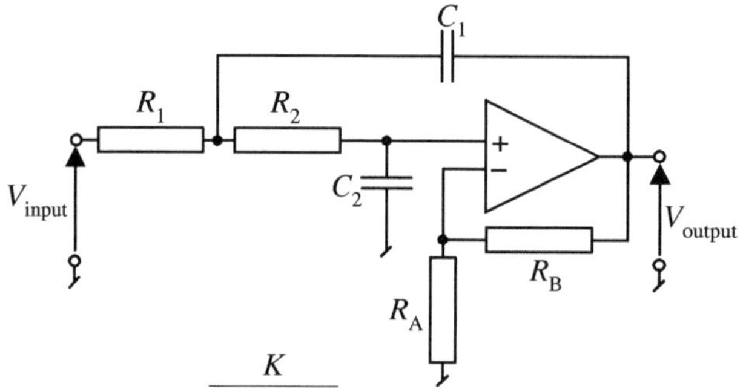

$$G(s) = \frac{\dfrac{K}{R_1R_2C_1C_2}}{s^2 + \left[\dfrac{1}{R_1C_1} + \dfrac{1}{R_2C_1} + \dfrac{1-K}{R_2C_2}\right]s + \dfrac{1}{R_1R_2C_1C_2}} \qquad K = \frac{R_A + R_B}{R_A}$$

Fig. 5.31 Second-order Sallen–Key low-pass circuit

Using the first-order (Fig. 5.30) and the second-order (Fig. 5.31) Sallen–Key low-pass circuits one can build higher-order anti-aliasing analog filters. Another option in this respect is use of the switched capacitor filters, which are composed of capacitors and electronic switches.

Design of the analog anti-aliasing filter with the given cut-off frequency requires assuming the respective standard approximation of the transfer function and its order. Three types of analog filters are commonly used: *Butterworth*, *Chebyshev*, and *Bessel* (also called a *Thompson* filter) [298]. A transfer function of each of them is obtained as a result of optimizing a different performance parameter. There is a wide literature on this subject.

6 One-end Impedance-based Fault-location Algorithms

Abstract In this chapter, a variety of one-end impedance-based fault-location algorithms are presented. The algorithms presented are designed for locating faults on single-circuit lines, double-circuit lines, and series-compensated lines. Both transposed and untransposed lines are taken into consideration. The algorithms are formulated for the lumped-line models. The chapter is concluded with the introduction of a distributed-parameter line model for improving fault-location accuracy.

6.1 Introduction

The one-end fault-location algorithms are simple and economical, when compared with two-end methods and those based on the traveling-wave and high-frequency component techniques. These algorithms are designed for estimating the location of transmission-line faults with use of three-phase current and voltage measured at one terminal of a line. Also, there are some algorithms that utilize three-phase voltage or current alone. The fundamental frequency phasors of the measured signals or the sampled data are processed in the algorithms. Also the impedance parameters of the network are required for determining a distance to fault.

6.2 Fault Location Based on Impedance Measurement

The impact of fault resistance on one-end impedance measurement is a key factor in deriving the majority of one-end fault-location algorithms. Let us start with considering a single-phase line (A–B) connected to a source at one end (A) only, i.e., the line that supplies no load at the end B (Fig. 6.1). The line is affected by a fault (F), which is at unknown distance d (p.u.) from the bus A, where the fault locator (FL) is installed. If the line charging current is neglected, then the current at the fault locator ($\underline{I}_A$) is equal to the current at the fault ($\underline{I}_F$). The impedance seen from the fault-locator terminal, calculated using the voltage $\underline{U}_A$ and current $\underline{I}_A$, is:

$$\underline{Z}_{\mathrm{FL}} = \frac{\underline{U}_{\mathrm{A}}}{\underline{I}_{\mathrm{A}}} = d\underline{Z}_{\mathrm{L}} + R_{\mathrm{F}} \tag{6.1}$$

Taking the imaginary part of (6.1) one obtains the distance to fault as:

$$d = \frac{\mathrm{imag}(\underline{Z}_{\mathrm{FL}})}{\mathrm{imag}(\underline{Z}_{\mathrm{L}})} \tag{6.2}$$

This formula is a predecessor of the one-end impedance-based fault-location algorithms [291]. It allows accurate determination of the distance to fault in the case of one-end supply (Fig. 6.1). This is so since the fault resistance (R_{F}) is seen from the fault-locator terminal as pure resistance, as shown in Fig. 1.2a.

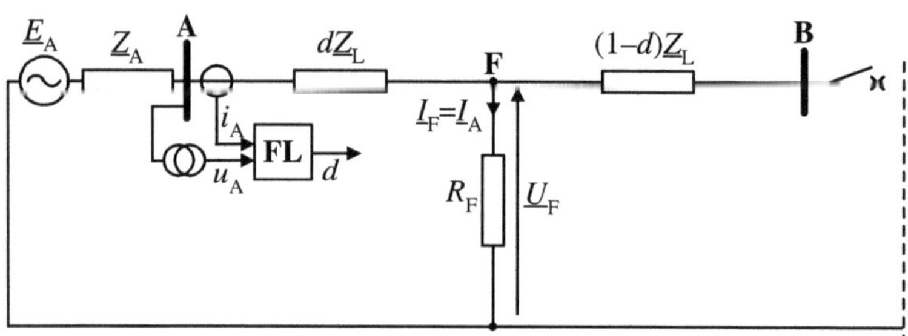

Fig. 6.1 Fault location based on impedance measurement for faulted line connected to the source at one end

However, if there is a two-end supply (Fig. 6.2), the current at the fault ($\underline{I}_{\mathrm{F}}$) is not equal to the current at the fault locator ($\underline{I}_{\mathrm{A}}$) since the remote current ($\underline{I}_{\mathrm{B}}$) also contributes to the total fault current ($\underline{I}_{\mathrm{F}} = \underline{I}_{\mathrm{A}} + \underline{I}_{\mathrm{B}}$). As a result, there is a contribution of the reactance in the impedance seen from the fault-locator terminal (the reactance effect), as shown in Chap. 1 in Figs. 1.2b and c.

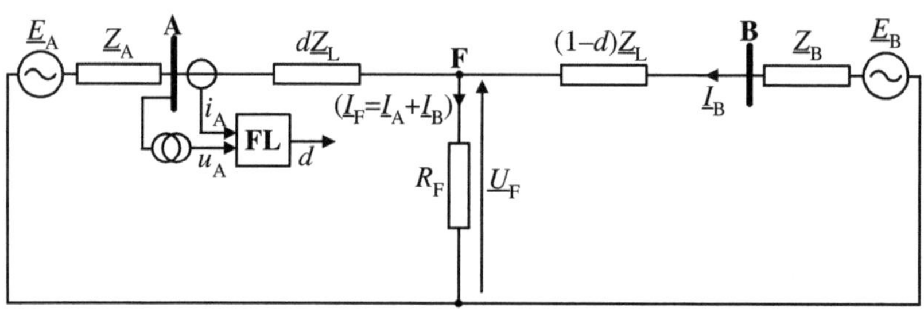

Fig. 6.2 Fault location based on impedance measurement for faulted line connected to the sources at two ends

Formulating Kirchoff's voltage law for the fault loop seen from the terminal A, i.e., the loop containing the faulted-line segment ($d\underline{Z}_L$) and fault-path resistance (R_F), one obtains the complex scalar equation:

$$\underline{U}_A - d\underline{Z}_L \underline{I}_A - R_F \underline{I}_F = 0 \tag{6.3}$$

This equation can be resolved into the real and imaginary parts (two equations), however, there are four unknowns: d, R_F, Re($\underline{I}_F$), Im($\underline{I}_F$) and thus the number of unknowns exceeds the number of equations. In order to assure solvability of the fault-location problem, Takagi *et al.* in their original work [309] proposed to decompose, using the *Thevenin theorem*, the faulted network (Fig. 6.3a) into the pre-fault network (Fig. 6.3b) and 'pure fault', i.e., superimposed network (Fig. 6.3c). They started the fault-location derivation by considering the distributed-parameter line model. However, they finally introduced some simplifications, which correspond to use of the simple lumped-line model. Therefore, in what follows the lumped-line model will be taken into account.

The superimposed circuit (Fig. 6.3c) is a current divider of the fault current and thus:

$$\Delta\underline{I}_A = \frac{(1-d)\underline{Z}_L + \underline{Z}_B}{\underline{Z}_A + \underline{Z}_L + \underline{Z}_B}\underline{I}_F \tag{6.4}$$

where $\Delta\underline{I}_A = \underline{I}_A - \underline{I}_A^{pre}$ – incremental current determined from the moment of the fault-inception occurrence (thus in the fault interval), and obtained by taking the fault current and subtracting the pre-fault current (present before fault inception). Note that the recordings of the pre-fault current have to be available.

This allows the total fault current to be determined as:

$$\underline{I}_F = \frac{\Delta\underline{I}_A}{\underline{k}_F} \tag{6.5}$$

where the fault current distribution factor ($\underline{k}_F$) [62, 106, 109, 330] is determined as follows:

$$\underline{k}_F = |\underline{k}_F|\,e^{j\gamma} = \frac{-d\underline{Z}_L + \underline{Z}_L + \underline{Z}_B}{\underline{Z}_A + \underline{Z}_L + \underline{Z}_B} \tag{6.6}$$

Substituting (6.5) and (6.6) into (6.3) results in:

$$\underline{U}_A - d\underline{Z}_L \underline{I}_A - \frac{R_F}{|\underline{k}_F|\,e^{j\gamma}}\Delta\underline{I}_A = 0 \tag{6.7}$$

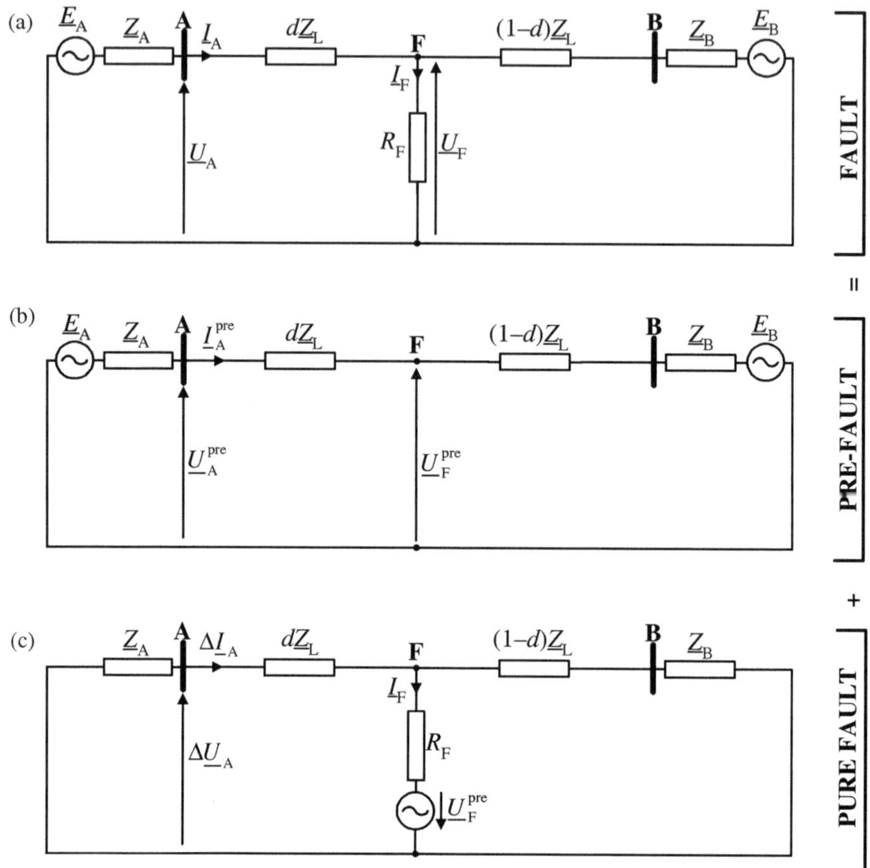

Fig. 6.3 Application of the Thevenin theory to a faulted network: (a) faulted network, (b) pre-fault network, and (c) superimposed component network

Multiplying (6.7) by the element ($e^{j\gamma}\Delta \underline{I}_A^*$) and taking the imaginary part yields the following formula for the distance to fault:

$$d = \frac{\mathrm{Im}(\underline{U}_A \Delta \underline{I}_A^* e^{j\gamma})}{\mathrm{Im}(\underline{Z}_L \underline{I}_A \Delta \underline{I}_A^* e^{j\gamma})} \tag{6.8}$$

where x^* denotes the conjugate of x.

It was assumed in [309] that the current distribution factor is a real number (γ=0), which facilitates calculations. This simplification is applied since all the impedances involved in (6.6) have approximately the same phases. Otherwise, iterative calculations, which require knowing all the impedances from (6.6), have to be performed.

The fault-location algorithm (6.8) was derived for a single-phase line. For three-phase lines, the symmetrical-components or phase-coordinates approaches are considered in successive sections of this chapter.

6.3 Fault Current Distribution Factors

6.3.1 Transmission Network with Single-circuit Line

Current distribution factors for symmetrical components were introduced in [62]. According to the fault model described in Chap. 3 in (3.1), the total fault current ($\underline{I}_{\mathrm{F}}$) is expressed as the weighted sum of its symmetrical components ($\underline{I}_{\mathrm{F1}}$, $\underline{I}_{\mathrm{F2}}$, $\underline{I}_{\mathrm{F0}}$), which can be determined with use of the fault current distribution factors:

$$\underline{I}_{\mathrm{F1}}=\frac{\Delta\underline{I}_{\mathrm{A1}}}{\underline{k}_{\mathrm{F1}}}=\frac{\underline{I}_{\mathrm{A1}}-\underline{I}_{\mathrm{A1}}^{\mathrm{pre}}}{\underline{k}_{\mathrm{F1}}} \tag{6.9}$$

$$\underline{I}_{\mathrm{F2}}=\frac{\underline{I}_{\mathrm{A2}}}{\underline{k}_{\mathrm{F2}}} \tag{6.10}$$

$$\underline{I}_{\mathrm{F0}}=\frac{\underline{I}_{\mathrm{A0}}}{\underline{k}_{\mathrm{F0}}} \tag{6.11}$$

In (6.9)–(6.11) and in Fig. 6.4, the respective subscript denotes: 1 – positive-, 2 – negative-, 0 – zero-sequences. For determining the positive-sequence component ($\underline{I}_{\mathrm{F1}}$), (6.9), the superimposed positive-sequence circuit (Fig. 6.4a) is considered. For the remaining sequences ((6.10)–(6.11)), the pre-fault sequence currents are not involved in the formulae since they are equal to zero for the completely symmetrical network before the fault occurrence. This is the condition of using the symmetrical-components approach.

From the analysis of the circuits presented in Fig. 6.4, and taking into account the fact that the respective network impedances for positive- and negative-sequences are basically identical, one obtains the fault current distribution factors:

$$\underline{k}_{\mathrm{F1}}=\underline{k}_{\mathrm{F2}}=\frac{\underline{K}_1 d+\underline{L}_1}{\underline{M}_1} \tag{6.12}$$

$$\underline{k}_{\mathrm{F0}}=\frac{\underline{K}_0 d+\underline{L}_0}{\underline{M}_0} \tag{6.13}$$

where:
$\underline{K}_1$, $\underline{L}_1$, $\underline{M}_1$ – complex coefficients dependent on positive-sequence impedances of the network (Table 6.1),
$\underline{K}_0$, $\underline{L}_0$, $\underline{M}_0$ – complex coefficients dependent on zero-sequence impedances of the network, having analogous forms as for the positive-sequence but the positive-sequence impedances are exchanged by the respective zero-sequence impedances.

In the case of a transmission network with a single line, the complex coefficients used for determining fault current distribution factors for positive- and negative-sequence (6.12) are determined by analysis of the fault currents flow in the circuits presented in Figs. 6.4a and b.

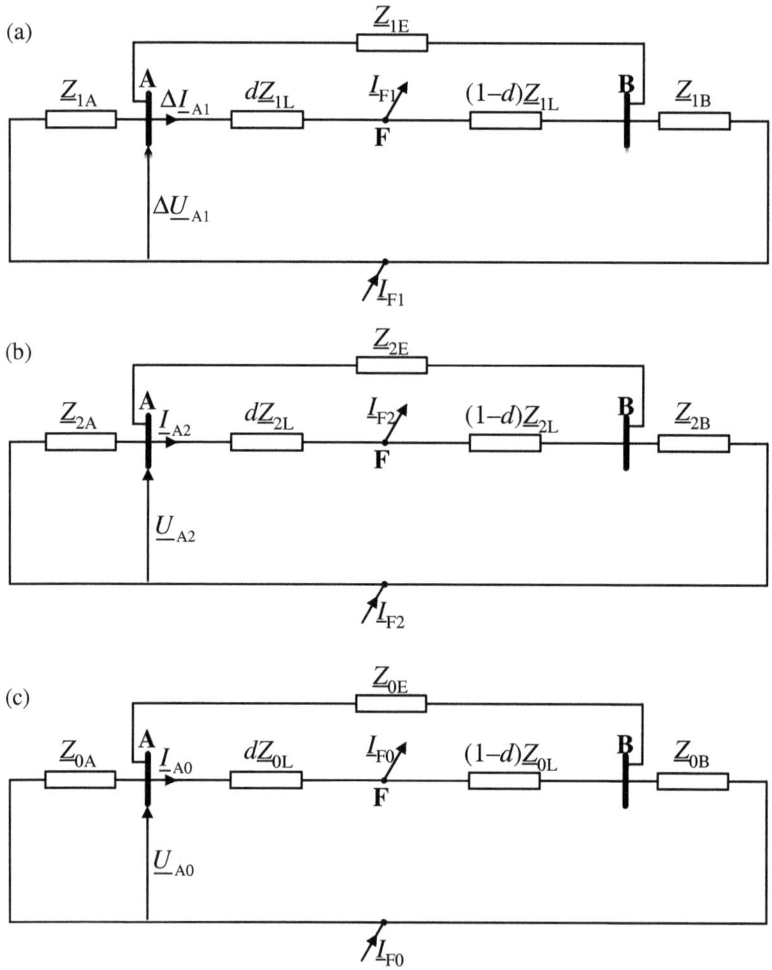

Fig. 6.4 Equivalent circuit diagrams of transmission network with single-circuit line ($\underline{Z}_L$) and extra link between the end buses ($\underline{Z}_E$) for: (a) incremental positive-, (b) negative-, and (c) zero-sequence

Table 6.1 Transmission network with single line (Figs. 6.4a and b) – complex coefficients used for determining fault current distribution factor for positive- and negative-sequence (6.12)

Extra link	Coefficients
$\underline{Z}_{1E} \neq \infty$ (extra link exists)	$\underline{K}_1 = -\underline{Z}_{1L}\underline{Z}_{1E} - (\underline{Z}_{1A} + \underline{Z}_{1B})\underline{Z}_{1L}$ $\underline{L}_1 = \underline{Z}_{1L}(\underline{Z}_{1A} + \underline{Z}_{1B}) + \underline{Z}_{1E}(\underline{Z}_{1L} + \underline{Z}_{1B})$ $\underline{M}_1 = (\underline{Z}_{1A} + \underline{Z}_{1B})(\underline{Z}_{1E} + \underline{Z}_{1L}) + \underline{Z}_{1L}\underline{Z}_{1E}$
$\underline{Z}_{1E} \to \infty$ (lack of extra link)	$\underline{K}_1 = -\underline{Z}_{1L}$ $\underline{L}_1 = \underline{Z}_{1L} + \underline{Z}_{1B}$ $\underline{M}_1 = \underline{Z}_{1A} + \underline{Z}_{1B} + \underline{Z}_{1L}$

6.3.2 Transmission Network with Double-circuit Line

Figure 6.5 presents equivalent circuit diagrams of transmission network with a double-circuit line ($\underline{Z}_{LI}$, $\underline{Z}_{LII}$) for determining the fault current distribution factors for the incremental positive-sequence (it is analogous for the negative-sequence). Different modes of buses connection at the sending and receiving ends are reflected with the switches: w_A, w_B (switch status: 0 – sections of buses are separated, 1 – sections of buses are connected). Usually, the sections of buses are connected ($w_A w_B = 11$).

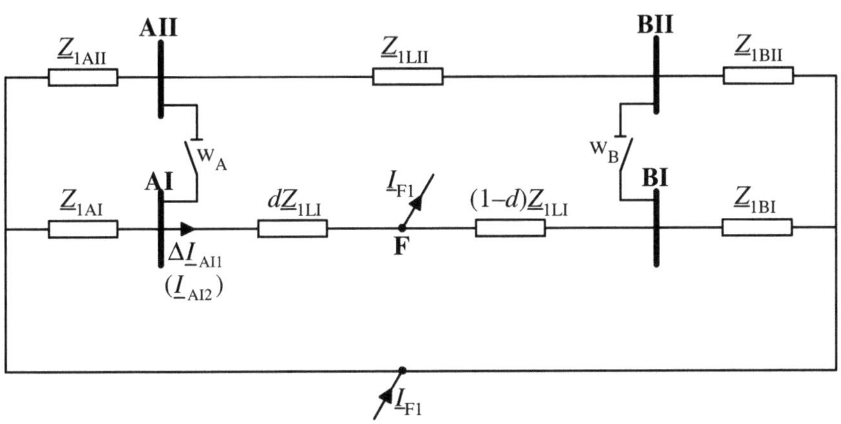

Fig. 6.5 Equivalent circuit diagram of transmission network with double-circuit line ($\underline{Z}_{LI}$, $\underline{Z}_{LII}$) with different modes of buses connection for incremental positive- or negative-sequence

From the analysis of the circuits presented in Fig. 6.5 the fault current distribution factor (6.12) can be derived. In Table 6.2, the coefficients involved in this factor are gathered.

The considerations for the network with a double-circuit line (Fig. 6.5, Table 6.2) have been performed under the assumption that the current from the faulted circuit (connected to the bus AI) only, i.e., $\Delta\underline{I}_{\mathrm{AI1}}$ – Fig. 6.5a, $\underline{I}_{\mathrm{AI2}}$ – Fig. 6.5b, is measured. However, there are some applications [121, 286, 343, 344] in which the currents from both circuits of the parallel lines (connected to the bus AI and to the bus AII) are utilized. Such an arrangement is shown in Fig. 6.6.

Comparing the voltage drops across two different routes in the circuit from Fig. 6.6: Route 1, Route 2, one obtains the following formula for the positive- and negative-sequence component of the total fault current:

$$\underline{I}_{\mathrm{F1}} = \frac{\underline{I}_{\mathrm{AI1}} - \dfrac{\underline{Z}_{\mathrm{1LII}}}{\underline{Z}_{\mathrm{1LI}}}\underline{I}_{\mathrm{AII1}}}{1-d} \tag{6.14}$$

$$\underline{I}_{\mathrm{F2}} = \frac{\underline{I}_{\mathrm{AI2}} - \dfrac{\underline{Z}_{\mathrm{2LII}}}{\underline{Z}_{\mathrm{2LI}}}\underline{I}_{\mathrm{AII2}}}{1-d} \tag{6.15}$$

Note that in (6.15) the negative-sequence impedances can be replaced by the respective positive-sequence quantities since they are identical.

Table 6.2 Transmission network with double-circuit line (Fig. 6.5) – complex coefficients used for determining fault current distribution factor for positive- and negative-sequence (6.12)

Status of switches	Coefficients
$\mathrm{w_A w_B} = 00$	$\underline{K}_1 = -\underline{Z}_{\mathrm{1LI}}$ $\underline{L}_1 = \underline{Z}_{\mathrm{1BI}} + \underline{Z}_{\mathrm{1LI}}$ $\underline{M}_1 = \underline{Z}_{\mathrm{1AI}} + \underline{Z}_{\mathrm{1LI}} + \underline{Z}_{\mathrm{1BI}}$
$\mathrm{w_A w_B} = 01$	$\underline{K}_1 = -\underline{Z}_{\mathrm{1LI}}(\underline{Z}_{\mathrm{1AII}} + \underline{Z}_{\mathrm{1B}} + \underline{Z}_{\mathrm{1LII}})$ $\underline{L}_1 = \underline{Z}_{\mathrm{1LI}}(\underline{Z}_{\mathrm{1LII}} + \underline{Z}_{\mathrm{1SII}} + \underline{Z}_{\mathrm{1B}}) + \underline{Z}_{\mathrm{1LII}}(\underline{Z}_{\mathrm{1LII}} + \underline{Z}_{\mathrm{1AII}})$ $\underline{M}_1 = (\underline{Z}_{\mathrm{1LI}} + \underline{Z}_{\mathrm{1AI}})(\underline{Z}_{\mathrm{1LII}} + \underline{Z}_{\mathrm{1AII}}) + \underline{Z}_{\mathrm{1B}}(\underline{Z}_{\mathrm{1LI}} + \underline{Z}_{\mathrm{1AI}} + \underline{Z}_{\mathrm{1LII}} + \underline{Z}_{\mathrm{1AII}})$
$\mathrm{w_A w_B} = 10$	$\underline{K}_1 = -\underline{Z}_{\mathrm{1LI}}(\underline{Z}_{\mathrm{1A}} + \underline{Z}_{\mathrm{1BII}} + \underline{Z}_{\mathrm{1LII}})$ $\underline{L}_1 = (\underline{Z}_{\mathrm{1BI}} + \underline{Z}_{\mathrm{1LI}})(\underline{Z}_{\mathrm{1LII}} + \underline{Z}_{\mathrm{1A}} + \underline{Z}_{\mathrm{1BII}})$ $\underline{M}_1 = (\underline{Z}_{\mathrm{1LI}} + \underline{Z}_{\mathrm{1BI}})(\underline{Z}_{\mathrm{1A}} + \underline{Z}_{\mathrm{1LII}} + \underline{Z}_{\mathrm{1BII}}) + \underline{Z}_{\mathrm{1A}}(\underline{Z}_{\mathrm{1LII}} + \underline{Z}_{\mathrm{1BII}})$
$\mathrm{w_A w_B} = 11$	$\underline{K}_1 = -\underline{Z}_{\mathrm{1LI}}(\underline{Z}_{\mathrm{1A}} + \underline{Z}_{\mathrm{1B}} + \underline{Z}_{\mathrm{1LII}})$ $\underline{L}_1 = \underline{Z}_{\mathrm{1LI}}(\underline{Z}_{\mathrm{1A}} + \underline{Z}_{\mathrm{1B}} + \underline{Z}_{\mathrm{1LII}}) + \underline{Z}_{\mathrm{1LII}}\underline{Z}_{\mathrm{1B}}$ $\underline{M}_1 = \underline{Z}_{\mathrm{1LI}}\underline{Z}_{\mathrm{1LII}} + (\underline{Z}_{\mathrm{1LI}} + \underline{Z}_{\mathrm{1LII}})(\underline{Z}_{\mathrm{1A}} + \underline{Z}_{\mathrm{1B}})$
Where: $\underline{Z}_{\mathrm{1A}} = \dfrac{\underline{Z}_{\mathrm{1AI}}\underline{Z}_{\mathrm{1AII}}}{\underline{Z}_{\mathrm{1AI}} + \underline{Z}_{\mathrm{1AII}}}$, $\underline{Z}_{\mathrm{1B}} = \dfrac{\underline{Z}_{\mathrm{1BI}}\underline{Z}_{\mathrm{1BII}}}{\underline{Z}_{\mathrm{1BI}} + \underline{Z}_{\mathrm{1BII}}}$.	

For the network with a single line (Fig. 6.4a) and with a double-circuit line, but with the current from one circuit only (Fig. 6.5a), the positive-sequence component of the total fault current was determined using the measured incremental positive-sequence current. Under the availability of measurements such as in Fig. 6.6a, it is possible to determine the positive-sequence component of the total fault current with use of positive-sequence currents. However, it is possible to also use the incremental positive-sequence currents, obtaining the following formula, which is analogous to (6.14):

$$\underline{I}_{F1} = \frac{\Delta\underline{I}_{AI1} - \dfrac{\underline{Z}_{1LII}}{\underline{Z}_{1LI}}\Delta\underline{I}_{AII1}}{1-d} \tag{6.16}$$

The formula (6.16) can be utilized for fault location if the incremental currents can be determined, i.e., the pre-fault currents are registered. Use of (6.16), instead of (6.14), in the fault-location algorithm derivation is advantageous since for the incremental quantities the line shunt capacitances effect (not taken into account here) has less influence on fault-location accuracy, i.e., higher accuracy is achieved.

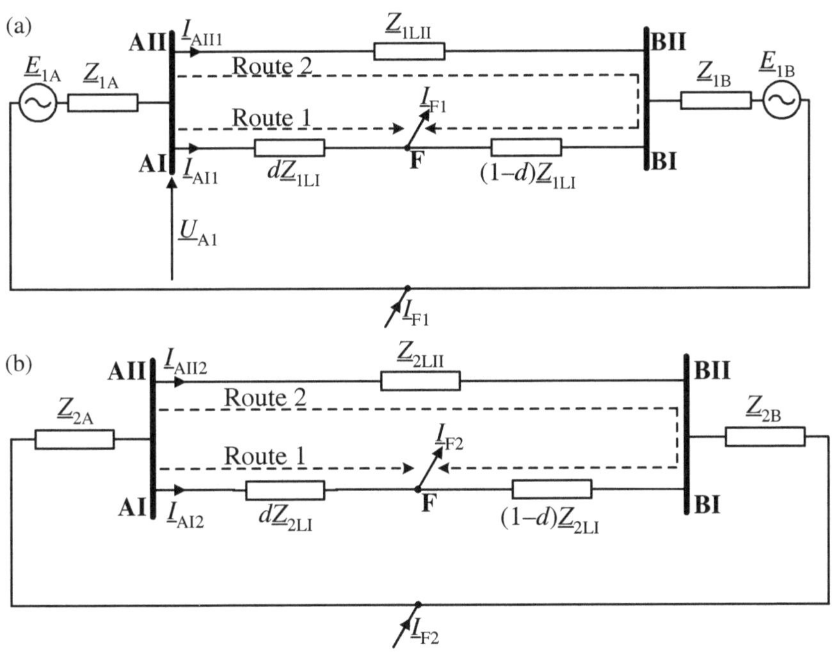

Fig. 6.6 Equivalent circuit diagrams of transmission network with double-circuit line for availability of measurements of currents from both line circuits: (a) for positive-sequence, and (b) for negative-sequence

The advantage of using measurement of currents from both line circuits relies on estimating the total fault current components (6.14)–(6.16) without involving source impedances, as in the case of measuring a current from one circuit only (Table 6.2).

In (6.14)–(6.16), impedances of both line circuits are involved, however, in practice they are identical, i.e.,: $\underline{Z}_{1LI}=\underline{Z}_{1LII}$.

The impedance of an overhead line for the zero-sequence since it is affected by the soil resistivity (difficult to measure and changeable) is considered as an uncertain parameter [62, 106, 109]. Therefore, the fault current distribution factors for the positive-(incremental positive-) and negative-sequence are basically used in fault-location algorithms. However, for example in the case of the complete lack of measurement of current from the healthy parallel line [114], there is a need for also using the fault current distribution factor for the zero-sequence. Equivalent circuits of double-circuit line for the zero-sequence, when both circuits are in operation, are presented in Fig. 2.18. The case where the parallel healthy line (line: LII) is switched off and grounded is presented in Fig. 2.19.

Considering the mesh of the circuit from Fig. 2.18b (*both line circuits in operation*) containing the elements: $\{\underline{Z}_{0LII}-\underline{Z}_{0m}\}$; $\{d(\underline{Z}_{0LI}-\underline{Z}_{0m})\}$; $\{(1-d)(\underline{Z}_{0LI}-\underline{Z}_{0m})\}$ yields the compact formula for the zero-sequence component of the total fault current:

$$\underline{I}_{F0}=\frac{\underline{I}_{AI0}-\underline{P}_0^{'}\underline{I}_{AII0}}{1-d} \tag{6.17}$$

where:

$$\underline{P}_0^{'}=\frac{\underline{Z}_{0LII}-\underline{Z}_{0m}}{\underline{Z}_{0LI}-\underline{Z}_{0m}} \tag{6.18}$$

For the case with the *parallel healthy line switched off and grounded* (Fig. 2.19) the consideration of the healthy line path (thus excluding impedances of the equivalent sources) appears to be advantageous. Taking into account that the sum of voltage drops defined in (2.14)–(2.17), across this path is:

$$\underline{U}_0^{E}+\underline{U}_0^{F}+\underline{U}_0^{G}+\underline{U}_0^{H}=0 \tag{6.19}$$

yields:

$$\underline{I}_{F0}=\frac{\underline{I}_{AI0}-\underline{P}_0^{''}\underline{I}_{AII0}}{1-d} \tag{6.20}$$

where:

$$\underline{P}_0^{''}=-\frac{\underline{Z}_{0LII}}{\underline{Z}_{0m}} \tag{6.21}$$

6.4 Models of Fault Loops

The majority of one-end fault-location algorithms are based on considering the fault loops composed according to the identified fault type, analogously as for the distance relays. The distance protective relay, say at the sending line end A, measures the apparent impedance of the fault loop under consideration:

$$\underline{Z}_{\mathrm{A\_P}} = \frac{\underline{U}_{\mathrm{A\_P}}}{\underline{I}_{\mathrm{A\_P}}} \tag{6.22}$$

where $\underline{U}_{\mathrm{A\_P}}$, $\underline{I}_{\mathrm{A\_P}}$ – protective (therefore the subscript: P) distance relay voltage and current signals, at the line end A (subscript A), which are composed as presented in Tables 6.3–6.6.

For solid faults (fault resistance $R_{\mathrm{F}} = 0$), the apparent impedance (6.22) is equal to the positive-sequence impedance of the line segment of the relative distance d (p.u.), i.e., from the measurement point to the fault. Thus, one obtains:

$$\underline{Z}_{\mathrm{A\_P}} = d\underline{Z}_{\mathrm{1L}} \tag{6.23}$$

Otherwise (for resistive faults), due to 'the reactance effect' the apparent impedance (6.22) is not a strict measure of the distance to fault. As opposed to protective distance relays, the one-end fault-location algorithms compensate for 'the reactance effect' by considering the fault-loop model, in which the term ($R_{\mathrm{F}}\underline{I}_{\mathrm{F}}$) represents the voltage drop across the fault-path resistance:

$$\underline{U}_{\mathrm{A\_P}} - d\underline{Z}_{\mathrm{1L}}\underline{I}_{\mathrm{A\_P}} - R_{\mathrm{F}}\underline{I}_{\mathrm{F}} = 0 \tag{6.24}$$

In Table 6.3, the composition of fault-loop signals (for the terminal A) for a single line is shown. Depending on the fault type, one assumes for the fault-loop voltage and current. Two examples (for a–g, a–b faults) show how these signals are composed.

a–g fault:

- voltage from faulted phase 'a': $\underline{U}_{\mathrm{Aa}}$;
- current from faulted phase 'a': $\underline{I}_{\mathrm{Aa}}$ and additionally the compensation for the zero-sequence component: $\underline{k}_0\underline{I}_{\mathrm{A0}}$.

a–b fault:

- difference of voltages from faulted phases 'a' and 'b': $\underline{U}_{\mathrm{Aa}} - \underline{U}_{\mathrm{Ab}}$;
- difference of currents from faulted phases 'a' and 'b': $\underline{I}_{\mathrm{Aa}} - \underline{I}_{\mathrm{Ab}}$.

Table 6.3 Composition of fault-loop voltage and current signals for single line

Fault type	Fault-loop voltage: $\underline{U}_{A\_P}$	Fault-loop current: $\underline{I}_{A\_P}$
a–g	$\underline{U}_{Aa}$	$\underline{I}_{Aa} + \underline{k}_0 \underline{I}_{A0}$
b–g	$\underline{U}_{Ab}$	$\underline{I}_{Ab} + \underline{k}_0 \underline{I}_{A0}$
c–g	$\underline{U}_{Ac}$	$\underline{I}_{Ac} + \underline{k}_0 \underline{I}_{A0}$
a–b, a–b–g a–b–c*, a–b–c–g*	$\underline{U}_{Aa} - \underline{U}_{Ab}$	$\underline{I}_{Aa} - \underline{I}_{Ab}$
b–c, b–c–g	$\underline{U}_{Ab} - \underline{U}_{Ac}$	$\underline{I}_{Ab} - \underline{I}_{Ac}$
c–a, c–a–g	$\underline{U}_{Ac} - \underline{U}_{Aa}$	$\underline{I}_{Ac} - \underline{I}_{Aa}$

Where: $\underline{k}_0 = \dfrac{\underline{Z}_{0L} - \underline{Z}_{1L}}{\underline{Z}_{1L}}$.

* – inter-phase fault loop (a–b) is considered, however, the other fault loops: (b–c), (c–a) can be taken as well.

In the case of single phase-to-earth faults on a double-circuit line (Table 6.4) additionally the compensation for the mutual coupling of the circuits: $\underline{k}_{0m} \underline{I}_{AII0}$ is included.

Table 6.4 Composition of fault-loop voltage and current signals for phase-to-ground faults on double-circuit line

Fault type	Fault-loop voltage: $\underline{U}_{A\_P}$	Fault-loop current: $\underline{I}_{A\_P}$
a–g	$\underline{U}_{Aa}$	$\underline{I}_{AIa} + \underline{k}_0 \underline{I}_{AI0} + \underline{k}_{0m} \underline{I}_{AII0}$
b–g	$\underline{U}_{Ab}$	$\underline{I}_{AIb} + \underline{k}_0 \underline{I}_{AI0} + \underline{k}_{0m} \underline{I}_{AII0}$
c–g	$\underline{U}_{Ac}$	$\underline{I}_{AIc} + \underline{k}_0 \underline{I}_{AI0} + \underline{k}_{0m} \underline{I}_{AII0}$

Where: $\underline{k}_0 = \dfrac{\underline{Z}_{0LI} - \underline{Z}_{1LI}}{\underline{Z}_{1LI}}$, $\underline{k}_{0m} = \dfrac{\underline{Z}_{0m}}{\underline{Z}_{1LI}}$.

For the remaining fault types the composition of fault-loop signals is analogous to the single-line case (Table 6.3).

In turn, in Tables 6.5 and 6.6, the fault-loop signals are expressed in terms of the respective symmetrical components of the measured voltages and currents (the last position of the subscript denotes the symmetrical component type):

$$\underline{U}_{A\_P} = \underline{a}_1 \underline{U}_{A1} + \underline{a}_2 \underline{U}_{A2} + \underline{a}_0 \underline{U}_{A0} \tag{6.25}$$

$$\underline{I}_{A\_P} = \underline{a}_1 \underline{I}_{A1} + \underline{a}_2 \underline{I}_{A2} + \underline{a}_0 \frac{\underline{Z}_{0L}}{\underline{Z}_{1L}} \underline{I}_{A0} \tag{6.26}$$

In the case of a double-circuit line the fault-loop current takes the following form:

$$\underline{I}_{\text{A\_P}} = \underline{a}_1 \underline{I}_{\text{AI1}} + \underline{a}_2 \underline{I}_{\text{AI2}} + \underline{a}_0 \left(\frac{\underline{Z}_{\text{0LI}}}{\underline{Z}_{\text{1LI}}} \underline{I}_{\text{AI0}} + \frac{\underline{Z}_{\text{0m}}}{\underline{Z}_{\text{1LI}}} \underline{I}_{\text{AII0}} \right) \tag{6.27}$$

where $\underline{a}_1$, $\underline{a}_2$, $\underline{a}_0$ – coefficients dependent on fault type, which are the same as for the single line.

The circuit diagrams with the indicated symmetrical component signals used in (6.25)–(6.27) are shown in Fig. 2.15 (single line) and in Figs. 2.17 and 2.18 (double-circuit line).

Table 6.5 Single-line circuit – composition of fault-loop voltage and current signals in terms of symmetrical components

Fault type	**Fault-loop voltage:** $\underline{U}_{\text{A\_P}} = \underline{a}_1 \underline{U}_{\text{A1}} + \underline{a}_2 \underline{U}_{\text{A2}} + \underline{a}_0 \underline{U}_{\text{A0}}$ **Fault-loop current:** $\underline{I}_{\text{A\_P}} = \underline{a}_1 \underline{I}_{\text{A1}} + \underline{a}_2 \underline{I}_{\text{A2}} + \underline{a}_0 \underline{I}_{\text{A0}}$		
	$\underline{a}_1$	$\underline{a}_2$	$\underline{a}_0$
a–g	1	1	1
b–g	$-0.5-\text{j}0.5\sqrt{3}$	$-0.5+\text{j}0.5\sqrt{3}$	1
c–g	$-0.5+\text{j}0.5\sqrt{3}$	$-0.5-\text{j}0.5\sqrt{3}$	1
a–b a–b–g a–b–c* a–b–c–g*	$1.5+\text{j}0.5\sqrt{3}$	$1.5-\text{j}0.5\sqrt{3}$	0
b–c b–c–g	$-\text{j}\sqrt{3}$	$\text{j}\sqrt{3}$	0
c–a c–a–g	$-1.5+\text{j}0.5\sqrt{3}$	$-1.5-\text{j}0.5\sqrt{3}$	0
* – inter-phase fault loop (a–b) is considered, however, the other fault loops: (b–c) and (c–a) can be taken as well.			

The notation used in (6.25)–(6.27) appears convenient if the compensation for line shunt capacitances is performed [121]. Of course, this notation is fully equivalent to the description traditionally used for distance protective relays (Tables 6.3 and 6.4).

Table 6.6 Double-circuit line – composition of the fault-loop voltage and current signals for phase-to-ground faults, in terms of symmetrical components

Fault type	Fault-loop voltage: $\underline{U}_{A\_P}=\underline{a}_1\underline{U}_{A1}+\underline{a}_2\underline{U}_{A2}+\underline{a}_0\underline{U}_{A0}$ Fault-loop current: $\underline{I}_{A\_P}=\underline{a}_1\underline{I}_{AI1}+\underline{a}_2\underline{I}_{AI2}+\underline{a}_0\left(\frac{\underline{Z}_{0LI}}{\underline{Z}_{1LI}}\underline{I}_{AI0}+\frac{\underline{Z}_{0m}}{\underline{Z}_{1LI}}\underline{I}_{AII0}\right)$		
	$\underline{a}_1$	$\underline{a}_2$	$\underline{a}_0$
a–g	1	1	1
b–g	$-0.5-\mathrm{j}0.5\sqrt{3}$	$-0.5+\mathrm{j}0.5\sqrt{3}$	1
c–g	$-0.5+\mathrm{j}0.5\sqrt{3}$	$-0.5-\mathrm{j}0.5\sqrt{3}$	1
For the remaining fault types the composition of fault-loop signals is analogous to the single-line case (as in Table 6.5).			

6.5 Fault Location with Use of Fault Current Distribution Factors

One of the earliest fault-location algorithms was presented in [309]. Its form for a single-phase line was presented by the formula (6.8). An extension to the three-phase application can be performed by utilizing the general fault-loop model (6.24) and the general formula for a total fault current (3.1). Combining both of them gives:

$$\underline{U}_{A\_P}-d\underline{Z}_{1L}\underline{I}_{A\_P}-R_F(\underline{a}_{F1}\underline{I}_{F1}+\underline{a}_{F2}\underline{I}_{F2}+\underline{a}_{F0}\underline{I}_{F0})=0 \qquad (6.28)$$

Taking into account such a set of weighting coefficients that for the zero-sequence: $\underline{a}_{F0}=0$ (Tables 3.3, 3.4 or 3.5) and expressing the symmetrical components of the total fault current with use of the fault current distribution factors (6.9) and (6.10) one obtains:

$$\underline{U}_{A\_P}-d\underline{Z}_{1L}\underline{I}_{A\_P}-R_F\left(\underline{a}_{F1}\frac{\Delta\underline{I}_{A1}}{\underline{k}_{F1}}+\underline{a}_{F2}\frac{\underline{I}_{A2}}{\underline{k}_{F2}}\right)=0 \qquad (6.29)$$

Considering that for the fault current distribution factors for the positive- and negative-sequence, with respect to their magnitude and angle, we have:

$$|\underline{k}_F|=|\underline{k}_{F1}|=|\underline{k}_{F2}| \qquad (6.30)$$

$$\gamma=\mathrm{angle}(\underline{k}_{F1})=\mathrm{angle}(\underline{k}_{F2}) \qquad (6.31)$$

the formula (6.29) transforms to:

$$\underline{U}_{\text{A\_P}} - d\underline{Z}_{1\text{L}}\underline{I}_{\text{A\_P}} - \frac{R_{\text{F}}}{|\underline{k}_{\text{F}}|\,\text{e}^{\text{j}\gamma}}(\underline{a}_{\text{F1}}\Delta\underline{I}_{\text{A1}} + \underline{a}_{\text{F2}}\underline{I}_{\text{A2}}) = 0 \tag{6.32}$$

Multiplying (6.32) by the term: $(\text{e}^{\text{j}\gamma}(\underline{a}_{\text{F1}}\Delta\underline{I}_{\text{A1}} + \underline{a}_{\text{F2}}\underline{I}_{\text{A2}})^*)$ and then rearranging, the resultant formula for the sought distance to fault (d (p.u.)) is obtained as follows:

$$d = \frac{\text{Im}(\underline{U}_{\text{A\_P}}(\underline{a}_{\text{F1}}\Delta\underline{I}_{\text{A1}} + \underline{a}_{\text{F2}}\underline{I}_{\text{A2}})^* \,\text{e}^{\text{j}\gamma})}{\text{Im}(\underline{Z}_{1\text{L}}\underline{I}_{\text{A\_P}}(\underline{a}_{\text{F1}}\Delta\underline{I}_{\text{A1}} + \underline{a}_{\text{F2}}\underline{I}_{\text{A2}})^* \,\text{e}^{\text{j}\gamma})} \tag{6.33}$$

where x^* denotes the conjugate of x.

The signals that are processed in the fault-location algorithm (6.33) are determined from measurements acquired at one line terminal (here, at the terminal A). Table 6.7 shows how to set the coefficients involved in the algorithm.

In formula (6.33), the angle of the current distribution factor (for the positive- or negative-sequence) is involved. In [309] it is proposed to assume that this angle equals zero ($\gamma = 0$), i.e., that the fault current distribution factor is a real number. In practice, this assumption is not completely fulfilled and thus there is a certain error due to this. However, in the case of a high-voltage network these additional errors are not substantial, as reported in [106, 330].

Table 6.7 Description of signals and coefficients of the fault-location algorithm (6.33)

Signals	$\underline{U}_{\text{A\_P}}$ – fault-loop voltage	Formula (6.25), and Table 6.3 or Table 6.5
	$\underline{I}_{\text{A\_P}}$ – fault-loop current	*Single line*: formula (6.26) and Table 6.3 or Table 6.5 *Double-circuit line*: formula (6.27) and Table 6.4 or Table 6.6
	$\Delta\underline{I}_{\text{A1}}$ – incremental positive-sequence current	*Single line*: $\Delta\underline{I}_{\text{A1}}$ in Fig. 6.4a *Double-circuit line*: $\Delta\underline{I}_{\text{AI1}}$ in Fig. 6.5
	$\underline{I}_{\text{A2}}$ – negative-sequence current	*Single line*: $\underline{I}_{\text{A2}}$ in Fig. 6.4b *Double-circuit line*: $\underline{I}_{\text{AI2}}$ in Fig. 6.5
Coefficients	$\underline{a}_{\text{F1}}$, $\underline{a}_{\text{F2}}$ – weighting coefficients	Tables 3.3–3.5, depending on the assumed preference with respect to using the respective sequences

6.6 Fault Location Related to Distance-protection Measurements

The other fundamental fault-location algorithm has been developed by Wiszniewski [330], which is somewhat similar to the algorithm from [309]. However, it is more related to the distance-protection technique. This is so since impedance measured by a distance relay is involved in the algorithm. The algorithm from [330] was derived utilizing the general fault-loop model (6.24) and the general formula for a total fault current (3.1). The same assumptions, as in [309] for obtaining formula (6.32), were assumed. Dividing (6.32) by the fault-loop current $\underline{I}_{\mathrm{A\_P}}$ yields:

$$\underline{Z}_{\mathrm{A\_P}} - d\underline{Z}_{\mathrm{1L}} - \frac{R_{\mathrm{F}}}{|\underline{k}_{\mathrm{F}}|} \frac{(\underline{a}_{\mathrm{F1}} \Delta \underline{I}_{\mathrm{A1}} + \underline{a}_{\mathrm{F2}} \underline{I}_{\mathrm{A2}})}{\underline{I}_{\mathrm{A\_P}} \mathrm{e}^{\mathrm{j}\gamma}} = 0 \tag{6.34}$$

where:

$\underline{Z}_{\mathrm{A\_P}} = \dfrac{\underline{U}_{\mathrm{A\_P}}}{\underline{I}_{\mathrm{A\_P}}} = R_{\mathrm{A\_P}} + \mathrm{j}X_{\mathrm{A\_P}}$ – fault-loop impedance (apparent impedance measured by a distance relay).

Resolving (6.34) into the real and imaginary parts results in:

$$R_{\mathrm{A\_P}} - dR_{\mathrm{1L}} - \frac{R_{\mathrm{F}}}{|\underline{k}_{\mathrm{F}}|} a = 0 \tag{6.35}$$

$$X_{\mathrm{A\_P}} - dX_{\mathrm{1L}} - \frac{R_{\mathrm{F}}}{|\underline{k}_{\mathrm{F}}|} b = 0 \tag{6.36}$$

where:

$$a = \mathrm{Re}\left(\frac{\underline{a}_{\mathrm{F1}} \Delta \underline{I}_{\mathrm{A1}} + \underline{a}_{\mathrm{F2}} \underline{I}_{\mathrm{A2}}}{\underline{I}_{\mathrm{A\_P}} \mathrm{e}^{\mathrm{j}\gamma}} \right) \tag{6.37}$$

$$b = \mathrm{Im}\left(\frac{\underline{a}_{\mathrm{F1}} \Delta \underline{I}_{\mathrm{A1}} + \underline{a}_{\mathrm{F2}} \underline{I}_{\mathrm{A2}}}{\underline{I}_{\mathrm{A\_P}} \mathrm{e}^{\mathrm{j}\gamma}} \right) \tag{6.38}$$

R_{1L}, X_{1L} – positive-sequence line resistance and reactance, respectively.

One of the possible forms of the solution of the set of two equations (6.35) and (6.36) is as follows [330]:

$$d = \frac{X_{\text{A\_P}}}{X_{\text{1L}}} - \frac{\frac{R_{\text{A\_P}}}{X_{\text{1L}}} \text{tg}(\varphi_{\text{1L}}) - \frac{X_{\text{A\_P}}}{X_{\text{1L}}}}{\frac{a}{b} \text{tg}(\varphi_{\text{1L}}) - 1} \tag{6.39}$$

where $\varphi_{\text{1L}} = \text{angle}(R_{\text{1L}} + \text{j}X_{\text{1L}})$ – angle of the positive-sequence line impedance.

The signals involved in the fault-location algorithm (6.39) are determined from measurements acquired at one line terminal (here, at terminal A) and the coefficients are identical with those used in the algorithm by Takagi *et al.* (Table 6.7).

The formula (6.39) allows for a simple interpretation of the reactance effect. The remarks concerning a need for making an assumption with respect to the angle of the fault current distribution factor are identical with those written in Sect. 6.5.

6.7 Fault Location with Use of Impedance Data of Network

An accurate one-end fault-location algorithm has been introduced by Eriksson *et al.* [62]. The fault-location algorithm uses novel compensation techniques for the errors resulting from remote source infeed into the fault resistance. This is accomplished by using a complete network model. The actual distribution of a fault current in the transmission network is taken into account. Compensation for fault resistance voltage drop utilizes pre-fault current and representative values for source resistance and reactance. This inclusion of source impedances is a novel approach.

In [62], it introduced into (6.34) the following form for the fault current distribution factor for the positive- (negative-) sequence:

$$\underline{k}_{\text{F}} = |\underline{k}_{\text{F}}| \text{e}^{\text{j}\gamma} = \frac{\underline{K}_1 d + \underline{L}_1}{\underline{M}_1} \tag{6.40}$$

where $\underline{K}_1$, $\underline{L}_1$, $\underline{M}_1$ – coefficients determined with use of impedances of the transmission network: Table 6.1 (network with a single line) and Table 6.2 (network with a double-circuit line).

Substituting (6.40) into (6.34) yields:

$$\underline{Z}_{\text{A\_P}} - d\underline{Z}_{\text{1L}} - R_{\text{F}} \frac{\underline{M}_1 (\underline{a}_{\text{F1}} \Delta \underline{I}_{\text{A1}} + \underline{a}_{\text{F2}} \underline{I}_{\text{A2}})}{(\underline{K}_1 d + \underline{L}_1) \underline{I}_{\text{A\_P}}} = 0 \tag{6.41}$$

where $\underline{Z}_{\text{A\_P}}$ – fault-loop impedance (which is measured by a distance relay).

One of the pioneering approaches in [62] is the extraction of the zero-sequence current from the fault current, since the zero-sequence distribution factor is not reliably known as positive- and negative-sequence distribution factors.

After performing the relevant rearrangements of (6.41) one gets:

$$\underline{K}_1\underline{Z}_{1L}d^2+(\underline{L}_1\underline{Z}_{1L}-\underline{K}_1\underline{Z}_{A\_P})d-\underline{L}_1\underline{Z}_{A\_P}+R_F\frac{(\underline{a}_{F1}\Delta\underline{I}_{A1}+\underline{a}_{F2}\underline{I}_{A2})\underline{M}_1}{\underline{I}_{A\_P}}=0 \tag{6.42}$$

Equation (6.42) can be rewritten to the following compact formula for complex numbers, with two unknowns: d (p.u.) – distance to fault, R_F – fault resistance:

$$\underline{A}_2d^2+\underline{A}_1d+\underline{A}_0+\underline{A}_{00}R_F=0 \tag{6.43}$$

where:

$\underline{A}_2=\underline{K}_1\underline{Z}_{1L}$,

$\underline{A}_1=\underline{L}_1\underline{Z}_{1L}-\underline{K}_1\underline{Z}_{A\_P}$

$\underline{A}_0=-\underline{L}_1\underline{Z}_{A\_P}$,

$\underline{A}_{00}=\dfrac{(\underline{a}_{F1}\Delta\underline{I}_{A1}+\underline{a}_{F2}\underline{I}_{A2})\underline{M}_1}{\underline{I}_{A\_P}}$

The complex coefficients $\underline{K}_1$, $\underline{L}_1$, $\underline{M}_1$ involved in (6.43) are gathered in the following tables:

- single line – Table 6.1; and
- double-circuit line – Table 6.2.

The formula (6.43) can be written down separately for the real and imaginary parts. Combining them in such a way that fault resistance is eliminated yields the quadratic formula for a sought distance to fault:

$$B_2d^2+B_1d+B_0=0 \tag{6.44}$$

where:

$B_2=\text{Re}(\underline{A}_2)\,\text{Im}(\underline{A}_{00})-\text{Im}(\underline{A}_2)\,\text{Re}(\underline{A}_{00})$,

$B_1=\text{Re}(\underline{A}_1)\,\text{Im}(\underline{A}_{00})-\text{Im}(\underline{A}_1)\,\text{Re}(\underline{A}_{00})$,

$B_0=\text{Re}(\underline{A}_0)\,\text{Im}(\underline{A}_{00})-\text{Im}(\underline{A}_0)\,\text{Re}(\underline{A}_{00})$.

There are two solutions of (6.44):

$$d_1=\frac{-B_1-\sqrt{B_1^2-4B_2B_0}}{2B_2} \tag{6.45}$$

$$d_2 = \frac{-B_1 + \sqrt{B_1^2 - 4B_2B_0}}{2B_2} \tag{6.46}$$

of which only one determines the real distance to fault (d), while the second solution usually lies outside the line range, i.e., outside the range: 0 to 1 (p.u.).

For the specific fault cases it may happen that both solutions (6.45) and (6.46) indicate the fault as occurring within the line range. In such rare cases one has to check the sign of the fault resistances R_{F1}, R_{F2}, which correspond to the calculated values of distance to fault: d_1, d_2. In a natural way, the solution that results in negative fault resistance has to be rejected. Fault resistance can be calculated by taking the real or imaginary part of the formula (6.43). With the real part, one obtains:

$$R_{F1} = \frac{-\mathrm{Re}(\underline{A}_2)d_1^2 - \mathrm{Re}(\underline{A}_1)d_1 - \mathrm{Re}(\underline{A}_0)}{\mathrm{Re}(\underline{A}_{00})} \tag{6.47}$$

$$R_{F2} = \frac{-\mathrm{Re}(\underline{A}_2)d_2^2 - \mathrm{Re}(\underline{A}_1)d_2 - \mathrm{Re}(\underline{A}_0)}{\mathrm{Re}(\underline{A}_{00})} \tag{6.48}$$

It has been suggested in [52] that selection of the valid solution for a distance to fault can also be performed by analyzing the relation between the symmetrical components of the total fault current, such as presented in (3.3) and Table 3.7.

The signals involved in (6.39) are determined from measurements performed at one line terminal (here, at terminal A) and the coefficients are identical to those of the algorithms from [309, 330]. It is important that additionally the positive-sequence source impedances (for the local source $\underline{Z}_{1A}$ and for the remote source $\underline{Z}_{1B}$) are involved in the complex coefficients $\underline{K}_1$, $\underline{L}_1$, $\underline{M}_1$ (see (6.43)), and thus required for solving the resultant quadratic formula (6.44).

The local source impedance can be determined from the incremental positive-sequence voltage and current:

$$\underline{Z}_{1A} = -\frac{\Delta\underline{U}_{A1}}{\Delta\underline{I}_{A1}} \tag{6.48}$$

or for all the faults, except the three-phase balanced ones, from the negative-sequence quantities:

$$\underline{Z}_{2A} = -\frac{\underline{U}_{A2}}{\underline{I}_{A2}} \tag{6.49}$$

The remote source impedance $\underline{Z}_{1B}$ as shown in Table 6.1 is the other parameter required by the fault-location formula (6.44). This impedance can be determined on condition that the exact network topology and parameters are known. Otherwise, its 'representative' value has to be provided as the input data [62]. Some mismatch between the actual impedance and the provided 'representative' can appear. However, in strong meshed modern networks the equivalent system configuration is fixed [233] and the expected mismatch is rather not too high. Moreover, it follows from the many years' of operational experiences [62] in more than 100 countries around the world that the mismatch causes no problem. So, the need for providing the remote source impedance ($\underline{Z}_{1B}$) cannot be considered as the algorithm limitation. On the contrary, the representative values of remote source impedance provide improved accuracy.

On the other hand, there is a possibility [93] for improving fault-location accuracy by sending the source impedance, which can be measured at the remote end devices (for example, in the recording device RD), with use of simple and even slow communication means (Fig. 6.7).

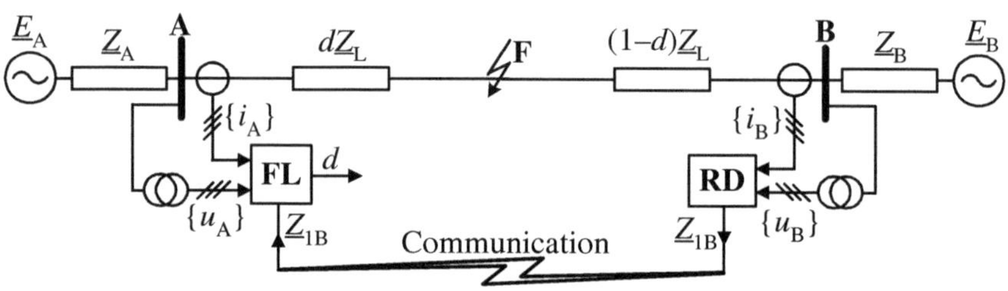

Fig. 6.7 Improving of the fault-location accuracy by measurement of remote source impedance ($\underline{Z}_{1B}$) in a recording device (RD) and use of communication for sending it to a fault locator (FL)

Example 6.1 Fault location according to the fault-location algorithm from [62]
Figure 6.8 presents an example of fault location on a single 400-kV, 300-km line for the following fault specifications:

- fault type: a–g;
- fault resistance: 10 Ω;
- fault location: 0.7 p.u.

The estimated fault distance (averaged over the time interval lasting from 30 ms up to 50 ms after the fault inception) equals 0.7079 p.u., and thus the error is 0.8%. The accuracy obtained can be improved by introducing the compensation for the line shunt capacitances (Sect. 6.15).

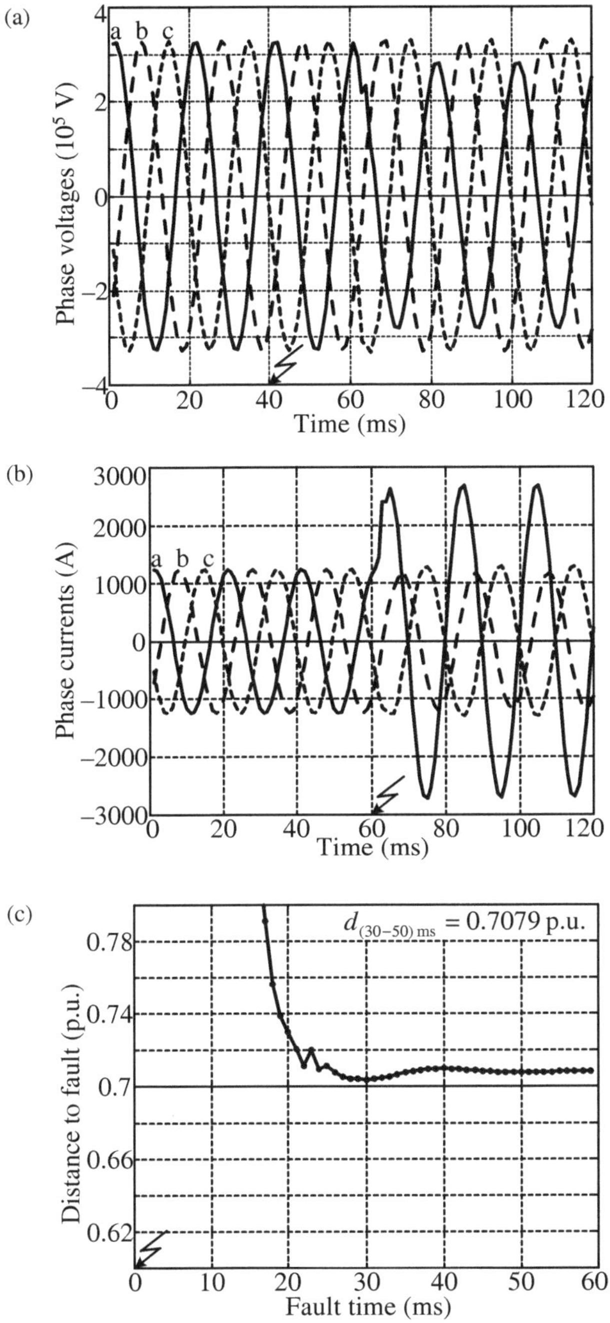

Fig. 6.8 Example 6.1 – (a) phase voltages, (b) phase currents, (c) estimated distance to fault, and (d) estimated fault resistance

(d)

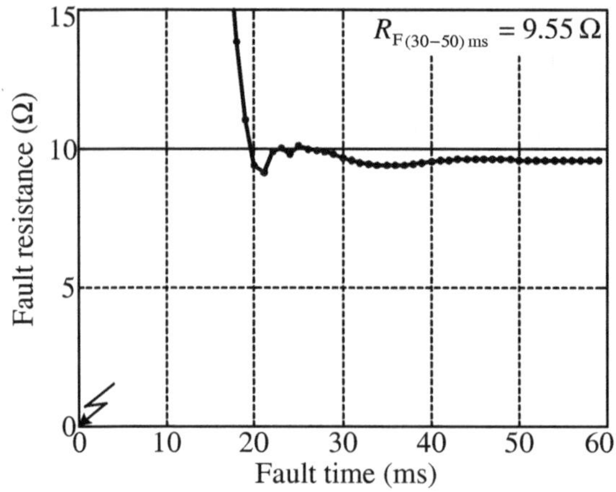

Fig. 6.8 *(continued)*

6.8 Fault Location on Double-circuit Line Using Complete Measurements at One End

A schematic diagram of fault location (FL) on a double-circuit line and distance protection (DP) with measurements of three-phase voltage $\{u_A\}$ and three-phase current from both faulted $\{i_{AI}\}$ and healthy $\{i_{AII}\}$ lines is shown in Fig. 6.9 [121, 286]. There is greater availability of measured signals in comparison to the case when only the zero-sequence current is measured in the healthy parallel line.

The fault current distribution factors for the considered availability of measurements were determined on the basis of the equivalent circuit diagrams from Fig. 6.6 and presented by the formulae (6.14)–(6.16). Considering the fault model (3.1), with zero-sequence component being eliminated ($\underline{a}_{F0}$=0), as in Tables 3.3–3.5, and substituting (6.14) and (6.15) into the fault-loop model (6.24) yields:

$$\underline{U}_{A\_P} - d\underline{Z}_{1LI}\underline{I}_{A\_P} - \frac{R_F}{1-d}\underline{N}_{12} = 0 \tag{6.50}$$

where:

$$\underline{N}_{12} = \underline{a}_{F1}\left(\underline{I}_{AI1} - \frac{\underline{Z}_{1LII}}{\underline{Z}_{1LI}}\underline{I}_{AII1}\right) + \underline{a}_{F2}\left(\underline{I}_{AI2} - \frac{\underline{Z}_{2LII}}{\underline{Z}_{2LI}}\underline{I}_{AII2}\right).$$

Resolving (6.50) into the real and imaginary parts and then eliminating the component ($R_F/(1-d)$) yields the following formula for a sought distance to fault:

$$d = \frac{\mathrm{Im}(\underline{U}_{A\_P})\mathrm{Re}(\underline{N}_{12}) - \mathrm{Re}(\underline{U}_{A\_P})\mathrm{Im}(\underline{N}_{12})}{\mathrm{Im}(\underline{Z}_{1LI}\underline{I}_{A\_P})\mathrm{Re}(\underline{N}_{12}) - \mathrm{Re}(\underline{Z}_{1LI}\underline{I}_{A\_P})\mathrm{Im}(\underline{N}_{12})} \tag{6.51}$$

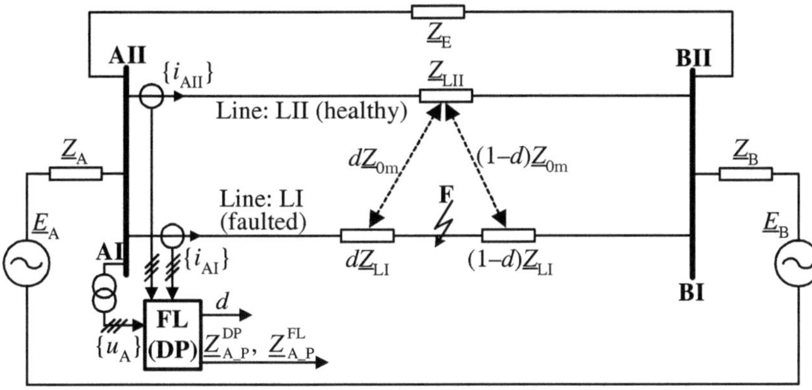

Fig. 6.9 Schematic diagram of fault location (FL) on a double-circuit line and distance protection (DP) with complete measurements at one end

The fault-location formula (6.51) can be presented in an even more compact alternative form:

$$d = \frac{\mathrm{Im}(\underline{U}_{\mathrm{A\_P}}\underline{N}_{12}^{*})}{\mathrm{Im}(\underline{Z}_{1\mathrm{LI}}\underline{I}_{\mathrm{A\_P}}\underline{N}_{12}^{*})} \tag{6.52}$$

where $\underline{N}_{12}^{*}$ – conjugate of $\underline{N}_{12}$, which is defined in (6.50).

Obtaining such compact formulae (as (6.51) or (6.52)) appears to be very attractive for application to adaptive distance protection. Neither impedances of the equivalent sources need be known nor are the pre-fault measurements required.

In Table 6.8 the signals and coefficients involved in (6.51) or (6.52) are specified. This facilitates writing down the detailed form of the considered algorithm.

Table 6.8 Description of signals and coefficients of the fault-location algorithm (6.51) or (6.52)

Signals	$\underline{U}_{\mathrm{A\_P}}$ – fault-loop voltage	Formula (6.25), and Table 6.3 or Table 6.5
	$\underline{I}_{\mathrm{A\_P}}$ – fault-loop current	Formula (6.27), and Table 6.4 or Table 6.6
	$\underline{I}_{\mathrm{AI1}}$ – positive-sequence current from faulted line	Fig. 6.6a
	$\underline{I}_{\mathrm{AII1}}$ – positive-sequence current from healthy line	Fig. 6.6a
	$\underline{I}_{\mathrm{AI2}}$ – negative-sequence current from faulted line	Fig. 6.6b
	$\underline{I}_{\mathrm{AII2}}$ – negative-sequence current from healthy line	Fig. 6.6b
Coefficients	$\underline{a}_{\mathrm{F1}}$, $\underline{a}_{\mathrm{F2}}$ – weighting coefficients	Tables 3.3–3.5, depending on the assumed preference for using the respective sequence components

It is worthwhile to note that the classic distance protection (the superscript DP is used) determines the fault-loop impedance from the fault-loop signals (6.24) and (6.26):

$$\underline{Z}_{\mathrm{A\_P}}^{\mathrm{DP}} = R_{\mathrm{A\_P}}^{\mathrm{DP}} + \mathrm{j}X_{\mathrm{A\_P}}^{\mathrm{DP}} = \frac{\underline{U}_{\mathrm{A\_P}}}{\underline{I}_{\mathrm{A\_P}}} \tag{6.53}$$

The impedance measurement (6.53) is affected by the *reactance effect*, relevant for resistive faults, and presence of pre-fault power flow. In consequence, the quality of protection can be adversely influenced. However, the fault-loop impedance measurement can be accomplished with the fault-location algorithm derived according to:

$$\underline{Z}_{\mathrm{A\_P}}^{\mathrm{FL}} = R_{\mathrm{A\_P}}^{\mathrm{FL}} + \mathrm{j}X_{\mathrm{A\_P}}^{\mathrm{FL}} = \frac{\mathrm{Im}(\underline{U}_{\mathrm{A\_P}}\underline{N}_{12}^{*})}{\mathrm{Im}(\underline{Z}_{\mathrm{1LI}}\underline{I}_{\mathrm{A\_P}}\underline{N}_{12}^{*})}\underline{Z}_{\mathrm{1LI}} \tag{6.54}$$

Examples 6.2 and 6.3 (Figs. 6.10 and 6.11) present the effectiveness of the fault-location algorithm utilizing complete measurements at one end of a double-circuit line.

Example 6.2 Fault location on 400-kV 300-km double-circuit line applying (6.51) The main specifications of the fault conditions are as follows – fault type: a–g, fault resistance: 15 Ω, fault location: 0.8 p.u., pre-fault power flow: from the bus B to A ($\underline{E}_{\mathrm{B}}$ leads $\underline{E}_{\mathrm{A}}$ by 35°).

(a)

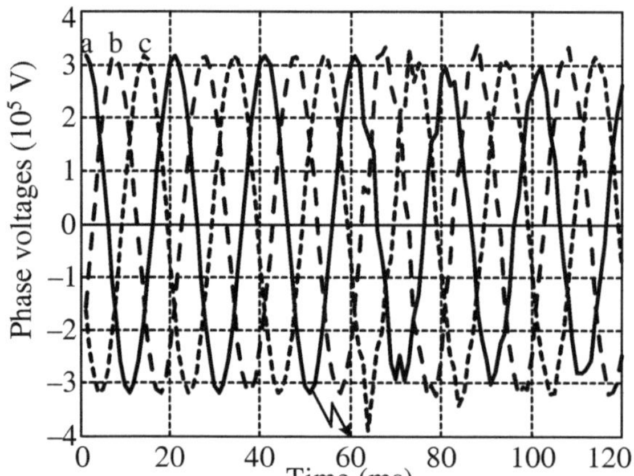

Fig. 6.10 Example 6.2 – (a) three-phase voltage, (b) faulted-line three-phase current, (c) healthy-line three-phase current, and (d) estimated distance to fault

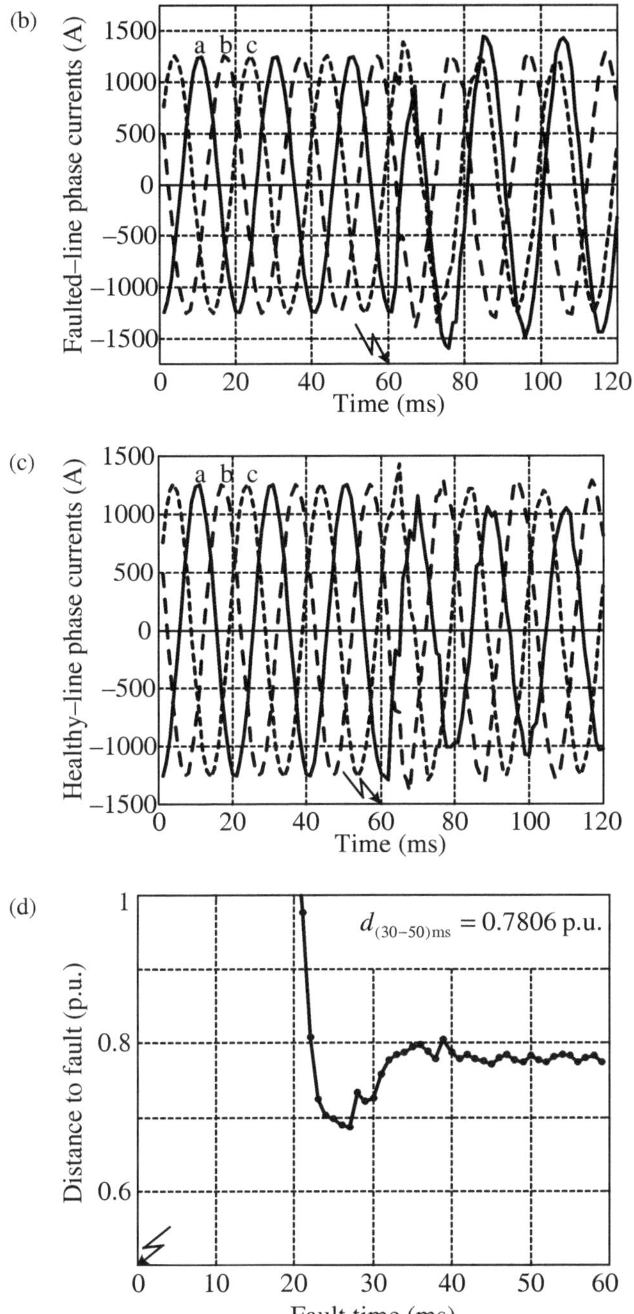

Fig. 6.10 (*continued*)

The estimated value of the distance to fault (Fig. 6.10d) is: d = 0.7806 p.u. (note: d_{actual} = 0.8 p.u.), and thus around 2% error is obtained. This result can be improved by compensating for the line shunt capacitances (Sect. 6.15).

Example 6.3 Comparison of fault-loop impedance measurements
Figure 6.11 presents the fault-loop resistance and reactance measured in two ways:

- classic distance protection principle (6.53) – resistance $R^{\text{DP}}_{\text{A\_P}}$, reactance $X^{\text{DP}}_{\text{A\_P}}$;
- fault-location-based method (6.54) – resistance $R^{\text{FL}}_{\text{A\_P}}$, reactance $X^{\text{FL}}_{\text{A\_P}}$.

The impedance measurement of the classic distance protection ($R^{\text{DP}}_{\text{A\_P}}$, $X^{\text{DP}}_{\text{A\_P}}$) departs greatly from the real values (dR_{1LI}, dX_{1LI}). By contrast, using the fault-location algorithm presented, one obtains the fault-loop impedance components ($R^{\text{FL}}_{\text{A\_P}}$, $X^{\text{FL}}_{\text{A\_P}}$), which coincide with the real values of resistance and reactance of the line segment from the measuring point (A) to the fault place (F). This indicates that the distance protection for a double-circuit power line, which is immune to 'the reactance effect', can be readily achieved using complete measurements from one line end. It is worth mentioning that this applies only if both lines are operating on common buses at both line ends. However, this is the basic configuration of double-circuit lines.

(a)

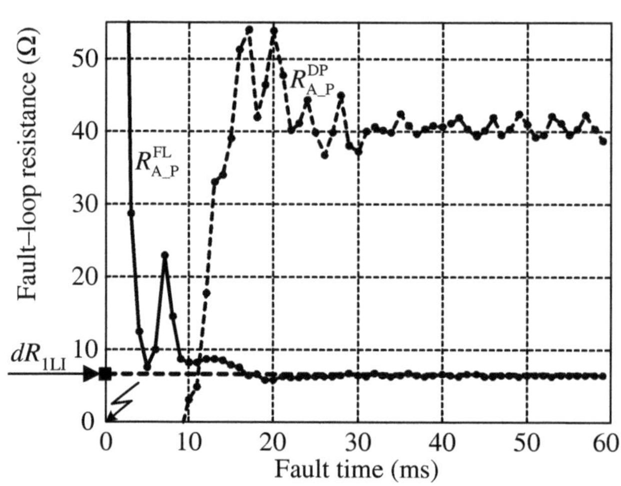

Fig. 6.11 Example 6.3 – fault location on a double-circuit transmission line, with the measurement of the fault-loop being performed according to the classic distance protection (6.53) and fault-location-based (6.54) principles: (a) resistance, and (b) reactance

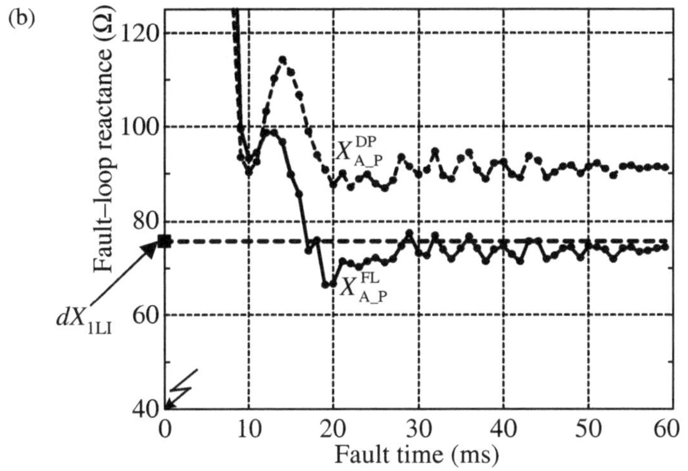

Fig. 6.11 (*continued*)

6.9 Fault Location on Double-circuit Line with Limited Measurements at One End

Figure 6.12 presents a schematic diagram of fault location on a double-circuit line with measurement of three-phase voltage $\{u_A\}$ and current $\{i_{AI}\}$ from a faulted line only [114]. It is considered that the healthy parallel line is *in operation* or *switched off and grounded at both ends.*

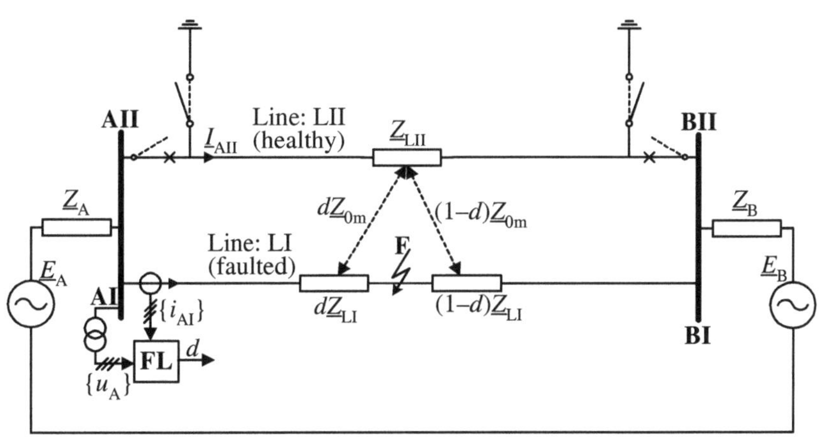

Fig. 6.12 Schematic diagram of fault location on a double-circuit line with measurements of three-phase voltage and current from the faulted line only, when the healthy parallel line is *in operation* (status of switches marked with the solid line) or *switched off and grounded* (status of switches marked with the dotted line)

Taking the fault-loop model (6.23) and expressing the total fault current with use of the symmetrical components of this current and the fault current distribution factors (6.9)–(6.11) one obtains:

$$\underline{U}_{\text{A\_P}} - d\underline{Z}_{\text{1LI}}\underline{I}_{\text{A\_P}} - R_{\text{F}}\left(\frac{\underline{a}_{\text{F1}}\Delta\underline{I}_{\text{AI1}}}{\underline{k}_{\text{F1}}} + \frac{\underline{a}_{\text{F2}}\underline{I}_{\text{AI2}}}{\underline{k}_{\text{F2}}} + \frac{\underline{a}_{\text{F0}}\underline{I}_{\text{AI0}}}{\underline{k}_{\text{F0}}}\right) = 0 \tag{6.55}$$

In (6.55), the fault-loop voltage ($\underline{U}_{\text{A\_P}}$) is determined by (6.24), while the fault-loop current can be expressed as:

$$\underline{I}_{\text{A\_P}} = \underline{I}_{\text{A\_P}}^{\text{SL}} + \underline{a}_0 \frac{\underline{Z}_{\text{0m}}}{\underline{Z}_{\text{1LI}}} \underline{I}_{\text{AII0}} \tag{6.56}$$

where the first component represents the fault-loop current without taking into account the mutual coupling effect, i.e., as applied for the single line (hence the superscript SL is used):

$$\underline{I}_{\text{A\_P}}^{\text{SL}} = \underline{a}_1 \underline{I}_{\text{AI1}} + \underline{a}_2 \underline{I}_{\text{AI2}} + \underline{a}_0 \frac{\underline{Z}_{\text{0LI}}}{\underline{Z}_{\text{1LI}}} \underline{I}_{\text{AI0}} \tag{6.57}$$

Since the zero-sequence current from the healthy line ($\underline{I}_{\text{AII0}}$), which is required for making the compensation for the mutual coupling (6.56), is considered here as unavailable from the measurement, it has to be estimated. For this purpose, one takes the formula determining the relations between the symmetrical components of the total fault current ((3.3) and Table 3.7), which can be expressed as:

$$\frac{\underline{I}_{\text{AI0}} - \underline{P}_0 \underline{I}_{\text{AII0}}}{1-d} = \frac{(\underline{b}_{\text{F1}}\Delta\underline{I}_{\text{AI1}} + \underline{b}_{\text{F2}}\underline{I}_{\text{AI2}})\underline{M}_1}{\underline{K}_1 d + \underline{L}_1} \tag{6.58}$$

which is obtained after substituting:

- formulae (6.17) and (6.18) or (6.20) and (6.21) for the zero-sequence current ($\underline{I}_{\text{F0}}$), depending on the mode of operation of the healthy parallel line;
- formula (6.12) for the fault current distribution factor for the positive- and negative-sequence.

The unavailable zero-sequence current from the healthy line ($\underline{I}_{\text{AII0}}$) can be determined from (6.58) as equal to:

$$\underline{I}_{\text{AII0}} = \frac{1}{\underline{P}_0}\left(\underline{I}_{\text{AI0}} - \frac{(1-d)(\underline{b}_{\text{F1}}\Delta\underline{I}_{\text{AI1}} + \underline{b}_{\text{F2}}\underline{I}_{\text{AI2}})\underline{M}_1}{\underline{K}_1 d + \underline{L}_1}\right) \tag{6.59}$$

Substituting (6.59) into the fault-loop current (6.56) and then into the fault-loop model (6.55) one obtains the following quadratic complex formula:

$$\underline{A}_2 d^2 + \underline{A}_1 d + \underline{A}_0 + \underline{A}_{00} R_F = 0 \tag{6.60}$$

where:

$$\underline{A}_2 = -\underline{Z}_{1LI}\underline{K}_1\underline{I}^{SL}_{A\_P} - \frac{\underline{Z}_{0m}}{\underline{P}_0}(\underline{K}_1\underline{I}_{AI0} - (\underline{b}_{F1}\Delta\underline{I}_{AI1} + \underline{b}_{F2}\underline{I}_{AI2})\underline{M}_1),$$

$$\underline{A}_1 = \underline{K}_1\underline{U}_{A\_P} - \underline{Z}_{1LI}\underline{L}_1\underline{I}^{SL}_{A\_P} - \frac{\underline{Z}_{0m}}{\underline{P}_0}\underline{L}_1\underline{I}_{AI0},$$

$$\underline{A}_0 = \underline{L}_1\underline{U}_{A\_P},$$

$$\underline{A}_{00} = -\underline{M}_1(\underline{a}_{F1}\Delta\underline{I}_{AI1} + \underline{a}_{F2}\underline{I}_{AI2}).$$

Solution of (6.60) is identical to the solution of the quadratic complex formula (6.43). For this purpose one needs to resolve (6.60) into the real and imaginary parts and then to eliminate the unknown fault resistance R_F.

In Table 6.9 the signals and coefficients involved in the fault-location algorithm (6.60) are specified.

Table 6.9 Description of signals and coefficients of the fault-location algorithm (6.60)

Signals	$\underline{U}_{A\_P}$ – fault-loop voltage	Formula (6.25), and Table 6.3 or Table 6.5
	$\underline{I}_{A\_P}$ – fault-loop current (complete)	Formula (6.56)
	$\underline{I}^{SL}_{A\_P}$ – fault-loop current without taking into account the mutual coupling effect, being the part of $\underline{I}_{A\_P}$	Formula (6.57), and Table 6.6
	$\Delta\underline{I}_{AI1}$ – incremental positive-sequence current from faulted line	Fig. 6.5
	$\underline{I}_{AI2}$ – negative-sequence current from faulted line	Fig. 6.5
	$\underline{I}_{AI0}$ – zero-sequence current from faulted line	Figs. 2.18 and 2.19
Coefficients	*Line LII in operation*	*Line LII switched off and grounded*
	$\underline{K}_1 = -\underline{Z}_{1LI}(\underline{Z}_{1A} + \underline{Z}_{1B} + \underline{Z}_{1LII})$	$\underline{K}_1 = -\underline{Z}_{1LI}$
	$\underline{L}_1 = -\underline{K}_1 + \underline{Z}_{1LII}\underline{Z}_{1B}$	$\underline{L}_1 = \underline{Z}_{1LI} + \underline{Z}_{1B}$
	$\underline{M}_1 = \underline{Z}_{1LI}\underline{Z}_{1LII} + (\underline{Z}_{1LI} + \underline{Z}_{1LII})(\underline{Z}_{1A} + \underline{Z}_{1B})$	$\underline{M}_1 = \underline{Z}_{1A} + \underline{Z}_{1B} + \underline{Z}_{1LI}$
	$\underline{P}_0 = \underline{P}'_0 = \frac{\underline{Z}_{0LII} - \underline{Z}_{0m}}{\underline{Z}_{0LI} - \underline{Z}_{0m}}$	$\underline{P}_0 = \underline{P}''_0 = -\frac{\underline{Z}_{0LII}}{\underline{Z}_{0m}}$
	$\underline{a}_1$, $\underline{a}_2$, $\underline{a}_0$: Table 6.6	
	$\underline{a}_{F1}$, $\underline{a}_{F2}$, $\underline{a}_{F0}$: The sets with $\underline{a}_{F0} = 0$ – Tables 3.3–3.5, according to the assumed preference for using the respective sequence components	
	$\underline{b}_{F1}$, $\underline{b}_{F2}$: (3.3) and Table 3.7	

Example 6.4 Fault location on double-circuit line according to the algorithm (6.60) In Fig. 6.13 the results of fault location on a 120-km double-circuit line are shown. The main specifications are as follows: a–g fault, d_{actual} =0.9 p.u., R_F=10 Ω, both lines are in operation [114]. Without using the zero-sequence current from the healthy parallel line, the compensation for mutual coupling of the lines is performed accurately, ensuring exact location of the fault (Fig. 6.13c).

The zero-sequence current from the healthy parallel line (real and imaginary parts), estimated according to (6.59), is shown in Fig. 6.14, together with the actual current. After completing the data window of the filters (within 20 ms) the difference between the estimated and the actual currents is very small.

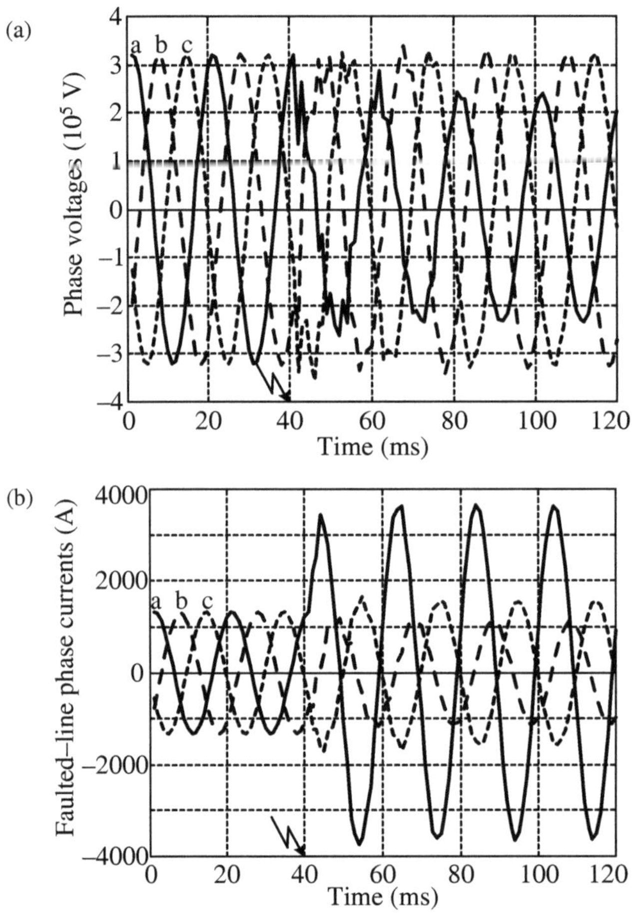

Fig. 6.13 Example 6.4 – (a) three-phase voltage, (b) three-phase current from faulted line, and (c) estimated distance to fault

(c)

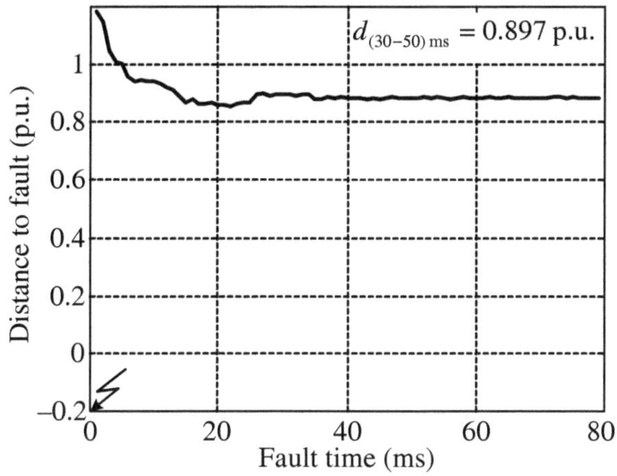

Fig. 6.13 (*continued*)

(a)

(b)

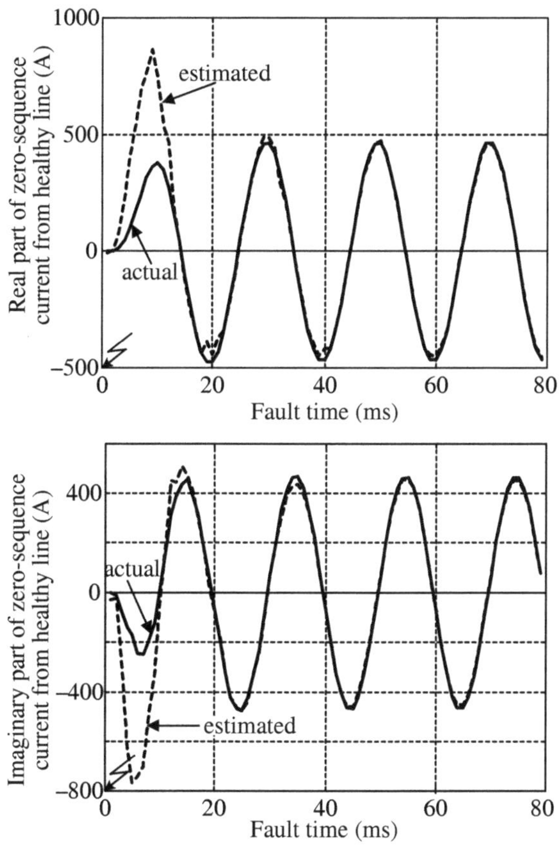

Fig. 6.14 Example 6.4 – estimated zero-sequence current from the healthy parallel line: (a) real part, and (b) imaginary part

Figure 6.15 shows errors of estimation of the distance to fault for the parallel lines of 120 km in length under single phase-to-ground faults applied at different locations 0.1, 0.2, ..., 0.9 p.u.) with fault resistance of 10 Ω. The errors in the case of no compensation for mutual coupling between the lines are big, which is very well known. For far-end faults the error for such location exceeds 10% (Fig. 6.15a). On the other hand, using the fault-location algorithm (6.60), the errors are quite small, especially for the case with the parallel line being in operation, for which the error does not exceed 0.3% (Fig. 6.15b). If the parallel line is switched off and earthed at both ends the errors are slightly bigger, but still acceptable.

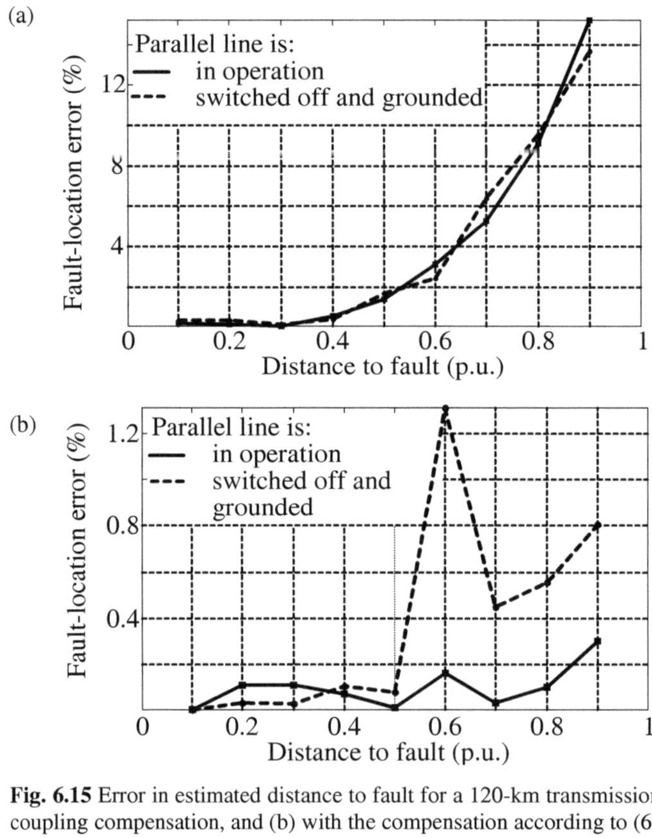

Fig. 6.15 Error in estimated distance to fault for a 120-km transmission line: (a) under no mutual coupling compensation, and (b) with the compensation according to (6.60)

6.10 Fault Location Utilizing Only Phase-current Phasors

Two fault-location algorithms, which use only current signals from one end of a single transmission line as the input data are presented in [52]. The first algorithm utilizes the relation between symmetrical components of the total fault current. As the algorithm covers only the phase-to-earth faults, in order to extend it to the phase-to-phase-to-earth faults, (3.3) and its coefficients gathered in Table 3.7 are recommended. After substituting (6.9)–(6.13) into (3.3) one obtains:

$$\frac{\underline{M}_0 \underline{I}_{A0}}{\underline{K}_0 d + \underline{L}_0} = (\underline{b}_{F1} \Delta \underline{I}_{A1} + \underline{b}_{F2} \underline{I}_{A2}) \frac{\underline{M}_1}{\underline{K}_1 d + \underline{L}_1} \tag{6.61}$$

The distance to fault can be determined from (6.61) as:

$$d = \frac{\underline{L}_1 \underline{M}_0 \underline{I}_{A0} - \underline{L}_0 \underline{M}_1 (\underline{b}_{F1} \Delta \underline{I}_{A1} + \underline{b}_{F2} \underline{I}_{A2})}{\underline{K}_0 \underline{M}_1 (\underline{b}_{F1} \Delta \underline{I}_{A1} + \underline{b}_{F2} \underline{I}_{A2}) - \underline{K}_1 \underline{M}_0 \underline{I}_{A0}} \tag{6.62}$$

where:
$\Delta \underline{I}_{A1}$, $\underline{I}_{A2}$, $\underline{I}_{A0}$ – incremental positive-, negative- and zero-sequence components of current from the terminal A,
$\underline{b}_{F1}$, $\underline{b}_{F2}$ – complex coefficients dependent on fault type (Table 3.7),
$\underline{K}_1$, $\underline{L}_1$, $\underline{M}_1$ – complex coefficients determined by the positive-sequence impedances of the network (Table 6.1),
$\underline{K}_0$, $\underline{L}_0$, $\underline{M}_0$ – complex coefficients determined by the zero-sequence impedances of the network (in coefficients from Table 6.1 the positive-sequence impedances have to be replaced by the respective zero-sequence impedances: $\underline{Z}_{1L}$ replaced by $\underline{Z}_{0L}$, etc.).

The first algorithm from [52] covers only the faults for which an earth is involved and thus the zero-sequence current is present. For the phase-to-phase fault one can formulate a relation between the incremental positive- and negative-sequence currents – the second algorithm from [52]. However, the distance to fault can be determined from this relation only for specific conditions with impedances of the network for the positive- and negative-sequence being not identical. Normally, this is not so and the fault current distribution factors for the incremental positive- and negative-sequence involved in the relation, being identical are cancelled and the relation becomes useless.

Evaluation of the fault-location accuracy performed in [52] has shown the presence of relatively big errors. Fortunately, the errors do not depend on fault resistance and pre-fault power-flow direction. Therefore, one can apply the correction factors, calculated in advance, for correcting the results obtained for the particular line and network topology. The other possibility is to derive the relation formula (3.3) taking into account the distributed-parameter line model.

The fault-location method discussed does not cover all faults, and thus cannot be considered as the only one suitable for implementing into the fault locator. It can be used as the supplement of other methods. Due to the simplicity of the algorithm, it can be utilized not only for fault location (off-line application), but also in the field of distance protection (on-line application).

6.11 Fault Location with Limited Use of Current Phasors

The need for limiting use of current phasors is considered as one of the remedies for assuring high accuracy of fault location under CT saturation.

The relation between the symmetrical components of the total fault current (6.61) can be used for formulating the fault-location algorithm utilizing only voltage phasors [109]. For this purpose, the symmetrical components of the measured three-phase current are expressed as follows:

$$\Delta \underline{I}_{A1} = \frac{-\Delta \underline{U}_{A1}}{\underline{Z}_{1A}} \tag{6.63}$$

$$\underline{I}_{A2} = \frac{-\underline{U}_{A2}}{\underline{Z}_{1A}} \tag{6.64}$$

$$\underline{I}_{A0} = \frac{-\underline{U}_{A0}}{\underline{Z}_{0A}} \tag{6.65}$$

Substituting (6.63)–(6.65) into (6.61) results in the following formula for the distance to fault:

$$d = \frac{\underline{Z}_{1A}\underline{L}_1\underline{M}_0\underline{U}_{A0} - \underline{Z}_{0A}\underline{L}_0\underline{M}_1(\underline{b}_{F1}\Delta\underline{U}_{A1} + \underline{b}_{F2}\underline{U}_{A2})}{\underline{Z}_{0A}\underline{K}_0\underline{M}_1(\underline{b}_{F1}\Delta\underline{U}_{A1} + \underline{b}_{F2}\underline{U}_{A2}) - \underline{Z}_{1A}\underline{K}_1\underline{M}_0\underline{U}_{A0}} \tag{6.66}$$

Again, as for the fault-location algorithm [52], all impedances of the network, for the positive- and zero-sequence are required to be known and only the faults involving earth are covered.

Non-standard use of one-terminal measurements of voltage and current for fault location was presented in [233]. Voltage phasors from the fault interval only are required, while the current phasors only from the pre-fault time, when CT saturation does not occur. When use is made of such input signals, the CT saturation does not affect the fault-location accuracy. The algorithm covers all faults. It requires impedances of equivalent systems at both line terminals to be known, as well as performing the fault classification, assuming at the same time that the fault impedance is purely resistive.

The solution in [233] consists in comparing the voltage measured at the local terminal (A) with the calculated one, taking into account the objective function of the sum of errors modules:

$$F(d, R_F) = \sum_{i=a,b,c} |\underline{U}_{Ai}^{\text{calc.}} - \underline{U}_{Ai}^{\text{measur.}}| \tag{6.67}$$

Node voltages under fault condition are calculated [233] using three-phase injected currents in the nodes (line terminals A, B and at the fault F), and a three-phase nodal admittance matrix.

Unknown fault distance (d) and fault resistance (R_F) are obtained through an optimization algorithm at the point of minimum of function (6.67).

6.12 Fault Location with Arc-voltage Estimation

The arcing character of faults is reflected in the algorithms presented in [51, 54]. This numerical algorithm is based on one terminal data and is derived in the time domain. The fault location and its nature, in terms of arcing or arc-less fault, are estimated using the least error squares technique. The faulted phase voltage (Fig. 6.16) is modeled as a serial connection of fault resistance (R_F) and arc voltage (u) – represented by the model defined in formula (3.11). The algorithm is derived for the most frequent case of a single phase-to-ground fault.

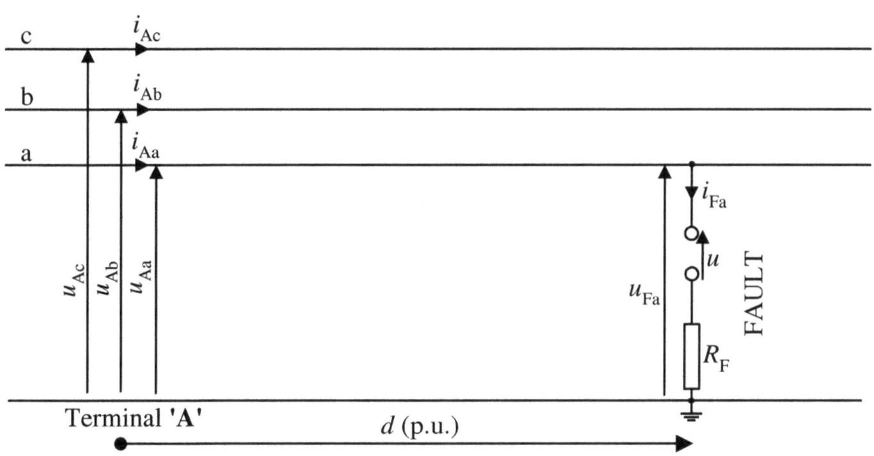

Fig. 6.16 Schematic diagram of three-phase line under phase ‘a’ to-ground fault with electrical arc and linear resistance

Considering the fault loop for the phase ‘a’ one obtains [51, 54]:

$$u_{Aa}(t) = \left[R_{1L}(i_{Aa}(t) - i_{A0}(t)) + \frac{X_{1L}}{\omega_1}\left(\frac{di_{Aa}(t)}{dt} + k_0 \frac{di_{A0}(t)}{dt} \right) \right] d \\ + U_a \,\mathrm{signum}(i_{A0}(t)) + R_e i_{A0}(t) + \xi(t) \tag{6.68}$$

where:

$k_0 = \dfrac{X_{0L} - X_{1L}}{X_{1L}}$, $R_e = R_{0L}d + k_a R_F$, $k_a = \dfrac{i_{Fa}(t)}{i_{A0}(t)}$,

R_{1L}, X_{1L} – positive-sequence resistance and reactance of the line,
R_{0L}, X_{0L} – zero-sequence resistance and reactance of the line,
ω_1 – fundamental radian frequency,
$i_{A0}(t)$ – zero-sequence current,
U_a – magnitude of the rectangular voltage wave: u (see (3.11)),
$\xi(t)$ – Gaussian noise with zero average value.

Equation (6.68) involves three unknowns:

- d (p.u.) – distance to fault,
- U_a – magnitude of the rectangular voltage wave,
- the quantity R_e.

After transforming (6.68) into the discrete form, an estimation with the use of the least error squares technique [51, 54] is carried out.

Besides the basic feature of the algorithm relying on determining the nature of the fault the distance to fault is also obtained. Moreover, the approach does not require the line zero-sequence resistance as input data.

6.13 Fault Location on Untransposed Lines

The symmetrical components approach can be effectively used for locating faults on completely symmetrical lines, i.e., on transposed lines. However, the symmetry of a line can be substantially disturbed, which is the case when there are long segments of a line without transposition of the conductors (untransposed line). The other causes are related with such installations as, for example, series-compensating capacitors equipped with MOVs for overvoltage protection, which introduce asymmetry upon the occurrence of unsymmetrical faults.

In order to take into account the asymmetry of a line, the phase coordinates approach [109, 216, 274, 284] will be applied for representing a faulted line in the fault-location algorithm. The voltage drop across a three-phase element, represented by a column matrix of three-phase voltage **U**, can be expressed as a product of an impedance matrix (**Z**) and a column matrix of measured three-phase current (**I**):

$$\mathbf{U} = \mathbf{Z}\mathbf{I} \tag{6.69}$$

where: $\mathbf{U} = \begin{bmatrix} \underline{U}_a \\ \underline{U}_b \\ \underline{U}_c \end{bmatrix}$, $\mathbf{I} = \begin{bmatrix} \underline{I}_a \\ \underline{I}_b \\ \underline{I}_c \end{bmatrix}$, $\mathbf{Z} = \begin{bmatrix} \underline{Z}_{aa} & \underline{Z}_{ab} & \underline{Z}_{ac} \\ \underline{Z}_{ab} & \underline{Z}_{bb} & \underline{Z}_{bc} \\ \underline{Z}_{ac} & \underline{Z}_{bc} & \underline{Z}_{cc} \end{bmatrix}$.

For a power line, which is perfectly transposed, the diagonal components of the impedance matrix **Z** (self-impedances: subscript 's') as well as all the off-diagonal elements (mutual impedances: subscript 'm') are accordingly equal to each other:

$$\underline{Z}_{s} = \underline{Z}_{aa} = \underline{Z}_{bb} = \underline{Z}_{cc} \tag{6.70}$$

$$\underline{Z}_{m} = \underline{Z}_{ab} = \underline{Z}_{bc} = \underline{Z}_{ac} \tag{6.71}$$

As a result, one obtains the following relations in which the impedances of a line for the positive- and zero-sequence are involved:

$$\underline{Z}_{s} = \frac{\underline{Z}_{0L} + 2\,\underline{Z}_{1L}}{3} \tag{6.72}$$

$$\underline{Z}_{m} = \frac{\underline{Z}_{0L} - \underline{Z}_{1L}}{3} \tag{6.73}$$

Models of a transmission network with a single line, for pre-fault and fault conditions, respectively, are presented in Figs. 6.17 and 6.18. In Figs. 6.19 and 6.20, a double-circuit line transmission network arrangement is presented. The voltages induced due to mutual coupling of the line circuits are denoted by voltage-source symbols, under which the value of the particular voltage induced (within a dashed rectangle) is specified.

In the models from Figs. 6.17 through 6.20, only the longitudinal line parameters are taken into account, while the shunt line parameters are neglected. Such simplification is applied with the aim of obtaining compact formulae for the distance to fault. Then, in order to improve fault-location accuracy, as required for the lines stretching over long distances, the compensation for the shunt line capacitances (analogously as in Sect. 6.15, where the compensation for the symmetrical components approach is presented) can be performed.

The transmission network with a double-circuit line, as a more general circuit, is taken further for deriving the fault-location algorithm. Location is considered as performed on a faulted line LI, utilizing the three-phase currents: $\mathbf{I}_{AI}^{pre}$, $\mathbf{I}_{AII}^{pre}$, $\mathbf{I}_{AI}$, $\mathbf{I}_{AII}$ and three-phase voltage $\mathbf{U}_{A}$ (Figs. 6.19 and 6. 20). So, it is assumed that the one-end fault locator is installed at the terminal A. Besides these input signals of the fault locator, the total fault current $\mathbf{I}_{F}$ and the current $\mathbf{I}_{BI}$ from the remote terminal BI are also marked in Fig. 6.20a. These currents are immeasurable for the one-end fault locator installed at the bus AI–AII. However, the currents $\mathbf{I}_{F}$, $\mathbf{I}_{BI}$ will be involved in the fault-location-algorithm derivation. During the derivation they will be eliminated as a result of being expressed by means of the measurable quantities and network parameters.

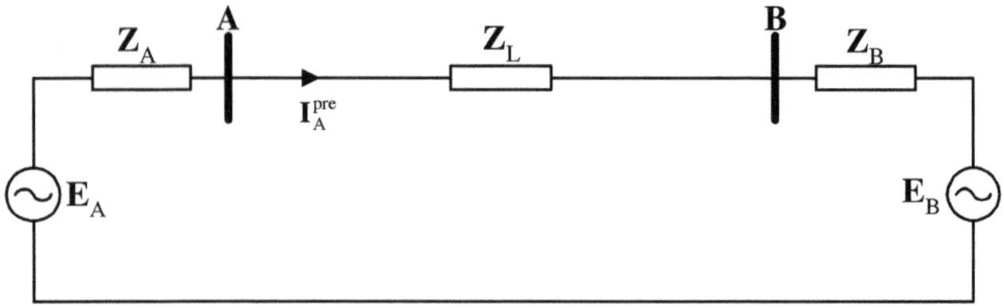

Fig. 6.17 Model of transmission network with single line for pre-fault conditions

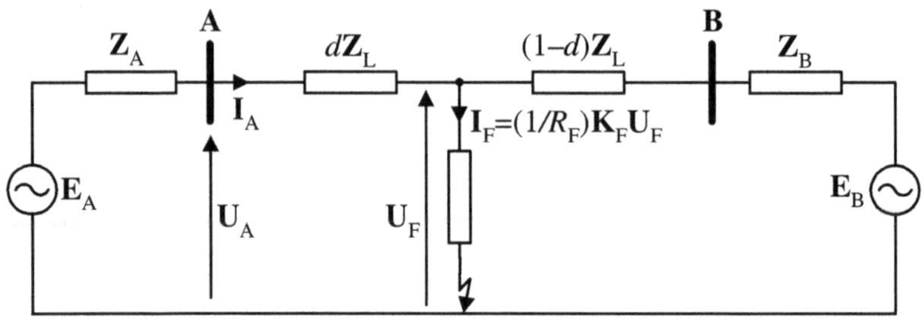

Fig. 6.18 Model of transmission network with single line for fault conditions

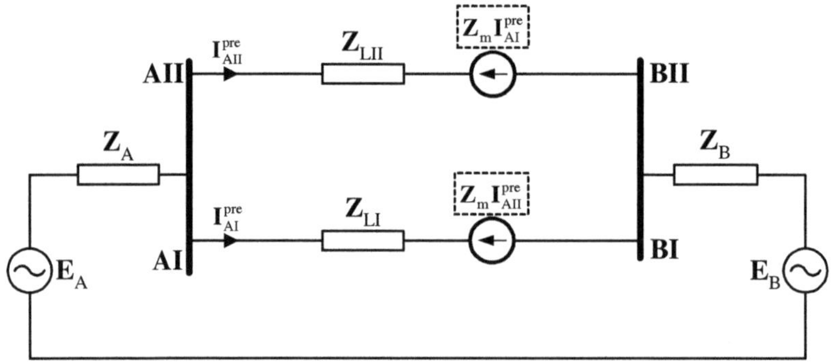

Fig. 6.19 Model of a transmission network with double-circuit line for pre-fault conditions

Considering the path formed by emfs: $\mathbf{E}_A$, $\mathbf{E}_B$, and impedances: $\mathbf{Z}_A$, $d\mathbf{Z}_{LI}$, $(1-d)\mathbf{Z}_{LI}$, $\mathbf{Z}_B$ in the circuit of Fig. 6.20a, one obtains the following matrix formula:

$$\Delta\mathbf{E} = \mathbf{E}_A - \mathbf{E}_B = (\mathbf{Z}_A + d\mathbf{Z}_{LI})\mathbf{I}_{AI} - ((1-d)\mathbf{Z}_{LI} + \mathbf{Z}_B)\mathbf{I}_{BI} + (\mathbf{Z}_A + \mathbf{Z}_B + \mathbf{Z}_m)\mathbf{I}_{AII} \quad (6.74)$$

in which the involved impedance matrices for the line LI ($\mathbf{Z}_{LI}$) and for the mutual coupling between line circuits ($\mathbf{Z}_m$) are:

$$\mathbf{Z}_{\mathrm{LI}} = \begin{bmatrix} \underline{Z}_{\mathrm{LI\_aa}} & \underline{Z}_{\mathrm{LI\_ab}} & \underline{Z}_{\mathrm{LI\_ac}} \\ \underline{Z}_{\mathrm{LI\_ab}} & \underline{Z}_{\mathrm{LI\_bb}} & \underline{Z}_{\mathrm{LI\_bc}} \\ \underline{Z}_{\mathrm{LI\_ac}} & \underline{Z}_{\mathrm{LI\_bc}} & \underline{Z}_{\mathrm{LI\_cc}} \end{bmatrix}, \quad \mathbf{Z}_{\mathrm{m}} = \begin{bmatrix} \underline{Z}_{\mathrm{m\_aa}} & \underline{Z}_{\mathrm{m\_ab}} & \underline{Z}_{\mathrm{m\_ac}} \\ \underline{Z}_{\mathrm{m\_ab}} & \underline{Z}_{\mathrm{m\_bb}} & \underline{Z}_{\mathrm{m\_bc}} \\ \underline{Z}_{\mathrm{m\_ac}} & \underline{Z}_{\mathrm{m\_bc}} & \underline{Z}_{\mathrm{m\_cc}} \end{bmatrix}.$$

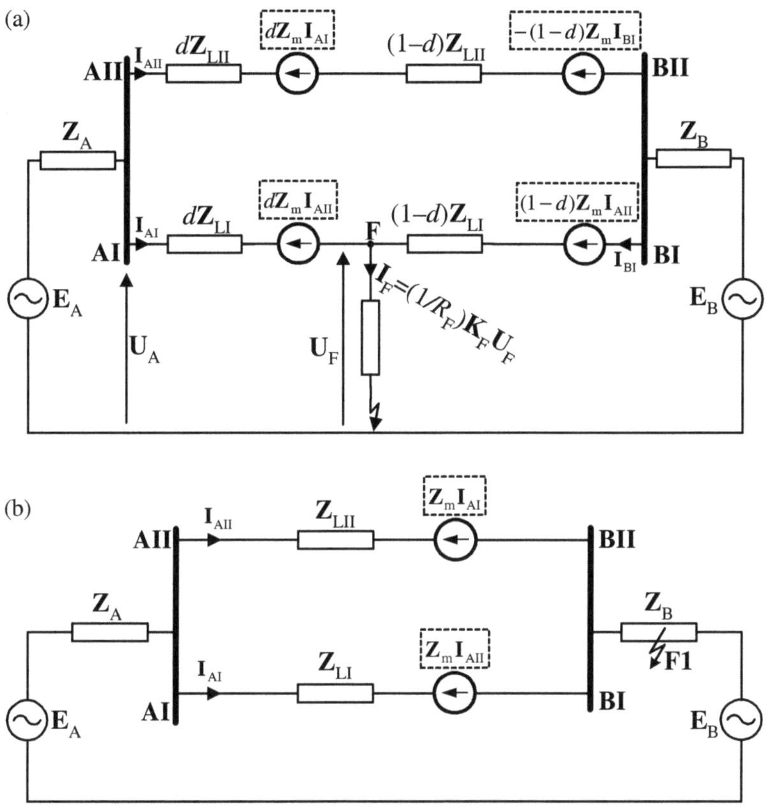

Fig. 6.20 Model of a transmission network with double-circuit line for: (a) fault F on line LI, and (b) fault F1 overreaching line length

For a completely symmetrical line, the elements of the impedance matrix $\mathbf{Z}_{\mathrm{LI}}$ are determined as in (6.72), while all the components of the mutual coupling impedance $\mathbf{Z}_{\mathrm{m}}$ are identical:

$$\underline{Z}_{\mathrm{m\_aa}} = \underline{Z}_{\mathrm{m\_bb}} = \underline{Z}_{\mathrm{m\_cc}} = \underline{Z}_{\mathrm{m\_ab}} = \underline{Z}_{\mathrm{m\_ac}} = \underline{Z}_{\mathrm{m\_bc}} = \frac{\underline{Z}_{0\mathrm{m}}}{3} \tag{6.75}$$

where $\underline{Z}_{0\mathrm{m}}$ – mutual coupling impedance for the zero-sequence.

Assuming that the emfs of the sources do not change due to a fault, the column matrix $\Delta\mathbf{E}$ determined in (6.74) can be expressed based on the pre-fault model (Fig. 6.19) as [284]:

$$\Delta\mathbf{E} = \mathbf{E}_{\mathrm{A}} - \mathbf{E}_{\mathrm{B}} = (\mathbf{Z}_{\mathrm{A}} + \mathbf{Z}_{\mathrm{B}} + \mathbf{Z}_{\mathrm{LI}})\mathbf{I}_{\mathrm{AI}}^{\mathrm{pre}} + (\mathbf{Z}_{\mathrm{A}} + \mathbf{Z}_{\mathrm{m}} + \mathbf{Z}_{\mathrm{B}})\mathbf{I}_{\mathrm{AII}}^{\mathrm{pre}} \tag{6.76}$$

The current at the remote substation B flowing in the line LI ($\mathbf{I}_{\mathrm{BI}}$), which is immeasurable, can be determined from (6.74) as:

$$\mathbf{I}_{\mathrm{BI}} = ((1-d)\mathbf{Z}_{\mathrm{LI}} + \mathbf{Z}_{\mathrm{B}})^{-1}((\mathbf{Z}_{\mathrm{A}} + d\mathbf{Z}_{\mathrm{LA}})\mathbf{I}_{\mathrm{AI}} + (\mathbf{Z}_{\mathrm{A}} + \mathbf{Z}_{\mathrm{B}} + \mathbf{Z}_{\mathrm{m}})\mathbf{I}_{\mathrm{AII}} - \Delta\mathbf{E}) \tag{6.77}$$

Column matrices of the voltage across a fault path and total fault current (Fig. 6.20a) are determined accordingly:

$$\mathbf{U}_{\mathrm{F}} = \mathbf{U}_{\mathrm{A}} - d\mathbf{Z}_{\mathrm{LI}}\mathbf{I}_{\mathrm{AI}} - d\mathbf{Z}_{\mathrm{m}}\mathbf{I}_{\mathrm{AII}} \tag{6.78}$$

$$\mathbf{I}_{\mathrm{F}} = \mathbf{I}_{\mathrm{AI}} + \mathbf{I}_{\mathrm{BI}} \tag{6.79}$$

A general fault model with use of the matrix notation was described in Chap. 3: (3.4) and Table 3.8. Taking into account the general fault model (3.4) and (6.78) and (6.79) one obtains [284]:

$$\frac{1}{R_{\mathrm{F}}}\mathbf{K}_{\mathrm{F}}(\mathbf{U}_{\mathrm{A}} - d\mathbf{Z}_{\mathrm{LI}}\mathbf{I}_{\mathrm{AI}} - d\mathbf{Z}_{\mathrm{m}}\mathbf{I}_{\mathrm{AII}}) = \mathbf{I}_{\mathrm{AI}} + \mathbf{I}_{\mathrm{BI}} \tag{6.80}$$

Combining (6.77) and (6.80), yields, after the rearrangements, the following matrix equation:

$$\mathbf{A}d^2 - \mathbf{B}d + \mathbf{C} - \mathbf{D}R_{\mathrm{F}} = \mathbf{0} \tag{6.81}$$

where:
$\mathbf{A} = \mathbf{Z}_{\mathrm{LI}}\mathbf{K}_{\mathrm{F}}\mathbf{Z}_{\mathrm{LI}}\mathbf{I}_{\mathrm{AI}} + \mathbf{Z}_{\mathrm{LI}}\mathbf{K}_{\mathrm{F}}\mathbf{Z}_{\mathrm{m}}\mathbf{I}_{\mathrm{AII}}$,
$\mathbf{B} = \mathbf{Z}_{\mathrm{LI}}\mathbf{K}_{\mathrm{F}}(\mathbf{U}_{\mathrm{A}} + \mathbf{Z}_{\mathrm{LI}}\mathbf{I}_{\mathrm{AI}}) + \mathbf{Z}_{\mathrm{B}}\mathbf{K}_{\mathrm{F}}\mathbf{Z}_{\mathrm{LI}}\mathbf{I}_{\mathrm{AI}} + \mathbf{Z}_{\mathrm{m}}\mathbf{K}_{\mathrm{F}}(\mathbf{Z}_{\mathrm{LI}} + \mathbf{Z}_{\mathrm{B}})\mathbf{I}_{\mathrm{AII}}$,
$\mathbf{C} = (\mathbf{Z}_{\mathrm{LI}} + \mathbf{Z}_{\mathrm{B}})\mathbf{K}_{\mathrm{F}}\mathbf{U}_{\mathrm{A}}$,
$\mathbf{D} = (\mathbf{Z}_{\mathrm{A}} + \mathbf{Z}_{\mathrm{LI}} + \mathbf{Z}_{\mathrm{B}})(\mathbf{I}_{\mathrm{AI}} - \mathbf{I}_{\mathrm{AI}}^{\mathrm{pre}}) + (\mathbf{Z}_{\mathrm{A}} + \mathbf{Z}_{\mathrm{B}} + \mathbf{Z}_{\mathrm{m}})(\mathbf{I}_{\mathrm{AII}} - \mathbf{I}_{\mathrm{AII}}^{\mathrm{pre}})$.

Transforming (6.81) into the scalar form one obtains the following quadratic formula for complex numbers:

$$\underline{A}_2 d^2 - \underline{A}_1 d + \underline{A}_0 - R_{\mathrm{F}} = 0 \tag{6.82}$$

where:

$$\underline{A}_2 = \mathbf{PA}, \quad \underline{A}_1 = \mathbf{PB}, \quad \underline{A}_0 = \mathbf{PC}, \quad \mathbf{P} = \frac{\mathbf{D}^{\mathrm{T}}}{\mathbf{D}^{\mathrm{T}}\mathbf{D}},$$

superscript T denotes transposition of the matrix.

The scalar quadratic equation (6.82) can be resolved into the real and imaginary parts, from which one calculates the distance to fault (d) and fault resistance (R_{F}), analogously as was presented for the formula (6.43).

In the case of a double-circuit line arrangement one ought to discriminate faults overreaching a total line length, i.e., occurring in a remote system (Fig. 6.20b: fault F1). This can be performed by analyzing the following column matrix:

$$\mathbf{F}_1 = (\mathbf{Z}_{\mathrm{LI}} - \mathbf{Z}_{\mathrm{m}})\mathbf{I}_{\mathrm{AI}} - (\mathbf{Z}_{\mathrm{LII}} - \mathbf{Z}_{\mathrm{m}})\mathbf{I}_{\mathrm{AII}} \tag{6.83}$$

The column matrix (6.83) for such faults has to possess all the components equal to zero. However, in practice, due to the presence of measurement errors, some threshold has to be applied.

Adaptation of the fault-location formula (6.82) (derived for a double-circuit line) to the case of a single line requires deleting all the components relevant to mutual coupling of the line circuits ($\mathbf{Z}_{\mathrm{m}}$) as well as all the components involving currents from the sound line ($\mathbf{I}_{\mathrm{AII}}$, $\mathbf{I}_{\mathrm{AII}}^{\mathrm{pre}}$).

In [109], a quantitative evaluation of the fault-location accuracy, with use of ATP-EMTP [56] generated fault data has been performed. It was shown that for the transmission system studied there, the presented algorithm allows the fault-location accuracy to be improved up to 2%, compared to the symmetrical components approach (applied after averaging the diagonal and off-diagonal elements of the line-impedance matrix). Note that such an improvement was obtained for comparatively small asymmetry of the line under consideration [109]. In the presence of larger asymmetry of the line, the improvement can be considerably higher.

6.14 Fault Location on Series-compensated Lines

Installation of SC and MOV banks (series capacitors (SCs) and their overvoltage protective devices MOVs) in series-compensated lines causes some difficulties for fault location. First, the presence of SC and MOV banks has to be reflected in the fault-location algorithms. Also, the detection at which line section (in front or behind the compensating bank) a fault occurred appears as an important issue.

6.14.1 Representation of Compensating Bank

Series capacitors equipped with MOVs (Fig. 6.21) create certain problems for transmission-line protective devices and fault locators. In order to cope with them, adequate representations of SCs and MOVs have to be developed, being required for both distance-protection and fault-location purposes.

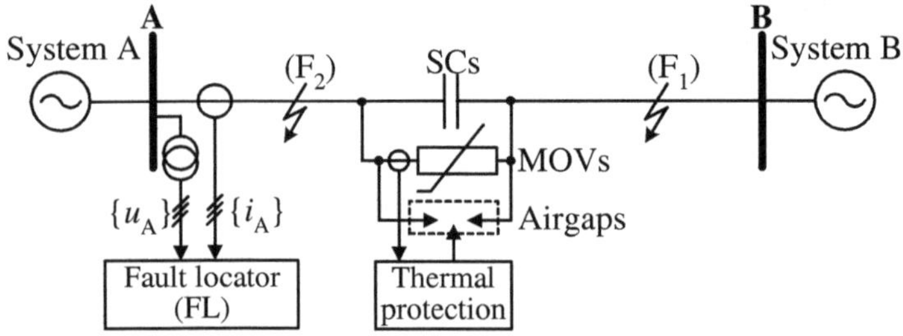

Fig. 6.21 Scheme of transmission line with series compensation in midline

The fundamental frequency concept has been utilized for representing SCs and MOVs for fault-location purposes in [111, 274]. In turn, a digital algorithm for estimating a voltage drop across the bank of SC and MOV has been considered for application to fault location in [254] and for protective relaying purposes in [277]. This estimation algorithm is based on the on-line solving of the strongly non-linear differential equation and is of recursive form.

Fundamental Frequency Equivalenting of SC and MOV

Figure 6.22 presents the equivalenting principle [111, 274]. The parallel connection of a fixed series capacitor (C) and its non-linear protecting resistor MOV (Fig. 6.22a) is represented for the steady state by the fundamental frequency equivalent (Fig. 6.22b). The equivalent is of the form of a series branch with the resistance (R_v) and the capacitive reactance (X_v), both dependent on the amplitude of a current ($|\underline{I}_v|$) entering the SC and MOV bank. Fundamental frequency currents and voltage drops denoted in the original scheme (Fig. 6.22a) and in the equivalent (Fig. 6.22b) must match each other. The equivalenting has to be done by scanning through different amplitudes of the fault current entering the parallel connection of the SC and MOV. This can be achieved, for example, by changing the voltage magnitude of the supplying source (Fig. 6.23).

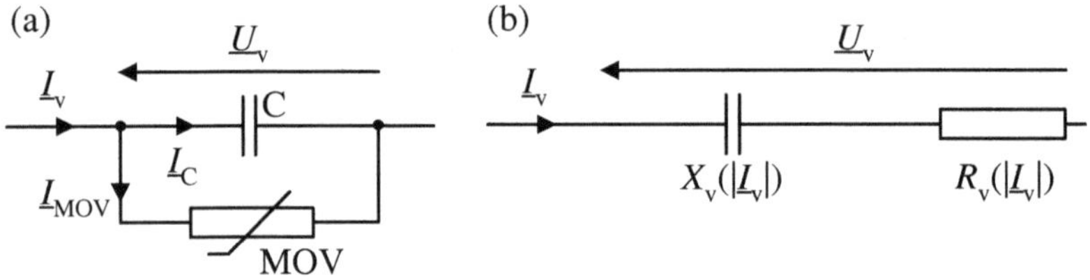

Fig. 6.22 Principle of SC and MOV equivalenting: (a) original circuit, and (b) scheme of the fundamental frequency equivalent

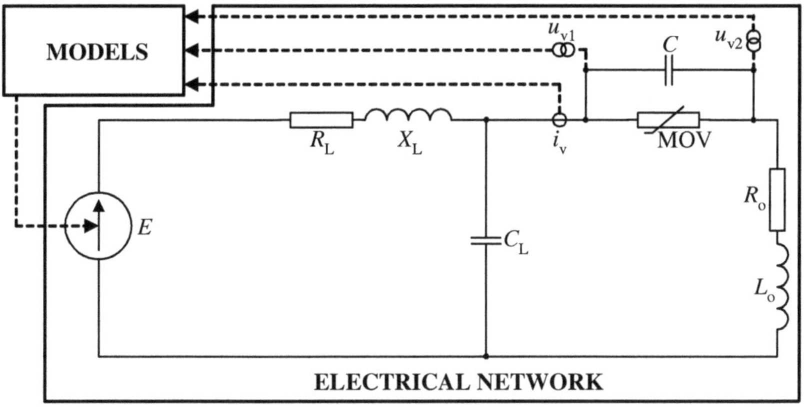

Fig. 6.23 Principle of using ATP-EMTP for equivalenting

Figure 6.23 presents the principle of equivalenting performed with use of ATP-EMTP [56] simulations. The circuit considered is supplied by a source for which the voltage magnitude (E) is controlled in the MODELS unit. The inductive impedance (R_L, X_L) represents the resultant impedance of the source and the line segment from the measuring point up to the SC and MOV installation point.

(a)

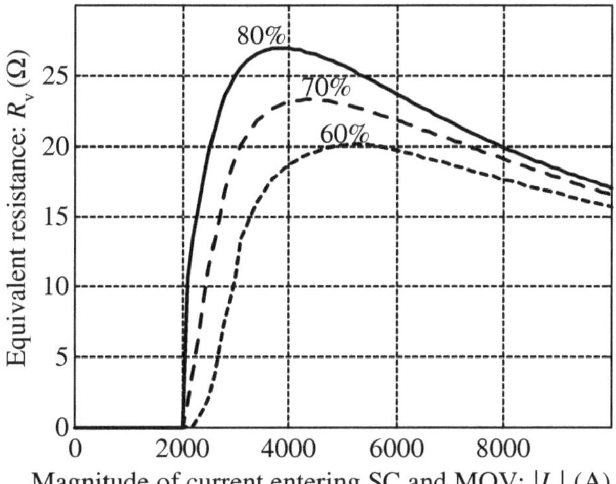

Fig. 6.24 Fundamental frequency equivalents under different compensation rates for a 400-kV, 300-km line: (a) equivalent resistance, and (b) equivalent reactance

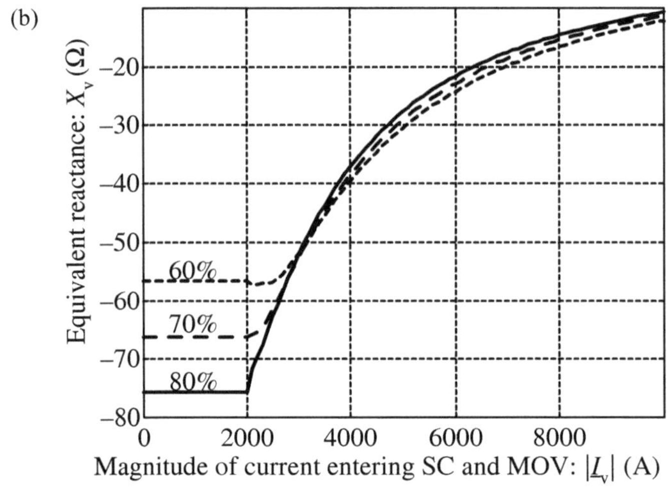

Fig. 6.24 *(continued)*

The capacitance (C_L) represents the shunt capacitance of this line segment, while R_o and X_o – the equivalent impedance for the remote faulted-line segment, together with the remote supplying system. Exchange of the signals, between the units of ATP-EMTP: the MODELS and the ELECTRICAL NETWORK, is shown in Fig. 6.23. The simulation time interval is subdivided into subintervals with different magnitudes of the voltage source (the number of subintervals is equal to the required number of points on the equivalent characteristics). The voltage magnitude is determined in the MODELS unit and sent to the ELECTRICAL NETWORK. The length of the simulation subintervals is set in such a way that steady-state measurement is achieved in each subinterval. This requires setting wider subintervals for smaller voltage magnitudes when long-lasting transients are present due to linear operation of MOVs. On the other hand, for higher supplying voltage magnitudes, the MOVs operate in a non-linear range and the transients are damped faster, and thus shorter subintervals can be designated. Voltages at both terminals of the compensating capacitor (u_{v1}, u_{v2}) and the current entering the SC and MOV bank (i_V) are picked up from the Electrical Network and sent to the MODELS unit. Then, these signals are transferred to the Matlab program for determining fundamental frequency phasors of the processed signals: $\underline{U}_v$, $\underline{I}_v$. From these phasors the equivalent characteristics (Fig. 6.24) are determined.

Results of the equivalenting for different rates of the capacitor compensation, namely for 60%, 70% and 80% rates, applied to a 400-kV, 300-km transmission line are shown in Fig. 6.24. When determining the equivalents presented in Fig. 6.24, it was assumed that the MOVs have identical characteristics, while different SCs are applied for providing 60%, 70% or 80% compensation of the line, respectively. It is seen that the capacitance of the SC influences mainly the equivalent resistance (R_v). For the analyzed compensation rates the equivalent reactance (X_v) differs only for low amplitudes of the fault current (when the MOVs operate linearly or almost linearly).

Using the phase coordinates approach, the three-phase equivalent of SCs and MOVs can be presented with the following matrix [274]:

$$\mathbf{Z}_v(|\mathbf{I}|) = \begin{bmatrix} \underline{Z}_{va}(|\underline{I}_a|) & 0 & 0 \\ 0 & \underline{Z}_{vb}(|\underline{I}_b|) & 0 \\ 0 & 0 & \underline{Z}_{vc}(|\underline{I}_c|) \end{bmatrix} \tag{6.84}$$

Estimation of Voltage Drop Across SC and MOV

Let us consider a parallel connection of the series capacitor SC, and the MOV shown in Fig. 2.13a and also in Fig. 6.21. Assuming the analytical approximation of the MOV characteristic in the form of (2.4), the non-linear circuit of SC and MOV can be described by the following non-linear differential equation [277]:

$$C\frac{du_v}{dt} + P\left(\frac{u_v}{V_{REF}}\right)^q - i_v = 0 \tag{6.85}$$

In this equation, all the parameters are known and constant; the current i_v entering the bank is available (since neglecting the shunt parameters of the line, this is the current in the substation where the fault locator is installed); while the voltage drop v_v is to be calculated. Thus, one needs to transform the continuous-time differential equation (6.85) into its algebraic discrete-time form. The 2nd-order Gear differentiation rule has been taken for this purpose. The following substitutions apply to (6.85):

$$i_v(t) \rightarrow i_{v(n)}, \quad u_v(t) \rightarrow u_{v(n)} \tag{6.86}$$

$$\frac{du_v(t)}{dt} \rightarrow \mathrm{D}\,(3u_{v(n)} - 4u_{v(n-1)} + u_{v(n-2)}) \tag{6.87}$$

where:

$$\mathrm{D} = \frac{2\pi f_1}{\sqrt{26 - 32\cos(\alpha) + 6\cos(2\alpha)}},$$

$\alpha = 2\pi f_1 T_s$,

f_1 – rated fundamental frequency,

T_s – sampling period,

n – discrete time index.

Substituting (6.86) and (6.87) into (6.85) yields the discrete-time equation:

$$F(x_{(n)}) = A_q x_{(n)}^q + A_1 x_{(n)} - A_0 = 0 \tag{6.88}$$

in which:

$$x_{(n)} = \frac{u_{v(n)}}{V_{REF}}, \qquad A_q = P, \qquad A_1 = 3DCU_{REF},$$

$$A_0 = i_{v(n)} + \frac{A_1}{3}(4x_{(n-1)} - x_{(n-2)}).$$

Equation (6.88) is to be solved for $x_{(n)}$ (the p.u. value of the sought voltage drop $u_{v(n)}$ at the current sampling instant n). The two parameters of this equation: A_q and A_1 are the constants, while A_0 depends on the sample of the current ($i_{v(n)}$) entering the bank and the two historical samples of the p.u. voltage drop $(x_{(n-1)}, x_{(n-2)})$.

In order to ensure good convergence of the algorithm, an appropriately modified Newton–Raphson method has been used. The form (6.88) of the equation is numerically efficient for '*small*' values of A_0, while for '*large*' values of A_0, it should be rewritten to:

$$F(y_{(n)}) = A_q y_{(n)} + A_1 (y_{(n)})^{\frac{1}{q}} - A_0 = 0 \tag{6.89}$$

(where: $y_{(n)} = x_{(n)}^q$) and solved for $y_{(n)}$.

The threshold value of A_0 alternating the two optimal formulae (6.88) and (6.89) is:

$$A_0^{\#} = A_q \left(\left(\frac{A_1}{qA_q} \right)^{\frac{1}{q-1}} \right)^q + A_1 \left(\left(\frac{A_1}{qA_q} \right)^{\frac{1}{q-1}} \right) \tag{6.90}$$

The formula (6.88) is solved iteratively with the Newton–Raphson method by applying the following algorithm:

$$x_{(n)\mathrm{new}} = x_{(n)\mathrm{old}} - \frac{A_q x_{(n)\mathrm{old}}^{\ q} + A_1 x_{(n)\mathrm{old}} - A_0}{qA_q x_{(n)\mathrm{old}}^{\ q-1} + A_1} \tag{6.91}$$

The formula (6.89) is solved iteratively with the Newton–Raphson method by applying:

$$y_{(n)\text{new}} = y_{\text{old}} - \frac{A_q y_{(n)\text{old}} + A_1 y_{\text{old}(n)}^{\frac{1}{q}} - A_0}{A_q + \frac{A_1}{q} y_{\text{old}(n)}^{\left(\frac{1-q}{q}\right)}} \tag{6.92}$$

Certainly, if (6.88) is applied, the sought voltage drop (in V) is eventually computed from:

$$u_{\text{v}(n)} = x_{(n)} U_{\text{REF}} \tag{6.93}$$

while, if (6.89) is applied, the voltage drop (in V) is obtained from:

$$u_{\text{v}(n)} = (y_{(n)})^{(q^{-1})} U_{\text{REF}} \tag{6.94}$$

The algorithm is accurate and numerically efficient owing to the following factors [277]:

- The difference in the signal levels (voltage is in thousands while current is in tens or hundreds) is removed by applying the p.u. value of the voltage drop.
- The strong non-linearity of the equation is moderated by using either of the two optimal formulae: (6.88) or (6.89), depending on the operating point on the characteristic of the MOV.
- The algorithm ensures satisfactory accuracy for time steps as large as 1/20th of the fundamental frequency cycle (it needs 2–3 iterations to find a solution). For shorter time steps (sampling frequency over 1000 Hz) the algorithm performs even better.

It is assumed that the pre-fault current is available and the estimation algorithm is started *a posteriori*, a few samples before detecting the fault. Thus, the algorithm is initiated in the pre-fault steady state using the zero conditions, i.e., assuming:

$$x_{(n-1)} = 0, \qquad x_{(n-2)} = 0 \tag{6.95}$$

In the pre-fault conditions, the compensating bank is a pure capacitance (since the MOV almost does not conduct a current) and when using the algorithm with zero initial conditions (6.95), a certain constant offset to the accurate solution is added. In order to remove this offset, it is recommended to apply the following digital filter:

$$\lambda_{(n)} = \frac{1}{2(\cos(\alpha) - 1)} (x_{(n)} - 2x_{(n-1)} + x_{(n-2)}) \tag{6.96}$$

where angle α, as in (6.87).

Implementation of the estimation algorithm, for the case where only the first form (6.88) is utilized, into the Matlab function [202] follows. The task of incorporating the second form (6.89) into this Matlab function is left for the reader.

Matlab Function for Estimating Voltage Drop Across SC and MOV

```
function vMOVa=MOV_fun(ia);        % 'ia' is the input current
n_iter=2;           % here fixed number (2) of iterations is set
C=4.8119e-005;      % capacitance of series capacitor
P=1000.0;           % reference current of MOV characteristic
q=23;               % exponent of MOV characteristic
V_REF=150000.0;     % reference voltage of MOV characteristic
f=50;               % rated power system frequency
fs=1000;            % sampling frequency
omega1=2*pi*f;      % rated power system angular frequency
a=omega1/fs;        % angle corresp. to single sampling interval
kF=0.5/(cos(a)-1);  % gain of filter for DC rejecting
D=omega1/sqrt(26-32*cos(a)+6*cos(2*a)); % const. of GEAR rule
A_q=P;                  % constant
A_1=3*D*C*V_REF;        % constant
l=size(ia);             % size of input current
x=[0;0;0];              % initial conditions
       for n=1:l(1), % processing with time
A_0=ia(n)+A_1*[0 4 -1]*x/3;        % from numeric formula
              for k=1:n_iter,      % perform iter. calculations
              z=abs(x(1));
                     if z,
                     z=log(z);
                     z=A_q*exp(q*z);
                            if x(1)<0, z=-z; end;
                       z1=q*z/x(1);
                     else,
                     z1=0;
                     end;
                     x(1)=x(1)-(z+A_1*x(1)-A_0)/(z1+A_1);
       end,
       if n<4,              % reject the constant offset
       xF(n)=kF*[1 -2 1]*x;
       end,
```

```
        if n==3,
        xF(n)=kF*[1 -2 1]*x;
        x(1)=ia(n)/A_1+(4*xF(n)-xF(n-1))/3;
        x(2)=xF(n);
        end,
        x(3)=x(2);
        x(2)=x(1);
        vMOVa_pu(n,1)=x(1);           % result in per units
        end;
vMOVa=vMOVa_pu*V_REF;                 % result in Volts
```

Figure 6.25 illustrates operation of the presented estimation algorithm for a 400-kV, 300-km line under a single phase-to-earth fault just behind the compensating bank (as seen from the bus A), with 5 Ω fault resistance.

(a)

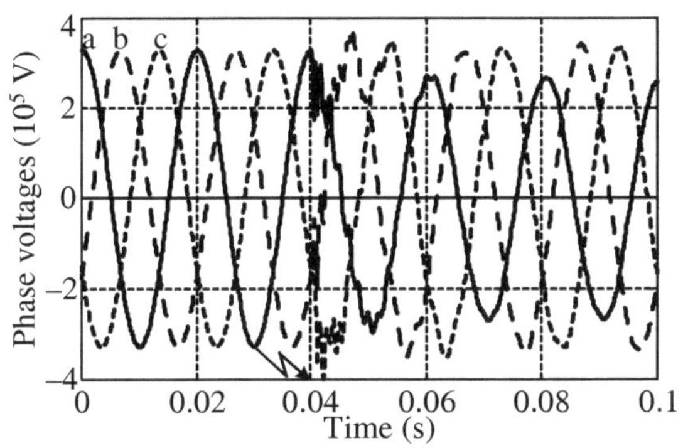

(b)

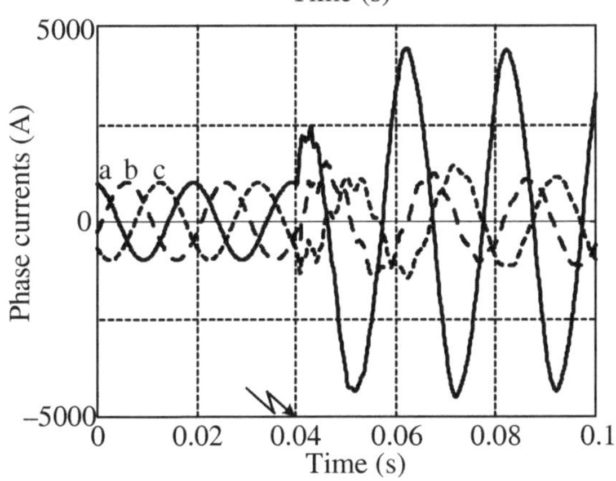

Fig. 6.25 Sample fault: (a) measured phase voltages, (b) measured phase currents (input to the algorithm), (c) voltage drops across SCs and MOVs from simulation, and (d) voltage drop across SCs and MOVs in the faulted phase – from simulation, and estimated (output from the algorithm)

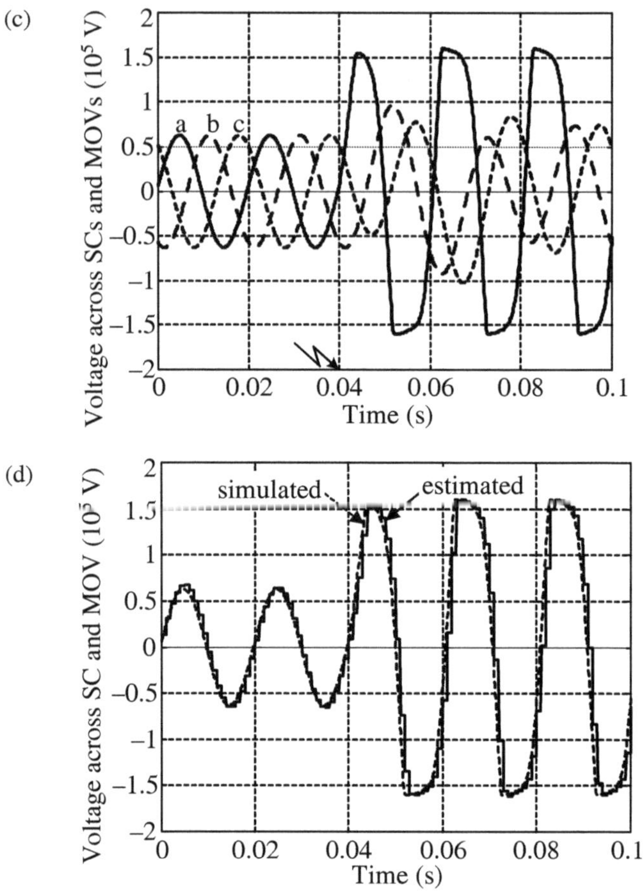

Fig. 6.25 (*continued*)

In the measured phase voltages (Fig. 6.25a) and phase currents (Fig. 6.25b) one observes the characteristic distortion of the waveforms. It is distinctive that, in contrast to the uncompensated-line case, there are no DC components in the currents. Limiting of the voltage drop across SC and MOV from the faulted phase to the level equal to around the reference voltage of MOV (V_{REF}) is observed (Fig. 6.25c). On the other hand, the MOVs from the sound phases operate linearly, which results in slowly decaying sub-harmonic oscillations (Fig. 6.25c). Figure 6.25d shows the actual voltage drop across the SC and MOV from the faulted phase (EMTP simulation) and the signal estimated with the algorithm presented. The accuracy of the voltage reconstruction is very high and the difference between the actual and calculated values is very small. Note that the voltage drop is calculated at the sampling instants only (1000 Hz sampling frequency was assumed here) and therefore the calculated voltage drop is presented as a stair waveform.

6.14.2 Fault Location on Single Series-compensated Line

Let us consider the one-end fault location on a single-circuit transmission line with series compensation inserted at the midline: at the distance d_{SC} (p.u.) from the terminal A. It is assumed that the fault locator is installed at the bus A (Fig. 6.21). For such a configuration of the series-compensated line there is a need to apply two subroutines, which are designed for locating faults:

- behind SCs and MOVs (fault F1, Fig. 6. 21) – the subroutine SUB_1;
- in front of SCs and MOVs (fault F2, Fig. 6. 21) – the subroutine SUB_2.

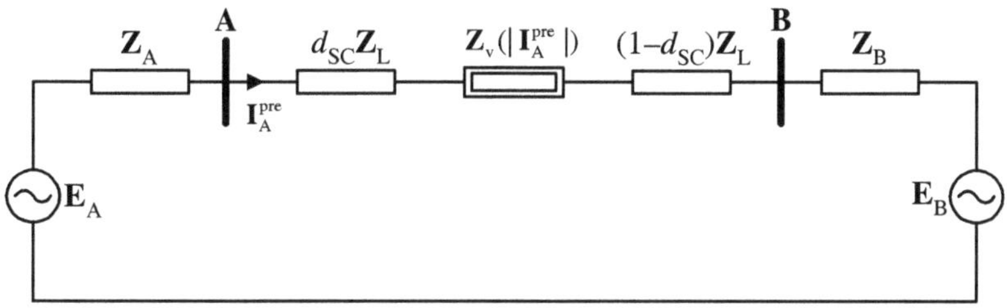

Fig. 6.26 Equivalent diagram for a transmission system with single line compensated at the midpoint, for pre-fault conditions (superscript: pre)

The position of a fault with respect to the SCs and MOVs (behind or in front of them) is of a random nature, therefore, in order to get the final result, the selection of the valid subroutine, i.e., the subroutine that yields the result consistent with the real fault, is necessary.

The fault-location algorithm presented here belongs to the phasor-based category [274]. In this algorithm, the SCs and MOVs are represented by the three-phase fundamental frequency equivalent (6.84). In the following circuit diagrams they are presented by the rectangle impedance symbol, but drawn with a double line, with the equivalent impedance being specified. For example, in the equivalent circuit diagram for pre-fault conditions (Fig. 6.26) the SCs and MOVs equivalent is denoted by $\mathbf{Z}_v(|\mathbf{I}_A^{pre}|)$.

Subroutine SUB_1 – for Faults Behind SCs

In this case (Fig. 6.27), the current flowing through the SCs and MOVs is directly measured by the fault locator at the terminal A. The following applies to the faulted network (Fig. 6.27):

$$\mathbf{E}_A - \mathbf{E}_B = (\mathbf{Z}_L + d_1\mathbf{Z}_L + \mathbf{Z}_v(|\mathbf{I}_A|))\mathbf{I}_A - ((1-d_1)\mathbf{Z}_L + \mathbf{Z}_B)\mathbf{I}_B \tag{6.97}$$

$$\mathbf{U}_A - \mathbf{U}_F = (d_1\mathbf{Z}_L + \mathbf{Z}_v(|\mathbf{I}_A|))\mathbf{I}_A \tag{6.98}$$

$$\mathbf{I}_F = \mathbf{I}_A + \mathbf{I}_B \tag{6.99}$$

where $d_{SC} < d_1 < 1$ is the sought fault distance (p.u.).

The pre-fault network (Fig. 6.26) is described by:

$$\mathbf{E}_A - \mathbf{E}_B = (\mathbf{Z}_A + \mathbf{Z}_L + \mathbf{Z}_v(|\mathbf{I}_A^{pre}|) + \mathbf{Z}_B)\mathbf{I}_A^{pre} \tag{6.100}$$

where the superscript 'pre' stands for the pre-fault values of three-phase current.

The fault equation (3.4) completes the model:

$$\mathbf{I}_F = \frac{1}{R_{F1}}\mathbf{K}_F\mathbf{U}_F \tag{6.101}$$

where R_{F1} – unknown resistance involved in the fault.

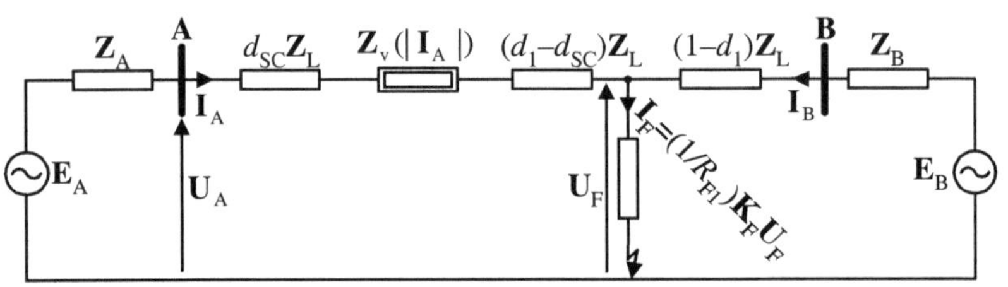

Fig. 6.27 Subroutine SUB_1 for faults occurring behind the compensating capacitors – equivalent circuit diagram for transmission network with single line compensated at midpoint

Combining (6.97)–(6.101) results in the following matrix formula:

$$\mathbf{A}d_1^2 - \mathbf{B}d_1 + \mathbf{C} - \mathbf{D}R_{F1} = \mathbf{0} \tag{6.102}$$

where:

$\mathbf{A} = \mathbf{Z}_L\mathbf{K}_F\mathbf{Z}_L\mathbf{I}_A$,

$\mathbf{B} = \mathbf{Z}_L\mathbf{K}_F(\mathbf{U}_A + (\mathbf{Z}_L - \mathbf{Z}_v(|\mathbf{I}_A|))\mathbf{I}_A) + \mathbf{Z}_B\mathbf{K}_F\mathbf{Z}_L\mathbf{I}_A$,

$\mathbf{C} = (\mathbf{Z}_L + \mathbf{Z}_B)\mathbf{K}_F(\mathbf{U}_A - \mathbf{Z}_v(|\mathbf{I}_A|)\mathbf{I}_A)$,

$\mathbf{D} = (\mathbf{Z}_A + \mathbf{Z}_L + \mathbf{Z}_B)(\mathbf{I}_A - \mathbf{I}_A^{pre}) + \mathbf{Z}_v(|\mathbf{I}_A|)\mathbf{I}_A - \mathbf{Z}_v(|\mathbf{I}_A^{pre}|)\mathbf{I}_A^{pre}$.

Transforming (6.102) into the scalar form yields the following quadratic formula (for complex numbers) in two unknowns: distance to fault (d_1), such that: $d_{SC} < d_1 < 1$ (p.u.) and fault resistance (R_{F1}):

$$\underline{A}_2 d_1^2 - \underline{A}_1 d_1 + \underline{A}_0 - R_{F1} = 0 \tag{6.103}$$

where:

$$\underline{A}_2 = \mathbf{PA}, \qquad \underline{A}_1 = \mathbf{PB}, \qquad \underline{A}_0 = \mathbf{PC}, \qquad \mathbf{P} = \frac{\mathbf{D}^{\mathrm{T}}}{\mathbf{D}^{\mathrm{T}}\mathbf{D}},$$

superscript T denotes transposition of the matrix.

The scalar quadratic equation (6.103) can be resolved into the real and imaginary parts, from which one calculates the distance to fault (d_1) and fault resistance (R_{F1}), analogously to that presented for the formula (6.43)

There are, certainly, two roots of the scalar quadratic formula, but one of them being constant (depending only on the system parameters) can be easily identified and rejected. Eventually, the first subroutine SUB_1 delivers the solution (d_1, R_{F1}) assuming the fault behind the SCs and MOVs.

Subroutine SUB_2 – for Faults in Front of SCs

The case of a fault between the substation and the SCs and MOVs is more involved because the current flowing through the SCs and MOVs ($\mathbf{I}_{\mathrm{B}}$) is not directly available to the locator (since here the one-end measurement at the terminal A is considered).

The following equations apply to the faulted network (Fig. 6.28):

$$\mathbf{E}_{\mathrm{A}} - \mathbf{E}_{\mathrm{B}} = (\mathbf{Z}_{\mathrm{A}} + d_2\mathbf{Z}_{\mathrm{L}})\mathbf{I}_{\mathrm{A}} - ((1-d_2)\mathbf{Z}_{\mathrm{L}} + \mathbf{Z}_{\mathrm{v}}(|\,\mathbf{I}_{\mathrm{B}}\,|) + \mathbf{Z}_{\mathrm{B}})\mathbf{I}_{\mathrm{B}} \tag{6.104}$$

$$\mathbf{U}_{\mathrm{A}} - \mathbf{U}_{\mathrm{F}} = d_2\mathbf{Z}_{\mathrm{L}}\mathbf{I}_{\mathrm{A}} \tag{6.105}$$

but now, $0 < d_2 < d_{\mathrm{SC}}$.

The model of the pre-fault network (6.100) and also the fault model (6.101), but with fault resistance R_{F2} instead of R_{F1}, remain valid.

In this case, one also obtains the quadratic matrix equation:

$$\mathbf{A}d_2^2 - \mathbf{B}d_2 + \mathbf{C} - \mathbf{D}R_{\mathrm{F2}} = \mathbf{0} \tag{6.106}$$

where:

$\mathbf{A} = \mathbf{Z}_{\mathrm{L}}\mathbf{K}_{\mathrm{F}}\mathbf{Z}_{\mathrm{L}}\mathbf{I}_{\mathrm{A}}$,

$\mathbf{B} = \mathbf{Z}_{\mathrm{L}}\mathbf{K}_{\mathrm{F}}\mathbf{U}_{\mathrm{A}} + (\mathbf{Z}_{\mathrm{L}} + \mathbf{Z}_{\mathrm{v}}(|\,\mathbf{I}_{\mathrm{B}}\,|) + \mathbf{Z}_{\mathrm{B}})\mathbf{K}_{\mathrm{F}}\mathbf{Z}_{\mathrm{L}}\mathbf{I}_{\mathrm{A}}$,

$\mathbf{C} = (\mathbf{Z}_{\mathrm{L}} + \mathbf{Z}_{\mathrm{B}} + \mathbf{Z}_{\mathrm{v}}(|\,\mathbf{I}_{\mathrm{B}}\,|))\mathbf{K}_{\mathrm{F}}\mathbf{U}_{\mathrm{A}}$,

$\mathbf{D} = (\mathbf{Z}_{\mathrm{A}} + \mathbf{Z}_{\mathrm{L}} + \mathbf{Z}_{\mathrm{B}})(\mathbf{I}_{\mathrm{A}} - \mathbf{I}_{\mathrm{A}}^{\mathrm{pre}}) + \mathbf{Z}_{\mathrm{v}}(|\,\mathbf{I}_{\mathrm{B}}\,|)\mathbf{I}_{\mathrm{A}} - \mathbf{Z}_{\mathrm{v}}(|\,\mathbf{I}_{\mathrm{A}}^{\mathrm{pre}}\,|)\mathbf{I}_{\mathrm{A}}^{\mathrm{pre}}$.

Transforming matrix formula (6.106) into the scalar form is performed analogously as for the subroutine SUB_1. Then, the unknowns d_2, R_{F2} are calculated, but an iterative numerical solution (Fig. 6.29b) is required because the coefficients depend on the unknown current from the remote substation B.

Eventually, the second subroutine delivers the solution (d_2, R_{F2}) assuming the fault in front of the SCs and MOVs.

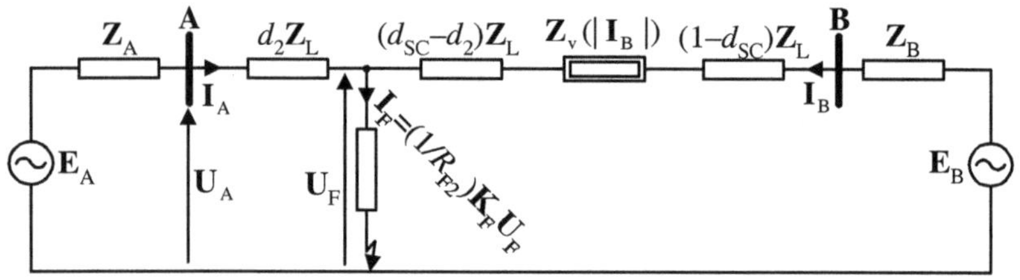

Fig. 6.28 Subroutine SUB_2 for faults in front of the compensating bank – equivalent circuit diagram for transmission network with single line compensated at midpoint

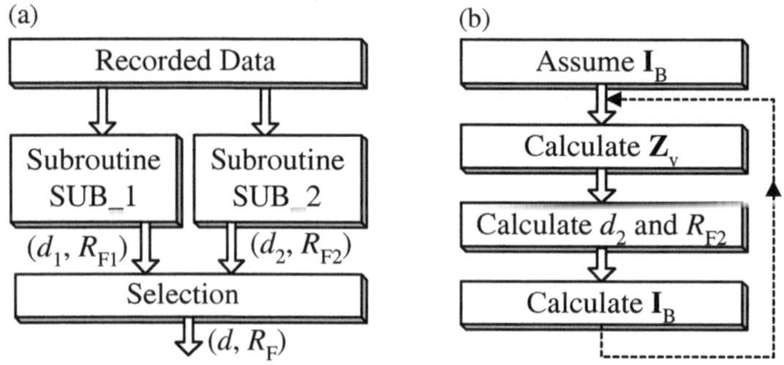

Fig. 6.29 Fault-location algorithm: (a) two subroutines compute two conditional fault locations and resistances, which undergo selection, and (b) iterative calculations of the subroutine SUB_2

Selection Algorithm

Locating a fault with respect to the SC and MOVs in the system of Fig. 6.21, i.e., selecting either SUB_1 or SUB_2 as a valid subroutine, is a separate issue. Here, however, the problem narrows to the selection of the correct pair (d, R_F) out of two alternatives (d_1, R_{F1}) and (d_2, R_{F2}), as depicted in Fig. 6.29a. The simple and straightforward algorithm:

if d_1 is out of $[d_{SC}, 1]$ and d_2 is in $[0, d_{SC}]$ then
  accept (d_2, R_{F2}) as the valid solution, otherwise
if d_1 is in $[d_{SC}, 1]$ and d_2 is out of $[0, d_{SC}]$ then
  accept (d_1, R_{F1}) as the valid solution, otherwise select the alternative with lower R_F as the valid solution.

This selecting algorithm works in most cases. In order to avoid false selection, which could happen in some rare cases, an additional selection criterion has been proposed. This criterion is based on calculating currents at the fault place, according to both subroutines: SUB_1, SUB_2, substituting the obtained distances to fault, d_1, d_2. Ideally, for the valid subroutine the calculated currents from healthy phases at fault have to be equal to zero (since in the healthy phases there is no current flow at all). In practice, due to the presence of the errors, a certain threshold has to be assumed for indicating that these currents are close to zero (the valid subroutine) or far away from zero (the false subroutine). This criterion can be applied except in the case of three-phase faults under which there are no healthy phases.

Summary of Data and Measurements

The algorithm needs the following data:

- line impedance (3×3 phase matrix);
- local source impedance (3×3 phase matrix);
- remote source impedance (3×3 phase matrix);
- location of the compensating bank (d_{SC}); and
- equivalenting branch for the SC and MOV (two functions for R_v, X_v or tables),

and the following measurements:

- type of fault;
- fault detection in terms of identifying the fault inception and clearance instants;
- three-phase local voltages (phasors); and
- three-phase local currents including their pre-fault values (phasors).

Improvements of the One-end Fault-location Algorithm

The fault algorithm has been presented in its very basic form. The important enhancements for its field implementation may include:

- the impedance of the local system should be traced on-line based on the sequence voltages and currents; owing to this, the local impedance is always perfectly matched;
- more involved techniques should be used for phasor estimation and dealing with the off-nominal frequencies;
- advanced post-processing techniques for output filtering and elaborated methods for selecting the final result from the available data window should be applied;
- the remote end source impedance may also be accurately measured by the remote end relays or fault locators and transmitted to the local substation; no sampling synchronization is needed (the voltage and current act as reference for each other); and
- the amplitude of the remote current may also be measured and transmitted enabling iterative calculations to be avoided for the subroutine SUB_2 and improving the accuracy.

Example 6.5 Fault-location on series-compensated line
The following specifications of the fault on a 400-kV, 300-km transmission line, compensated at 70% rate with the three-phase compensating bank installed at the midline (d_{SC} = 0.5 p.u.) were assumed:

- fault type: a–g;
- distance to fault: $d = 0.5^+$ p.u. (just behind SCs and MOVs – the subroutine SUB_1 is valid);
- fault resistance: $R_F = 10\ \Omega$.

The results for this example are shown in Figs. 6.30a–f.

(a)

(b)

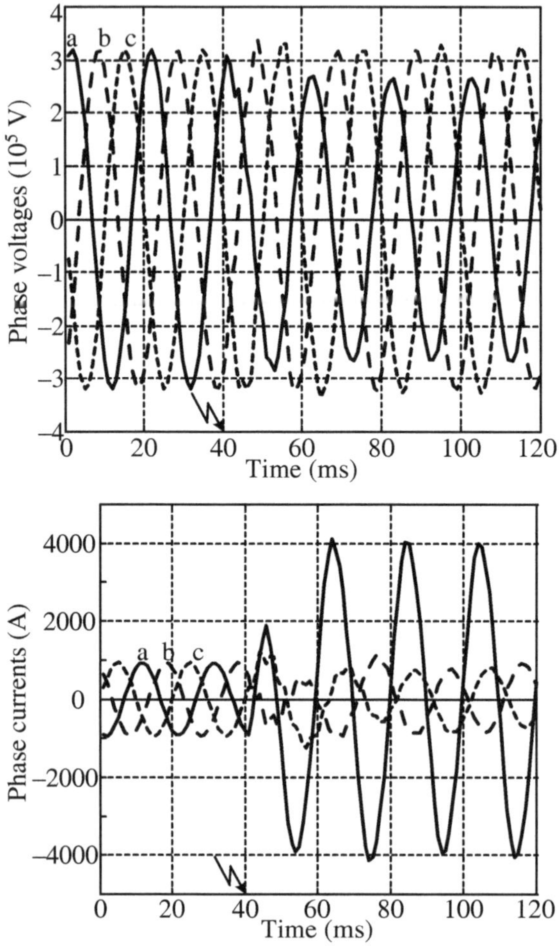

Fig. 6.30 Example 6.5: (a) three-phase input voltage, (b) three-phase input current, (c) distance to fault, (d) fault resistance, (e) magnitude of total fault current from healthy phase ‘b’ according to SUB_1 and SUB_2, and (f) fault-distance estimation error for valid subroutine (SUB_1)

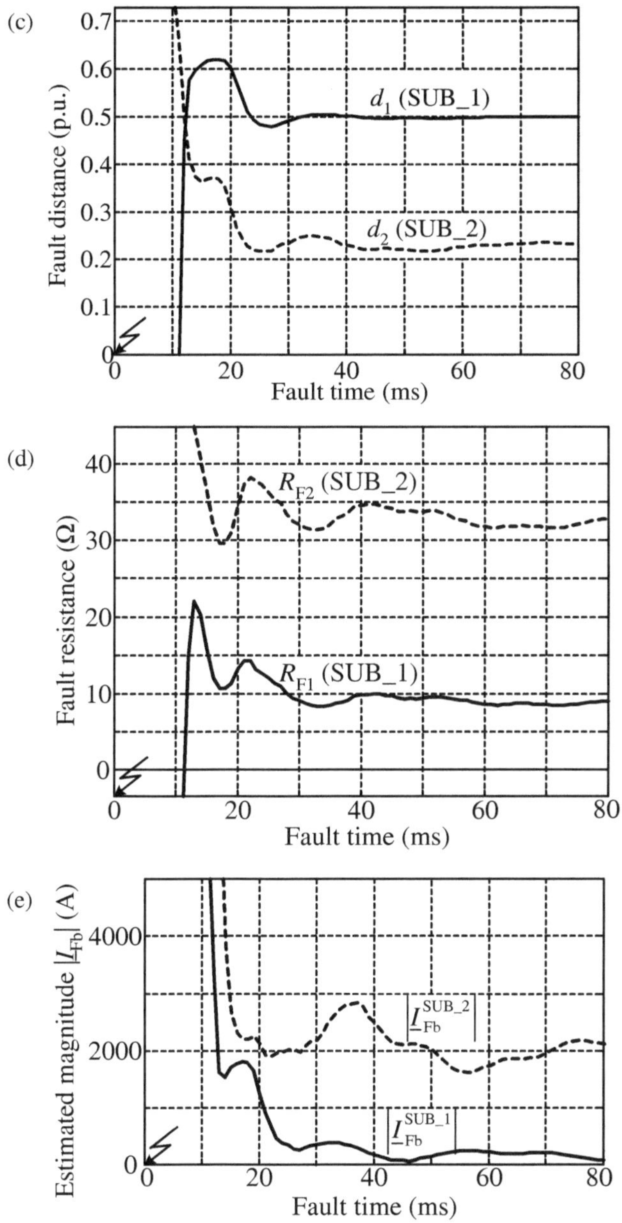

Fig. 6.30 (*continued*)

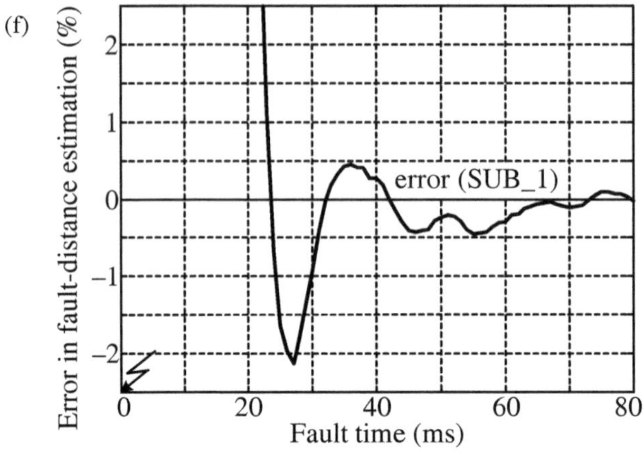

Fig. 6.30 (*continued*)

The fault-locator input signals are shown in Figs. 6.30a and b. Results for estimated distance to fault from both subroutines (Fig. 6.30c) could be accepted and thus selection of the valid result is required. A lower positive fault resistance (Fig. 6.30d) is yielded by the subroutine SUB_1 and this supports selection of it as the valid subroutine. Also, the selection criterion based on calculating the healthy phase total fault current (Fig. 6.30e) indicates the subroutine SUB_1 as the valid one. The estimation error for the fault distance (from the valid subroutine SUB_1) shown in Fig. 6.30f is definitely acceptable, as it does not exceed 0.5% after around 1.5 fundamental-frequency cycles since fault incipience. Note: full-cycle Fourier filtration was applied for determining the phasors of the processed signals.

6.14.3 Fault Location on Line Compensated at Both Ends

The presented earlier fault-location algorithm (for a line compensated at midline) can be easily adapted to a line compensated at both ends (Fig. 6.31), as presented in [272]. Taking into consideration the more difficult case with instrument transformers installed on the bus side (Fig. 2.30a), one can distinguish two characteristic fault positions (Fig. 6.31), as seen by the one-end fault locator installed at the bus A:

- fault F1 on the line between the compensation banks I, II: the subroutine SUB_1; and
- fault F2 overreaching the line length (in the remote system B): the subroutine SUB_2.

The remote compensating bank II has different equivalents, depending on which subroutine is considered. For the subroutine SUB_1 it is dependent on the magnitude of the remote current $\mathbf{I}_B$, which is unavailable for the considered one-end fault locator from the substation A. Therefore, analogously as for the subroutine SUB_2 of the fault-location algorithm for a line compensated with a single compensating bank installed at midline, iterative calculations have to be performed. Selection of the valid subroutine can be done utilizing the same criteria as for a line compensated with a single compensating bank installed at midline.

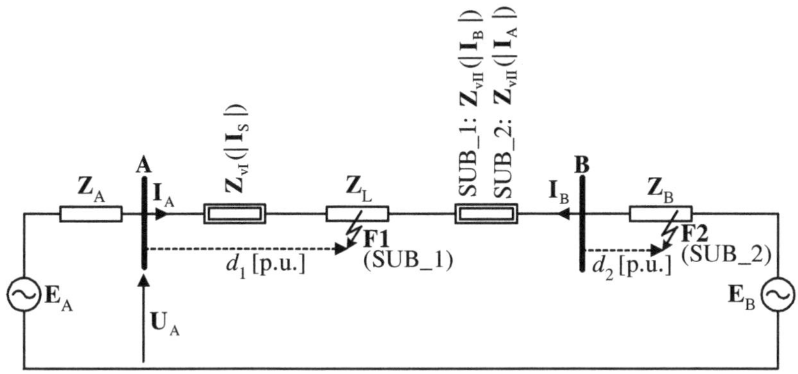

Fig. 6.31 Principle of fault location on single line compensated at both ends

6.14.4 Fault Location on Double-circuit Series-compensated Line

Fault location on a double-circuit line requires additionally taking into account a mutual coupling effect, when adapting the fault-location algorithm, which was designed for a single-circuit line. Two algorithms for application to a double-circuit series-compensated line follow:

- basic algorithm;
- improved algorithm.

6.14.4.1 Basic Algorithm

Analogously, as for the earlier considered configurations of series-compensated lines, there is a need for also applying two subroutines (SUB_1, SUB_2) for double-circuit series-compensated line [284]. The final result is obtained with the selection algorithm, used for indicating the valid subroutine.

Equivalent circuit diagram of double-circuit series-compensated line for the pre-fault state is shown in Fig. 6.32. The description of this circuit is utilized for deriving both subroutines.

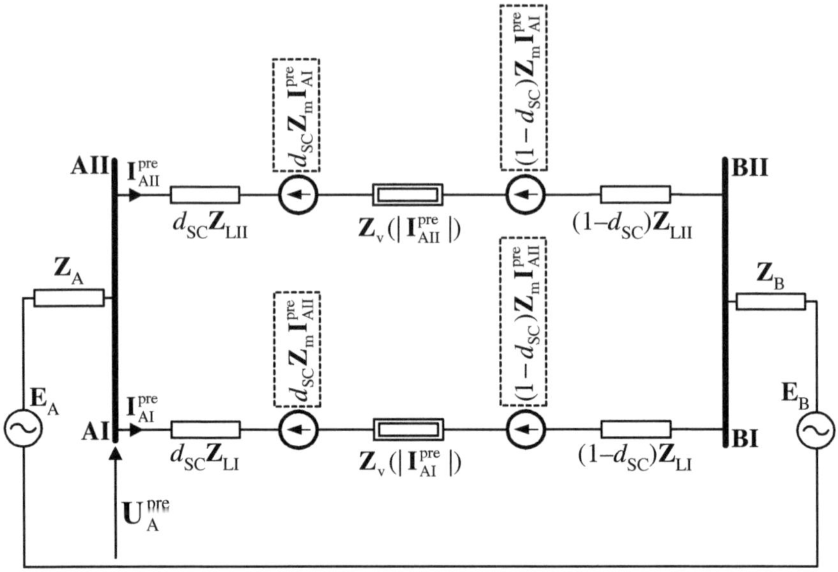

Fig. 6.32 Equivalent circuit diagram of double-circuit series-compensated line for pre-fault state (superscript: pre)

Subroutine SUB_1 – for Faults Behind SCs and MOVs

For faults behind SCs and MOVs (Fig. 6.33) one gets:

$$\mathbf{E}_A - \mathbf{E}_B = \left(\mathbf{Z}_A + d_1\mathbf{Z}_{LI} + \mathbf{Z}_v(|\mathbf{I}_{AI}|)\right)\mathbf{I}_{AI} - \left((1-d_1)\mathbf{Z}_{LI} + \mathbf{Z}_{LII}\right)\mathbf{I}_{BI} + \left(\mathbf{Z}_A + \mathbf{Z}_B + \mathbf{Z}_m\right)\mathbf{I}_{AII} \tag{6.107}$$

$$\mathbf{V}_A - \mathbf{V}_F = \left(d_1\mathbf{Z}_{LI} + \mathbf{Z}_v(|\mathbf{I}_{AI}|)\right)\mathbf{I}_{AI} + d_1\mathbf{Z}_m\mathbf{I}_{AII} \tag{6.108}$$

$$\mathbf{I}_F = \mathbf{I}_{AI} + \mathbf{I}_{BI} \tag{6.109}$$

$$\mathbf{I}_F = \frac{1}{R_{F1}}\mathbf{K}_F\mathbf{U}_F \tag{6.110}$$

where: $d_{SC} < d_1 < 1$ – sought distance to fault (p.u.).

Description for the pre-fault state (Fig. 6.32) supplements the above set of fault formulae (6.107)–(110):

$$\mathbf{E}_A - \mathbf{E}_B = \left(\mathbf{Z}_A + \mathbf{Z}_B + \mathbf{Z}_{LI} + \mathbf{Z}_v(|\mathbf{I}_{AI}^{pre}|)\right)\mathbf{I}_{AI}^{pre} + \left(\mathbf{Z}_A + \mathbf{Z}_B + \mathbf{Z}_m\right)\mathbf{I}_{AII}^{pre} \tag{6.111}$$

From (6.107)–(6.111) one gets to the following matrix formula:

$$\mathbf{A}d_1^{\;2} - \mathbf{B}d_1 + \mathbf{C} - \mathbf{D}R_{\mathrm{F1}} = \mathbf{0} \tag{6.112}$$

where:

$\mathbf{A} = \mathbf{Z}_{\mathrm{LI}}\mathbf{K}_{\mathrm{F}}\mathbf{Z}_{\mathrm{LI}}\mathbf{I}_{\mathrm{AI}} + \mathbf{Z}_{\mathrm{LI}}\mathbf{K}_{\mathrm{F}}\mathbf{Z}_{\mathrm{m}}\mathbf{I}_{\mathrm{AII}}$,

$\mathbf{B} = \mathbf{Z}_{\mathrm{LI}}\mathbf{K}_{\mathrm{F}}\left(\mathbf{U}_{\mathrm{A}} + \left(\mathbf{Z}_{\mathrm{LI}} - \mathbf{Z}_{\mathrm{v}}(|\,\mathbf{I}_{\mathrm{AI}}\,|)\right)\mathbf{I}_{\mathrm{AI}}\right) + \mathbf{Z}_{\mathrm{B}}\mathbf{K}_{\mathrm{F}}\mathbf{Z}_{\mathrm{LI}}\mathbf{I}_{\mathrm{AI}} + \mathbf{Z}_{\mathrm{m}}\mathbf{K}_{\mathrm{F}}\left(\mathbf{Z}_{\mathrm{LI}} + \mathbf{Z}_{\mathrm{B}}\right)\mathbf{I}_{\mathrm{AII}}$,

$\mathbf{C} = \left(\mathbf{Z}_{\mathrm{LI}} + \mathbf{Z}_{\mathrm{B}}\right)\mathbf{K}_{\mathrm{F}}\left(\mathbf{U}_{\mathrm{A}} - \mathbf{Z}_{\mathrm{v}}(|\,\mathbf{I}_{\mathrm{AI}}\,|)\mathbf{I}_{\mathrm{AI}}\right)$,

$\mathbf{D} = \left(\mathbf{Z}_{\mathrm{A}} + \mathbf{Z}_{\mathrm{B}}\right)\left(\mathbf{I}_{\mathrm{AI}} + \mathbf{I}_{\mathrm{AII}} - \mathbf{I}_{\mathrm{AI}}^{\mathrm{pre}} - \mathbf{I}_{\mathrm{AII}}^{\mathrm{pre}}\right) + \mathbf{Z}_{\mathrm{v}}(|\,\mathbf{I}_{\mathrm{AI}}\,|)\mathbf{I}_{\mathrm{AI}} - \mathbf{Z}_{\mathrm{v}}(|\,\mathbf{I}_{\mathrm{AI}}^{\mathrm{pre}}\,|)\mathbf{I}_{\mathrm{AI}}^{\mathrm{pre}}$
$\quad + \mathbf{Z}_{\mathrm{LI}}\left(\mathbf{I}_{\mathrm{AI}} - \mathbf{I}_{\mathrm{AI}}^{\mathrm{pre}}\right) + \mathbf{Z}_{\mathrm{m}}\left(\mathbf{I}_{\mathrm{AII}} - \mathbf{I}_{\mathrm{AII}}^{\mathrm{pre}}\right)$.

Transforming of (6.112) into the scalar form, to get the quadratic formula (for complex numbers) in two unknowns: distance to fault (d_1) and fault resistance (R_{F1}) can be performed identically as was presented for the matrix formula (6.102).

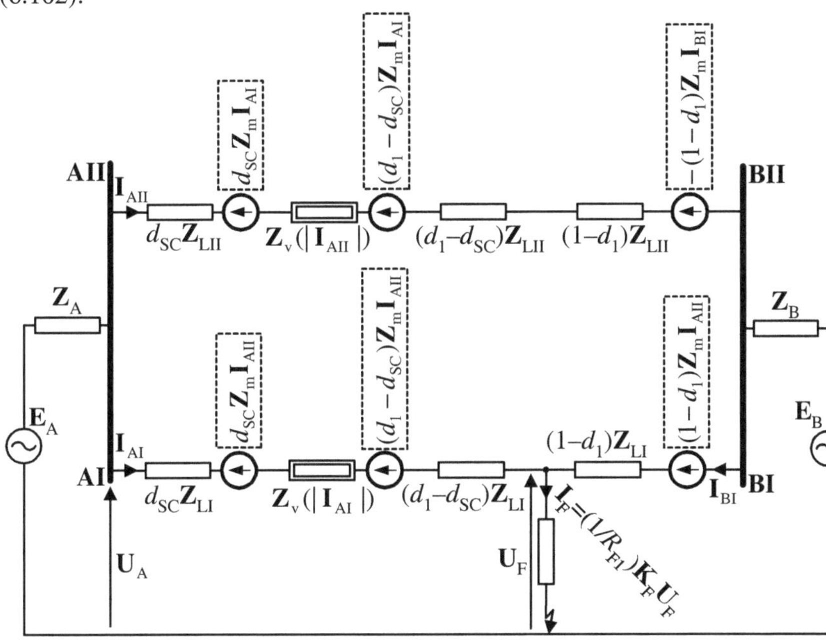

Fig. 6.33 Circuit diagram of double-circuit series-compensated line for fault behind SCs and MOVs

Subroutine SUB_2 – for Faults in Front of SCs and MOVs

For faults occurring in front of SCs and MOVs, one obtains from analysis of the circuit diagram from Fig. 6.34:

$$\begin{aligned}\mathbf{E}_{\mathrm{A}} - \mathbf{E}_{\mathrm{B}} = {} & \left(\mathbf{Z}_{\mathrm{A}} + d_2\mathbf{Z}_{\mathrm{LI}}\right)\mathbf{I}_{\mathrm{AI}} + \left(\mathbf{Z}_{\mathrm{A}} + \mathbf{Z}_{\mathrm{B}} + \mathbf{Z}_{\mathrm{m}}\right)\mathbf{I}_{\mathrm{AII}} \\ & - \left((1-d_2)\mathbf{Z}_{\mathrm{LI}} + \mathbf{Z}_{\mathrm{B}} + \mathbf{Z}_{\mathrm{v}}(|\,\mathbf{I}_{\mathrm{BI}}\,|)\right)\mathbf{I}_{\mathrm{BI}}\end{aligned} \tag{6.113}$$

$$\mathbf{U}_{\mathrm{A}} - \mathbf{U}_{\mathrm{F}} = d_2\mathbf{Z}_{\mathrm{LI}}\mathbf{I}_{\mathrm{AI}} + d_2\mathbf{Z}_{\mathrm{m}}\mathbf{I}_{\mathrm{AII}} \tag{6.114}$$

$$\mathbf{I}_{\mathrm{F}} = \mathbf{I}_{\mathrm{AI}} + \mathbf{I}_{\mathrm{BI}} \tag{6.115}$$

$$\mathbf{I}_{\mathrm{F}} = \frac{1}{R_{\mathrm{F2}}}\mathbf{K}_{\mathrm{F}}\mathbf{U}_{\mathrm{F}} \tag{6.116}$$

where: $0 < d_2 < d_{\mathrm{SC}}$ – sought distance to fault (p.u.).

Utilizing (6.113)–(6.116) together with the description for the pre-fault state (6.11) leads to the following matrix formula:

$$\mathbf{A}d_2^{\,2} - \mathbf{B}d_2 + \mathbf{C} - \mathbf{D}R_{\mathrm{F2}} = \mathbf{0} \tag{6.117}$$

where:

$\mathbf{A} = \mathbf{Z}_{\mathrm{LI}}\mathbf{K}_{\mathrm{F}}\mathbf{Z}_{\mathrm{LI}}\mathbf{I}_{\mathrm{AI}} + \mathbf{Z}_{\mathrm{LI}}\mathbf{K}_{\mathrm{F}}\mathbf{Z}_{\mathrm{m}}\mathbf{I}_{\mathrm{AII}}$,

$\mathbf{B} = \mathbf{Z}_{\mathrm{LI}}\mathbf{K}_{\mathrm{F}}\mathbf{U}_{\mathrm{A}} + (\mathbf{Z}_{\mathrm{LI}} + \mathbf{Z}_{\mathrm{v}}(|\mathbf{I}_{\mathrm{BI}}|) + \mathbf{Z}_{\mathrm{B}})\mathbf{K}_{\mathrm{F}}\mathbf{Z}_{\mathrm{LI}}\mathbf{I}_{\mathrm{AI}} + \mathbf{Z}_{\mathrm{m}}\mathbf{K}_{\mathrm{F}}(\mathbf{Z}_{\mathrm{LI}} + \mathbf{Z}_{\mathrm{B}} + \mathbf{Z}_{\mathrm{v}}(|\mathbf{I}_{\mathrm{BI}}|))\mathbf{I}_{\mathrm{AII}}$,

$\mathbf{C} = (\mathbf{Z}_{\mathrm{LI}} + \mathbf{Z}_{\mathrm{B}} + \mathbf{Z}_{\mathrm{v}}(|\mathbf{I}_{\mathrm{BI}}|))\mathbf{K}_{\mathrm{F}}\mathbf{U}_{\mathrm{A}}$,

$$\mathbf{D} = (\mathbf{Z}_{\mathrm{A}} + \mathbf{Z}_{\mathrm{B}})(\mathbf{I}_{\mathrm{AI}} + \mathbf{I}_{\mathrm{AII}} - \mathbf{I}_{\mathrm{AI}}^{\mathrm{pre}} - \mathbf{I}_{\mathrm{AII}}^{\mathrm{pre}}) + \mathbf{Z}_{\mathrm{v}}(|\mathbf{I}_{\mathrm{BI}}|)\mathbf{I}_{\mathrm{AI}} - \mathbf{Z}_{\mathrm{v}}(|\mathbf{I}_{\mathrm{AI}}^{\mathrm{pre}}|)\mathbf{I}_{\mathrm{AI}}^{\mathrm{pre}} + \mathbf{Z}_{\mathrm{LI}}(\mathbf{I}_{\mathrm{AI}} - \mathbf{I}_{\mathrm{AI}}^{\mathrm{pre}}) + \mathbf{Z}_{\mathrm{m}}(\mathbf{I}_{\mathrm{AII}} - \mathbf{I}_{\mathrm{AII}}^{\mathrm{pre}}).$$

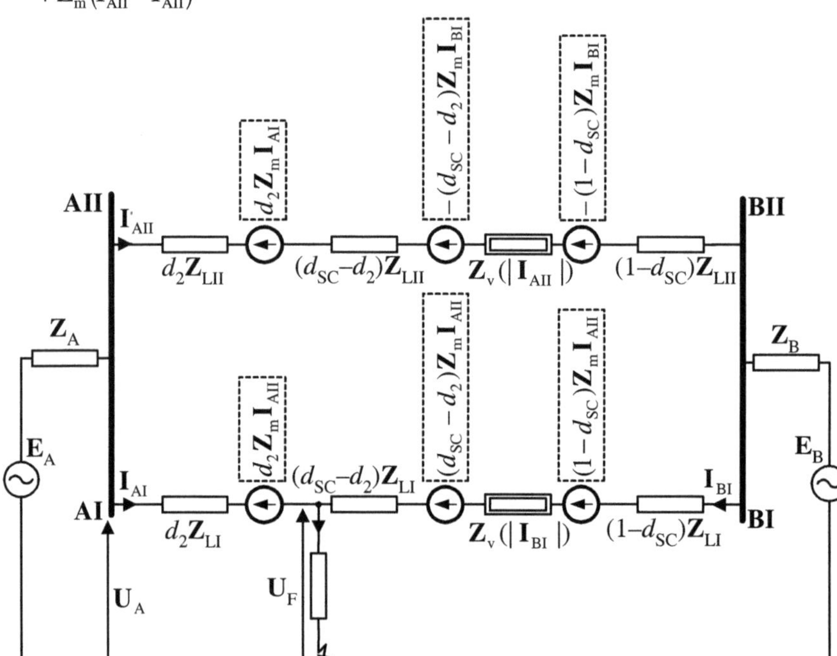

Fig. 6.34 Circuit diagram of double-circuit series-compensated line for fault in front of SCs and MOVs

Transformation of matrix formula (6.117) into the scalar form can be performed analogously as for the subroutine SUB_1. Then, the unknowns: d_2, R_{F2} are calculated, but an iterative numerical solution (as shown in Fig. 6.29b) is required because the coefficients depend on the unknown current from the remote substation B.

Having the results yielded by the subroutines SUB_1, SUB_2 one has to select the valid subroutine. This can be performed by applying the same approach as was presented for a line compensated with a single compensating bank installed at midline.

6.14.4.2 Improved Algorithm

The basic fault-location algorithm, for which the matrix formulae: (6.112) and (6.117) were derived for the subroutines SUB_1 and SUB_2, requires knowing the source impedances $\mathbf{Z}_A$, $\mathbf{Z}_B$. Also the pre-fault measurements of currents $\mathbf{I}_{AI}^{pre}$, $\mathbf{I}_{AII}^{pre}$ at the local buses AI, AII are required. Both requiring the source impedances and use of pre-fault measurements can be, in some cases, considered as the fault-location algorithm limitations. In order to assure the possibly highest accuracy of fault location, these limitations have to be overcome. This aim was achieved in [285], where the improved fault-location algorithm for a double-circuit series-compensated line has been introduced. The improved algorithm also applies two subroutines, which follow.

Subroutine SUB_1 of Improved Algorithm – for Faults Behind SCs and MOVs

Let us consider the case of faults behind the SCs and MOVs, but not overreaching the line length (Fig. 6.33). Again, the local ($\mathbf{I}_{AI}$) and remote current ($\mathbf{I}_{BI}$) from the faulted circuit contribute to the total fault current:

$$\mathbf{I}_F = \mathbf{I}_{AI} + \mathbf{I}_{BI} \tag{6.118}$$

In order to apply (6.118) together with the fault model (6.110), the remote current ($\mathbf{I}_{BI}$) and the voltage at fault ($\mathbf{U}_F$) have to be expressed in terms of the local measurements of the fault locator. For this purpose the faulted- and the healthy-line circuits are considered:

- Faulted-line circuit:

$$\mathbf{U}_F = \mathbf{U}_A - d_1\mathbf{Z}_{LI}\mathbf{I}_{AI} - \mathbf{Z}_v(|\,\mathbf{I}_{AI}\,|)\mathbf{I}_{AI} - d_1\mathbf{Z}_m\mathbf{I}_{AII} \tag{6.119}$$

$$\mathbf{U}_F = \mathbf{U}_\mathbf{B} - (1-d_1)\mathbf{Z}_{LI}\mathbf{I}_{BI} + (1-d_1)\mathbf{Z}_m\mathbf{I}_{AII} \tag{6.120}$$

- Healthy-line circuit:

$$\mathbf{U}_{\mathrm{A}}-\mathbf{U}_{\mathrm{B}}=\left(\mathbf{Z}_{\mathrm{LII}}+\mathbf{Z}_{\mathrm{v}}(|\,\mathbf{I}_{\mathrm{AII}}\,|)\right)\mathbf{I}_{\mathrm{AII}}+d_1\mathbf{Z}_{\mathrm{m}}\mathbf{I}_{\mathrm{AI}}-(1-d_1)\mathbf{Z}_{\mathrm{m}}\mathbf{I}_{\mathrm{BI}} \tag{6.121}$$

The set of formulae (6.110) and (6.118)–(6.121) yields the matrix equation for the fault distance (d_1) and aggregated fault resistance (R_{F1}):

$$\mathbf{A}d_1^{\,2}-\ \mathbf{B}d_1+\mathbf{C}-\mathbf{D}R_{\mathrm{F1}}=\mathbf{0} \tag{6.122}$$

in which the column matrices (3×1 matrices) are determined as follows:

$\mathbf{A}=\left(\mathbf{Z}_{\mathrm{m}}-\mathbf{Z}_{\mathrm{LI}}\right)\mathbf{K}_{\mathrm{F}}\left(\mathbf{Z}_{\mathrm{LI}}\mathbf{I}_{\mathrm{AI}}+\mathbf{Z}_{\mathrm{m}}\mathbf{I}_{\mathrm{AII}}\right)$,
$\mathbf{C}=\left(\mathbf{Z}_{\mathrm{m}}-\mathbf{Z}_{\mathrm{LI}}\right)\mathbf{K}_{\mathrm{F}}\left(\mathbf{U}_{\mathrm{A}}-\mathbf{Z}_{\mathrm{v}}(|\,\mathbf{I}_{\mathrm{AI}}\,|)\mathbf{I}_{\mathrm{AI}}\right)$,
$\mathbf{B}=\mathbf{A}+\mathbf{C}$,
$\mathbf{D}=\left(\mathbf{Z}_{\mathrm{m}}-\mathbf{Z}_{\mathrm{LI}}-\mathbf{Z}_{\mathrm{v}}(|\,\mathbf{I}_{\mathrm{AI}}\,|)\right)\mathbf{I}_{\mathrm{AI}}-\left(\mathbf{Z}_{\mathrm{m}}-\mathbf{Z}_{\mathrm{LII}}-\mathbf{Z}_{\mathrm{v}}(|\,\mathbf{I}_{\mathrm{AII}}\,|)\right)\mathbf{I}_{\mathrm{AII}}$.

In the obtained matrix formula (6.122) the source impedances are not involved and the measurements only from the fault interval are utilized. This is advantageous for assuring high accuracy of fault location.

Subroutine SUB_2 of Improved Algorithm – for Faults in Front of SCs and MOVs

The case of a fault between the considered fault locator (at the bus AI) and the SCs (Fig. 6.34) is more involved because the equivalent of SCs and MOVs from the faulted circuit is dependent on the current, which is not directly available to the one-end locator at the bus A and thus ought to be estimated.

In this case the formulae (6.110), (6.118) are valid but (6.119)–(6.121) have to be replaced by the following:

- Faulted-line circuit:

$$\mathbf{U}_{\mathrm{F}}=\mathbf{U}_{\mathrm{A}}-d_2\mathbf{Z}_{\mathrm{LI}}\mathbf{I}_{\mathrm{AI}}-d_2\mathbf{Z}_{\mathrm{m}}\mathbf{I}_{\mathrm{AII}} \tag{6.123}$$

$$\mathbf{U}_{\mathrm{F}}=\mathbf{U}_{\mathrm{B}}-(1-d_2)\left(\mathbf{Z}_{\mathrm{LI}}\mathbf{I}_{\mathrm{BI}}-\mathbf{Z}_{\mathrm{m}}\mathbf{I}_{\mathrm{AII}}\right)-\mathbf{Z}_{\mathrm{v}}(|\,\mathbf{I}_{\mathrm{BI}}\,|)\mathbf{I}_{\mathrm{BI}} \tag{6.124}$$

- Healthy-line circuit:

$$\mathbf{U}_{\mathrm{A}}-\mathbf{U}_{\mathrm{B}}=\left(\mathbf{Z}_{\mathrm{LII}}+\mathbf{Z}_{\mathrm{v}}(|\,\mathbf{I}_{\mathrm{AII}}\,|)\right)\mathbf{I}_{\mathrm{AII}}+d_2\mathbf{Z}_{\mathrm{m}}\mathbf{I}_{\mathrm{AI}}-(1-d_2)\mathbf{Z}_{\mathrm{m}}\mathbf{I}_{\mathrm{BI}} \tag{6.125}$$

The set of formulae (6.110), (6.118), (6.123)–(6.125) yields the following matrix formula:

$$\mathbf{A}d_2^{\,2}-\ \mathbf{B}d_2+\mathbf{C}-\mathbf{D}R_{\mathrm{F2}}=\mathbf{0} \tag{6.126}$$

in which the column matrices (3×1 matrices) are determined as follows:

$$\mathbf{A} = (\mathbf{Z}_{\text{m}} - \mathbf{Z}_{\text{LI}})\mathbf{K}_{\text{F}}(\mathbf{Z}_{\text{LI}}\mathbf{I}_{\text{AI}} + \mathbf{Z}_{\text{m}}\mathbf{I}_{\text{AII}}),$$

$$\mathbf{B} = (\mathbf{Z}_{\text{m}} - \mathbf{Z}_{\text{LI}})\mathbf{K}_{\text{F}}(\mathbf{U}_{\text{A}} + \mathbf{Z}_{\text{LI}}\mathbf{I}_{\text{AI}} + \mathbf{Z}_{\text{m}}\mathbf{I}_{\text{AII}}) - \mathbf{Z}_{\text{v}}(|\,\mathbf{I}_{\text{BI}}\,|)\mathbf{K}_{\text{F}}(\mathbf{Z}_{\text{LI}}\mathbf{I}_{\text{AI}} + \mathbf{Z}_{\text{m}}\mathbf{I}_{\text{AII}}),$$

$$\mathbf{C} = (\mathbf{Z}_{\text{m}} - \mathbf{Z}_{\text{LI}} - \mathbf{Z}_{\text{v}}(|\,\mathbf{I}_{\text{RI}}\,|)\mathbf{K}_{\text{F}}\mathbf{U}_{\text{A}},$$

$$\mathbf{D} = (\mathbf{Z}_{\text{m}} - \mathbf{Z}_{\text{LI}} - \mathbf{Z}_{\text{v}}(|\,\mathbf{I}_{\text{BI}}\,|))\mathbf{I}_{\text{AI}} - (\mathbf{Z}_{\text{m}} - \mathbf{Z}_{\text{LII}} - \mathbf{Z}_{\text{v}}(|\,\mathbf{I}_{\text{AII}}\,|))\mathbf{I}_{\text{AII}}.$$

Again, in the obtained matrix formula (6.126) the source impedances are not involved and the measurements only from the fault interval are utilized. This is advantageous for assuring high accuracy of fault location.

Transformation of the matrix formulae (6.122) and (6.126) can be performed identically as was presented for the matrix formula (6.102).

Having the results yielded by the subroutines SUB_1, SUB_2 one has to select the valid subroutine. This can be performed by applying the same approach as was presented for a line compensated with a single compensating bank installed at midline.

In [285] some fault examples showing the benefits of using the improved fault-location algorithm are delivered.

6.14.5 Differential-equation-based Algorithm

In [254, 279] the one-end fault-location algorithm for a series-compensated line, which is based on differential-equation approach has been delivered. This algorithm (SUB_1diff) has been developed with the aim of applying it as the supplementary subroutine to the subroutine SUB_1 (designated for locating faults behind SCs and MOVs), formulated with use of the fundamental-frequency phasor approach. The differential-equation approach has advantages for locating faults behind SCs and MOVs involving high fault resistances, i.e., for the fault cases when the MOVs operate linearly or almost linearly, which results in an increase in slowly decaying subharmonic components.

The fault-location algorithm based on the differential-equation approach [254, 279] utilizes a Clarke transformation (2.44) [25, 56, 254, 279].

Clarke Transformation

Transformation of the phase coordinates (in the time domain) of a 3-phase (a, b, c) symmetrical system into the 0, α, β coordinates can be performed according to the following matrix relation:

$$\mathbf{i}_{0\alpha\beta}(t) = \mathbf{C}\mathbf{i}_{\text{abc}}(t) \qquad (6.127)$$

where:

$$\mathbf{i}_{0\alpha\beta}(t)=\begin{bmatrix} i_0(t) \\ i_\alpha(t) \\ i_\beta(t) \end{bmatrix}, \quad \mathbf{i}_{\text{abc}}(t)=\begin{bmatrix} i_\text{a}(t) \\ i_\text{b}(t) \\ i_\text{c}(t) \end{bmatrix}, \quad \mathbf{C}=\frac{1}{3}\begin{bmatrix} 1 & 1 & 1 \\ 2 & -1 & -1 \\ 0 & \sqrt{3} & -\sqrt{3} \end{bmatrix}$$

It is also possible to use the Clarke transformation to phasor representation of a three-phase system. In this case the instantaneous 0, α, β coordinates can be obtained by taking into account only a real part of the complex transformation.

Considering the phasor representation of currents, the following relations between the Clarke and symmetrical components can be derived:

$$\mathbf{i}_{0\alpha\beta}=\mathbf{C}_\text{s}\mathbf{i}_{012} \tag{6.128}$$

$$\mathbf{i}_{012}=\mathbf{C}_\text{s}^{-1}\mathbf{i}_{0\alpha\beta} \tag{6.129}$$

where:

$$\mathbf{C}_\text{s}=\begin{bmatrix} 1 & 0 & 0 \\ 0 & 1 & 1 \\ 0 & -\text{j} & \text{j} \end{bmatrix}, \qquad \mathbf{C}_\text{s}^{-1}=\frac{1}{2}\begin{bmatrix} 2 & 0 & 0 \\ 0 & 1 & \text{j} \\ 0 & 1 & -\text{j} \end{bmatrix}$$

Analogously is for voltages, and as a result of that, one derives the relation between impedance parameters for the 0, α, β and symmetrical sequences:

$$\mathbf{Z}_{0\alpha\beta}=\mathbf{C}_\text{s}\mathbf{Z}_{012}\mathbf{C}_\text{s}^{-1}=\frac{1}{2}\begin{bmatrix} 2\underline{Z}_0 & 0 & 0 \\ 0 & \underline{Z}_1+\underline{Z}_2 & \text{j}(\underline{Z}_1-\underline{Z}_2) \\ 0 & -\text{j}(\underline{Z}_1-\underline{Z}_2) & \underline{Z}_1+\underline{Z}_2 \end{bmatrix} \tag{6.130}$$

The transmission network is a static network (positive- and negative-sequence parameters are identical: $\underline{Z}_1=\underline{Z}_2$), and therefore one gets that impedance parameters for the 0, α, β and symmetrical sequences are respectively equal to each other:

$$\{\underline{Z}_0\}_{0\alpha\beta}=\{\underline{Z}_0\}_{012}, \quad \underline{Z}_\alpha=\underline{Z}_1, \quad \underline{Z}_\beta=\underline{Z}_1 \tag{6.131}$$

Description of the Transmission Network for 0, α, β Modes

Taking into account the above *considerations* with respect to the Clarke transformation (6.131) one can represent the transmission network with the series-compensated line as depicted in Fig. 6.35. The equivalent diagram from Fig. 6.35 is general and is valid for each of 0, α, β modes.

It is assumed in Fig. 6.35 that the SCs and MOVs banks are installed at distance d_{SC} (p.u.) from the terminal A. The respective voltage drops across the SCs and MOVs (U_v) for particular 0, α, β modes represent the compensating banks. These voltage drops are obtained from the instantaneous voltage drops, which in this algorithm are considered as estimated by solving the non-linear differential equation stated for the SC and MOV circuit [277]. The estimation algorithm is based on Newton–Raphson method iterative calculations according to (6.91) and (6.192).

The respective resistances and inductances represent the transmission line (note: shunt parameters of the line are not taken into account at this stage). The respective impedance parameters and the emfs represent the equivalent systems behind the line terminals. For the balanced 3-phase sources their representation for the 0 mode involves the emf equal to zero.

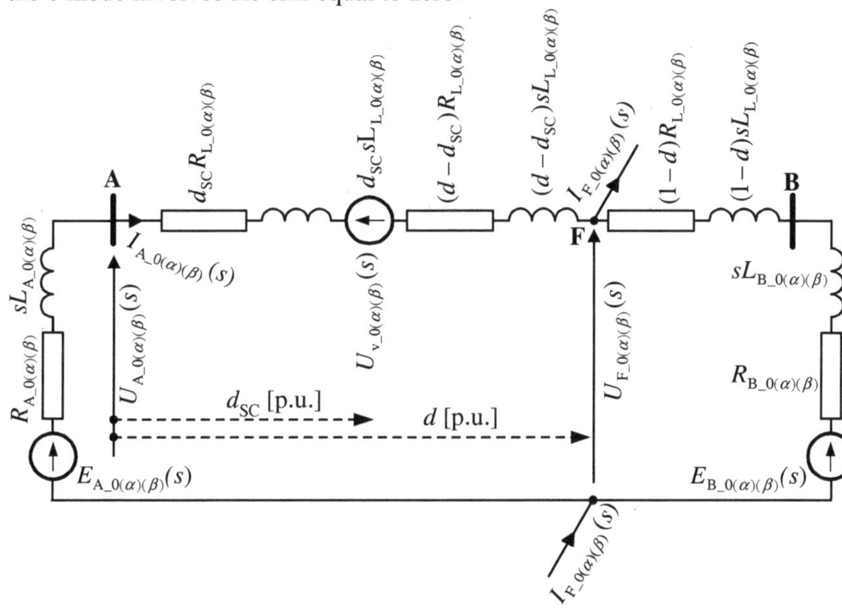

Fig. 6.35 Equivalent circuit diagram of the investigated transmission network for 0, α, β modal components

Generalized Fault-loop Model

A generalized model of the fault loop covering different fault types can be described in a similar way as for the symmetrical components [121]. Taking the Laplace transformation (s – Laplace operator) one obtains the following fault-loop model:

$$U_{A\_P}(s) - dZ_{1L}(s)I_{A\_P}(s) - U_{v\_P}(s) - R_F\left(a_{F0}I_{F0}(s) + a_{F\alpha}I_{F\alpha}(s) + a_{F\beta}I_{F\beta}(s)\right) = 0 \quad (6.132)$$

where:

d – sought distance to fault (p.u.),

$Z_{1L}(s) = R_{1L} + sL_{1L}$ – positive-sequence line impedance,

$U_{A\_P}(s)$, $I_{A\_P}(s)$ – fault-loop voltage and current, composed according to the fault type,

$U_{\mathrm{v\_P}}(s)$ – voltage drop across SC and MOV circuit for the considered fault loop (single phase or inter-phase loop),
R_{F} – fault resistance,
$I_{\mathrm{F}i}(s)$ – components of the total fault current for the Clarke 0, α, β coordinates,
$a_{\mathrm{F}i}$ – share coefficients dependent on fault type.

Fault-loop voltage and current from (6.132) are determined in the following way:

$$U_{\mathrm{A\_P}}(s) = a_0 U_{\mathrm{A0}}(s) + a_\alpha U_{\mathrm{A}\alpha} + a_\beta U_{\mathrm{A}\beta}(s) \tag{6.133}$$

$$I_{\mathrm{A\_P}}(s) = a_0 \frac{Z_{\mathrm{0L}}(s)}{Z_{\mathrm{1L}}(s)} I_{\mathrm{A0}}(s) + a_\alpha I_{\mathrm{A}\alpha} + a_\beta I_{\mathrm{A}\beta}(s) \tag{6.134}$$

where:
$Z_{\mathrm{0L}}(s) = R_{\mathrm{0L}} + sL_{\mathrm{0L}}$ – zero-sequence line impedance.

The weighting coefficients a_0, a_α, a_β in (6.133) and (6.134) and the share coefficients a_{F0}, $a_{\mathrm{F}\alpha}$, $a_{\mathrm{F}\beta}$ from (6.132) have to be determined in such a way that they take the real number values, which substantially facilitates the algorithm formulation. Using the transformation rules (6.128) and (6.129) one determines their values, which are gathered in Tables 6.10 and 6.11.

Table 6.10 Composition of fault-loop signals with use of 0, α, β components, according to (6.133) and (6.134)

Fault type	a_0	a_α	a_β
a–g	1	1	0
b–g	1	–0.5	$0.5\sqrt{3}$
c–g	1	–0.5	$-0.5\sqrt{3}$
a–b, a–b–g a–b–c, a–b–c–g	0	1.5	$-0.5\sqrt{3}$
b–c, b–c–g	0	0	$\sqrt{3}$
c–a, c–a–g	0	–1.5	$-0.5\sqrt{3}$

The share coefficients for the α, β modes can be determined as follows:

$$a_{\mathrm{F}\alpha} = 0.5(a_{\mathrm{F1}} + a_{\mathrm{F2}}) \tag{6.135}$$

$$a_{\mathrm{F}\beta} = 0.5\,\mathrm{j}(a_{\mathrm{F1}} - a_{\mathrm{F2}}) \tag{6.136}$$

The share coefficient for the zero mode is equal to the share coefficient for the zero-sequence.

There is some freedom in selecting the share coefficients [121]. In order to get the share coefficients (6.135) and (6.136) in the form of real numbers (Table 6.11), the special set of these coefficients for the symmetrical components has been selected, i.e., such as in Table 3.5, where the set of weighting coefficients with elimination of the zero-sequence and using both positive- and negative-sequences is contained.

Table 6.11 Sets of weighting coefficients for determining the voltage drop across the fault-path resistance

Fault type	a_{F0}	$a_{F\alpha}$	$a_{F\beta}$
a–g	0	1.5	0
b–g	0	–0.75	$0.75\sqrt{3}$
c–g	0	–0.75	$-0.75\sqrt{3}$
a–b	0	0.75	$-0.25\sqrt{3}$
b–c	0	0	$0.5\sqrt{3}$
c–a	0	–0.75	$-0.25\sqrt{3}$
a–b–g	0	1.5	$-0.5\sqrt{3}$
b–c–g	0	0	$\sqrt{3}$
c–a–g	0	–1.5	$-0.5\sqrt{3}$
a–b–c–g, a–b–c	0	1.5	$-0.5\sqrt{3}$

The next step in the derivation of the fault-location algorithm is related to the determination of the total fault current components for the respective 0, α, β modes. This requires considering the flow of fault currents in the circuits for particular, i.e., for the 0, α, β modes (Fig. 6.36 shows this circuit for the α mode). As a result of this one obtains:

$$I_{F\alpha}(s) = \frac{\Delta I_{A\alpha}(s)}{k_{F\alpha}(s)} + \frac{\Delta U_{v\alpha}(s)}{Z_{\Delta v\alpha}(s)} \tag{6.137}$$

$$I_{F\beta}(s) = \frac{\Delta I_{A\beta}(s)}{k_{F\beta}(s)} + \frac{\Delta U_{v\beta}(s)}{Z_{\Delta v\beta}(s)} \tag{6.138}$$

$$I_{F0}(s) = \frac{I_{A0}(s)}{k_{F0}(s)} + \frac{\Delta U_{v0}(s)}{Z_{\Delta v0}(s)} \tag{6.139}$$

where:

$I_{F0}(s)$, $I_{F\alpha}(s)$; $I_{F\beta}(s)$ – 0, α, β components of the total fault current,

$k_{F0}(s)$, $k_{F\alpha}(s)$, $k_{F\beta}(s)$ – fault current distribution factors for particular modal quantities,

$I_{A0}(s)$, $\Delta I_{A\alpha}(s)=I_{A\alpha}(s)-I_{A\alpha}^{pre}(s)$ and $\Delta I_{A\beta}(s)=I_{A\beta}(s)-I_{A\beta}^{pre}(s)$ – 0, α, β components of currents measured at the terminal A,

$U_{v0}(s)$, $\Delta U_{v\alpha}(s)=U_{v\alpha}(s)-U_{v\alpha}^{pre}(s)$ and $\Delta U_{v\beta}(s)=U_{v\beta}(s)-U_{v\beta}^{pre}(s)$ – 0, α, β components of incremental voltage drop across the SC and MOV, as seen from the terminal A,

$Z_{\Delta v0}(s)$, $Z_{\Delta v\alpha}(s)$, $Z_{\Delta v\beta}(s)$ – impedance of the equivalent SC and MOV circuit, which reflects participation of the adequate voltage components: $\Delta U_{v0}(s)$, $\Delta U_{v\alpha}(s)$ and $\Delta U_{v\beta}(s)$ in particular components of the total fault current.

Note that in the cases of the α and β modes the incremental quantity (fault current minus pre-fault quantity) is used.

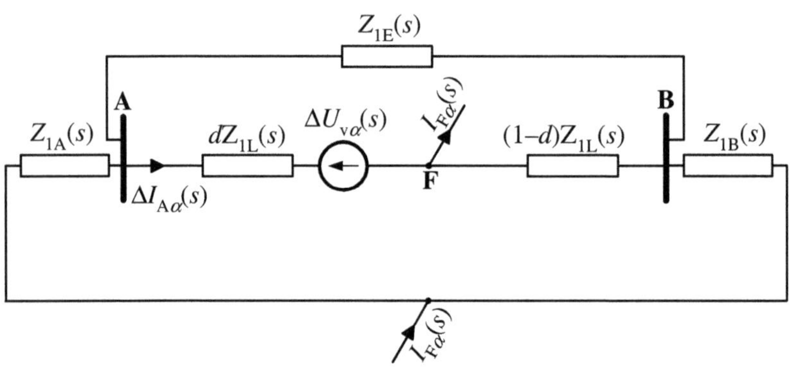

Fig. 6.36 Circuit diagram for α mode, for determining total fault current

Fault current distribution factors and the mentioned impedances of the equivalent circuits depend on the configuration of the transmission network (Fig. 6.36) and impedance parameters. Assuming that all impedances for the positive- and negative-sequence are equal to each other, one obtains analogously as for the symmetrical-components approach:

$$k_{F\alpha}(s)=k_{F\beta}(s)=\frac{dK_1(s)+L_1(s)}{M_1(s)} \tag{6.140}$$

$$Z_{\Delta v\alpha}(s)=Z_{\Delta v\beta}(s)=\frac{dK_1(s)+L_1(s)}{N_1(s)} \tag{6.141}$$

Coefficients, which are used for composing the fault current distribution factor (6.140) and impedance of the equivalent SC and MOV circuit (6.141) are gathered in Table 6.12. The case of a single line (Fig. 6.36) has been considered for this. Parameters for the zero-sequence circuit are not considered here because the share coefficient for the zero-sequence and thus for the 0 mode equals zero.

Table 6.12 Coefficients used for composing fault current distribution factors

$Z_{1E}(s) \neq \infty$	$K_1(s) = -Z_{1L}(s)Z_{1E}(s) - (Z_{1A}(s)+Z_{1B}(s))Z_{1L}(s)$ $L_1(s) = Z_{1L}(s)(Z_{1A}(s)+Z_{1B}(s)) + Z_{1E}(s)(Z_{1L}(s)+Z_{1B}(s))$ $M_1(s) = (Z_{1A}(s)+Z_{1B}(s))(Z_{1E}(s)+Z_{1L}(s)) + Z_{1L}(s)Z_{1E}(s)$ $N_1(s) = Z_{1A}(s)+Z_{1B}(s)+Z_{1E}(s)$
$Z_{1E}(s) \to \infty$	$K_1(s) = -Z_{1L}(s)$ $L_1(s) = Z_{1L}(s)+Z_{1B}(s)$ $M_1(s) = Z_{1A}(s)+Z_{1B}(s)+Z_{1L}(s)$ $N_1(s) = 1$

After introducing the fault current distribution factors (6.140) and (6.141) into the generalized fault-loop model (6.132), the quadratic formula with two unknowns (d (p.u.) – sought fault distance, R_F – fault resistance) is obtained:

$$A(s)d^2 - B(s)d + C(s) - R_F D(s) = 0 \tag{6.142}$$

where:

$A(s) = -K_1(s)Z_{1L}(s)I_{A\_P}(s)$,

$B(s) = (L_1(s)Z_{1L}(s)I_{A\_P}(s) - K_1(s)(U_{A\_P}(s) - U_{v\_P}(s)))$,

$C(s) = L_1(s)(U_{A\_P}(s) - U_{v\_P}(s))$,

$D(s) = M_1(s)(a_{F\alpha}\Delta I_{A\alpha}(s) + a_{F\beta}\Delta I_{A\beta}(s)) + N_1(s)(a_{F\alpha}\Delta U_{v\alpha}(s) + a_{F\beta}\Delta U_{v\beta}(s))$

Then, the bilinear transformation is used for transforming (6.142) from the Laplace (s) domain into the discrete time (k) domain:

$$s = \frac{\omega_1}{\tan(0.5\omega_1 T_s)} \frac{1 - z^{-1}}{1 + z^{-1}} \tag{6.143}$$

where:

T_s – sampling period, ω_1 – angular rated frequency.

Application of (6.143) to (6.142) yields:

$$A(k)d^2 - B(k)d + C(k) - R_F D(k) = 0 \tag{6.144}$$

The coefficients involved in (6.144) are determined by the network parameters, three consecutive samples of the measured signals and three consecutive samples of the estimated voltage drop across the SC and MOV.

Example 6.6 Fault location on series-compensated line with the fault-location algorithm (6.144)
This example is for the 400-kV, 300-km line, compensated at 70% rate by the compensating bank installed in the middle (d_{SC}=0.5 p.u.), a–g fault at d=0.5$^+$ p.u. (just behind SCs and MOVs). In Fig. 6.37 the results are shown.

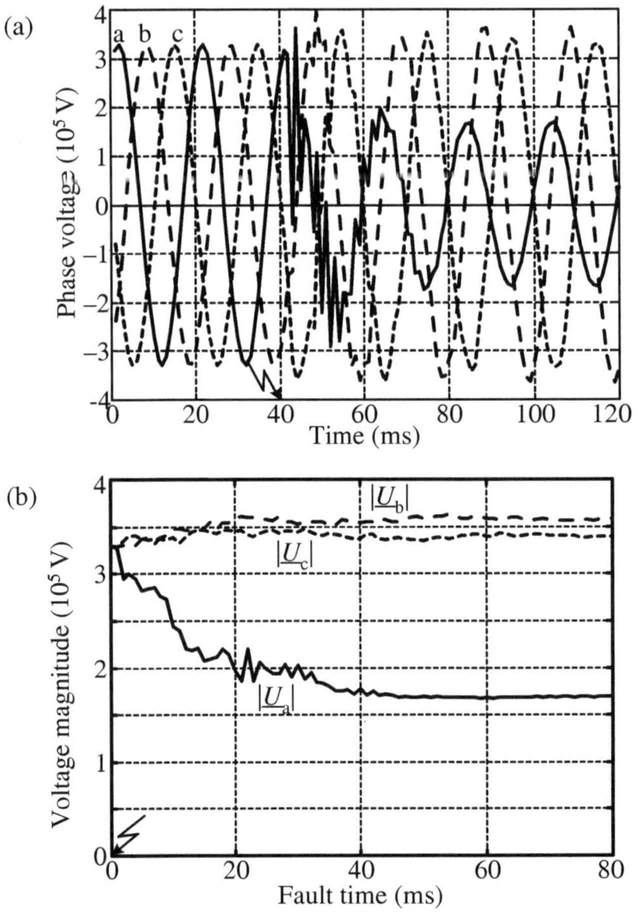

Fig. 6.37 Example 6.6: (a) phase voltage, (b) voltage magnitude, (c) phase current, (d) current magnitude, (e) estimated distance to fault, and (f) estimated fault resistance

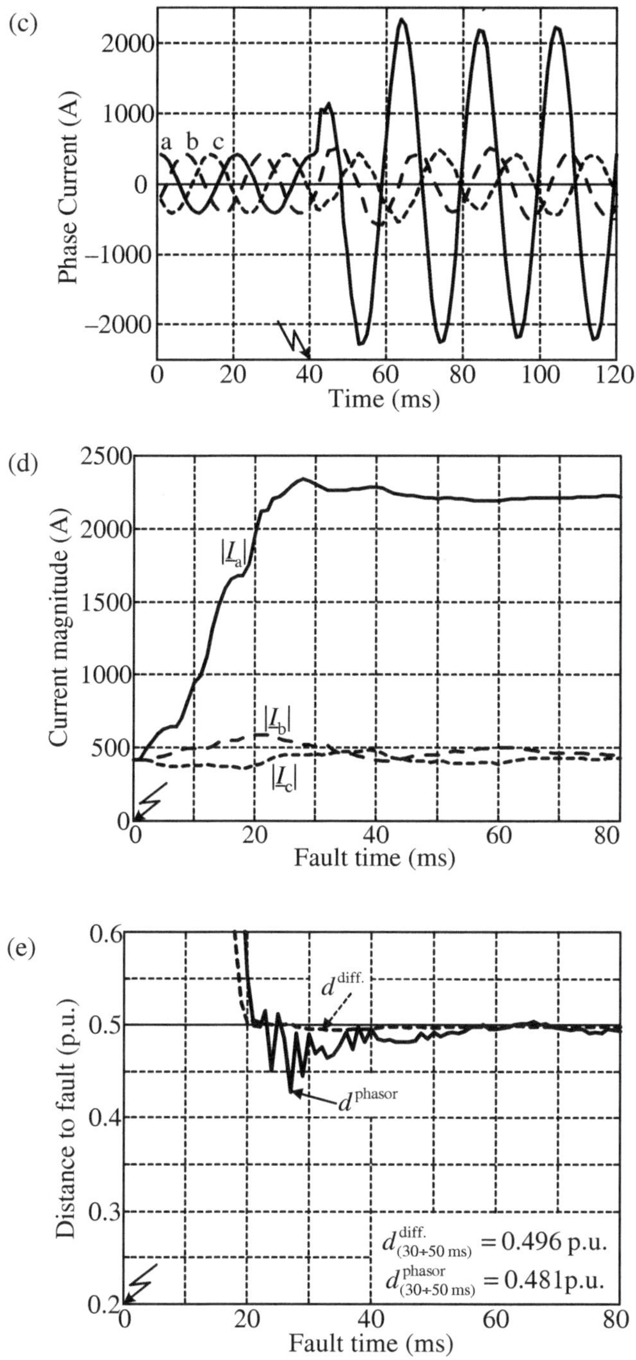

Fig. 6.37 (*continued*)

(f)

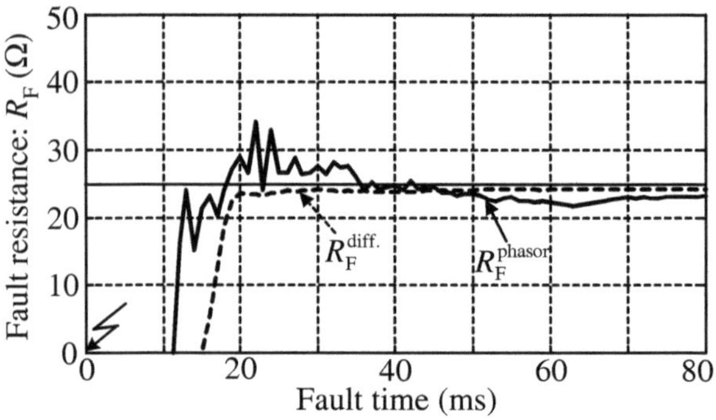

Fig. 6.37 (*continued*)

The fault-location results (Figs. 6.37e and f) obtained with the fundamental frequency phasor-based approach: d^{phasor}, R_F^{phasor} exhibit some transients. In contrast, for the results obtained with use of the differential-equation-based approach: d^{diff}, R_F^{diff} very good stabilization is observed.

6.15 Accuracy Improvement by Application of Distributed-parameter Line Model

The varieties of fault-location algorithms presented above (Sects. 6.5 and 6.14) were derived assuming the lumped model of a transmission line and neglecting the line shunt parameters (pre-dominantly the shunt capacitances). The fault-loop formula (6.28) was used for deriving those algorithms. Introducing in (6.28) the fault-loop current for the single line (6.26) one obtains the following form for the fault-loop formula [121]:

$$\underline{U}_{A\_P} - d\underline{Z}_{1L}\left(\underline{a}_1\underline{I}_{A1} + \underline{a}_2\underline{I}_{A2} + \underline{a}_0\frac{\underline{Z}_{0L}}{\underline{Z}_{1L}}\underline{I}_{A0}\right) - R_F(\underline{a}_{F1}\underline{I}_{F1} + \underline{a}_{F2}\underline{I}_{F2} + \underline{a}_{F0}\underline{I}_{F0}) = 0 \quad (6.145)$$

in which there are the coefficients dependent on fault type (gathered in Table 6.5 and Tables 3.3–3.5).

Using (6.145) accurate location of faults on transmission lines of comparatively short length can be performed. In order to keep the fault-location errors in the case of long lines (usually considered as exceeding 150 km) on the acceptable level, the distributed-parameter line model has to be incorporated.

The distributed-parameter model of faulted line for the *i*th symmetrical component from Fig. 2.25 will be utilized for simplified introduction of the distributed-parameter line model into the fault-loop formula (6.145). This model is simplified (not strict) since in order to avoid solving non-linear equations, iterative calculations are performed. Performing the current iteration (the iteration number n), the shunt parameters (capacitances) for the faulted-line section are determined not by an accurate value of the distance to fault but by the value obtained in the last but one iteration (the iteration number: $(n-1)$). Also, the longitudinal line impedance for the line section in the model from Fig. 2.25 is determined using the distance to fault value obtained in the previous iteration of the calculations. In order to start the iterative calculations, the distance to fault obtained for the lumped-line model has to be taken.

Introducing into (6.145) the distributed-parameter line model from Fig. 2.25 one obtains:

$$\underline{U}_{\mathrm{A\_P}} - d_{(n)}\underline{Z}_{1\mathrm{L}}\left(\underline{a}_1\underline{A}^{\mathrm{sh}}_{1_{(n-1)}}\underline{I}^{\mathrm{comp}}_{\mathrm{A1}} + \underline{a}_2\underline{A}^{\mathrm{sh}}_{2(n-1)}\underline{I}^{\mathrm{comp}}_{\mathrm{A2}} + \underline{a}_0\underline{A}^{\mathrm{sh}}_{0(n-1)}\frac{\underline{Z}_{0\mathrm{L}}}{\underline{Z}_{1\mathrm{L}}}\underline{I}^{\mathrm{comp}}_{\mathrm{A0}}\right)$$

$$-R_{\mathrm{F}_{(n)}}(\underline{a}_{\mathrm{F1}}\underline{I}^{\mathrm{comp}}_{\mathrm{F1}} + \underline{a}_{\mathrm{F2}}\underline{I}^{\mathrm{comp}}_{\mathrm{F2}} + \underline{a}_{\mathrm{F0}}\underline{I}^{\mathrm{comp}}_{\mathrm{F0}}) = 0 \qquad (6.146)$$

where:

$d_{(n)}$, $d_{(n-1)}$ – distance to fault from current and previous iterations,

$R_{\mathrm{F}_{(n)}}$ – fault resistance from current iteration,

$\underline{A}^{\mathrm{sh}}_{1_{(n-1)}} = \underline{A}^{\mathrm{sh}}_{2_{(n-1)}} = \dfrac{\sinh(\underline{\gamma}_1 d_{(n-1)}\ell)}{\underline{\gamma}_1 d_{(n-1)}\ell}$ – correction factor for determining the positive-sequence (negative-sequence) longitudinal impedance of line section having the length d (p.u.),

$\underline{A}^{\mathrm{sh}}_{0_{(n-1)}} = \dfrac{\sinh(\underline{\gamma}_0 d_{(n-1)}\ell)}{\underline{\gamma}_0 d_{(n-1)}\ell}$ – as above, but for the zero-sequence.

The positive-, negative- and zero-sequence currents after deducing the shunt capacitance currents equal:

$$\underline{I}^{\mathrm{comp}}_{\mathrm{A1}} = \underline{I}_{\mathrm{A1}} - 0.5d_{(n-1)}\ell\underline{Y}'_{1\mathrm{L}}\underline{A}^{\mathrm{th}}_{1_{(n-1)}}\underline{U}_{\mathrm{A1}}$$

$$\underline{I}^{\mathrm{comp}}_{\mathrm{A2}} = \underline{I}_{\mathrm{A2}} - 0.5d_{_{(n-1)}}\ell\underline{Y}'_{2\mathrm{L}}\underline{A}^{\mathrm{th}}_{2_{(n-1)}}\underline{U}_{\mathrm{A2}}$$

$$\underline{I}^{\mathrm{comp}}_{\mathrm{A0}} = \underline{I}_{\mathrm{A0}} - 0.5d_{(n-1)}\ell\underline{Y}'_{0\mathrm{L}}\underline{A}^{\mathrm{th}}_{0_{(n-1)}}\underline{U}_{\mathrm{A0}}$$

where:

$\underline{A}^{\mathrm{th}}_{1_{(n-1)}} = \underline{A}^{\mathrm{th}}_{2_{(n-1)}}$ – correction factor for the positive- (negative-)sequence shunt admittance of line section having the length d (p.u.),

$\underline{A}^{\mathrm{th}}_{0_{(n-1)}}$ – as above, but for the zero-sequence,

$\underline{I}^{\mathrm{comp}}_{\mathrm{F1}}$, $\underline{I}^{\mathrm{comp}}_{\mathrm{F2}}$, $\underline{I}^{\mathrm{comp}}_{\mathrm{F0}}$ – symmetrical components of total fault current determined considering the distributed-parameter line model, however, it is common practice [259, 260] that they are identical with the currents estimated using the lumped-line model.

The fault-loop formula (6.146) is solved iteratively for the unknowns: $d_{(n)}$, $R_{\mathrm{F}_{(n)}}$ with checking the convergence of iterative calculations (this requires setting the convergence threshold) or applying the pre-defined number of iterations. Usually, performance of two iterations appears to be sufficient.

7 Two-end and Multi-end Fault-location Algorithms

Abstract This chapter is focused on two-end and multi-end fault-location algorithms. First, the algorithms utilizing two-end synchronized measurements are presented for both phasor-based and time-domain approaches. Then, use of the unsynchronized measurements is considered in detail. Different options for measuring the synchronization angle are introduced. Use of both complete and incomplete two-end measurements is considered for two-end fault-location algorithms. Algorithms utilizing measurements of distance relays are presented. Fault location on three-terminal and multi-terminal lines is addressed.

7.1 Introduction

With the aim of overcoming the limitations of the one-end fault-location techniques and improving fault-location accuracy, two-end and multi-end fault-location algorithms have been proposed. Use of both synchronized [29, 65, 105, 107, 132, 133, 163, 164, 190, 191, 235, 236, 294, 323, 340] and unsynchronized [42, 116, 131, 185–187, 224–226, 297, 319, 320] measurements is considered for these algorithms. In the case of unsynchronized measurements, different ways of dealing with the unknown synchronization angle are presented. With regard to the utilized measurements, the following options are of interest for the two-end fault location:

- *complete two-end measurements*: three-phase voltages and currents from both line ends; and
- *incomplete two-end measurements*:
  - three-phase voltages from both line ends, with three-phase current from only one end;
  - three-phase currents from both line ends, with three-phase voltage from only one end;
  - three-phase voltages from both line ends.

Use of measurements of distance relays from the line terminals is considered for deriving fault-location algorithms, which are designed for application to post-fault analysis of operation of the relays [120, 130, 259, 260, 270].

Also, different options with respect to the availability of measurements from the terminals of three-terminal and multi-terminal lines are considered for use by the fault-location algorithms.

7.2 Use of Two-end Synchronized Measurements

7.2.1 Phasor-based Approach

The distributed-parameter model of a faulted line for the *i*th symmetrical component, with use of the correction factors for representing series and shunt parameters, as in Fig. 2.23, is taken into consideration. The voltage at fault point (F), viewed from the terminals A and B, respectively, is expressed as follows:

$$\underline{U}_{\mathrm{F}i}^{\mathrm{A}} = \cosh(\underline{\gamma}_i d\ell)\underline{U}_{\mathrm{A}i} - \sinh(\underline{\gamma}_i d\ell)\underline{Z}_{\mathrm{c}i}\underline{I}_{\mathrm{A}i} \tag{7.1}$$

$$\underline{U}_{\mathrm{F}i}^{\mathrm{B}} = \cosh(\underline{\gamma}_i (1-d)\ell)\underline{U}_{\mathrm{B}i} - \sinh(\underline{\gamma}_i (1-d)\ell)\underline{Z}_{\mathrm{c}i}\underline{I}_{\mathrm{B}i} \tag{7.2}$$

where: $\underline{U}_{\mathrm{A}i}$, $\underline{I}_{\mathrm{A}i}$, $\underline{U}_{\mathrm{B}i}$, $\underline{I}_{\mathrm{B}i}$ – phasors of the *i*th symmetrical component of voltages and currents obtained from synchronized measurements at the line terminals.

After taking into account the following trigonometric identities:

$$\cosh(\underline{\gamma}_i (1-d)\ell) = \cosh(\underline{\gamma}_i \ell)\cosh(\underline{\gamma}_i d\ell) - \sinh(\underline{\gamma}_i \ell)\sinh(\underline{\gamma}_i d\ell) \tag{7.3}$$

$$\sinh(\underline{\gamma}_i (1-d)\ell) = \sinh(\underline{\gamma}_i \ell)\cosh(\underline{\gamma}_i d\ell) - \cosh(\underline{\gamma}_i \ell)\sinh(\underline{\gamma}_i d\ell) \tag{7.4}$$

and performing further rearrangements, the formula (7.2) can be presented as:

$$\underline{U}_{\mathrm{F}i}^{\mathrm{B}} = \underline{G}_i \cosh(\underline{\gamma}_i d\ell) + \underline{H}_i \sinh(\underline{\gamma}_i d\ell) \tag{7.5}$$

where:

$\underline{G}_i = \underline{U}_{\mathrm{B}i}\cosh(\underline{\gamma}_i \ell) - \underline{Z}_{\mathrm{c}i}\underline{I}_{\mathrm{B}i}\sinh(\underline{\gamma}_i \ell)$,

$\underline{H}_i = -\underline{U}_{\mathrm{B}i}\sinh(\underline{\gamma}_i \ell) + \underline{Z}_{\mathrm{c}i}\underline{I}_{\mathrm{B}i}\cosh(\underline{\gamma}_i \ell)$.

The voltages (7.1) and (7.5), as present at the same fault point (F), are to be compared:

$$\underline{U}_{\mathrm{F}i}^{\mathrm{A}} = \underline{U}_{\mathrm{F}i}^{\mathrm{B}} \tag{7.6}$$

Performing such a comparison results in obtaining:

$$\underline{X}_i \sinh(\underline{\gamma}_i d\ell) - \underline{Y}_i \cosh(\underline{\gamma}_i d\ell) = 0 \tag{7.7}$$

where:
$\underline{X}_i = \sinh(\underline{\gamma}_i \ell)\underline{U}_{\mathrm{B}i} - \underline{Z}_{\mathrm{c}i} \cosh(\underline{\gamma}_i \ell)\underline{I}_{\mathrm{B}i} - \underline{Z}_{\mathrm{c}i}\underline{I}_{\mathrm{A}i}$,
$\underline{Y}_i = \cosh(\underline{\gamma}_i \ell)\underline{U}_{\mathrm{B}i} - \underline{Z}_{\mathrm{c}i} \sinh(\underline{\gamma}_i \ell)\underline{I}_{\mathrm{B}i} - \underline{U}_{\mathrm{A}i}$.

From (7.7) one obtains the following formula for the distance to fault, with use of the *i*th symmetrical component of voltages and currents – obtained from synchronized measurements at the line terminals [135]:

$$d = \frac{1}{\underline{\gamma}_i \ell} \tanh^{-1}\left(\frac{\cosh(\underline{\gamma}_i \ell)\underline{U}_{\mathrm{B}i} - \underline{Z}_{\mathrm{c}i} \sinh(\underline{\gamma}_i \ell)\underline{I}_{\mathrm{B}i} - \underline{U}_{\mathrm{A}i}}{\sinh(\underline{\gamma}_i \ell)\underline{U}_{\mathrm{B}i} - \underline{Z}_{\mathrm{c}i} \cosh(\underline{\gamma}_i \ell)\underline{I}_{\mathrm{B}i} - \underline{Z}_{\mathrm{c}i}\underline{I}_{\mathrm{A}i}} \right) \tag{7.8}$$

The obtained fault location (7.8) can be applied for the modal quantities as well.

7.2.2 Time-domain Approach

In Chap. 2, the *modified telegrapher's equations* of the long-line model were formulated. Such first-order partial differential equations can be solved using the *method of characteristics* [37]. This method introduces lines (called characteristic lines or characteristics) along which the partial differential equation degenerates into an ordinary differential equation.

Using the characteristics [37], the partial differential equations relating v and i (2.25) and (2.26) can be written as ordinary differential equations:

$$\frac{dv}{ds} - \chi \frac{di}{ds} = \frac{\eta i}{\sqrt{1+\chi^2}} \tag{7.9}$$

$$\frac{dv}{d\rho} + \chi \frac{di}{d\rho} = \frac{-\eta i}{\sqrt{1+\chi^2}} \tag{7.10}$$

where s, ρ – length along the two characteristics: $t \pm \chi x = \text{const}$ (straight lines in the distance–time plane).

The solution of (7.9) and (7.10) requires the discretization of the continuous-time system. Assuming [88] that the distance (x) axis is discretized by the index j and the time (t) axis by the index k, the following expressions for the voltage and current are obtained:

$$v_{j,k} = \frac{1}{2}(v_{j-1,k-1} + v_{j-1,k+1}) + \frac{Z_c}{2}(i_{j-1,k-1} - i_{j-1,k+1}) + \frac{R'_L \Delta x}{4}(i_{j-1,k-1} + i_{j-1,k+1}) - \frac{R'_L \Delta x}{2} i_{j,k} \tag{7.11}$$

$$i_{j,k} = \frac{1}{2Z_c}(v_{j-1,k-1} - v_{j-1,k+1}) + \frac{1}{2}(i_{j-1,k-1} + i_{j-1,k+1}) + \frac{R'_L \Delta x}{4Z_c}(i_{j-1,k+1} - i_{j-1,k-1}) \tag{7.12}$$

where $Z_c = \sqrt{L'_L / C'_L}$ – surge impedance of the transmission line.

Starting from the ends of the line, one can compute the voltage and current profile along the line, applying (7.11) and (7.12).

If in (7.11) and (7.12), the series resistance R'_L is neglected, they will reduce to the well-known *Bergeron equations* for a transmission line. Instead of computing the profile point by discrete point, the voltage and current at any point on the line can be determined knowing the voltage and current at the end of the line and the distance to the point.

The Bergeron equations for the voltage and current at a point x_j and time instant t_k are then given by:

$$v_{j,k} = \frac{1}{2}(v_{0,k-j} + v_{0,k+j}) + \frac{Z_c}{2}(i_{0,k-j} - i_{0,k+j}) \tag{7.13}$$

$$i_{j,k} = \frac{1}{2Z_c}(v_{0,k-j} - v_{0,k+j}) + \frac{1}{2}(i_{0,k-j} + i_{0,k+j}) \tag{7.14}$$

Steps in Fault Location

Using the voltage and current profiles, calculated by means of the method of characteristics, the fault-location procedure from [88] consists of the following steps:

- Modal decomposition used for obtaining three decoupled single-phase transmission lines (from a three-phase coupled system), to which the method of characteristics can be applied.

- Discretization of the transmission line: Based on the sampling frequency of the data acquisition, each mode of the transmission line is discretized into a finite number of points.
- Location of the approximate fault point: For each discrete point determined above, compute the voltage due to the sending end voltage and current. Then, the procedure using the receiving end data is repeated and the approximate fault point is found (the point with the least error).
- Refining the fault location performed with use of reconstructed voltage and current values transformed back to the phase domain.

7.3 Use of Two-end Unsynchronized Measurements

When the phasors of voltages and currents are obtained from digital measurements acquired asynchronously at the line terminals, there is no common time reference. This happens when the GPS facility is not used or for some reasons the signal from the GPS is not received. In order to make use of such measurements for fault location, a common time reference has to be provided. Different approaches for that follow in the next sections. In these approaches the measurements from the terminal B will be assumed as the base, while all phasors for voltage and current from the end A will be multiplied by the synchronization operator $e^{j\delta}$, where δ is the synchronization angle, which has to be determined or eliminated in mathematical manipulations.

7.3.1 Fault Location with Measurement of Synchronization Angle

In the case of utilizing unsynchronized measurements one may perform fault location with use of (7.8), which is valid for synchronized measurements. However, earlier determination of the synchronization operator $e^{j\delta}$ and making analytical synchronization of measurements are required.

The needed synchronization operator can be determined by utilizing the followings:

- relationships between pre-fault quantities – Sect. 7.3.1.1;
- boundary conditions of particular faults – Sect. 7.3.1.2.

In order to achieve high accuracy for determination of the synchronization operator, the distributed-parameter line model for both approaches will be utilized.

7.3.1.1 Measurement of Synchronization Angle with Use of Pre-fault Quantities

Figure 7.1 represents a distributed-parameter model of a line for a pre-fault positive-sequence. One can distinguish the following phasors of signals measured asynchronously, which after analytical synchronization (performed with use of the synchronization operator $e^{j\delta}$) are as follows:

- at the terminal B: $\underline{I}_{B1}^{pre}$, $\underline{U}_{B1}^{pre}$;
- at the terminal A: $\underline{I}_{A1}^{pre}e^{j\delta}$, $\underline{U}_{A1}^{pre}e^{j\delta}$.

Different options with regard to use of these phasors can be considered, as presented in the next sections. Note that in Fig. 7.1, the phasors that are to be calculated are marked with the dashed boxes. This is done with the aim of distinguishing them from those obtained from measurements.

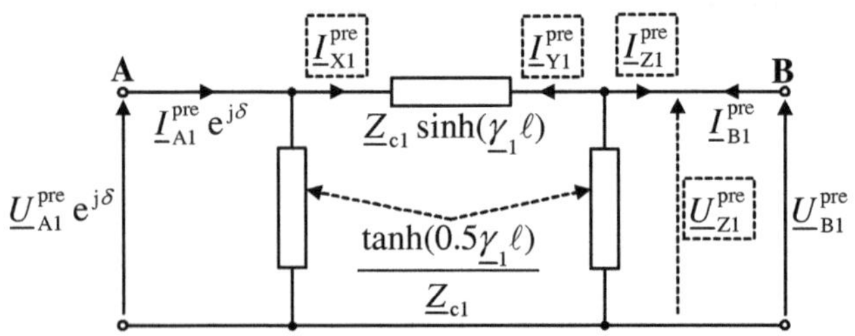

Fig. 7.1 Distributed-parameter model of line for the pre-fault positive-sequence (*the signals to be calculated are marked with dashed boxes*)

Use of Pre-fault Positive-sequence Voltage and Current from Two Ends

Using the equivalent π circuit model (Fig. 7.1), the current $\underline{I}_{X1}^{pre}$ is calculated by deducing the shunt current from the current measured at the bus A:

$$\underline{I}_{X1}^{pre} = \underline{I}_{A1}^{pre}e^{j\delta} - \frac{\tanh(0.5\underline{\gamma}_1\ell)}{\underline{Z}_{c1}}\underline{U}_{A1}^{pre}e^{j\delta} \tag{7.15}$$

Analogously, the current $\underline{I}_{Y1}^{pre}$ is calculated as:

$$\underline{I}_{Y1}^{pre} = \underline{I}_{B1}^{pre} - \frac{\tanh(0.5\underline{\gamma}_1\ell)}{\underline{Z}_{c1}}\underline{U}_{B1}^{pre} \tag{7.16}$$

The currents determined, (7.15) and (7.16), satisfy:

$$\underline{I}_{X1}^{pre} = -\underline{I}_{Y1}^{pre} \tag{7.17}$$

As a result of substituting (7.15) and (7.16) into (7.17), one derives the following formula for the synchronization operator:

$$\mathrm{e}^{\mathrm{j}\delta}=\frac{-\underline{Z}_{\mathrm{c1}}\underline{I}_{\mathrm{B1}}^{\mathrm{pre}}+\tanh(0.5\underline{\gamma}_1\ell)\underline{U}_{\mathrm{B1}}^{\mathrm{pre}}}{\underline{Z}_{\mathrm{c1}}\underline{I}_{\mathrm{A1}}^{\mathrm{pre}}-\tanh(0.5\underline{\gamma}_1\ell)\underline{U}_{\mathrm{A1}}^{\mathrm{pre}}} \tag{7.18}$$

Use of Pre-fault Positive-sequence Voltage from Two Ends and Current from One End

The positive-sequence pre-fault voltage $\underline{U}_{\mathrm{Z1}}^{\mathrm{pre}}$ can be calculated by transferring analytically the signals from the bus A towards the bus B:

$$\underline{U}_{\mathrm{Z1}}^{\mathrm{pre}}=\cosh(\underline{\gamma}_1\ell)\underline{U}_{\mathrm{A1}}^{\mathrm{pre}}\mathrm{e}^{\mathrm{j}\delta}-\underline{Z}_{\mathrm{c1}}\sinh(\underline{\gamma}_1\ell)\underline{I}_{\mathrm{A1}}^{\mathrm{pre}}\mathrm{e}^{\mathrm{j}\delta} \tag{7.19}$$

The transferred voltage (7.19) is equal to the voltage at the bus B:

$$\underline{U}_{\mathrm{Z1}}^{\mathrm{pre}}=\underline{U}_{\mathrm{B1}}^{\mathrm{pre}} \tag{7.20}$$

Substitution of (7.19) into (7.20) results in the following formula for the synchronization operator:

$$\mathrm{e}^{\mathrm{j}\delta}=\frac{\underline{U}_{\mathrm{B1}}^{\mathrm{pre}}}{\cosh(\underline{\gamma}_1\ell)\underline{U}_{\mathrm{A1}}^{\mathrm{pre}}-\underline{Z}_{\mathrm{c1}}\sinh(\underline{\gamma}_1\ell)\underline{I}_{\mathrm{A1}}^{\mathrm{pre}}} \tag{7.21}$$

Use of Pre-fault Positive-sequence Current from Two Ends and Voltage from One End

Transferring signals measured at the bus A towards the bus B one obtains:

$$\underline{I}_{\mathrm{Z1}}^{\mathrm{pre}}=-\frac{1}{\underline{Z}_{\mathrm{c1}}}\sinh(\underline{\gamma}_1\ell)\underline{U}_{\mathrm{A1}}^{\mathrm{pre}}\mathrm{e}^{\mathrm{j}\delta}+\cosh(\underline{\gamma}_1\ell)\underline{I}_{\mathrm{A1}}^{\mathrm{pre}}\mathrm{e}^{\mathrm{j}\delta} \tag{7.22}$$

The transferred current (7.22) and that measured at the bus B satisfy:

$$I_{\mathrm{Z1}}^{\mathrm{pre}}=-I_{\mathrm{B1}}^{\mathrm{pre}} \tag{7.23}$$

Substitution of (7.22) into (7.23) results in the following formula for the synchronization operator:

$$\mathrm{e}^{\mathrm{j}\delta}=\frac{-\underline{Z}_{\mathrm{c1}}\underline{I}_{\mathrm{B1}}^{\mathrm{pre}}}{-\sinh(\underline{\gamma}_1\ell)\underline{U}_{\mathrm{A1}}^{\mathrm{pre}}+\cosh(\underline{\gamma}_1\ell)\underline{Z}_{\mathrm{c1}}\underline{I}_{\mathrm{A1}}^{\mathrm{pre}}} \tag{7.24}$$

Summarizing, the unknown synchronization operator can be determined from pre-fault positive-sequence currents and voltages, measured asynchronously at the line terminals, by applying (7.18) or (7.21) or (7.24).

7.3.1.2 Measurement of Synchronization Angle with Use of Fault Quantities

Alternatively, the synchronization operator can be determined using only fault quantities (measured within the fault period). For this purpose, the boundary conditions of particular faults can be explored [123]. In the case of three-phase balanced faults, the positive-sequence components are the only ones, which are present in the measured currents and voltages. For this reason, the boundary conditions will be considered for all faults, except for three-phase balanced faults, for which the synchronization operator can be determined with use of pre-fault positive-sequence quantities, i.e., by applying (7.18) or (7.21) or (7.24).

Use of Two-end Voltages and Current from One End

For analysis of boundary conditions of particular faults the symmetrical components of the total fault current have to be determined. This will be done using the distributed-parameter line model as presented in Fig. 7.2.

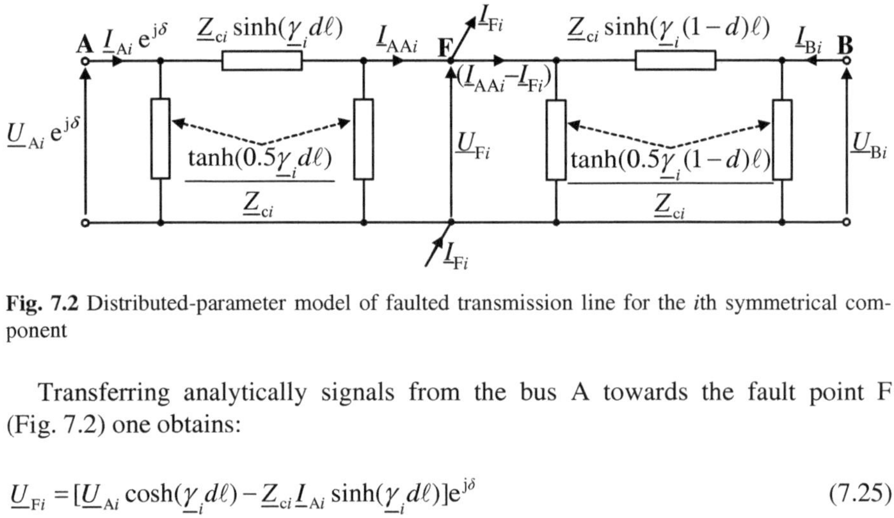

Fig. 7.2 Distributed-parameter model of faulted transmission line for the *i*th symmetrical component

Transferring analytically signals from the bus A towards the fault point F (Fig. 7.2) one obtains:

$$\underline{U}_{Fi} = [\underline{U}_{Ai}\cosh(\underline{\gamma}_i d\ell) - \underline{Z}_{ci}\underline{I}_{Ai}\sinh(\underline{\gamma}_i d\ell)]e^{j\delta} \tag{7.25}$$

$$\underline{I}_{AAi} = \left[-\frac{1}{\underline{Z}_{ci}}\underline{U}_{Ai}\sinh(\underline{\gamma}_i d\ell) + \underline{I}_{Ai}\cosh(\underline{\gamma}_i d\ell) \right] e^{j\delta} \tag{7.26}$$

Voltage at the remote terminal B (Fig. 7.2) takes the following form:

$$\underline{U}_{\mathrm{B}i} = \underline{U}_{\mathrm{F}i}\cosh(\underline{\gamma}_i(1-d)\ell) - \underline{Z}_{ci}(\underline{I}_{\mathrm{AA}i} - \underline{I}_{\mathrm{F}i})\sinh(\underline{\gamma}_i(1-d)\ell) \tag{7.27}$$

Substituting (7.25) and (7.26) into (7.27) and performing manipulations on hyperbolic functions (with use of the trigonometric identities (7.3) and (7.4)) one obtains the following formula for the *i*th symmetrical component of the total fault current [123]:

$$\underline{I}_{\mathrm{F}i} = \frac{\underline{U}_{\mathrm{B}i} + \underline{N}_{\mathrm{A}i}\mathrm{e}^{\mathrm{j}\delta}}{\underline{Z}_{ci}\sinh(\underline{\gamma}_i(1-d)\ell)} \tag{7.28}$$

where:

$\underline{N}_{\mathrm{A}i} = -\underline{U}_{\mathrm{A}i}\cosh(\underline{\gamma}_i\ell) + \underline{Z}_{ci}\underline{I}_{\mathrm{A}i}\sinh(\underline{\gamma}_i\ell)$,

$\underline{U}_{\mathrm{A}i}$, $\underline{U}_{\mathrm{B}i}$, $\underline{I}_{\mathrm{A}i}$ – the *i*th symmetrical component of the signals obtained from asynchronous measurements of two-end voltages and one-end current.

The total fault current can be presented as a composition of its respective components, using the share coefficients dependent on the fault type (the alternative sets of the share coefficients for different faults are gathered in Tables 3.3–3.6). Choice of the particular set depends on the preferred priority for using respective sequences. Two characteristic sets of the share coefficients for the phase-to-ground and phase-to-phase faults, which will be used, are collected in Table 7.1.

Table 7.1 Two alternative sets of share coefficients for phase-to-ground and phase-to-phase faults

Fault type	**I Set** (as in Table 3.3)		**II Set** (as in Table 3.4)	
	$\underline{a}_{\mathrm{F1}}^{\mathrm{I\,Set}}$	$\underline{a}_{\mathrm{F2}}^{\mathrm{I\,Set}}$	$\underline{a}_{\mathrm{F1}}^{\mathrm{II\,Set}}$	$\underline{a}_{\mathrm{F2}}^{\mathrm{II\,Set}}$
a–g	0	3	3	0
b–g	0	$-1.5+\mathrm{j}1.5\sqrt{3}$	$-1.5-\mathrm{j}1.5\sqrt{3}$	0
c–g	0	$-1.5-\mathrm{j}1.5\sqrt{3}$	$-1.5+\mathrm{j}1.5\sqrt{3}$	0
a–b	0	$1.5-\mathrm{j}0.5\sqrt{3}$	$1.5+\mathrm{j}0.5\sqrt{3}$	0
b–c	0	$\mathrm{j}\sqrt{3}$	$-\mathrm{j}\sqrt{3}$	0
c–a	0	$-1.5-\mathrm{j}0.5\sqrt{3}$	$-1.5+\mathrm{j}0.5\sqrt{3}$	0

Applying the share coefficients from Table 7.1, the total fault current can be expressed as:

$$\underline{I}_{\mathrm{F}} = \underline{a}_{\mathrm{F1}}^{\mathrm{I\,Set}}\underline{I}_{\mathrm{F1}} + \underline{a}_{\mathrm{F2}}^{\mathrm{I\,Set}}\underline{I}_{\mathrm{F2}} \tag{7.29}$$

or alternatively:

$$\underline{I}_{\mathrm{F}} = \underline{a}_{\mathrm{F1}}^{\mathrm{II\,Set}} \underline{I}_{\mathrm{F1}} + \underline{a}_{\mathrm{F2}}^{\mathrm{II\,Set}} \underline{I}_{\mathrm{F2}} \tag{7.30}$$

Considering (7.28) for $i = 1$ (positive-sequence) and $i = 2$ (negative-sequence) one can take into account that the propagation constants are identical, and also the surge impedances are identical for these sequences. As a result of comparing (7.29) with (7.30) one gets the following formula for the synchronization operator, valid for all faults listed in Table 7.1 (phase-to-ground and phase-to-phase faults):

$$[\mathrm{e}^{\mathrm{j}\delta}]_{\mathrm{ph-g,\,ph1-ph2}} = \frac{\underline{a}_{\mathrm{F2}}^{\mathrm{I\,Set}} \underline{U}_{\mathrm{B2}} - \underline{a}_{\mathrm{F1}}^{\mathrm{II\,Set}} \underline{U}_{\mathrm{B1}}}{\underline{a}_{\mathrm{F1}}^{\mathrm{II\,Set}} \underline{N}_{\mathrm{A1}} - \underline{a}_{\mathrm{F2}}^{\mathrm{I\,Set}} \underline{N}_{\mathrm{A2}}} \tag{7.31}$$

where:

$\underline{N}_{\mathrm{A1}} = -\underline{U}_{\mathrm{A1}} \cosh(\underline{\gamma}_1 \ell) + \underline{Z}_{\mathrm{c1}} \underline{I}_{\mathrm{A1}} \sinh(\underline{\gamma}_1 \ell)$,

$\underline{N}_{\mathrm{A2}} = -\underline{U}_{\mathrm{A2}} \cosh(\underline{\gamma}_1 \ell) + \underline{Z}_{\mathrm{c1}} \underline{I}_{\mathrm{A2}} \sinh(\underline{\gamma}_1 \ell)$.

In (7.31), the positive- and negative-sequence quantities from a fault period are involved. Thus, one avoids using the pre-fault measurements.

For phase-to-phase-to-ground faults the relation between the zero-sequence component of the total fault current and the remaining components ((3.3) and Table 3.7) is utilized. Substituting (7.28) into (3.3) gives:

$$\frac{\underline{U}_{\mathrm{B0}} + \underline{N}_{\mathrm{A0}} [\mathrm{e}^{\mathrm{j}\delta}]_{\mathrm{ph1-ph2-g}}}{\underline{Z}_{\mathrm{c0}} \sinh(\underline{\gamma}_0 (1-d)\ell)} = \frac{\underline{b}_{\mathrm{F1}} \underline{U}_{\mathrm{B1}} + \underline{b}_{\mathrm{F2}} \underline{U}_{\mathrm{B2}} + (\underline{b}_{\mathrm{F1}} \underline{N}_{\mathrm{A1}} + \underline{b}_{\mathrm{F2}} \underline{N}_{\mathrm{A2}}) [\mathrm{e}^{\mathrm{j}\delta}]_{\mathrm{ph1-ph2-g}}}{\underline{Z}_{\mathrm{c1}} \sinh(\underline{\gamma}_1 (1-d)\ell)} \tag{7.32}$$

where:

$\underline{b}_{\mathrm{F1}}, \underline{b}_{\mathrm{F2}}$ – coefficients from Table 3.7,

$\underline{N}_{\mathrm{A0}} = -\underline{U}_{\mathrm{A0}} \cosh(\underline{\gamma}_0 \ell) + \underline{Z}_{\mathrm{c0}} \underline{I}_{\mathrm{A0}} \sinh(\underline{\gamma}_0 \ell)$,

$\underline{N}_{\mathrm{A1}}$, $\underline{N}_{\mathrm{A2}}$ as in (7.31).

Direct determination of the synchronization operator from (7.32), i.e., without knowing the distance to fault (d), cannot be accomplished. Therefore, the solution of (7.32) can be considered together with the fault-location algorithm. Approximate determination of the synchronization operator from (7.32) can be performed by linearizing this formula, i.e., by applying the substitution for the positive- and zero-sequence ($i = 1$ and $i = 0$):

$$\sinh(\underline{\gamma}_i (1-d)\ell) \rightarrow (\underline{\gamma}_i (1-d)\ell) \tag{7.33}$$

Use of Two-end Currents and One-end Voltage

Considering the line section F–B (Fig. 7.2), one obtains the following formula for the *i*th sequence of the bus B current:

$$\underline{I}_{\mathrm{B}i} = \frac{1}{\underline{Z}_{\mathrm{c}i}} \underline{U}_{\mathrm{F}i} \sinh(\underline{\gamma}_i (1-d)\ell) - (\underline{I}_{\mathrm{AA}i} - \underline{I}_{\mathrm{F}i}) \cosh(\underline{\gamma}_i (1-d)\ell) \tag{7.34}$$

Substituting (7.25) and (7.26) into (7.34) and performing manipulations on hyperbolic functions, with use of the trigonometric identities (7.3) and (7.4), one obtains the following formula for the *i*th symmetrical component of the total fault current:

$$\underline{I}_{\mathrm{F}i} = \frac{\underline{I}_{\mathrm{B}i} + \underline{N}_{\mathrm{A}i} \mathrm{e}^{\mathrm{j}\delta}}{\cosh(\underline{\gamma}_i (1-d)\ell)} \tag{7.35}$$

where:

$$\underline{N}_{\mathrm{A}i} = \underline{I}_{\mathrm{A}i} \cosh(\underline{\gamma}_i \ell) - \frac{1}{\underline{Z}_{\mathrm{c}i}} \underline{U}_{\mathrm{A}i} \sinh(\underline{\gamma}_i \ell),$$

$\underline{I}_{\mathrm{A}i}$, $\underline{I}_{\mathrm{B}i}$, $\underline{U}_{\mathrm{A}i}$ – phasors of the *i*th symmetrical component of the signals obtained from asynchronous measurements of two-end currents and one-end voltage.

Using (7.35) one can determine the synchronization operator, analogously to the method presented in the previous section, where (7.28) was used for determining the *i*th symmetrical component of the total fault current.

7.3.2 Elimination of Synchronization Angle

The algorithm presented here applies elimination of the synchronization angle as a result of the applied mathematical manipulations. It starts with considering the lumped model of the faulted line while disregarding the shunt parameters (Fig. 7.3). Then, in order to improve accuracy the distributed-parameter line model (Fig. 7.4) is taken into consideration.

Comparison of the *i*th symmetrical component of voltage at fault point ($\underline{U}_{\mathrm{F}i}$), viewed from both A and B ends gives:

$$\underline{U}_{\mathrm{A}i} \mathrm{e}^{\mathrm{j}\delta} - d\underline{Z}_{i\mathrm{L}} \underline{I}_{\mathrm{A}i} \mathrm{e}^{\mathrm{j}\delta} = \underline{U}_{\mathrm{B}i} - (1-d)\underline{Z}_{i\mathrm{L}} \underline{I}_{\mathrm{B}i} \tag{7.36}$$

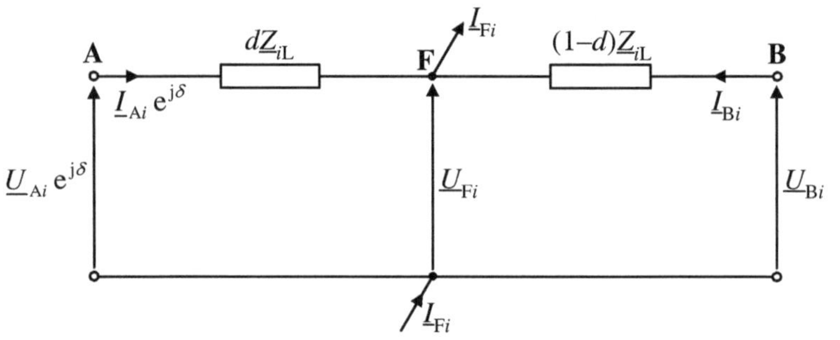

Fig. 7.3 Lumped model of faulted line for the *i*th symmetrical component

From (7.36) one obtains the following formula for the synchronization operator:

$$e^{j\delta} = \frac{U_{Bi} - Z_{iL}I_{Bi} + dZ_{iL}I_{Bi}}{U_{Ai} - dZ_{iL}I_{Ai}} \tag{7.37}$$

Taking into account that the absolute value of the synchronization operator applies:

$$|e^{j\delta}| = 1 \tag{7.38}$$

after using (7.37) and rearrangements one obtains:

$$A_{2i}d^2 + A_{1i}d + A_{0i} = 0 \tag{7.39}$$

where:

$A_{2i} = |Z_{iL}I_{Ai}|^2 - |Z_{iL}I_{Bi}|^2$,

$A_{1i} = -2\,\mathrm{Re}[U_{Ai}(Z_{iL}I_{Ai})^* + (U_{Bi} - Z_{iL}I_{Bi})(Z_{iL}I_{Bi})^*]$,

$A_{0i} = |U_{Ai}|^2 - |U_{Bi} - Z_{iL}I_{Bi}|^2$,

X^* – conjugate of X.

Solution of (7.39), formulated with the *i*th type of the symmetrical components, yields two results for the distance to fault (p.u.): (d_{i_1}, d_{i_2}). Usually, one of them lies in the line range: from 0 to 1 p.u., and in a natural way such a result is selected as the valid one, i.e., which complies with the actual fault. However, it can happen for some rare cases that both results (d_{i_1}, d_{i_2}) lie within the line range, and in order to select the valid result, additional calculations, with use of another type of symmetrical components, have to be performed:

$$A_{2j}d^2 + A_{1j}d + A_{0j} = 0 \tag{7.40}$$

where: j – subscript denoting that the jth type symmetrical components ($j \neq i$) are utilized, i.e., in (7.39) the ith type is replaced by the jth type of symmetrical components.

Thus, from (7.39) and (7.40) one gets four solutions: (d_{i_1}, d_{i_2}) and (d_{j_1}, d_{j_2}). For the pair of valid solutions (d_{i_k}, d_{j_m}), where the indices: k=1 or 2, m=1 or 2, one gets the smallest difference: abs($d_{i\_k} - d_{j\_m}$)=min., while larger for the others.

In order to improve the fault-location accuracy achieved using (7.39) or (7.40), the distributed-parameter line model has to be included into (7.36), which is used for deriving (7.39). Applying the distributed-parameter line model suitable for simple iterative calculations (as in Fig. 2.25), the circuit diagram as in Fig. 7.4 is obtained and the following substitutions apply:

$$d\underline{Z}_{i\mathrm{L}} \rightarrow d_{(n)}\ell \underline{Z}_{i\mathrm{L}}' \underline{A}_{i(n-1)}^{\mathrm{sh}} \tag{7.41}$$

$$\underline{I}_{\mathrm{A}i} \rightarrow \underline{I}_{\mathrm{A}i}^{\mathrm{comp}} = \underline{I}_{\mathrm{A}i} - 0.5 d_{(n-1)}\ell \underline{Y}_{i\mathrm{L}}' \underline{A}_{i(n-1)}^{\mathrm{th}} \underline{U}_{\mathrm{A}i} \tag{7.42}$$

$$(1-d)\underline{Z}_{i\mathrm{L}} \rightarrow (1-d_{(n)})\ell \underline{Z}_{i\mathrm{L}}' \underline{B}_{i(n-1)}^{\mathrm{sh}} \tag{7.43}$$

$$\underline{I}_{\mathrm{B}i} \rightarrow \underline{I}_{\mathrm{B}i}^{\mathrm{comp}} = \underline{I}_{\mathrm{B}i} - 0.5(1-d_{(n-1)})\ell \underline{Y}_{i\mathrm{L}}' \underline{B}_{i(n-1)}^{\mathrm{th}} \underline{U}_{\mathrm{B}i} \tag{7.44}$$

where:
$d_{(n)}$, $d_{(n-1)}$ – distance to fault in the 'current' and 'previous' iterations,
$\underline{A}_{i(n-1)}^{\mathrm{sh}}$, $\underline{A}_{i(n-1)}^{\mathrm{th}}$, $\underline{B}_{i(n-1)}^{\mathrm{sh}}$, $\underline{B}_{i(n-1)}^{\mathrm{th}}$ – values determined in the 'previous' iteration using (2.39)–(2.42).

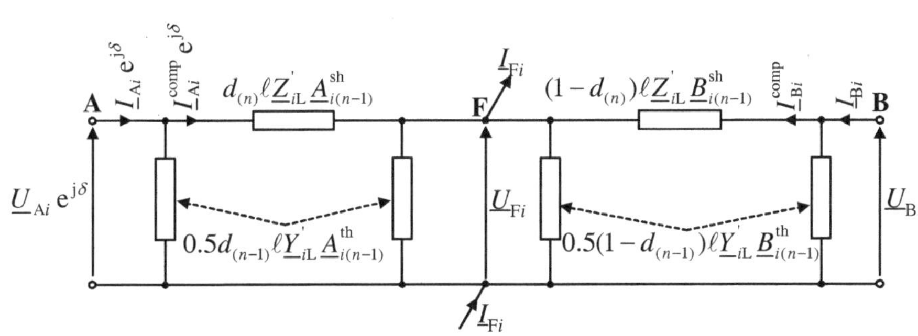

Fig. 7.4 Distributed-parameter model of faulted line for the ith symmetrical component suitable for iterative calculations of distance to fault

7.3.3 Analytical Determination of Synchronization Angle

In [225], a fundamental fault-location algorithm, which utilizes two-end unsynchronized measurement has been introduced. It was formulated in terms of symmetrical components of the measured currents and voltages. Since different types of components can be utilized, the algorithm will be presented in general form, i.e., for the *i*th symmetrical component (subscript *i*). The algorithm from [225] is derived starting with substitution of the synchronization operator:

$$e^{j\delta_i} = \cos(\delta_i) + j\sin(\delta_i) \tag{7.45}$$

into (7.36), thus obtaining:

$$[\underline{U}_{Ai} - d\underline{Z}_{iL}\underline{I}_{Ai}]\cdot[\cos(\delta_i) + j\sin(\delta_i)] - \underline{U}_{Bi} + (1-d)\underline{Z}_{iL}\underline{I}_{Bi} = 0 \tag{7.46}$$

Resolving (7.46) into the real/imaginary parts, and eliminating the unknown distance to fault, results in obtaining a simple formula for the unknown synchronization angle [116]:

$$A_i \cos(\delta_i) + B_i \sin(\delta_i) = C_i \tag{7.47}$$

where:

$A_i = \mathrm{Im}[\underline{Z}_{iL}^*((\underline{U}_{Bi} - \underline{Z}_{iL}\underline{I}_{Bi})\underline{I}_{Ai}^* - \underline{U}_{Ai}\underline{I}_{Bi}^*)]$,

$B_i = \mathrm{Re}[-\underline{Z}_{iL}^*((\underline{U}_{Bi} - \underline{Z}_{iL}\underline{I}_{Bi})\underline{I}_{Ai}^* + \underline{U}_{Ai}\underline{I}_{Bi}^*)]$,

$C_i = \mathrm{Im}[-\underline{Z}_{iL}^*((\underline{U}_{Bi} - \underline{Z}_{iL}\underline{I}_{Bi})\underline{I}_{Bi}^* - \underline{U}_{Ai}\underline{I}_{Ai}^*)]$,

x^* – conjugate of x.

Note that (7.47) is completely equivalent to the original formula from reference [225]. However, it contains a more compact description of the involved terms: A_i, B_i, C_i.

In [225], (7.47) is solved iteratively, applying the Newton–Raphson method. The iterative calculations are stopped when the difference between the last two values of the synchronization angle is smaller than a pre-assigned limit. An initial value for the unknown synchronization angle is required. In [225], setting the initial value to zero has been recommended. This is due to the fact that in practice, the synchronization angle is contained within a limited range around zero. The calculations have been observed to converge rapidly in numerous cases.

Once the synchronization angle is known, the distance to fault can be calculated from the real or imaginary part of (7.46).

7.3.4 Optimal Two-end Unsynchronized Fault-location Algorithm

In order to avoid iterative calculations at the stage of using the lumped-line model (which are applied in the approach from [225]), in [116] rewriting of (7.46) to the following form has been proposed:

$$\sin(\delta_{i\_1(2)} + \alpha_i) = \frac{C_i}{\sqrt{A_i^2 + B_i^2}} \tag{7.48}$$

where:

$$\sin(\alpha_i) = \frac{A_i}{\sqrt{A_i^2 + B_i^2}}, \quad \cos(\alpha_i) = \frac{B_i}{\sqrt{A_i^2 + B_i^2}}.$$

Two solutions of the trigonometric equation (7.48) from the range $(-\pi, \pi)$ are as follows (Fig. 7.5):

$$\delta_{i\_1} = \operatorname{asin}\left(\frac{C_i}{\sqrt{A_i^2 + B_i^2}}\right) - \operatorname{atan2}[\sin(\alpha_i),\ \cos(\alpha_i)] \tag{7.49}$$

$$\delta_{i\_2} = -\operatorname{asin}\left(\frac{C_i}{\sqrt{A_i^2 + B_i^2}}\right) - \operatorname{atan2}[\sin(\alpha_i),\ \cos(\alpha_i)] - \pi \tag{7.50}$$

where ‘atan2’ – four-quadrant arctangent function for calculating the angle with setting the sine and cosine values of this angle (note: this is a function that is used in the Matlab program [202]).

Performing calculations according to (7.49) and (7.50) one needs to consider the following:

- take the real part from the arcsine function, since in the presence of certain transient errors the term: $C_i\big/\sqrt{A_i^2 + B_i^2}$ could fall outside the permissible range: (–1, 1);
- limit the values for the synchronization angle $\delta_{i\_1(2)}$ to the range $(-\pi, \pi)$, by adding or subtracting 2π, should it happen that the values $\delta_{i\_1(2)}$ fall outside the range of interest $(-\pi, \pi)$.

Limitation of the synchronization angle to the range $(-\pi, \pi)$ is reasonable, since the synchronization angle is basically considered to have a small value [4].

In Fig. 7.5, an illustration of the solution of the trigonometric formula (7.48) is given. In the case when only one solution is close to zero, while the latter is far away from zero, the selection of the valid synchronization angle becomes straightforward. The solution being close to zero is selected as the valid one (in Fig. 7.5a, the solution δ_{i_1}), and further taken for determining the distance to fault. In contrast, if both solutions lie close to zero (Fig. 7.5b), none of them can be simply rejected and there is a need for performing the selection. This will be explained when presenting *Example 7.1*.

After determination of the synchronization angle, the distance to fault can be calculated, as for example, from the real part of (7.46):

$$d_{i\_1(2)} = \mathrm{Re}\left(\frac{\underline{U}_{\mathrm{A}i}\,\mathrm{e}^{\mathrm{j}\delta_{i\_1(2)}} - (\underline{U}_{\mathrm{B}i} - \underline{Z}_{i\mathrm{L}}\underline{I}_{\mathrm{B}i})}{\underline{Z}_{i\mathrm{L}}(\underline{I}_{\mathrm{A}i}\,\mathrm{e}^{\mathrm{j}\delta_{i\_1(2)}} + \underline{I}_{\mathrm{B}i})} \right) \tag{7.51}$$

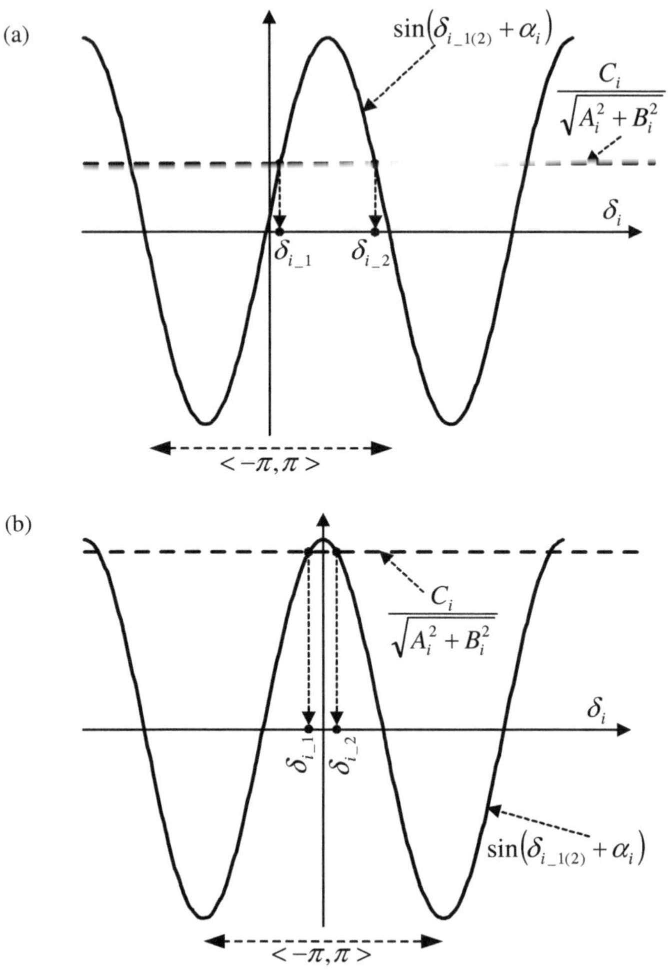

Fig. 7.5 Illustration of the solution of trigonometric formula (7.48): (a) only one solution is close to zero, and (b) both solutions are close to zero

Again, there are two solutions for the distance to fault (d_{i_1}, d_{i_2}). The solution, which is obtained for the earlier selected valid solution of the synchronization angle, is the valid solution for the distance to fault.

Optimal Two-end Fault-location Algorithm – Use of Distributed-parameter Line Model

Applying the distributed-parameter line model (Fig. 7.2), the voltage at the fault point F for the ith symmetrical component, viewed from the terminals A and B, is determined as follows:

$$\underline{U}_{\mathrm{F}i}^{\mathrm{A}}(d,\delta)=[\underline{U}_{\mathrm{A}i}\cosh(\underline{\gamma}_i d\ell)-\underline{Z}_{\mathrm{c}i}\underline{I}_{\mathrm{A}i}\sinh(\underline{\gamma}_i d\ell)]\cdot[\cos(\delta)+\mathrm{j}\sin(\delta)] \tag{7.52}$$

$$\underline{U}_{\mathrm{F}i}^{\mathrm{B}}(d)=\underline{U}_{\mathrm{B}i}\cosh(\underline{\gamma}_i(1-d)\ell)-\underline{Z}_{\mathrm{c}i}\underline{I}_{\mathrm{B}i}\sinh(\underline{\gamma}_i(1-d)\ell) \tag{7.53}$$

Comparison of (7.52) with (7.53) does not result in direct solution for the synchronization angle and the distance to fault, as was the case for the lumped-line model. This can be obtained by performing iterative calculations utilizing the Newton–Raphson method. For this purpose, the following function of the sought unknowns is considered:

$$\underline{F}(d,\delta)=\underline{U}_{\mathrm{F}i}^{\mathrm{A}}(d,\delta)-\underline{U}_{\mathrm{F}i}^{\mathrm{B}}(d)=0 \tag{7.54}$$

The iterative calculations are performed according to the following matrix formula:

$$\mathbf{X}_{\mathrm{new}}=\mathbf{X}_{\mathrm{old}}-\mathbf{J}^{-1}(\mathbf{F}_{\mathrm{old}})*\mathbf{F}_{\mathrm{old}} \tag{7.55}$$

where:

$$\mathbf{X}_{\mathrm{new}}=\begin{bmatrix} d_{\mathrm{new}} \\ \delta_{\mathrm{new}} \end{bmatrix},\quad \mathbf{X}_{\mathrm{old}}=\begin{bmatrix} d_{\mathrm{old}} \\ \delta_{\mathrm{old}} \end{bmatrix},$$

$$\mathbf{F}_{\mathrm{old}}=\begin{bmatrix} F_{\mathrm{R}}(d_{\mathrm{old}},\delta_{\mathrm{old}}) \\ F_{\mathrm{I}}(d_{\mathrm{old}},\delta_{\mathrm{old}}) \end{bmatrix},$$

$$\mathbf{J}(\mathbf{F}_{\mathrm{old}})=\begin{bmatrix} \dfrac{\partial F_{\mathrm{R}}(d_{\mathrm{old}},\delta_{\mathrm{old}})}{\partial d_{\mathrm{old}}} & \dfrac{\partial F_{\mathrm{R}}(d_{\mathrm{old}},\delta_{\mathrm{old}})}{\partial \delta_{\mathrm{old}}} \\ \dfrac{\partial F_{\mathrm{I}}(d_{\mathrm{old}},\delta_{\mathrm{old}})}{\partial d_{\mathrm{old}}} & \dfrac{\partial F_{\mathrm{I}}(d_{\mathrm{old}},\delta_{\mathrm{old}})}{\partial \delta_{\mathrm{old}}} \end{bmatrix}.$$

The respective elements of the vector $\mathbf{F}_{\text{old}}$ are defined as follows:

$$F_{\text{R}}(d_{\text{old}},\delta_{\text{old}})=\text{Re}[\underline{U}_{\text{F}i}^{\text{A}}(d_{\text{old}},\delta_{\text{old}})-\underline{U}_{\text{F}i}^{\text{B}}(d_{\text{old}})] \tag{7.56}$$

$$F_{\text{I}}(d_{\text{old}},\delta_{\text{old}})=\text{Im}[\underline{U}_{\text{F}i}^{\text{A}}(d_{\text{old}},\delta_{\text{old}})-\underline{U}_{\text{F}i}^{\text{B}}(d_{\text{old}})] \tag{7.57}$$

After taking into account (7.52) and (7.54) one obtains:

$$F_{\text{R}}(d_{\text{old}},\delta_{\text{old}})=\text{Re}[\underline{U}_{\text{F}i}^{\text{A\_unsynchr.}}(d_{\text{old}})]\cos(\delta_{\text{old}})-\text{Im}[\underline{U}_{\text{F}i}^{\text{A\_unsynchr.}}(d_{\text{old}}]\sin(\delta_{\text{old}}) \\ -\text{Re}[\underline{U}_{\text{F}i}^{\text{B}}(d_{\text{old}})] \tag{7.58}$$

$$F_{\text{I}}(d_{\text{old}},\delta_{\text{old}})=\text{Im}[\underline{U}_{\text{F}i}^{\text{A\_unsynchr.}}(d_{\text{old}})]\cos(\delta_{\text{old}})+\text{Re}[\underline{U}_{\text{F}i}^{\text{A\_unsynchr.}}(d_{\text{old}})]\sin(\delta_{\text{old}}) \\ -\text{Im}[\underline{U}_{\text{F}i}^{\text{B}}(d_{\text{old}})] \tag{7.59}$$

Applying (7.58) and (7.59), the respective components of the Jacobian matrix $\mathbf{J}(\mathbf{F}_{\text{old}})$ can be determined.

The results for the synchronization angle (7.49) and (7.50) and the distance to fault (7.51), obtained considering the lumped model of the line, are used as the initial values for $\delta_{\text{old}}, d_{\text{old}}$ in the calculations according to (7.55), (7.58) and (7.59). The iterative calculations are performed for the pre-defined fixed number of iterations or until the required convergence for the sought results is achieved. It has been checked that performing three iterations is sufficient.

Example 7.1 Fault location with use of the algorithm [116]
In Figs. 7.6 through 7.8, the results for the fault location performed according to the fault-location algorithm from [116] are presented. The specifications of the fault on a 400-kV, 300-km line are as follows:

- a–g fault;
- distance to fault $d = 0.8$ p.u.;
- fault resistance $R_{\text{F}} = 10\ \Omega$;
- actual synchronization angle $\delta = 36°$ (obtained by introducing a delay of signals from the side A by two samples at the sampling frequency $f_s = 1000$ Hz and the fundamental frequency $f_1 = 50$ Hz).

The digital values for the synchronization angle (Fig. 7.7) and distance to fault (Fig. 7.8) are singled out by averaging (subscript: av.) within the interval lasting from 30 to 50 ms after the fault inception.

Using positive-sequence quantities the following averaged values of the synchronization angle have been obtained: 94.28°, 37.42° (Fig. 7.7a), while using the negative-sequence components: 36.11°, –143.78° (Fig. 7.7a). There are four possible pairs of the calculated angles: (94.28°, 36.11°); (94.28°, –143.78°); (37.42°, 36.11°); (37.42°, –143.78°).

The pair (37.42°, 36.11°) consists of the synchronization angle values that are the closest to each other and do not differ too much. Such a pair with coincident results indicates the valid solution for the synchronization angle. Taking the values 37.42° (*for the positive-sequence*) or 36.11° (*for the negative-sequence*), respectively, the distance to fault (Fig. 7.8) was determined. In the case of using the *positive-sequence* quantities one gets:

- for the lumped-line model: 0.8964 p.u.;
- for the distributed-parameter line model: 0.8003 p.u.

Applying the *negative-sequence* components as the fault-locator input signals, the following results have been obtained:

- for the lumped-line model: 0.7948 p.u.;
- for the distributed-parameter line model: 0.7998 p.u.

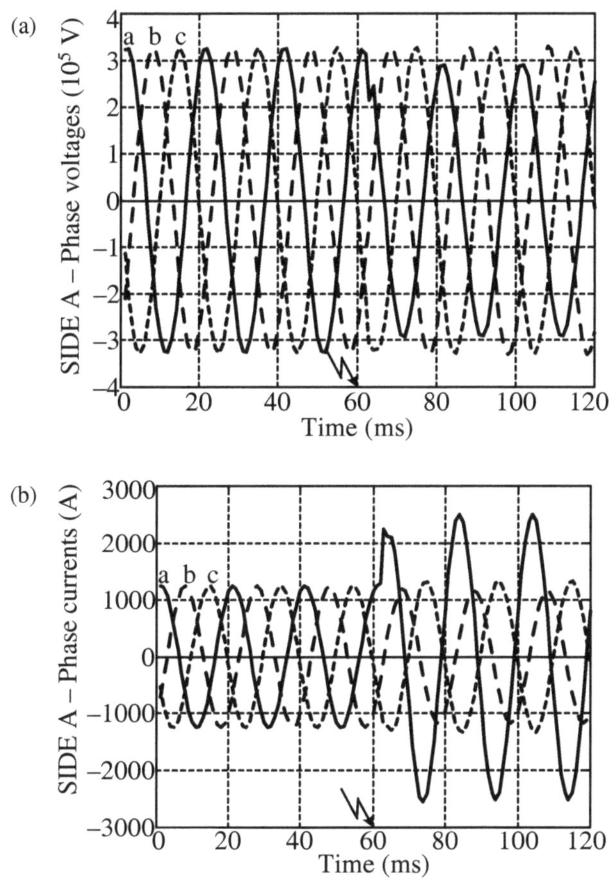

Fig. 7.6 Example 7.1 – input signals of the fault locator: (a) side A voltage, (b) side A current, (c) side B voltage, and (d) side B current

(c)

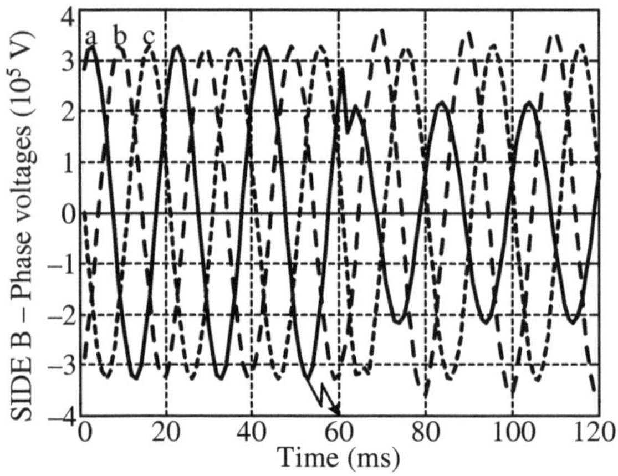

(d)

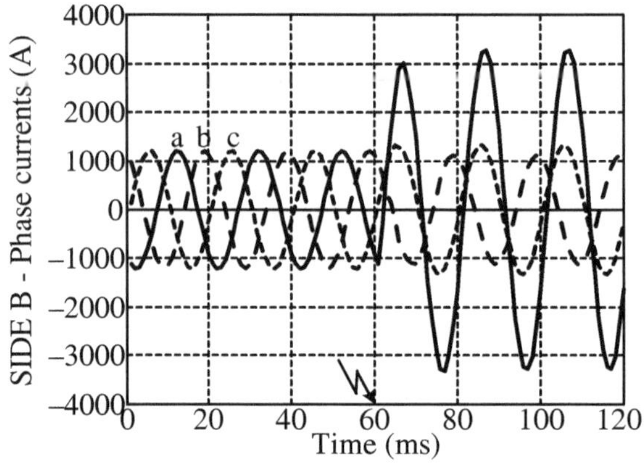

Fig. 7.6 (*continued*)

(a)

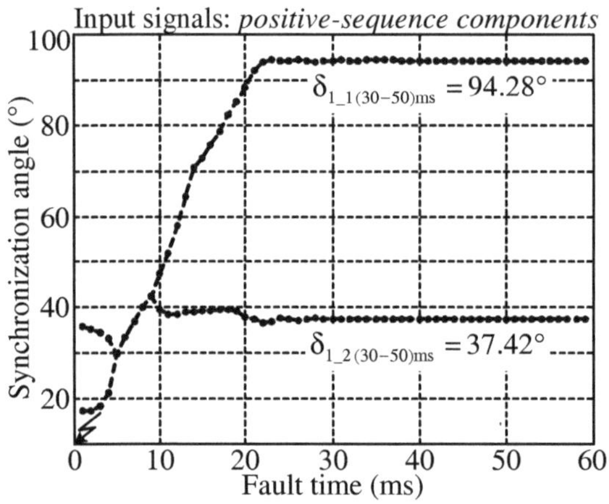

Fig. 7.7 Example 7.1 – determination of the synchronization angle using: (a) positive-sequence components, and (b) negative-sequence components

(b)

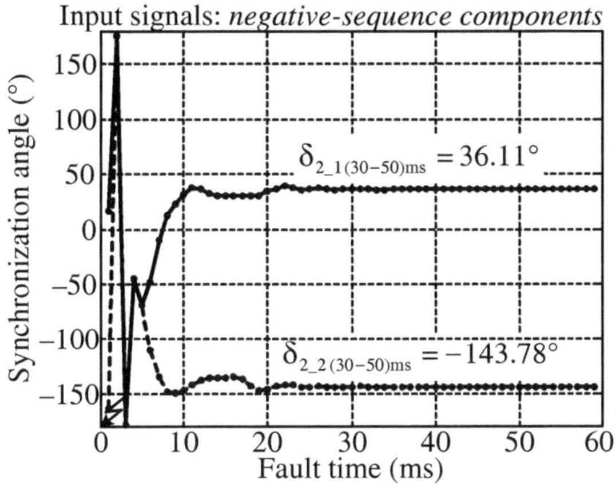

Fig. 7.7 (*continued*)

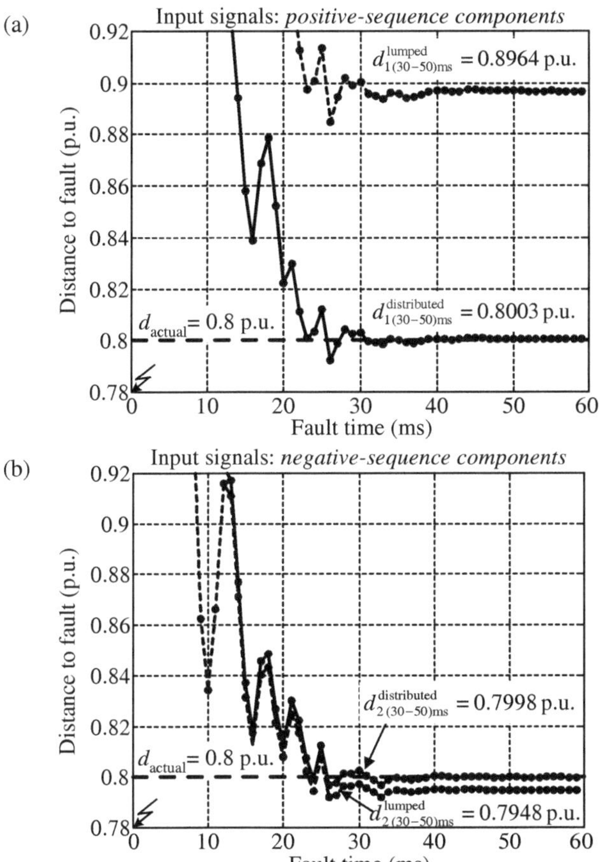

Fig. 7.8 Example 7.1 – determination of the distance to fault applying lumped- and distributed-parameter line models, using: (a) positive-sequence components, and (b) negative-sequence components

Applying the distributed-parameter line model for both types of symmetrical components, accurate results for the distance to fault are obtained. When using the lumped-line model, application of the negative-sequence components as the fault-locator input signals assures a much more accurate result than for the positive-sequence components.

7.3.5. Synchronization of Measurements of Distance Relays

Figure 7.9 presents the idea for the post-fault analysis of protective distance relays (REL_A, REL_B) installed at the terminals (AI, BI) of a double-circuit line (similarly as for a single line). The considerations will be carried out for a double-circuit line operating at common buses (AI/AII and BI/BII) at both ends.

It is considered that the relays are supplied with the measured currents and voltages, as for example the relay REL_A: $\{u_A\}$ – three-phase voltage, $\{i_A\}$ – three-phase current from the faulted-line circuit I, i_{AII0} – zero-sequence current from the healthy-line circuit II.

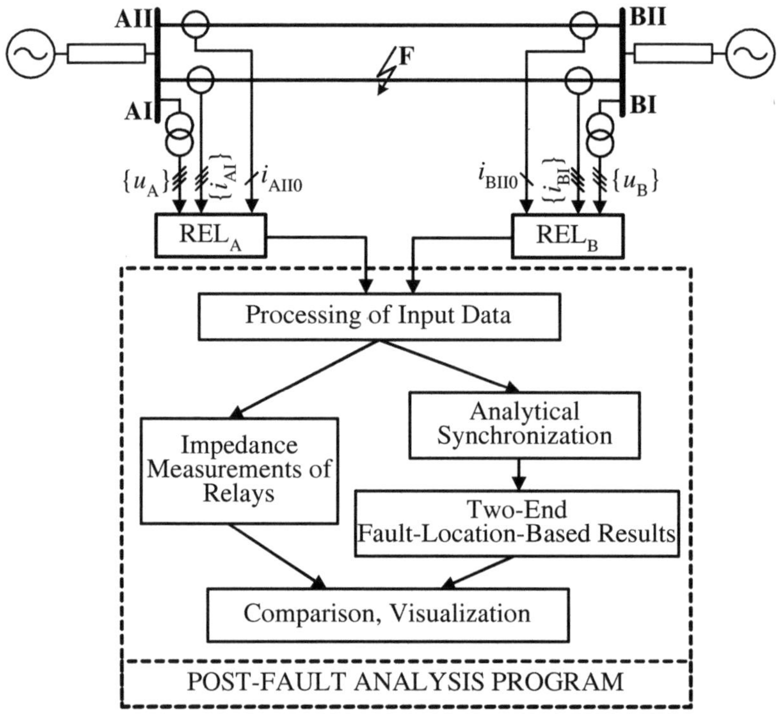

Fig. 7.9 Structure of post-fault analysis program for protective distance relays

The voltage and current signals measured by the relays are sent to the computer performing the analysis. Versatile methods for processing the input signals have been included in the post-fault analysis program developed [120]. In addition, the fault-loop impedance measurement performed in the distance relays is taken into account. The other part of the program utilizes the fault-loop impedance measurements that are based on the two-end fault-location principle. Prior to the fault location, the processed two-end signals are analytically synchronized.

Distance relays measure the apparent impedance of the appropriate fault loop using the relaying voltage ($\underline{U}_{A\_P}$) and current ($\underline{I}_{A\_P}$) signals. These signals are composed according to the identified fault type, as shown in:

- Tables 6.3 and 6.5: for single-circuit line; and
- Tables 6.4 and 6.6: for double-circuit line.

For the side A relay the fault loop can be described with the following formula for the phasor notation:

$$\underline{U}_{A\_P} - d\underline{Z}_{1L}\underline{I}_{A\_P} - R_F\underline{I}_F = 0 \tag{7.60}$$

where:
d – distance to fault (p.u.), measured from the bus A up to fault point F,
R_F – fault-path resistance,
$\underline{I}_F$ – total fault current.

The measurements of distance relays from both sides (A and B) are considered as performed asynchronously. Assuming the measurements of the relay B as the reference, the phasors of s the relay A signals are multiplied by the synchronization operator $e^{j\delta}$, in order to provide a common base.

In the models from Fig. 7.10a (for the relay A) and Fig. 7.10b (for the relay B) there are longitudinal branches represented by the positive-sequence impedances of the line sections A–F and F–B, respectively. The transverse branch in these models represents the fault path through which the total fault current flows. Merging the models of measurements of these two relays (from Figs. 7.10a and b), one obtains a general model with a *fictitious transverse branch* as in Fig. 7.11. The following sum of the relaying currents flows through this fictitious branch:

$$\underline{I}_{AB} = \underline{I}_{A\_P}e^{j\delta} + \underline{I}_{B\_P} \tag{7.61}$$

In [120, 270], it has been proposed to define the impedance of this fictitious branch $\underline{Z}_{FLT}$ as a function of the real fault-path resistance R_F (see fault models in Fig. 3.1) and the fault-type coefficient $\underline{P}_{FLT}$:

$$\underline{Z}_{FLT} = \frac{R_F}{\underline{P}_{FLT}} \tag{7.62}$$

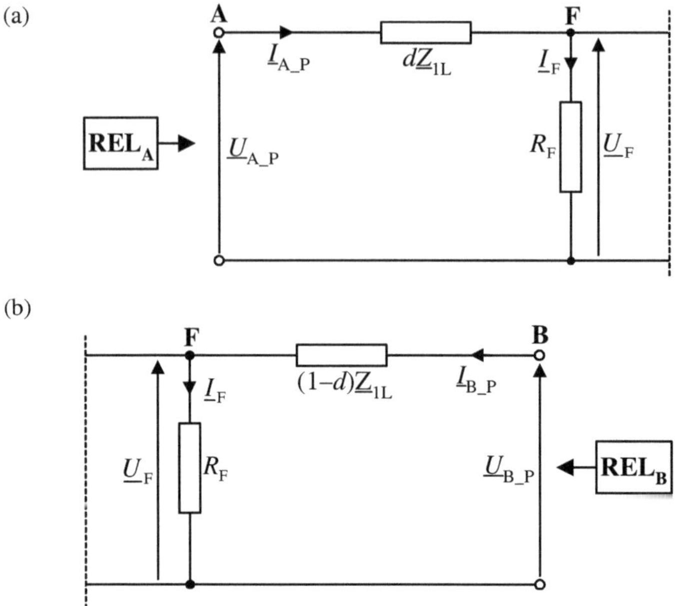

Fig. 7.10 Models of fault-loop measurement for: (a) relay at bus A, and (b) relay at bus B

The fault-type coefficient $\underline{P}_{FLT}$ has to be set to such a value that the voltage drop ($\underline{U}_F$) across the fault path in the model from Fig. 7.10a or Fig. 7.10b is identical with the drop across the impedance of the fictitious branch $\underline{Z}_{FLT}$ (7.62) in the model from Fig. 7.11. This condition results in the following formula:

$$R_F \underline{I}_F = \frac{R_F}{\underline{P}_{FLT}} (\underline{I}_{A\_P} e^{j\delta} + \underline{I}_{B\_P}) \tag{7.63}$$

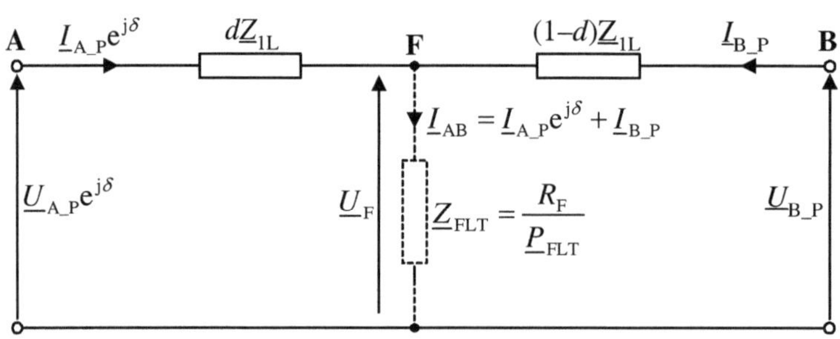

Fig. 7.11 General model of measurements of relays obtained by merging the models from Figs. 7.10a and b, assuming the time reference of measurements of relay REL$_B$ as the base

Table 7.2 Coefficient $\underline{P}_{\mathrm{FLT}}$ for different fault types

Fault type (fault models as in Fig. 3.1)	$\underline{P}_{\mathrm{FLT}}$
a–g, b–g, c–g	$\dfrac{2\underline{Z}_{1\mathrm{L}}+\underline{Z}_{0\mathrm{L}}}{3\underline{Z}_{1\mathrm{L}}}$
a–b, b–c, c–a	2
a–b–g, b–c–g, c–a–g, a–b–c, a–b–c–g	1

From (7.63) one obtains the fault-type coefficient as a function of currents:

$$\underline{P}_{\mathrm{FLT}}=\frac{\underline{I}_{\mathrm{A\_P}}\mathrm{e}^{\mathrm{j}\delta}+\underline{I}_{\mathrm{B\_P}}}{\underline{I}_{\mathrm{F}}} \tag{7.64}$$

Expressing the relaying currents from the nominator of (7.64) and the total fault current from the denominator of (7.64) in terms of the respective symmetrical components allows the coefficient $\underline{P}_{\mathrm{FLT}}$ to be determined for different fault types, as collected in Table 7.2.

Example 7.2 Determination of $\underline{P}_{\mathrm{FLT}}$ for a–g fault
According to Table 6.5, the relaying currents for the single-circuit line and a–g fault are expressed as follows:

$$[\underline{I}_{\mathrm{A\_P}}\mathrm{e}^{\mathrm{j}\delta}]_{(\mathrm{a-g})}=\left(\underline{I}_{\mathrm{A1}}+\underline{I}_{\mathrm{A2}}+\frac{\underline{Z}_{0\mathrm{L}}}{\underline{Z}_{1\mathrm{L}}}\underline{I}_{\mathrm{A0}}\right)\mathrm{e}^{\mathrm{j}\delta}$$

$$[\underline{I}_{\mathrm{B\_P}}]_{(\mathrm{a-g})}=\underline{I}_{\mathrm{B1}}+\underline{I}_{\mathrm{B2}}+\frac{\underline{Z}_{0\mathrm{L}}}{\underline{Z}_{1\mathrm{L}}}\underline{I}_{\mathrm{B0}}$$

The total fault current for a–g fault can be expressed as the following sum of its symmetrical components (see: *Example 3.1*):

$$\underline{I}_{\mathrm{F}}=\underline{I}_{\mathrm{F1}}+\underline{I}_{\mathrm{F2}}+\underline{I}_{\mathrm{F0}}$$

Substitution of the above three formulae into (7.64) yields:

$$[\underline{P}_{\mathrm{FLT}}]_{(\mathrm{a-g})}=\frac{\left(\underline{I}_{\mathrm{A1}}+\underline{I}_{\mathrm{A2}}+\frac{\underline{Z}_{0\mathrm{L}}}{\underline{Z}_{1\mathrm{L}}}\underline{I}_{\mathrm{A0}}\right)\mathrm{e}^{\mathrm{j}\delta}+\left(\underline{I}_{\mathrm{B1}}+\underline{I}_{\mathrm{B2}}+\frac{\underline{Z}_{0\mathrm{L}}}{\underline{Z}_{1\mathrm{L}}}\underline{I}_{\mathrm{B0}}\right)}{\underline{I}_{\mathrm{F1}}+\underline{I}_{\mathrm{F2}}+\underline{I}_{\mathrm{F0}}}$$

Taking into account that the symmetrical components of the total fault current can be expressed as the respective sums of components of currents from both ends (with the line shunt capacitances neglected):

$$\underline{I}_{F1} = \underline{I}_{A1} e^{j\delta} + \underline{I}_{B1}$$

$$\underline{I}_{F2} = \underline{I}_{A2} e^{j\delta} + \underline{I}_{B2}$$

$$\underline{I}_{F0} = \underline{I}_{A0} e^{j\delta} + \underline{I}_{B0}$$

the fault-type coefficient equals:

$$[\underline{P}_{FLT}]_{(a-g)} = \frac{\underline{I}_{F1} + \underline{I}_{F2} + \frac{\underline{Z}_{0L}}{\underline{Z}_{1L}} \underline{I}_{F0}}{\underline{I}_{F1} + \underline{I}_{F2} + \underline{I}_{F0}}$$

Consideration of the boundary conditions for a–g fault:

$$\underline{I}_{Fb} = \underline{I}_{Fc} = 0 \text{ and } \underline{I}_{F} = \underline{I}_{Fa} \neq 0$$

results in the following relation between the symmetrical components of the total fault current:

$$\underline{I}_{F1} = \underline{I}_{F2} = \underline{I}_{F0} .$$

Therefore, the fault-type coefficient for an a–g fault is equal to:

$$[\underline{P}_{FLT}]_{(a-g)} = \frac{2\underline{Z}_{1L} + \underline{Z}_{0L}}{3\underline{Z}_{1L}}$$

The same value of $\underline{P}_{FLT}$ one also obtains for the other single phase-to-ground faults (Table 7.2):

$$[\underline{P}_{FLT}]_{(a-g)} = [\underline{P}_{FLT}]_{(b-g)} = [\underline{P}_{FLT}]_{(c-g)} = \frac{2\underline{Z}_{1L} + \underline{Z}_{0L}}{3\underline{Z}_{1L}}$$

Since impedances of transmission lines for the positive- and zero-sequence, with respect to both magnitude and phase angle, are not identical, the fault-type coefficient $\underline{P}_{FLT}$ for phase-to-ground faults is a complex number.

In the case of double-circuit transmission lines the fault-type coefficient $\underline{P}_{FLT}$ for single phase-to-ground faults is determined identically as for the single-circuit line. This results from the fact that, when neglecting shunt capacitances of the line, the sum of the zero-sequence currents from both ends of the parallel healthy line equals zero:

$$\frac{\underline{Z}_{0m}}{\underline{Z}_{1L}}\underline{I}_{AII0}e^{j\delta}+\frac{\underline{Z}_{0m}}{\underline{Z}_{1L}}\underline{I}_{BII0}=0$$

The coefficient $\underline{P}_{FLT}$ for other fault types can be determined analogously (Table 7.2). Also, one obtains identical values for both single-circuit and double-circuit lines.

Example 7.3 Fault location with analytical synchronization of measurements of distance relays

Figures 7.12 through 7.17 present the results for the a–g fault on a 400-kV, 300-km single-circuit transmission line. The emfs at the bus B were assumed as leading the emfs of the side A by 30°; distance to fault from the terminal A: d = 0.8 p.u., fault resistance: $R_F = 10\ \Omega$.

Processing the input signals as shown in Fig. 7.12, the distance relay at the terminal A determines the fault-loop resistance (Fig. 7.13a) and reactance (Fig. 7.13b). Due to the presence of the reactance effect, the determined values differ from the actual resistance and reactance of the transmission-line section from the measuring point (A) up to the fault point (F). A full-cycle Fourier filtering was applied to the measurement. Polar plots of the impedance measurements are shown in Fig. 7.14a (for the first 20 samples) and Fig. 7.14b (for the next 20 samples). One observes that due to the reactance effect the trajectory of the measured fault-loop impedance does not enter the MHO impedance characteristic, for which the reach of 85% of the line positive-sequence impedance was set. The reactance effect causes the fault occurring at 80% of the line length to be not detected by the first zone MHO element set at 85%.

In turn, application of the two-end fault-location algorithm allows accurate location of the fault. It starts with determination of the synchronization angle using the relaying signals (fault-loop voltage and current) – Fig. 7.15a, or the negative-sequence signals – Fig. 7.15b. This allows us to select the valid values for the synchronization angle (18.47°, 18.15°), which are very close to the introduced intentional desynchronization (δ_{actual} = 18°). Performing analytical synchronization with use of the determined synchronization angle, the distance to fault (d [p.u.]) was calculated. Then, the fault-loop impedance using the fault-location principle (Fig. 7.16) was calculated: $\underline{Z}_{FL}=d\underline{Z}_{1L}$, the trajectory of which surely enters the interior of the MHO characteristic. Figure 7.17 shows a comparison of the fault-loop resistance (Fig. 7.17a) and reactance (Fig. 7.17b) determined in two different ways, according to the following principles:

- classic distance relay measurement; and
- using the two-end fault-location principle.

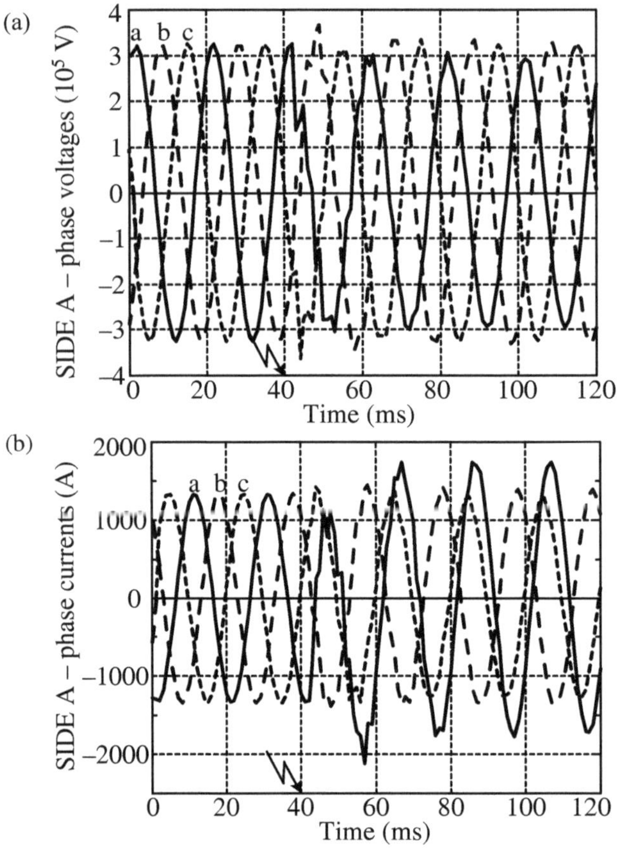

Fig. 7.12 Example 7.3 – (a) three-phase voltage at side A, and (b) three-phase current at side A

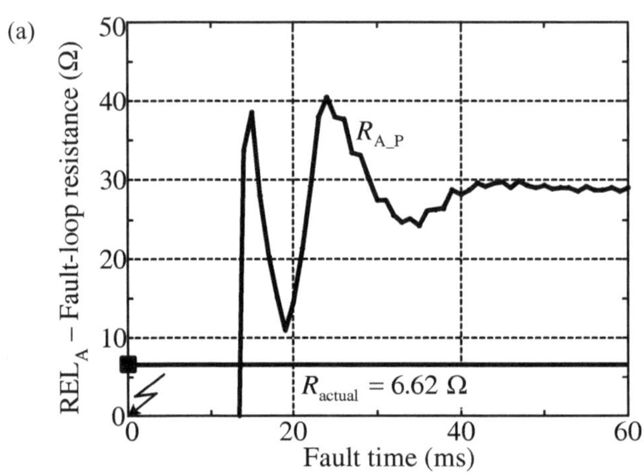

Fig. 7.13 Example 7.3 – (a) fault-loop resistance measured by relay at A, and (b) fault-loop reactance measured by relay at A

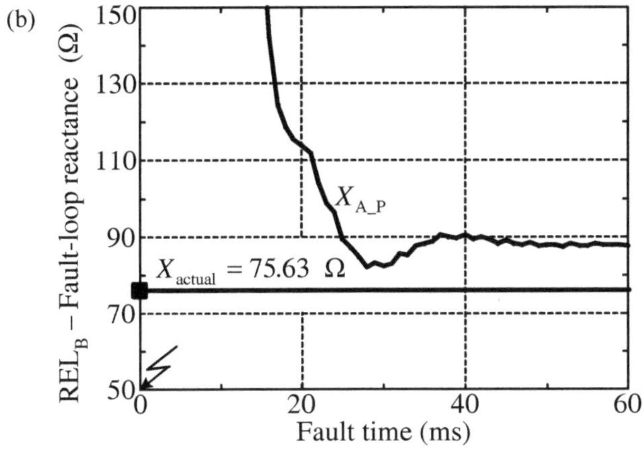

Fig. 7.13 (*continued*)

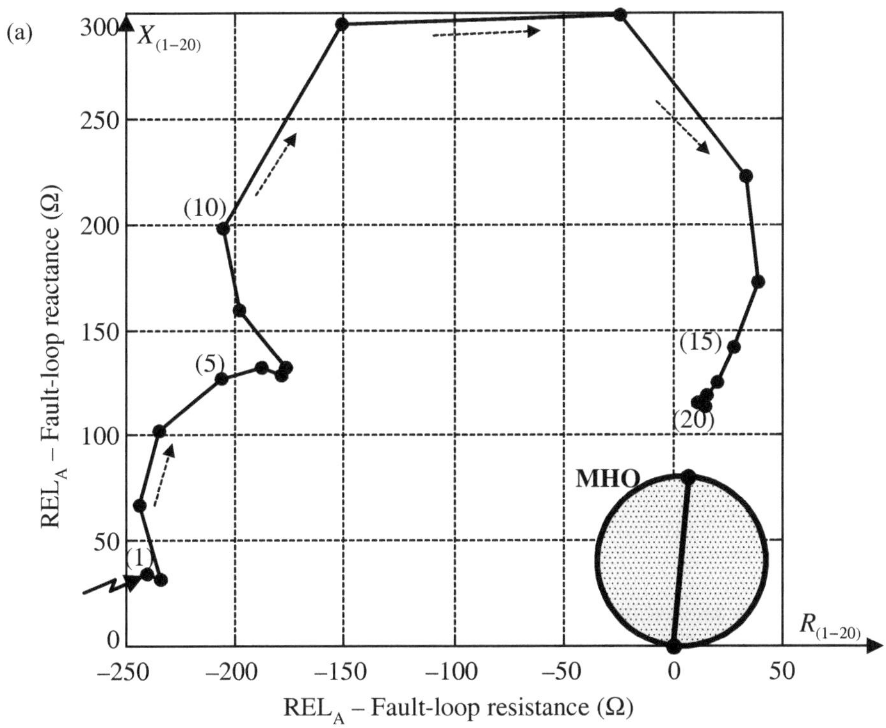

Fig. 7.14 Example 7.3 –(a) polar plot of fault-loop impedance measured by relay at A – first 20 samples since fault inception, and (b) polar plot of fault-loop impedance measured by relay at A - next 20 samples (the samples from 21up to 40 ms since fault inception)

(b)

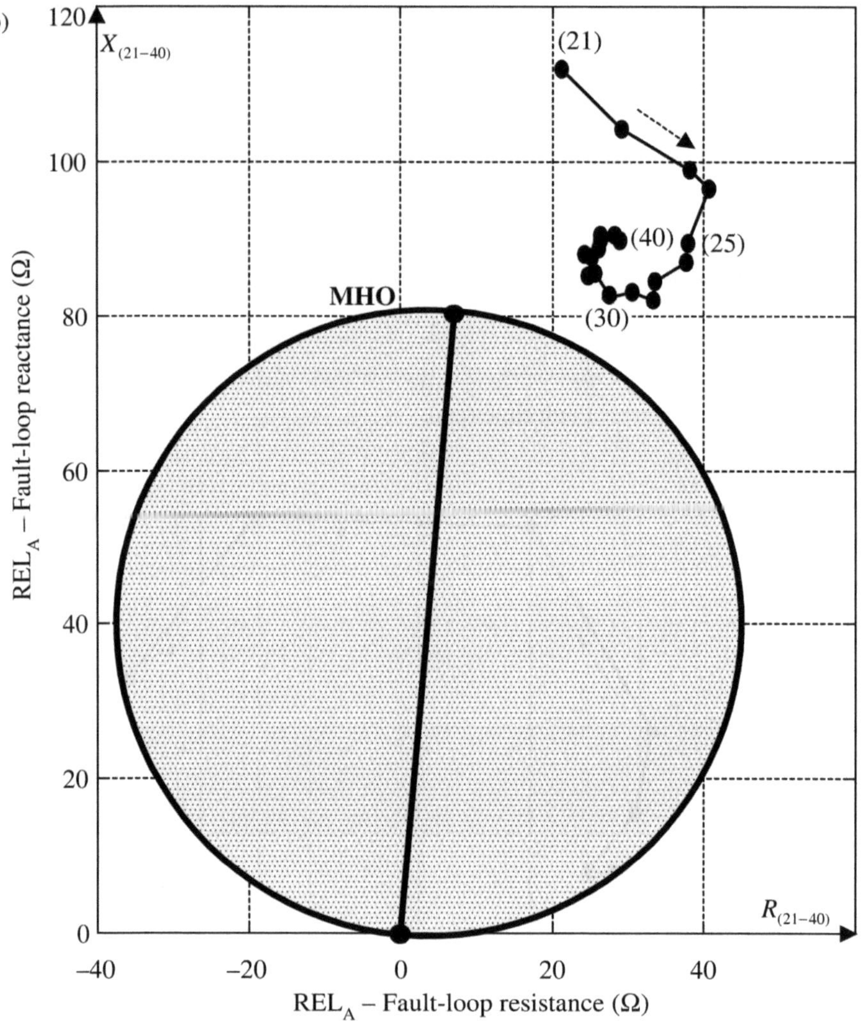

Fig. 7.14 (*continued*)

It is clearly shown in Fig. 7.16 and Fig. 7.17 that using the two-end fault-location principle, the influence of the reactance effect is fully compensated for.

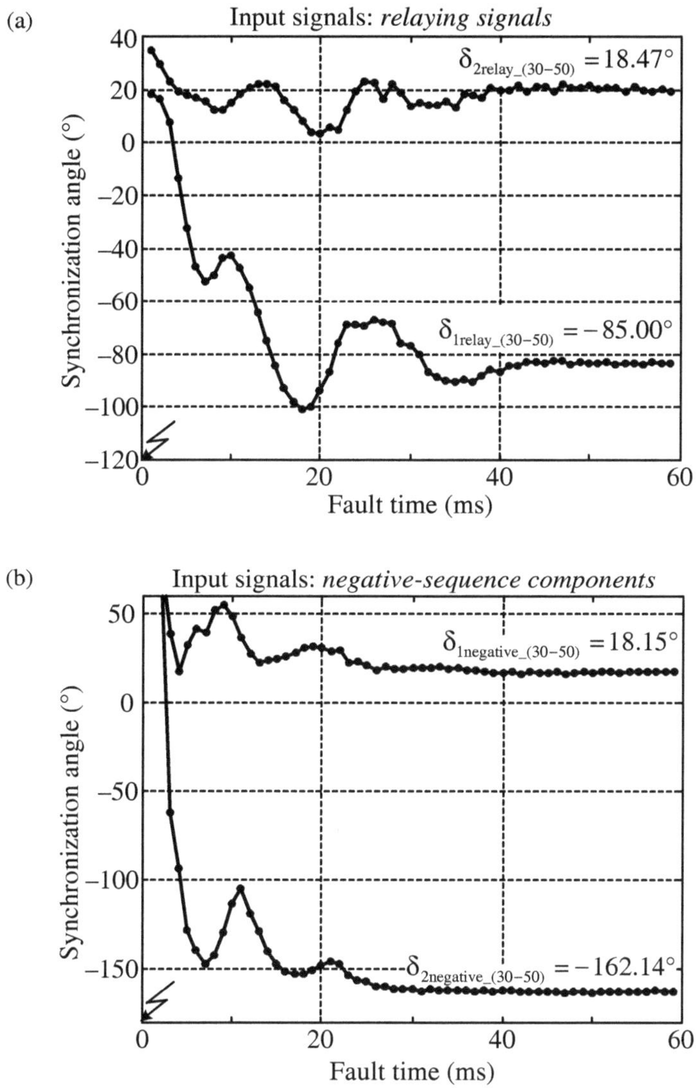

Fig. 7.15 Example 7.3 – determination of synchronization angle using: (a) relaying signals, and (b) negative-sequence components

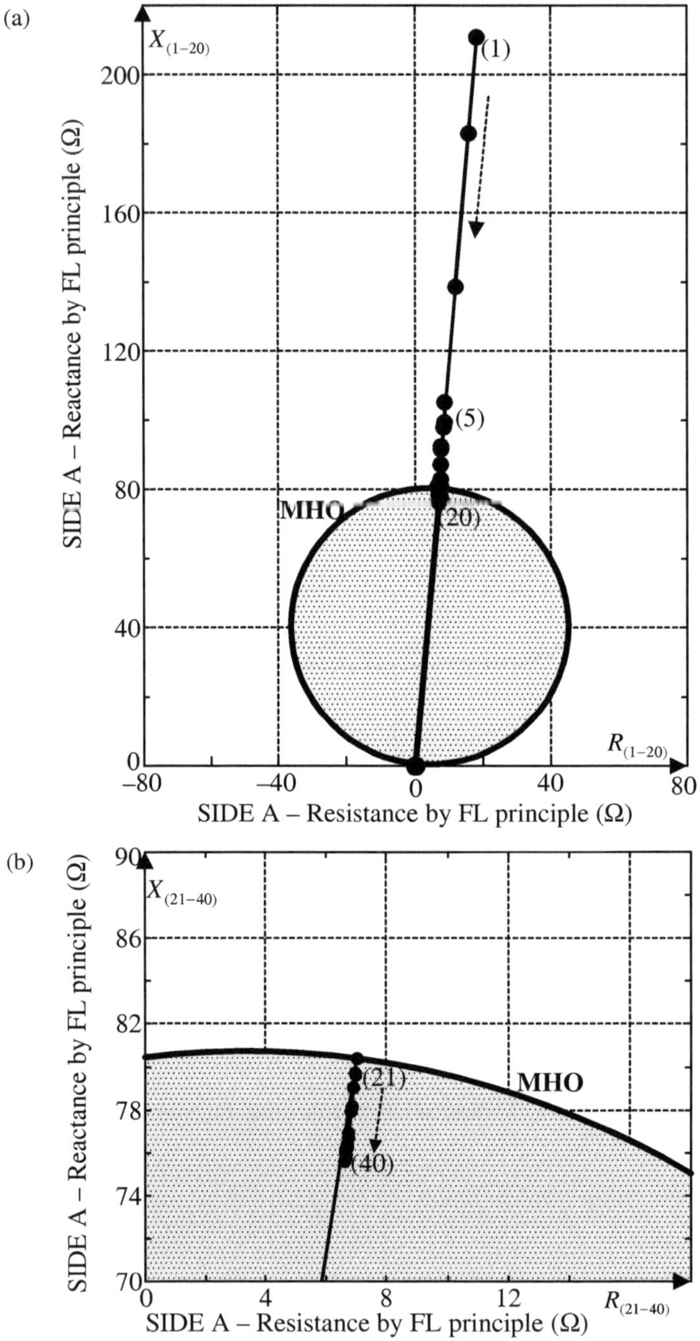

Fig. 7.16 Example 7.3 – determination of fault-loop impedance by applying the fault-location principle: (a) for the first 20 samples since fault inception, and (b) for the next 20 samples (the samples from 21up to 40 ms since fault inception)

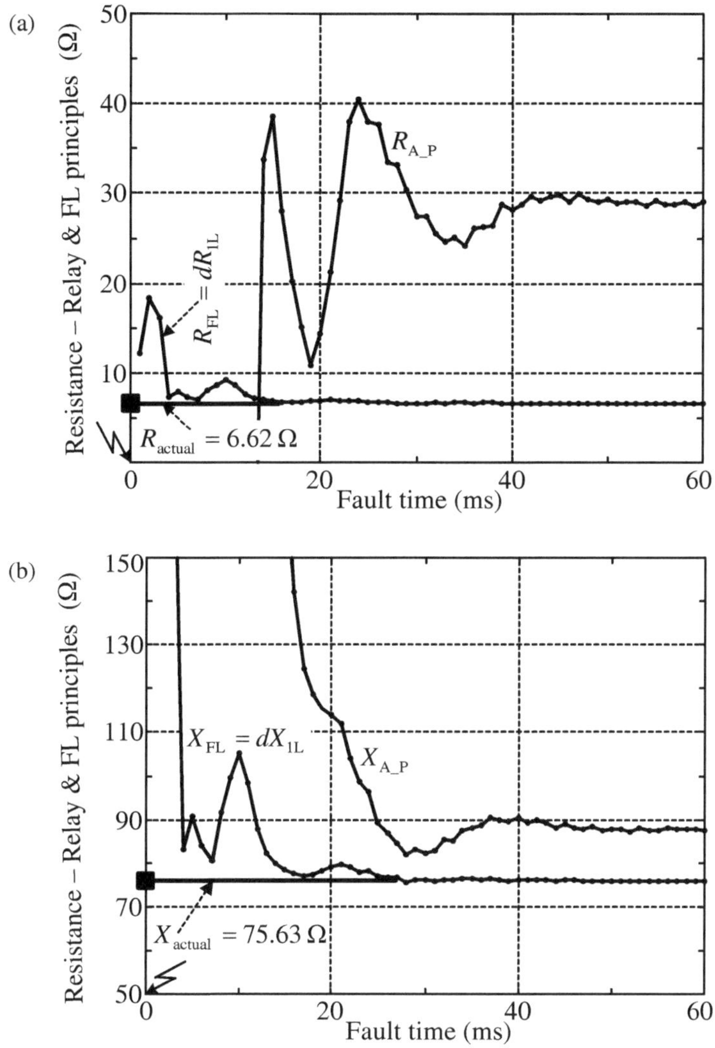

Fig. 7.17 Example 7.3 – comparison of fault-loop impedance determined by distance relay measurement and by applying the fault-location principle: (a) fault-loop resistance, and (b) fault-loop reactance

7.3.6 Use of Unsynchronized Measurements of Distance Relays

Measurements performed by impedance relays from both terminals of the faulted line can be described by applying the general model derived, depicted in Fig. 7.11, as follows:

$$\underline{U}_{\mathrm{A\_P}}\,\mathrm{e}^{\mathrm{j}\delta}-d\underline{Z}_{1\mathrm{L}}\underline{I}_{\mathrm{A\_P}}\,\mathrm{e}^{\mathrm{j}\delta}-\frac{R_{\mathrm{F}}}{\underline{P}_{\mathrm{FLT}}}(\underline{I}_{\mathrm{A\_P}}\,\mathrm{e}^{\mathrm{j}\delta}+\underline{I}_{\mathrm{B\_P}})=0 \tag{7.65}$$

$$\underline{U}_{\mathrm{B\_P}}-(1-d)\underline{Z}_{1\mathrm{L}}\underline{I}_{\mathrm{B\_P}}-\frac{R_{\mathrm{F}}}{\underline{P}_{\mathrm{FLT}}}(\underline{I}_{\mathrm{A\_P}}\,\mathrm{e}^{\mathrm{j}\delta}+\underline{I}_{\mathrm{B\_P}})=0 \tag{7.66}$$

Introducing apparent impedances measured by the impedance relays:

$$\underline{Z}_{\mathrm{A\_P}}=\frac{\underline{U}_{\mathrm{A\_P}}\,\mathrm{e}^{\mathrm{j}\delta}}{\underline{I}_{\mathrm{A\_P}}\,\mathrm{e}^{\mathrm{j}\delta}} \tag{7.67}$$

$$\underline{Z}_{\mathrm{B\_P}}=\frac{\underline{U}_{\mathrm{B\_P}}}{\underline{I}_{\mathrm{B\_P}}} \tag{7.68}$$

(7.65) and (7.66) transform to:

$$\underline{Z}_{\mathrm{A\_P}}-d\underline{Z}_{1\mathrm{L}}-\frac{R_{\mathrm{F}}}{\underline{P}_{\mathrm{FLT}}}\frac{(\underline{I}_{\mathrm{A\_P}}\,\mathrm{e}^{\mathrm{j}\delta}+\underline{I}_{\mathrm{B\_P}})}{\underline{I}_{\mathrm{A\_P}}\,\mathrm{e}^{\mathrm{j}\delta}}=0 \tag{7.69}$$

$$\underline{Z}_{\mathrm{B\_P}}-(1-d)\underline{Z}_{1\mathrm{L}}-\frac{R_{\mathrm{F}}}{\underline{P}_{\mathrm{FLT}}}\frac{(\underline{I}_{\mathrm{A\_P}}\,\mathrm{e}^{\mathrm{j}\delta}+\underline{I}_{\mathrm{B\_P}})}{\underline{I}_{\mathrm{B\_P}}}=0 \tag{7.70}$$

Combining (7.69) and (7.70), and eliminating the fault-loop currents, one obtains the following quadratic formula for the complex numbers with the following unknowns d, R_{F}:

$$\underline{D}_2 d^2+\underline{D}_1 d+\underline{D}_0+\frac{R_{\mathrm{F}}}{\underline{P}_{\mathrm{FLT}}}\underline{N}_0=0 \tag{7.71}$$

where:
$\underline{D}_2=\underline{Z}_{1\mathrm{L}}\underline{Z}_{1\mathrm{L}}$,
$\underline{D}_1=\underline{Z}_{1\mathrm{L}}\underline{Z}_{\mathrm{B\_P}}-\underline{Z}_{1\mathrm{L}}\underline{Z}_{1\mathrm{L}}-\underline{Z}_{1\mathrm{L}}\underline{Z}_{\mathrm{A\_P}}$,
$\underline{D}_0=\underline{Z}_{1\mathrm{L}}\underline{Z}_{\mathrm{A\_P}}-\underline{Z}_{\mathrm{A\_P}}\underline{Z}_{\mathrm{B\_P}}$,
$\underline{N}_0=\underline{Z}_{\mathrm{A\_P}}+\underline{Z}_{\mathrm{B\_P}}-\underline{Z}_{1\mathrm{L}}$,
$\underline{P}_{\mathrm{FLT}}$ – fault-type coefficient as in Table 7.2.

Analysis of the quantity $\underline{N}_0$ from (7.71) allows distinguishing the solid faults (or involving comparatively low fault resistances) from those evidently resistive. In the second case (resistive faults), after resolving (7.71) into the real and imaginary components, respectively, the following quadratic formula for a sought fault distance is obtained:

$$\mathrm{F_A}(d) = A_2 d^2 + A_1 d + A_0 = 0 \tag{7.72}$$

where A_2, A_1, A_0 – real number coefficients (derived accordingly from (7.71)), involving only the apparent impedances – defined in (7.67) and (7.68), and the fault-type coefficient $\underline{P}_{\mathrm{FLT}}$ (Table 7.2).

One of the roots of (7.72) gives the valid solution for a distance to fault. In some cases it may happen that both roots are within the line length. In order to cope with such troublesome cases, an extra formula has to be derived. This can be accomplished by combining (7.69) and (7.70) with elimination of the quantity $R_{\mathrm{F}}/\underline{P}_{\mathrm{FLT}}$. As a result, one obtains:

$$d = \frac{1 - \dfrac{\underline{Z}_{\mathrm{B\_P}}}{\underline{Z}_{\mathrm{1L}}} + \dfrac{\underline{Z}_{\mathrm{A\_P}}}{\underline{Z}_{\mathrm{1L}}}\underline{\vartheta}_{\mathrm{AB}}}{\underline{\vartheta}_{\mathrm{AB}} + 1} \tag{7.73}$$

where:

$$\underline{\vartheta}_{\mathrm{AB}} = \frac{\underline{I}_{\mathrm{A\_P}} \mathrm{e}^{\mathrm{j}\delta}}{\underline{I}_{\mathrm{B\_P}}} = |\underline{\vartheta}_{\mathrm{AB}}| \mathrm{e}^{\mathrm{j}\theta},$$

$$|\underline{\vartheta}_{\mathrm{AB}}| = \left| \frac{\underline{I}_{\mathrm{A\_P}} \mathrm{e}^{\mathrm{j}\delta}}{\underline{I}_{\mathrm{B\_P}}} \right| = \frac{|\underline{I}_{\mathrm{A\_P}}|}{\underline{I}_{\mathrm{B\_P}}},$$

$$\theta = \mathrm{angle}\left(\frac{\underline{I}_{\mathrm{A\_P}} \mathrm{e}^{\mathrm{j}\delta}}{\underline{I}_{\mathrm{B\_P}}} \right) = \mathrm{angle}\left(\frac{\underline{I}_{\mathrm{A\_P}}}{\underline{I}_{\mathrm{B\_P}}} \right) + \delta .$$

A sought distance to fault can be calculated from (7.73) if besides the apparent impedances, defined in (7.67) and (7.68), the ratio of the phasors of the fault-loop currents of both relays ($\underline{\vartheta}_{\mathrm{AB}}$) is provided. The magnitude of this ratio ($|\underline{\vartheta}_{\mathrm{AB}}|$) is not affected by the synchronization angle (δ). In contrast, the angle of this ratio (θ) involves the synchronization angle. There are different possibilities of measuring the angle (θ), such as for example, utilizing the pre-fault measurements. However, setting the angle θ in (7.73) by a trial and error method, so that the right-hand side of (7.73) is the real number, may be a good alternative.

7.4 Use of Incomplete Two-end Measurements

7.4.1 Use of Two-end Voltages

A fault-location method utilizing synchronized measurements of two-end voltages only was proposed in [23, 24]. By complete rejection of currents from the input signals, the fault-location algorithm free from CT errors has been obtained. Thus, these algorithms appear to be completely immune to CT saturation, which is an important advantage of them. On the other hand, the need for providing source impedances from both line ends appears as a particular limitation of this algorithm. This is so since a mismatch between the data provided and the actual source impedances may happen and thus the fault-location accuracy can be degraded. The algorithms from [23, 24] are formulated using a matrix description of the transmission network.

In [186], several algorithms utilizing only the voltages, dispensing with current transformers and thus eliminating the errors caused by saturation of current transformers are presented. The algorithms are applicable to phase-to-ground faults, phase-to-phase faults, and phase-to-phase-to-ground faults, thus not to three-phase balanced faults. The algorithms utilize unsynchronized fault voltage measurements from two ends of a line and do not require pre-fault data. Shunt capacitances of the line are fully considered, which is advantageous from the fault-location accuracy point of view.

In turn, in [184] Liao introduced an interesting fault-location approach for single-circuit lines utilizing only voltage measurements from one or two buses, which may be distant from the faulted line. An advantage of the bus-impedance matrix technique is utilized for that. With the addition of a fictitious bus where the fault occurs, the transfer impedances of this bus and other buses are revealed as a function of the fault location. It is shown in [184] that based on the relationship between the bus-voltage change due to the fault and the transfer impedance, the fault location can be derived. The shunt capacitance of the line is ignored first and then fully considered based on a distributed-parameter line model.

A method for locating faults on two-terminal transmission lines based on the fundamental components of fault and pre-fault voltage measured at the two ends of a transmission line has been introduced in [341]. This methodology allows one to establish a direct calculation procedure that is independent of fault and pre-fault currents, fault type, fault resistance, synchronization condition of register devices located on line ends, and pre-fault condition, either balanced or not. This is achieved by defining a new concept called the '*distance factor*' (K_V) in relation to the circuit diagram of the transmission network for the incremental positive-sequence:

$$K_{\mathrm{V}} = \left| \frac{\Delta \underline{U}_{\mathrm{A1}} \mathrm{e}^{\mathrm{j}\delta}}{\Delta \underline{U}_{\mathrm{B1}}} \right| \tag{7.74}$$

where $\Delta \underline{U}_{\mathrm{A1}}$, $\Delta \underline{U}_{\mathrm{B1}}$ – incremental positive-sequence voltage obtained from asynchronous measurements of voltages at both line ends (A, B).

Taking into account that for the synchronization operator we have $|\mathrm{e}^{\mathrm{j}\delta}| = 1$, one can derive that the distance factor (7.74) is a real-number function of the following parameters (Fig. 7.18):

- positive-sequence impedance ($\underline{Z}_{1\mathrm{L}}^{'}$, per unit length) and admittance ($\underline{Y}_{1\mathrm{L}}^{'}$, per unit length) of the transmission line,;
- positive-sequence impedances of the sources at both line ends ($\underline{Z}_{1\mathrm{A}}$, $\underline{Z}_{1\mathrm{B}}$) and extra link between the buses ($\underline{Z}_{1\mathrm{E}}$);
- distance to fault (d).

Thus, the distance factor can be expressed as the following function:

$$K_{\mathrm{V}} = f(\underline{Z}_{1\mathrm{L}}^{'}, \underline{Y}_{1\mathrm{L}}^{'}, \underline{Z}_{1\mathrm{A}}, \underline{Z}_{1\mathrm{B}}, \underline{Z}_{1\mathrm{E}}, d) \tag{7.75}$$

Knowing the impedance/admittance data involved in (7.75) one can draw the plot for the distance factor as in Fig. 7.19: $(K_{\mathrm{V}})_{\mathrm{theoretical}}$. The same can be done with use of a simulation tool, by applying faults at subsequent locations and calculating the distance factor according to the definition (7.74).

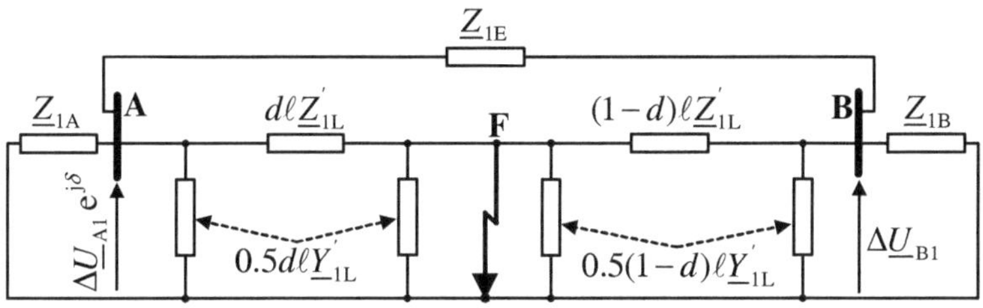

Fig. 7.18 Equivalent circuit diagram for two-end unsynchronized measurement of incremental positive-sequence voltage

The distance to fault can be determined by finding the intersection of the distance factor obtained from theoretical (or simulation) considerations ($(K_{\mathrm{V}})_{\mathrm{theoretical}}$) and that measured according to the definition (7.74), as shown in Fig. 7.19.

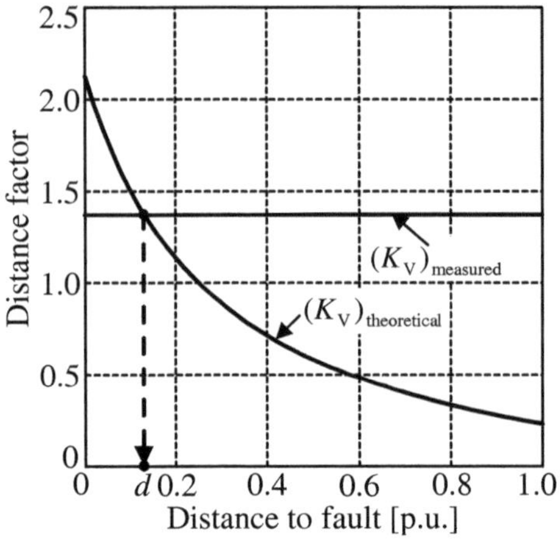

Fig. 7.19 Plots of the distance factors determined theoretically and obtained from measurement

The fault-location algorithm from [341] is completely immune to CT saturation, similarly as the other algorithms utilizing voltage signals. However, the source impedances are also required to be provided as the fault locator-input data.

7.4.2 Use of Two-end Voltages and One-end Current

It is well known from both experience and the literature [292] that if the CT saturation occurs, it occurs at one line side only. Therefore, it seems to be advantageous if the secondary currents of the CTs from the line side with no CT saturation are included in the fault-locator input signals, besides the voltages acquired at both line sides (utilized alone by the methods introduced in [23, 24, 184, 186, 341]). As a result, still incomplete two-end signals are utilized, but more information can be gained from the signals. The fault-location algorithm utilizing such incomplete two-end signals measured asynchronously was introduced in [123, 271].

Let us derive the fault-location algorithm for the case with CT saturation occurring at the line side B (Fig. 7.20). The other case: saturation of CTs at the side A can be resolved analogously. The samples of three-phase currents and voltages from the measurement units at both line ends (MU_A, MU_B) are transferred via the communication channels to the fault locator. First, the currents from both line ends are processed to recognize saturation. Algorithms known from the literature [143, 144, 152] can be applied for this. As a result of identifying the CT saturation at the line end B (in the considered case – Fig. 7.20), the three-phase current measured at the side B: $\{i_B(n)\}=\{i_{Ba}(n), i_{Bb}(n), i_{Bc}(n)\}$ are further rejected, regardless of how many CTs (only one, or two, or all three CTs) have become saturated.

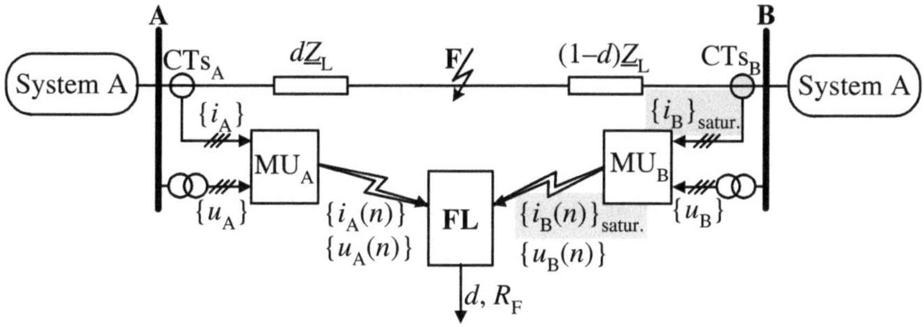

Fig. 7.20 Schematic diagram for considering two-end fault location immune to CT saturation (the case of CT saturation at the line end B)

Finally, the following three-phase phasors are applied for fault location:

- voltage at the bus A: $\{\underline{U}_A\}=\{\underline{U}_{Aa}, \underline{U}_{Ab}, \underline{U}_{Ac}\}$;
- voltage at the bus B: $\{\underline{U}_B\}=\{\underline{U}_{Ba}, \underline{U}_{Bb}, \underline{U}_{Bc}\}$;
- current at the bus A: $\{\underline{I}_A\}=\{\underline{I}_{Aa}, \underline{I}_{Ab}, \underline{I}_{Ac}\}$.

Since asynchronous measurements are considered in this algorithm [123, 271], an analytical synchronization is performed by multiplying all phasors from the side A (the side B measurement is assumed here as the base) by the synchronization operator $e^{j\delta}$.

The fault-location algorithm derivation is related to Fig. 7.2, where the distributed-parameter model of a faulted transmission line for the *i*th symmetrical component is presented. The following generalized model, describing the fault loop seen from the terminal A can be formulated as follows:

$$\underline{U}_{F\_P}(d, e^{j\delta}) - R_F \underline{I}_F = 0 \tag{7.76}$$

where:

$\underline{U}_{F\_P}(d, e^{j\delta})$ – fault-loop voltage after having been transferred from the measuring point A to the fault point F (therefore, the first index of the subscript is F instead of A),

d – unknown distance to fault (p.u.), as measured from the terminal A to fault point F,

R_F – unknown fault-path resistance,

$\underline{I}_F$ – fault-path current (total fault current).

The fault-loop voltage at the fault point F can be composed in terms of its symmetrical components:

$$\underline{U}_{F\_P}(d, e^{j\delta}) = \underline{a}_1 \underline{U}_{F1} + \underline{a}_2 \underline{U}_{F2} + \underline{a}_0 \underline{U}_{F0} \tag{7.77}$$

where: $\underline{a}_1$, $\underline{a}_2$, $\underline{a}_0$ – weighting coefficients dependent on fault type (Table 6.5).

Applying the distributed-parameter model of the line (Fig. 7.2), the *i*th symmetrical component of voltages at the fault point, involved in (7.77), is:

$$\underline{U}_{\mathrm{F}i} = [\underline{U}_{\mathrm{A}i}\cosh(\underline{\gamma}_i d\ell) - \underline{Z}_{\mathrm{c}i}\underline{I}_{\mathrm{A}i}\sinh(\underline{\gamma}_i d\ell)]\cdot \mathrm{e}^{\mathrm{j}\delta} \tag{7.78}$$

Then, the total fault current ($\underline{I}_{\mathrm{F}}$) is determined as a composition of its positive- and negative-sequence components, i.e., eliminating the zero-sequence component:

$$\underline{I}_{\mathrm{F}} = \frac{\underline{M}_{12}(\mathrm{e}^{\mathrm{j}\delta})}{\underline{Z}_{\mathrm{c}1}\sinh(\underline{\gamma}_1(1-d)\ell)} \tag{7.79}$$

where:

$\underline{M}_{12}(\mathrm{e}^{\mathrm{j}\delta}) = \underline{a}_{\mathrm{F}1}(\underline{U}_{\mathrm{B}1} + \underline{N}_{\mathrm{A}1}\mathrm{e}^{\mathrm{j}\delta}) + \underline{a}_{\mathrm{F}2}(\underline{U}_{\mathrm{B}2} + \underline{N}_{\mathrm{A}2}\mathrm{e}^{\mathrm{j}\delta})$,

$\underline{N}_{\mathrm{A}1}$, $\underline{N}_{\mathrm{A}2}$ – as in (7.28),

$\underline{a}_{\mathrm{F}1}$, $\underline{a}_{\mathrm{F}2}$ – share coefficients dependent on fault type (any set taken from Tables 3.3–3.5 can be applied).

Substituting the total fault current (7.79) into the general fault-loop model (7.76) yields:

$$\underline{U}_{\mathrm{F\_P}}(d, \mathrm{e}^{\mathrm{j}\delta}) - R_{\mathrm{F}}\frac{\underline{M}_{12}(\mathrm{e}^{\mathrm{j}\delta})}{\underline{Z}_{\mathrm{c}1}\sinh(\underline{\gamma}_1(1-d)\ell)} = 0 \tag{7.80}$$

In order to solve (7.80) for the unknown distance to fault (d) and fault resistance (R_{F}), the synchronization operator $\mathrm{e}^{\mathrm{j}\delta}$ has to be determined as well. It has been proposed in [123] to determine this operator for particular faults as follows:

- *single-phase* and *phase-to-phase faults* – by utilizing formula (7.31), obtained from analysis of boundary conditions of the faults;
- *phase-to-phase-to-ground faults* – with use of formula (7.32), obtained by considering the relation between the zero-sequence component of the total fault current and the remaining components;
- *three-phase balanced faults* – by utilizing formula (7.21) involving pre-fault positive-sequence voltages from both line sides and one-end current (in this case from the bus A).

Note that the most complex solution appears for the case of phase-to-phase-to-ground faults since here iterative calculations (for example applying the Newton–Raphson method) have to be carried out for (7.80) and (7.32). Resolving them into the real and imaginary parts one obtains four real-number equations, which have to be solved for the following unknowns: d, R_{F}, $\sin(\delta)$, $\cos(\delta)$.

Example 7.4 Fault location under CT saturation
Figures 7.21–7.25 present the results for fault location of the a–b fault on a 400-kV, 300-km single-circuit transmission line. The emfs at the bus A were assumed as leading the emfs of the side B by 30°; distance to fault from the terminal A: $d_A = 0.1$ p.u., fault resistance: $R_F = 0.5\ \Omega$.

Three-phase voltages from both line ends are shown in Fig. 7.21, while three-phase currents in Fig. 7.22. One can observe that the CTs from the phases 'a', 'b' at the line side A (Fig. 7.22a) become saturated. This is clearly visible in Fig. 7.23 where secondary currents from the saturated CTs (recalculated to the primary level) are drawn together with the primary currents. In Fig. 7.24 the synchronization angle – calculated with use of two-end voltages and currents from the line end B. The value of the calculated synchronization angle (–18.0099°) is very close to the intentionally introduced delay (–18°).

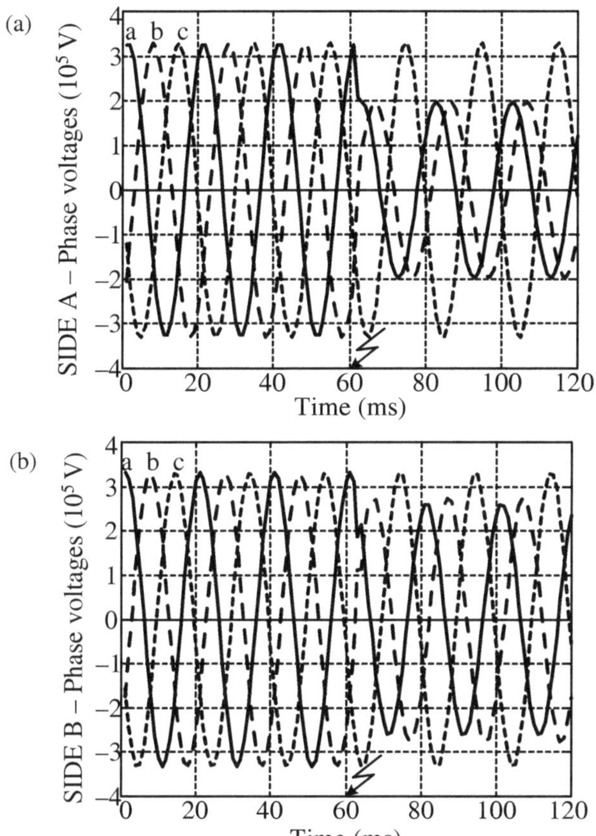

Fig. 7.21 Example 7.4 – three-phase voltage from: (a) side A, and (b) side B

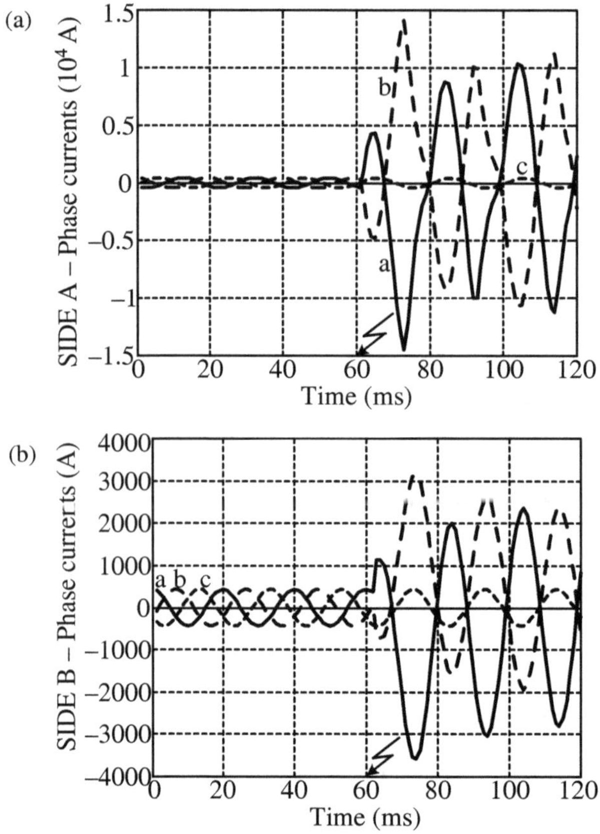

Fig. 7.22 Example 7.4 – three-phase current from: (a) side A, and (b) side B

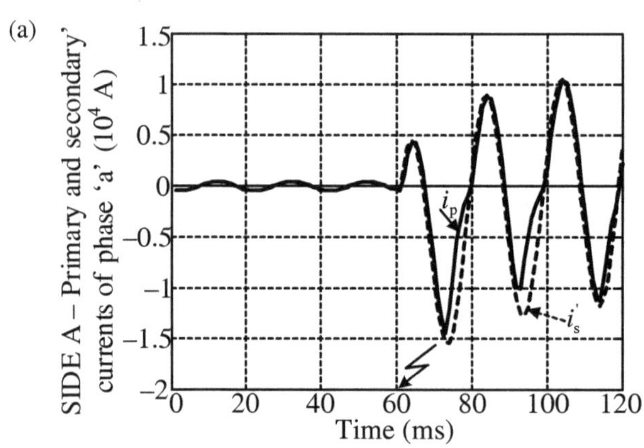

Fig. 7.23 Example 7.4 – currents (primary and secondary recalculated to primary level) of CTs that become saturated: (a) from phase 'a' at line side A, and (b) from phase 'b' at line side A

(b)

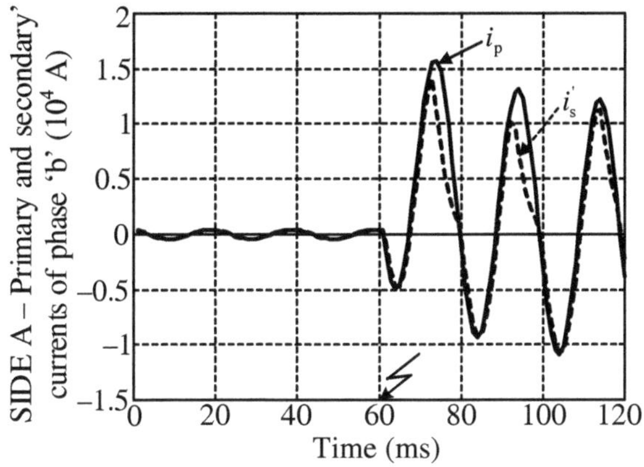

Fig. 7.23 (*continued*)

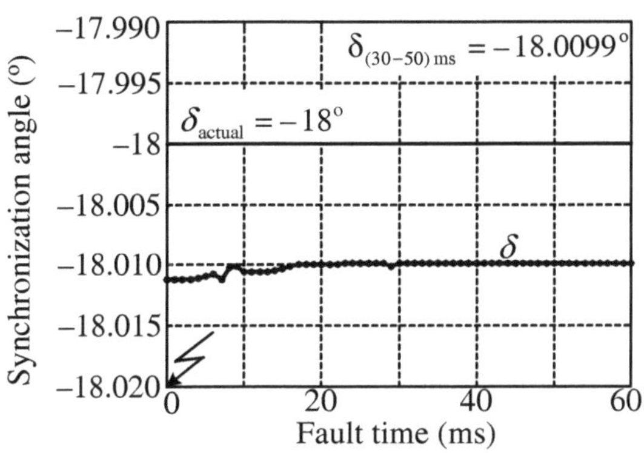

Fig. 7.24 Example 7.4 – determination of the synchronization angle according to (7.31)

In Fig. 7.24 the synchronization angle is calculated with use of two-end voltages and currents from the line end B. The value of the calculated synchronization angle (–18.0099°) is very close to the intentionally introduced delay (–18°).

Results of the distance to fault calculations are presented in Fig. 7.25. When applying a typical two-end unsynchronized fault-location algorithm [225], which is non-immune to saturation, one gets the distance to fault (from the bus B to the fault point F) equal to: 0.8754 p.u. (Fig. 7.25a). Thus, the fault-location error is around 2.5%, and can be even higher in the other CT saturation cases. In contrast with this result, fault location with use of the presented CT saturation immune algorithm the location accuracy is considerably improved: for the distributed-parameter line model the distance from bus B to fault equals: 0.8958 p.u.

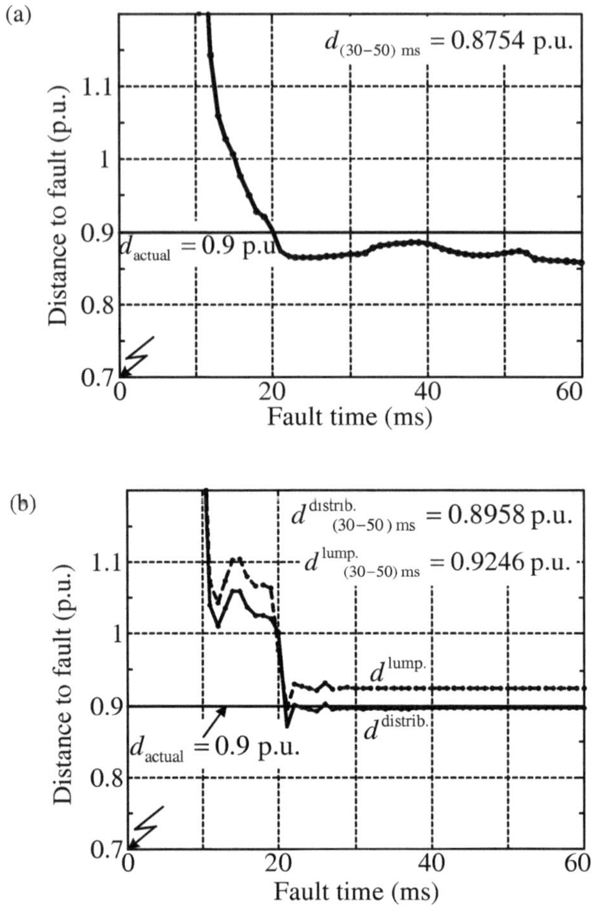

Fig. 7.25 Example 7.4 – determination of the distance to fault according to a fault-location algorithm that is: (a) non-immune to CT saturation [225], and (b) immune to CT saturation [123]

7.4.3 Use of Two-end Currents and One-end Voltage

Modern microprocessor-based current differential relays exchange locally measured current phasors over long distances. For this purpose different forms of communication means are utilized. The current differential protection principle requires synchronization of digital measurements performed at different line terminals. This is accomplished using the well-known global positioning system (GPS) or other techniques (Sect. 4.3.3), [107].

This fault-location method is aimed at incorporating it into current differential protective relays of the two-terminal power transmission or distribution line [119], analogously as in [124] for a three-terminal line. It is assumed that such a fault locator apart from three-phase currents from both ends of the line is additionally provided with three-phase voltages from the local line terminal (Fig. 7.26). The signals provided from both ends of the line are considered as synchronized. When this is not so, the synchronization angle can be determined with use of the already-known algorithms (Sect. 7.3.1).

Distinctive availability of the measurement signals for this method has been assumed since the fault-location algorithm (being of an off-line regime nature) is designed as the added feature for the current differential protection relay. The differential relay when applied to protect the two-terminal line utilizes phase currents measured synchronously at both line terminals. Therefore, in order to incorporate the fault-location function additionally to the protection function itself, the local phase voltages (from the terminal A in further considerations) have to be supplied to the relay. In this way, the differential relay, with the fault location added, can identify the fault not only indicating whether it has occurred within the zone or outside it (which is performed by the differential relay principle itself), but also stating precisely at what distance from the particular line terminal it occurred.

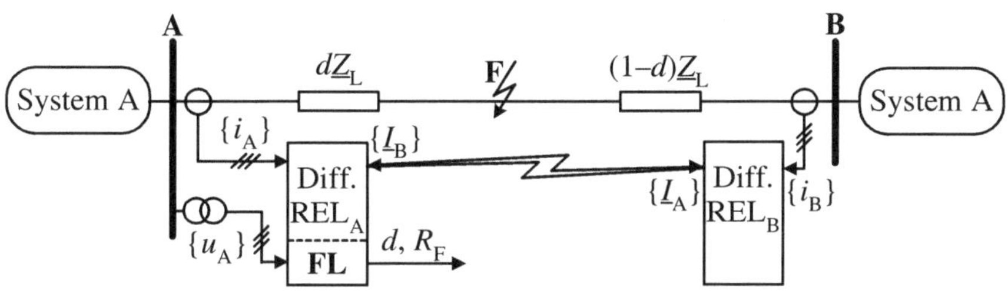

Fig. 7.26 Fault locator added to the current differential protective relays of two-terminal line

The generalized fault-loop model, expressed in terms of phasors, can be utilized for deriving the fault-location algorithm. This is a single universal formula covering different fault types involving coefficients dependent on a fault type:

$$\underline{U}_{A\_P} - d\underline{Z}_{1L}\underline{I}_{A\_P} - R_F\underline{I}_F = 0 \tag{7.81}$$

where:

$\underline{U}_{A\_P}$, $\underline{I}_{A\_P}$ – fault-loop voltage and current, composed as shown in Tables 6.5 and 6.6,

d – unknown distance to fault (p.u.), measured from the bus A,

R_F – unknown fault resistance,

$\underline{I}_F$ – total fault current.

Since currents at both line ends are available (measured synchronously), the total fault current ($\underline{I}_F$) can be easily determined by adding, in particular phases (a, b, c), currents flowing from both ends towards the fault:

$$\underline{I}_{Fa} = \underline{I}_{Aa} + \underline{I}_{Ba} \tag{7.82}$$

$$\underline{I}_{Fb} = \underline{I}_{Ab} + \underline{I}_{Bb} \tag{7.83}$$

$$\underline{I}_{Fc} = \underline{I}_{Ac} + \underline{I}_{Bc} \tag{7.84}$$

The respective phase current from (7.82)–(7.84) (for single phase-to-ground faults) or respective difference of phase currents (for inter-phase faults) is then substituted into (7.81).

As an alternative to the direct addition, as presented in (7.82)–(7.84), the total fault current can be composed using the respective sequence currents from both line ends.

For all the faults, but with the exception of three-phase balanced faults, it is advantageous to utilize the negative- and zero-sequence components, while avoiding the positive-sequence (choosing a set of the share coefficients, for which $a_{F1} = 0$):

$$[\underline{I}_F]_{ph-g,\, ph1-ph2,\, ph1-ph2-g} = \underline{a}_{F2}(\underline{I}_{A2} + \underline{I}_{B2}) + \underline{a}_{F0}(\underline{I}_{A0} + \underline{I}_{B0}) \tag{7.85}$$

where $\underline{a}_{F2}$, $\underline{a}_{F0}$ – share coefficients for all the faults except three-phase balanced ones, see Table 3.6.

For three-phase balanced faults, there are no negative- nor zero-sequence quantities, and the total fault current can be determined from the positive-sequence components, or alternatively and more accurately from the incremental positive-sequence components:

$$[\underline{I}_F]_{3-ph} = \underline{a}_{F1}(\Delta\underline{I}_{A1} + \Delta\underline{I}_{B1}) \tag{7.86}$$

where:

$\underline{a}_{F1}$ – share coefficient for three-phase balanced faults from Table 3.6.

Having determined the total fault current (7.85) and (7.86), the generalized fault-loop model can be solved for the unknowns d and R_F.

The fault-location accuracy of the method presented can be improved by introducing the distributed-parameter line model, similarly as for the algorithm from Sect. 7.4.2 (fault location with use of two-end voltages and one-end current). The total fault current ($\underline{I}_F$) can be determined as the following composition of its positive- and negative-sequence components:

$$\underline{I}_F = \frac{\underline{M}_{12}}{\cosh(\underline{\gamma}_1(1-d)\ell)} \tag{7.87}$$

where:

$\underline{M}_{12} = \underline{a}_{F1}(\underline{I}_{B1} + \underline{N}_{S1}) + \underline{a}_{F2}(\underline{I}_{B2} + \underline{N}_{S2})$

$\underline{N}_{S1}$, $\underline{N}_{S2}$ – as in (7.35),

$\underline{a}_{F1}$, $\underline{a}_{F2}$ – share coefficients dependent on fault type (any set from Tables 3.3–3.5 can be applied).

7.4.4 Fault Location with Exchange of Limited Information

In [319, 320], a fault-location system for two-terminal, and for also multi-terminal transmission lines, has been introduced. The algorithm used by this system is suitable for inclusion in a numerical protection relay, which communicates with remote relay(s) and exchanges information over a protective relaying channel. The data volume communicated between relays is sufficiently small to be easily transmitted using a digital protection channel. The algorithm does not require data alignment, pre-fault load-flow information, phase-selection information, and does not perform iterations to calculate the distance to the fault. Pre-fault load flow, zero-sequence mutual coupling, fault resistance, power system non-homogeneity, and current infeed from other line terminals or tapped loads do not affect the fault-location accuracy.

In relation to Fig. 7.27, one can write down the formula, which comes from comparison of the negative-sequence voltage at the fault point F, viewed from both line terminals A, B, respectively:

$$(\underline{Z}_{2A} + d\underline{Z}_{2L})\underline{I}_{A2}e^{j\delta} = (\underline{Z}_{2B} + (1-d)\underline{Z}_{2L})\underline{I}_{B2} \tag{7.88}$$

Application of negative-sequence components in (7.88) implies that this algorithm is for all faults, except three-phase balanced ones, for which the incremental positive-sequence can be applied.

The absolute value of the negative-sequence current from the remote bus R is determined from (7.88) as:

$$|\underline{I}_{B2}| = \frac{|\underline{Z}_{2A}\underline{I}_{A2} + d\underline{Z}_{2L}\underline{I}_{A2}|}{|\underline{Z}_{2B} + \underline{Z}_{2L} - d\underline{Z}_{2L}|} \tag{7.89}$$

This leads to cancelling of the synchronization operator. Finally, the following quadratic formula for the distance to fault is obtained:

$$A_2d^2 + A_1d + A_0 = 0 \tag{7.90}$$

where:

A_2, A_1, A_0 – coefficients dependent on negative-sequence quantities measured by each relay and impedance data of the network.

It is distinctive that the relay at each line terminal of the two-terminal line must transmit a minimal amount of information. The information sent by the relay at side A for a two-terminal application includes:

- magnitude of the negative-sequence current: $|\underline{I}_{A2}|$;
- magnitude of the negative-sequence source impedance: $|\underline{Z}_{2A}|$;
- angle of negative-sequence source impedance: angle($\underline{Z}_{2A}$).

Using the above information combined with the negative-sequence quantities, which are measured by relays, we can determine the fault location at each terminal without iterations. However, the lumped-line model is applied here, and thus acceptable accuracy is achieved for lines being not too long.

This algorithm is also extended to three-terminal line application, which was presented in [319, 320].

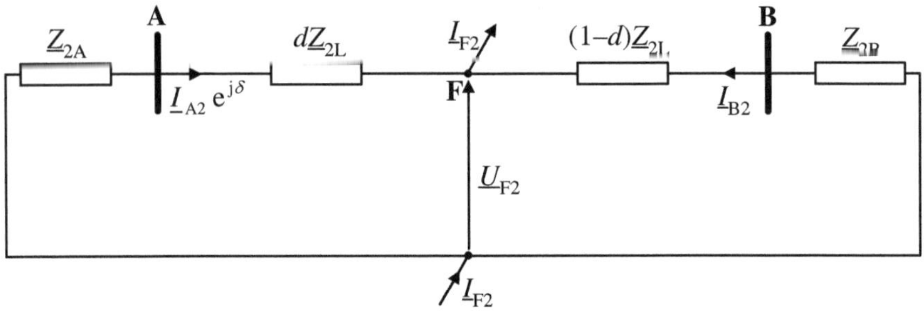

Fig. 7.27 Circuit diagram of transmission network for negative-sequence

7.5 Two-end Fault Location on Series-compensated Lines

7.5.1 Introduction

In Sect. 6.14 the one-end fault-location algorithms for both the single- and double-circuit series-compensated transmission lines were presented in detail. Those algorithms require representing the compensating bank (banks) in the form of the fundamental-frequency equivalent or by the estimated voltage drop across the bank. This can be considered as a certain limitation of the one-end algorithms. With the aim of overcoming such limitation, use of two-end measurements to series-compensated line fault location has been proposed in [273, 340]. As a result, representing the compensating bank is avoided, under the condition that only one compensating bank is installed in the line. Such algorithms suit the application to transmission lines compensated with fixed series capacitors (FSC) or thyristor-controlled series capacitors (TCSC).

It is considered in Fig. 7.28 that the fault locator (FL) is installed at the line end A and is supplied with local three-phase current $\{i_A\}$ and voltage $\{u_A\}$. Three-phase current $\{i_B\}$ and voltage $\{u_B\}$ at the line end B are recorded with use of a digital fault recorder (DFR) and sent to the fault locator.

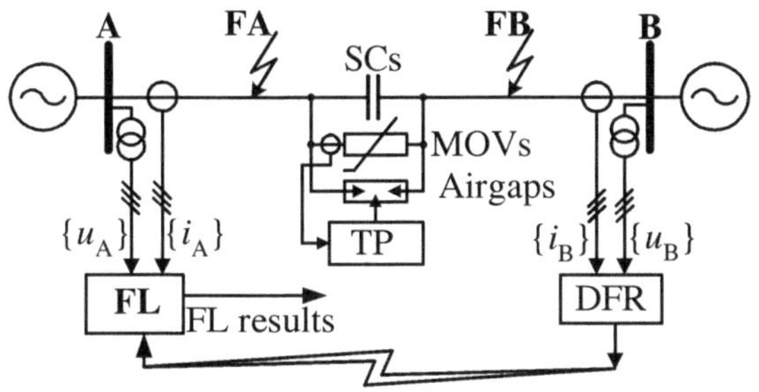

Fig. 7.28 Schematic diagram of two-end location of faults on series-compensated line

7.5.2 Fault-location Subroutines

A fault is of a random nature and thus can appear at any line section: fault FA and fault FB – Fig. 7.28. Therefore, two subroutines: SUB_A (considered circuit diagram as in Fig. 7.29a), SUB_B (considered circuit diagram as in Fig. 7.29b) are utilized for locating those hypothetical faults FA and FB, respectively. The final result has to be selected with use of a special selection procedure.

The three-phase SC and MOV bank divides the line of the length ℓ [km] into two line segments having the length: d_{SC} [km] and $(\ell - d_{SC})$ [km]. The subroutines SUB_A, SUB_B determine the per unit distance to fault: d_{FA}, d_{FB}, each related to the length of the particular line segment (d_{SC} or $(\ell - d_{SC})$). Finally, one recalculates those relative distances: d_{FA}, d_{FB} into the distances: d_A, d_B, which are related to the common base, i.e., to the whole line length (ℓ).

As in [273], use of unsynchronized measurements is considered here. Common phase reference for the phasors from two line ends is assured by mathematical synchronization as follows:

- subroutine SUB_A – the synchronization operator: $\exp(j\delta_A)$ is introduced into the phasors from the bus A; and
- subroutine SUB_B – the synchronization operator: $\exp(j\delta_B)$ is introduced into the phasors from the bus B.

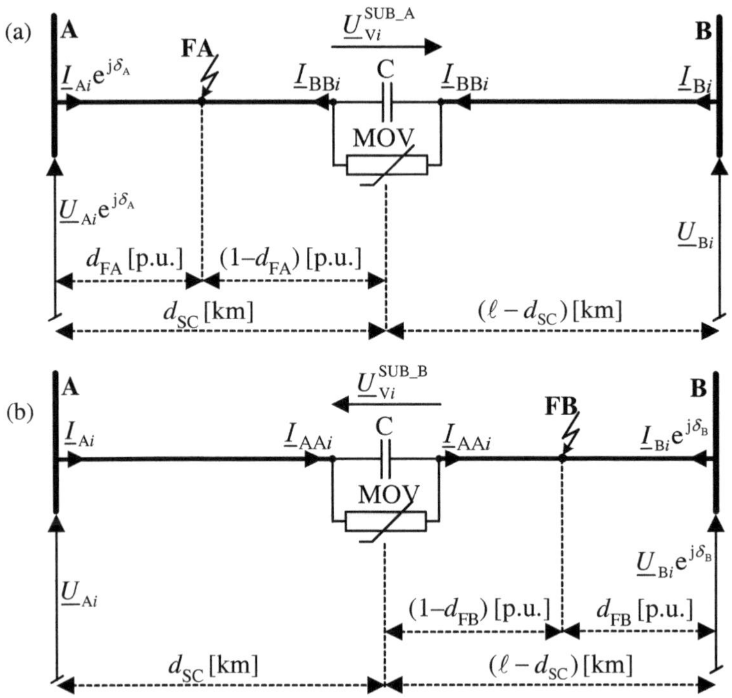

Fig. 7.29 Equivalent circuit diagrams considered for deriving fault-location subroutines: (a) subroutine SUB_A, and (b) subroutine SUB_B

The line segment adjacent to the bus B (Fig. 7.29a) is healthy and thus one can transfer analytically the current ($\underline{I}_{Bi}$) from the bus B towards the SC and MOV bank, obtaining:

$$\underline{I}_{BBi} = \frac{-\underline{U}_{Bi}}{\underline{Z}_{ci}} \sinh(\underline{\gamma}_i(\ell - d_{SC})) + \underline{I}_{Bi} \cosh(\underline{\gamma}_i(\ell - d_{SC})) \tag{7.91}$$

where:
i – index denoting a type of the symmetrical component (i=1: positive-, i=2: negative-, i=0: zero-sequence),
$\underline{Z}_{ci}$, $\underline{\gamma}_i$ – characteristic impedance and propagation constant of the line for the ith sequence, respectively.

Analogously, in relation to the subroutine SUB_B (Fig. 7.29b) one transfers the current ($\underline{I}_{Ai}$) from the bus A towards the SC and MOV bank:

$$\underline{I}_{AAi} = \frac{-\underline{U}_{Ai}}{\underline{Z}_{ci}} \sinh(\underline{\gamma}_i d_{SC}) + \underline{I}_{Ai} \cosh(\underline{\gamma}_i d_{SC}) \tag{7.92}$$

After transferring currents for the individual sequences (7.91) and (7.92), the phasors of three-phase currents $\{\underline{I}_{BB}\}$, $\{\underline{I}_{AA}\}$ can be calculated, using the symmetrical-components theory.

Note that the transferred currents (7.91) and (7.92) are identical at both sides of the compensating device. In contrast, due to the presence of a voltage drop across the bank, the voltages at both sides of the compensating bank differ and transfer of voltage across it requires knowing the bank parameters. Avoiding the necessity for providing such parameters as the fault-location algorithm input data, we have the case of using the incomplete two-end measurements (currents from two ends, while voltage only from one end of the faulted line section) for fault location:

- subroutine SUB_A – phasors of three-phase signals: $\{\underline{U}_A\}$, $\{\underline{I}_A\}$, $\{\underline{I}_{BB}\}$;
- subroutine SUB_B – phasors of three-phase signals: $\{\underline{U}_B\}$, $\{\underline{I}_B\}$, $\{\underline{I}_{AA}\}$.

Analytical synchronization of measurements can be performed with use of the determined synchronization angles δ_A, δ_B, however, only one of them has to be calculated since there is the following relation between them:

$$\delta_B = -\delta_A \tag{7.93}$$

The synchronization angle δ_A can be determined from:

- pre-fault positive-sequence phasors of the transferred currents: $\underline{I}_{AA1}^{pre}$, $\underline{I}_{BB1}^{pre}$;
- fault quantities, as presented in Sect. 7.3.1.2 – applicable for single phase-to-ground and phase-to-phase faults.

7.5.3 Selection of Valid Subroutine

The applied subroutines SUB_A, SUB_B yield the results for a distance to fault and fault resistance: (d_A, R_{FA}); (d_B, R_{FB}), respectively. The results from only one subroutine are consistent with the actual fault.

First, the subroutine yielding the distance to fault outside the section range and/or the fault resistance of negative value is rejected. Next, the equivalent impedance of the series capacitor and its MOV, in terms of the impedance for the phase involved in a fault, is determined. For the subroutine SUB_A this impedance is determined with the calculated signals: $\underline{I}_{BBi}$, $\underline{V}_{Vi}^{SUB\_A}$ (Fig. 7.29a):

$$\underline{Z}_{\mathrm{V}}^{\mathrm{SUB\_A}} = \frac{\underline{a}_1^{'}\underline{V}_{\mathrm{V1}}^{\mathrm{SUB\_A}} + \underline{a}_2^{'}\underline{V}_{\mathrm{V2}}^{\mathrm{SUB\_A}} + \underline{a}_0^{'}\underline{V}_{\mathrm{V0}}^{\mathrm{SUB\_A}}}{\underline{a}_1^{'}\underline{I}_{\mathrm{BB1}} + \underline{a}_2^{'}\underline{I}_{\mathrm{BB2}} + \underline{a}_0^{'}\underline{I}_{\mathrm{BB2}}} \tag{7.94}$$

where:
$\underline{a}_i^{'}$ – coefficients allowing us to compose the phase voltage and current, as taken from the first phase involved in a fault (Table 7.3).

Table 7.3 Coefficients for composing phase quantities in (7.94)

Fault type	$\underline{a}_1^{'}$	$\underline{a}_2^{'}$	$\underline{a}_0^{'}$
a–g, a–b, a–b–g, a–b–c, a–b–c–g	1	1	1
b–g, b–c, b–c–g	$-0.5-\mathrm{j}0.5\sqrt{3}$	$-0.5+\mathrm{j}0.5\sqrt{3}$	1
c–g, c–a, c–a–g	$-0.5+\mathrm{j}0.5\sqrt{3}$	$-0.5-\mathrm{j}0.5\sqrt{3}$	1

Example 7.5 Two-end unsynchronized fault location on series-compensated line Figures 7.30–7.32 present the results for fault location of the a–g fault on a 400-kV, 300-km single-circuit transmission line compensated with the compensating bank installed in the middle (d_{SC}=150 km), with the compensation rate of 70% [273]. The emfs at the bus A were assumed as leading the emfs of the side B by 30°; distance to fault from the terminal A: $d_{\mathrm{A}} = 0.5^{-}$ p.u. (just in front of the compensating bank, as seen from the bus A), fault resistance: $R_{\mathrm{FA}} = 25\ \Omega$.

Figure 7.30 presents waveforms of the fault-locator input signals acquired at both line sides. In Fig. 7.31 the estimated distance to fault and fault resistance are shown. The fault distances of both subroutines are within the line sections and the fault resistances are of positive values. Therefore the selection has to be continued. In Fig. 7.32 the estimated real/imaginary parts of the equivalent impedance for the SC and MOV from the faulted phase 'a' are shown. The subroutine SUB_A yields the equivalent impedance of the *RC* circuit character, which is consistent with the reality. In contrast, the subroutine SUB_B estimates the equivalent resistance of evidently negative value. Finally, the subroutine SUB_A remains selected as valid, giving an estimation of the fault distance with around 0.2% error.

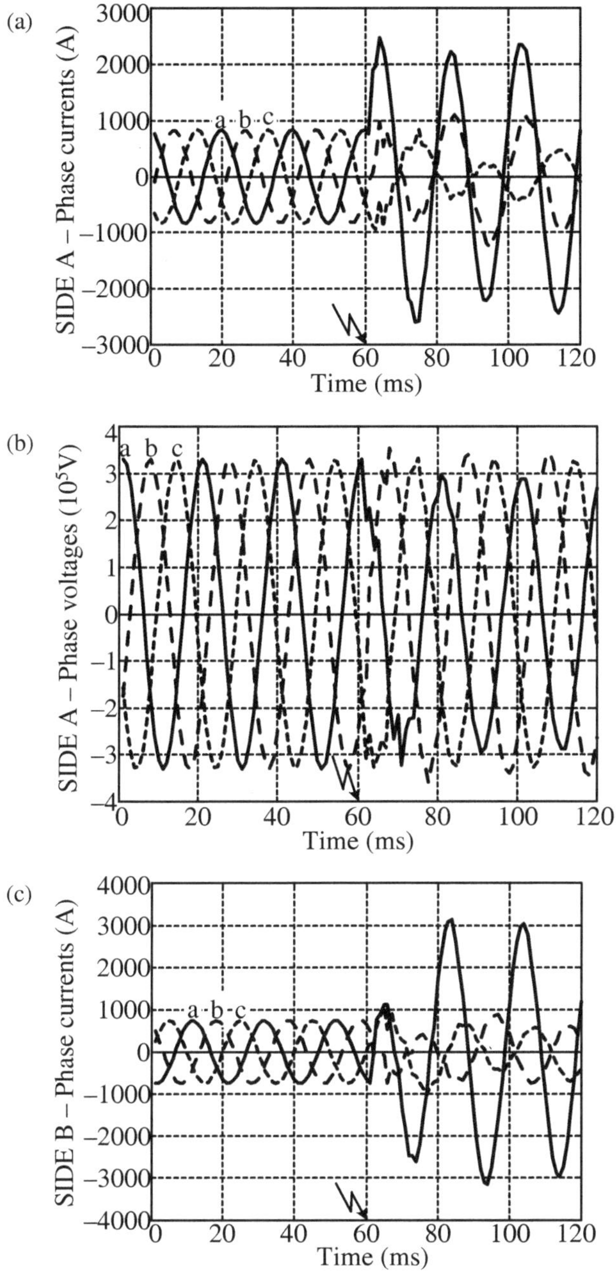

Fig. 7.30 Example 7.5 – input signals (three-phase): (a) current of side A, (b) voltage of side A, (c) current of side B, and (d) voltage of side B

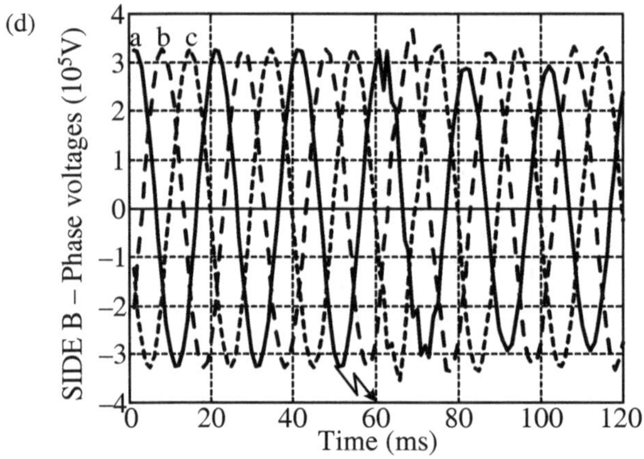

Fig. 7.30 (*continued*)

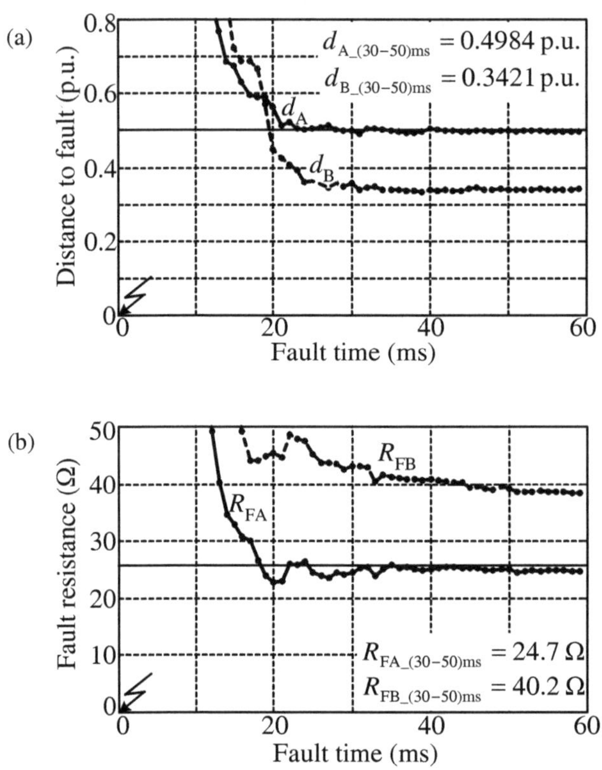

Fig. 7.31 Example 7.5 – calculation results: (a) distance to fault, and (b) fault resistance

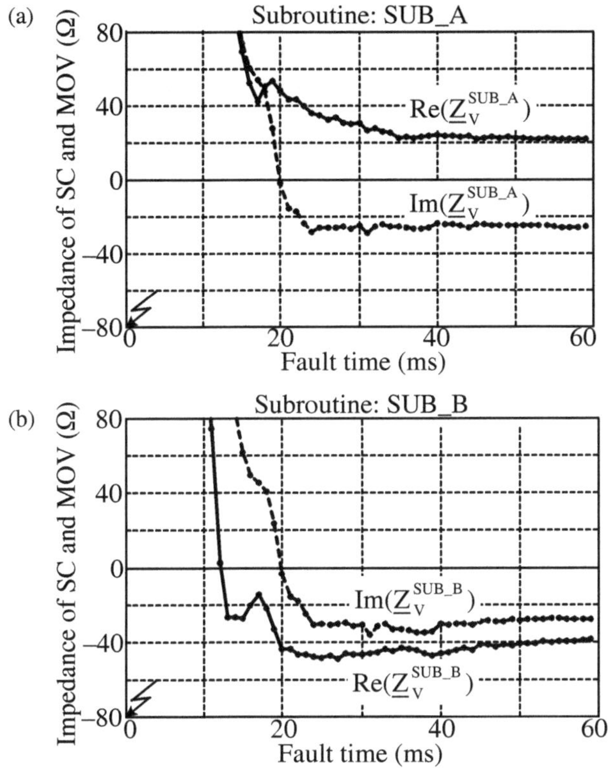

Fig. 7.32 Example 7.5 – estimation of the compensating bank impedance components according to: (a) subroutine SUB_A, and (b) subroutine SUB_B

7.6 Recent Developments of Two-end Fault Location

When considering the solution of a problem with a particular number of unknowns it is important to determine how many equations can be formulated. If the number of equations is smaller, then obtaining the solution requires making certain simplifying assumptions. In the case of the same number of equations as the number of unknowns, the means to their solution is of interest. There is also a possibility for the case with redundancy, i.e., when it is possible to formulate more equations than the number of unknowns. Such a case occurs if two-end synchronized or unsynchronized measurements are utilized for fault location. The resulting redundancy can be utilized to obtain certain improvements in fault location. This is observed in recent developments in this subject, as for example presented in [187–189, 255].

In [187, 188], the fault-location technique not requiring line parameters, obtained by utilizing the redundancy of information brought by the fault-locator input signals, has been delivered. In reality, values of line parameters may not always be available, and such an algorithm appears highly desired.

An optimal estimate of transmission-line fault location considering measurement errors utilizing the redundancy has been proposed in [189]. It has been shown that it is feasible to design an approach for detecting, identifying and removing the possible bad measurements. In turn, when synchronized measurements are utilized, possible synchronizing errors can also be detected. As a result of using this approach, improvement of the fault-location estimation accuracy can be achieved.

A new accurate algorithm for locating faults on power transmission lines with use of the two-end unsynchronized measurements of voltages and currents has been presented in [255]. Simultaneous usage of two kinds of symmetrical components, has allowed utilization of the redundancy with respect to the fault-location equations. As a result, the synchronization operator is determined by applying simple non-iterative calculations, even the distributed-parameter line model is considered. This appears as very attractive from the practical point of view, since it can be easily implemented into contemporary microprocessor-based protective relays. Moreover, simple, but at the same time quite accurate determination of the synchronization angle has been obtained when using the lumped-line model. This angle was utilized for selecting the valid solution for the angle determined with use of the distributed-parameter line model. However, apart from that purpose, it can be considered also for simple checking of the synchronism of the distributed measurements. Such checking appears to be attractive for differential protective relay applications. Also, potential synchronization errors due to synchronizing device failures, such as communication problems or loss of the signal from the GPS, could be easily detected.

7.7 Fault Location on Three-terminal Lines

Different availability of the fault-locator input signals can be considered in relation to application to three-terminal lines.

In Sect. 7.7.1 use of complete three-terminal measurements is considered, focusing on utilizing the unsynchronized measurements. Then, application of incomplete measurements is taken into account. In particular, association of the fault locator with current differential relays is proposed in Sect. 7.7.2. Then, use of measurements acquired at only two line ends is considered and optimized. Finally, the problem of defining the minimal measurements to provide for the fault-locator input signals is addressed.

Apart from presenting the procedures for calculating the distance to fault, the selection algorithms, aimed at finding which section of the line is faulted, are also considered here. In order to achieve high accuracy of fault location, the distributed-parameter line model (the line sections are drawn with a thick line and not with use of the rectangle as an impedance symbol) will be utilized.

7.7.1 Use of Three-end Measurements

A circuit diagram of a three-terminal line for considering a fault location using three-end measurements is presented in Fig. 7.33 (synchronized measurements) and in Fig. 7.34 (unsynchronized measurements). Different types of signals can be processed in the fault-location algorithms and in Figs. 7.33 and 7.34 the type is marked by the subscript *i*.

In the case of unsynchronized measurements [118], the measurements at one end have to be assumed as the basis (in Fig. 7.34: from the end B), while those from the buses A and C have to be analytically synchronized, with use of the synchronization operators $e^{j\delta_A}$ and $e^{j\delta_C}$, respectively. Thus, two additional unknowns: synchronization angles δ_A and δ_C appear in comparison to the case of synchronized measurements.

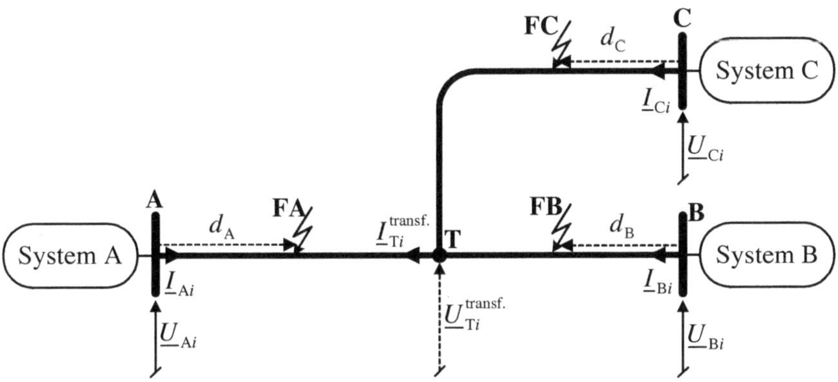

Fig. 7.33 Circuit diagram of three-terminal line for considering fault location using three-end synchronized measurements

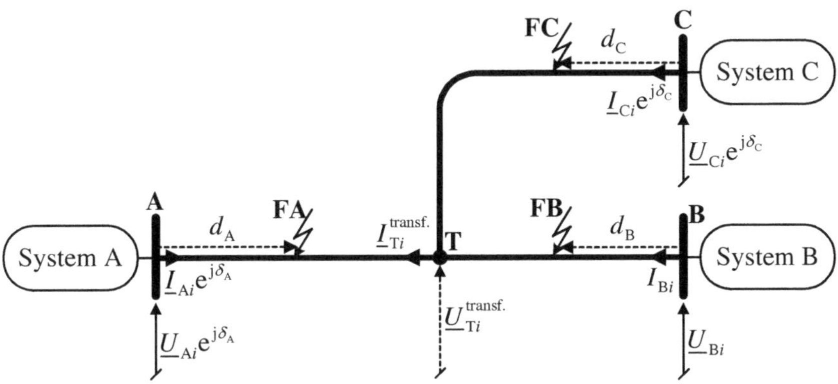

Fig. 7.34 Circuit diagram of three-terminal line for considering fault location using three-end unsynchronized measurements: the measurements from the bus B are assumed as the base, while analytical synchronization is performed for the measurements from the buses A and C

The fault may occur at any of the three line sections (faults: FA, FB or FC in Figs. 7.33 and 7.34). The distance to the particular fault, measured from the particular bus (A, B or C) up to the fault point is denoted by d_A, d_B, d_C, respectively. So, there are three hypotheses regarding placement of the fault in a particular line section: A–T, T–B or T–C. In general, one can distinguish the following approaches:

- all three hypotheses regarding indication of the faulted-line section are considered and after calculating the distances to fault (d_A, d_B, d_C) with use of three subroutines, the judgment on selecting only one solution (the valid solution), which corresponds to the actual fault position, is performed;
- first, the faulted-line section is identified and then the distance to fault for this section is determined.

In [80], a fault-location method for two- and three-terminal lines has been presented. As the input signals of the fault locator, complete two- and three-terminal measurements, both synchronized and unsynchronized, have been considered. Three-phase voltage and current phasors from the line terminals are processed for determining the distance to fault (d) and the synchronization angle (δ) if unsynchronized measurements are considered. So, there is one unknown (d) and two unknowns (d and δ) in the case of synchronized and unsynchronized measurements, respectively. However, on using complete three-phase voltage and current measurement at all line terminals, a redundancy of the fault-location equations is assured. For example, in the case of two-terminal line and synchronized measurements, the distance to fault is calculated applying the least-squares estimates from three complex equations or six real-number equations in one unknown, d. Similarly, the redundancy of the fault-location equations is assured for fault location on a three-terminal line.

In the fault-location method from [80], the line sections are represented with the line series impedance matrix only. Thus, the line capacitances are neglected, which could affect the calculation of the fault location. However, the authors of this method state that the fault-location error is minimized due to the redundancy of the fault-location equations, which leads to a squares estimate of the fault location. Important advantages of the fault-location algorithm rely on the fact that it is independent of fault type, insensitive to fault resistance, and does not require any information on source impedance. Neither does it assume transposition.

Another important contribution to fault location on three-terminal lines is presented in [4]. The technique described in [4] makes use of incremental, modal components (the phase values are transformed into three modes: an earth mode and two aerial modes) of voltages and currents rather than total, phase values, as it was for the method presented in [80]. The contribution in [4] is based on the approach of the fault-location method developed earlier for a two-terminal line [135]. Thus, the distributed-parameter line model is fully utilized and the fault location in (7.8), developed for a two-terminal line is applied.

The fault-location algorithm [4] is designed for both synchronized and unsynchronized three-end measurement of voltage and current. In the case of unsynchronized measurement, the shifting of the data at the unsynchronized end(s) is considered. For this purpose, the voltage at the tee point (T), based on the knowledge of the pre-fault power frequency voltage and current phasors at all the line terminals is determined, applying the distributed-parameter models of all line sections.

It is important to identify the faulted leg of the tee (the faulted-line section) before determining the distance to fault [4]. The technique consists in evaluating the voltages at the tee point based on the knowledge of the fault current and voltage phasors at the three ends. With reference to Fig. 7.33, one can state that the fault is in the section A–T if the voltage at the T point transferred analytically from the buses B and C (superscript: B and C):

$$\underline{U}_{\mathrm{T}i}^{\mathrm{transf.\,B}} = \cosh(\underline{\gamma}_{i\mathrm{LB}}\ell_{\mathrm{LB}})\underline{U}_{\mathrm{B}i} - \underline{Z}_{\mathrm{c}i\mathrm{LB}}\sinh(\underline{\gamma}_{i\mathrm{LB}}\ell_{\mathrm{LB}})\underline{I}_{\mathrm{B}i} \tag{7.95}$$

$$\underline{U}_{\mathrm{T}i}^{\mathrm{transf.\,C}} = \cosh(\underline{\gamma}_{i\mathrm{LC}}\ell_{\mathrm{LC}})\underline{V}_{\mathrm{C}i} - \underline{Z}_{\mathrm{c}i\mathrm{LC}}\sinh(\underline{\gamma}_{i\mathrm{LC}}\ell_{\mathrm{LC}})\underline{I}_{\mathrm{C}i} \tag{7.96}$$

is identical (in practice are close each other), whereas that transferred from the line end A:

$$\underline{U}_{\mathrm{T}i}^{\mathrm{transf.\,A}} = \cosh(\underline{\gamma}_{i\mathrm{LA}}\ell_{\mathrm{LA}})\underline{U}_{\mathrm{A}i} - \underline{Z}_{\mathrm{c}i\mathrm{LA}}\sinh(\underline{\gamma}_{i\mathrm{LA}}\ell_{\mathrm{LA}})\underline{I}_{\mathrm{A}i} \tag{7.97}$$

is very different, indicating that the faulted-line section is A–T.

Note that in (7.95)–(7.97), the distributed parameters of the respective line sections (having length: ℓ_{LA}, ℓ_{LB}, ℓ_{LC}) are involved:

- $\underline{\gamma}_{i\mathrm{LA}}$, $\underline{\gamma}_{i\mathrm{LB}}$, $\underline{\gamma}_{i\mathrm{LC}}$ – propagation constants;
- $\underline{Z}_{\mathrm{c}i\mathrm{LA}}$, $\underline{Z}_{\mathrm{c}i\mathrm{LB}}$, $\underline{Z}_{\mathrm{c}i\mathrm{LC}}$ – surge impedances.

The other faulted-line sections can be similarly identified. If, however, there is no discernible difference in the voltages transferred from all three ends, then it can be safely assumed that the fault is at the tee point itself [4]. Having identified the faulted-line section, the distance to fault can be calculated. As for example, for the faulted-line section A–T, the distance to fault (d_{A}) is calculated using the signals from the bus A ($\underline{U}_{\mathrm{A}i}$, $\underline{I}_{\mathrm{A}i}$) and at the tee point T ($\underline{U}_{\mathrm{T}i}^{\mathrm{transf.}}$, $\underline{I}_{\mathrm{T}i}^{\mathrm{transf.}}$), obtained after the analytical transfer (Fig. 7.33):

$$d_{\mathrm{A}} = \frac{\tanh^{-1}\left(\dfrac{\underline{U}_{\mathrm{T}i}^{\mathrm{transf.}}\cosh(\underline{\gamma}_{i\mathrm{LA}}\ell_{\mathrm{LA}}) - \underline{Z}_{\mathrm{c}i\mathrm{LA}}\underline{I}_{\mathrm{T}i}^{\mathrm{transf.}}\sinh(\underline{\gamma}_{i\mathrm{LA}}\ell_{\mathrm{LA}}) - \underline{U}_{\mathrm{A}i}}{\underline{U}_{\mathrm{T}i}^{\mathrm{transf.}}\sinh(\underline{\gamma}_{i\mathrm{LA}}\ell_{\mathrm{LA}}) - \underline{Z}_{\mathrm{c}i\mathrm{LA}}\underline{I}_{\mathrm{T}i}^{\mathrm{transf.}}\cosh(\underline{\gamma}_{i\mathrm{LA}}\ell_{\mathrm{LA}}) - \underline{Z}_{\mathrm{c}i\mathrm{LA}}\underline{I}_{\mathrm{A}i}}\right)}{\underline{\gamma}_{i\mathrm{LA}}\ell_{\mathrm{LA}}} \tag{7.98}$$

where:

$\underline{U}_{\mathrm{T}i}^{\mathrm{transf.}}$ – voltage transferred analytically to the tee point T, from the bus B (7.95) or the bus C (7.96),

$$\underline{I}_{\mathrm{T}i}^{\mathrm{transf.}} = -\frac{\sinh(\underline{\gamma}_{i\mathrm{LB}}\ell_{\mathrm{LB}})\underline{U}_{\mathrm{B}i}}{\underline{Z}_{\mathrm{c}i\mathrm{LB}}} + \cos(\underline{\gamma}_{i\mathrm{LB}}\ell_{\mathrm{LB}})\underline{I}_{\mathrm{B}i} - \frac{\sinh(\underline{\gamma}_{i\mathrm{LC}}\ell_{\mathrm{LC}})\underline{U}_{\mathrm{C}i}}{\underline{Z}_{\mathrm{c}i\mathrm{LC}}} + \cosh(\underline{\gamma}_{i\mathrm{LC}}\ell_{\mathrm{LC}})\underline{I}_{\mathrm{C}i} \tag{7.99}$$

Analogously to (7.98) and (7.99), the distance to fault in the section T–B and in the section T–C is calculated.

In the case of unsynchronized measurements, the synchronization operators ($\mathrm{e}^{\mathrm{j}\delta_\mathrm{A}}$, $\mathrm{e}^{\mathrm{j}\delta_\mathrm{C}}$) have to be determined. In [4], it is proposed to determine them by processing the pre-fault quantities. The synchronization operator $\mathrm{e}^{\mathrm{j}\delta_\mathrm{A}}$ can be determined by comparing the pre-fault voltage at the tee point T, transferred analytically from the bus A and the bus B:

$$\underline{U}_{\mathrm{T}i}^{\mathrm{pre\_transf.\,A}} = [\cosh(\underline{\gamma}_{i\mathrm{LA}}\ell_{\mathrm{LA}})\underline{U}_{\mathrm{A}i}^{\mathrm{pre}} - \underline{Z}_{\mathrm{c}i\mathrm{LA}}\sinh(\underline{\gamma}_{i\mathrm{LA}}\ell_{\mathrm{LA}})\underline{I}_{\mathrm{A}i}^{\mathrm{pre}}]\mathrm{e}^{\mathrm{j}\delta_\mathrm{A}} \tag{7.100}$$

$$\underline{U}_{\mathrm{T}i}^{\mathrm{pre\_transf.\,B}} = \cosh(\underline{\gamma}_{i\mathrm{LB}}\ell_{\mathrm{LB}})\underline{U}_{\mathrm{B}i}^{\mathrm{pre}} - \underline{Z}_{\mathrm{c}i\mathrm{LB}}\sinh(\underline{\gamma}_{i\mathrm{LB}}\ell_{\mathrm{LB}})\underline{I}_{\mathrm{B}i}^{\mathrm{pre}} \tag{7.101}$$

Analogously, the other synchronization operator $\mathrm{e}^{\mathrm{j}\delta_\mathrm{C}}$ can be determined by comparing the pre-fault voltage at the tee point T, transferred analytically from the bus C and the bus B.

7.7.2 Fault Location Associated with Current Differential Protective Relays

In Sect. 7.4.3, the fault-location method using two-end currents and one-end voltage, aimed at application to current differential protective relays of the two-terminal power transmission or distribution line, was presented. Analogously, this method can be applied in association with current differential relays protecting a three-terminal line [124], as shown in Fig. 7.35. In this case, the fault-location (FL) function is supplemented to the current differential relay from the substation A (Fig. 7.35). For this purpose, phasors of three-phase current from all line terminals: $\underline{I}_\mathrm{A}$, $\underline{I}_\mathrm{B}$, $\underline{I}_\mathrm{C}$ – exchanged by the current differential relays A, B, C, together with the locally measured three-phase voltage phasor ($\underline{U}_\mathrm{A}$) are taken as the fault-locator input signals. In a natural way, these measurements are considered as synchronized.

The fault-location technique proposed [124] is based on using three subroutines: SUB_A, SUB_B, SUB_C. They are designed for locating faults: FA, FB,

FC at hypothetical fault spots within particular line sections: A–T, T–B, T–C, respectively. Note that any of the sections may be faulted at random. Therefore, the position of a fault is a random factor, and thus, the faulted-line section is not known in advance. The faulted section will be indicated using a special selection procedure.

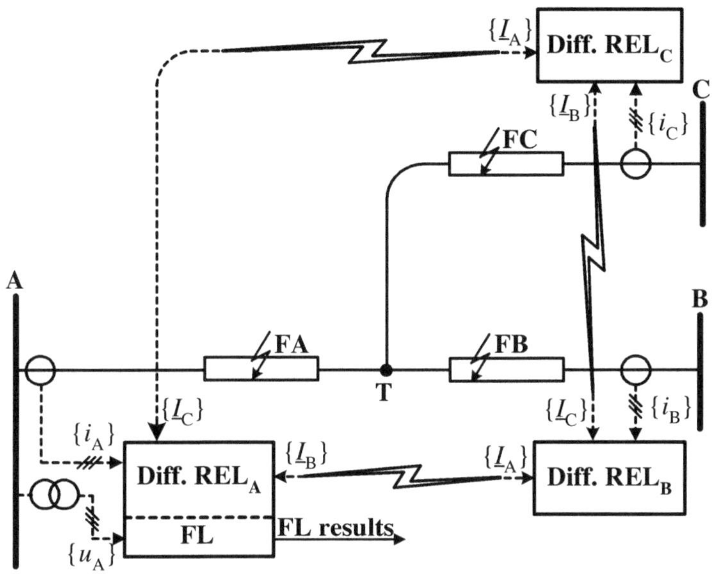

Fig. 7.35 Fault location on three-terminal line included into one of the current differential relays

The subroutine SUB_A, designed for locating faults (FA) within the line section A–T (Fig. 7.36), is based on the following generalized fault-loop model:

$$\underline{U}_{\text{A\_P}} - d_{\text{A}} \underline{Z}_{\text{1LA}} \underline{I}_{\text{A\_P}} - R_{\text{FA}} \underline{I}_{\text{F}} = 0 \tag{7.102}$$

where:

$\underline{U}_{\text{A\_P}}$, $\underline{I}_{\text{A\_P}}$ – fault-loop voltage and current, composed as shown in Tables 6.5 and 6.6,

$\underline{I}_{\text{F}}$ – total fault current (fault-path current).

Before calculating from (7.102) the unknowns: d_{A} – distance to fault, R_{FA} – fault-path resistance, one has to determine the total fault current $\underline{I}_{\text{F}}$. This can be accomplished analogously as for the two-terminal line (7.85) and (7.86), but instead of adding currents from two line ends, the respective symmetrical components of currents from three line ends are added.

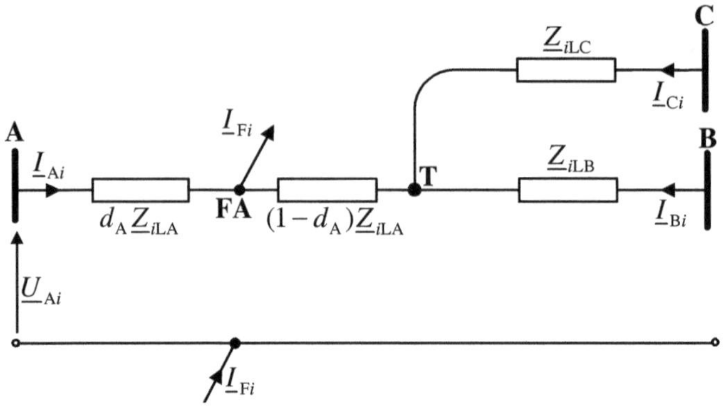

Fig. 7.36 Circuit diagram of three-terminal line under fault in section A–T, for the *i*th sequence

The remaining subroutines SUB_B, SUB_C complete the location procedures. An illustration for deriving the subroutine SUB_B is shown in Fig. 7.37. An analytic transfer of three-phase measurements: $\underline{U}_A$, $\underline{I}_A$, $\underline{I}_C$ to the beginning of the section LB, with the distributed-parameter line model being strictly taken into account, is performed. The superscript 'transf.' is used to distinguish the analytically transferred signals from the measured ones. Certainly, such a transfer has to be performed separately for each of the *i*th type symmetrical components of three-phase voltage and current, as presented in (7.97) and (7.99).

Using the transferred signals $\underline{U}_{Ti}^{\text{transf.}}$ and $\underline{I}_{TBi}^{\text{transf.}}$, the fault-loop voltage $\underline{U}_{T\_P}$ and current $\underline{I}_{TB\_P}$ are composed. Then, as in (7.100), the fault-loop model for the subroutine SUB_B can be formulated, and the unknown distance to fault d_B and fault resistance R_{FB} can be determined. The subroutine SUB_C, designed for locating faults on the line section T–C, can be derived analogously to the subroutine SUB_B.

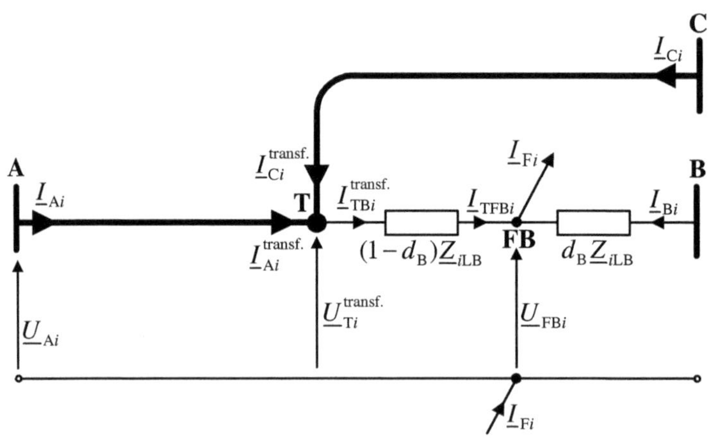

Fig. 7.37 Circuit diagram of three-terminal line under fault in section T–B, for the *i*th sequence

Three fault estimates are calculated assuming the fault to be on the A–T, T–B or T–C segment of the line, respectively.

A selection procedure is required to indicate the faulted-line segment, i.e., which of the subroutines (SUB_A, SUB_B or SUB_C) is valid, i.e., corresponds to the real fault. In the first step of the selection procedure, the results of the fault distance and resistance calculations are utilized. The subroutine that yields the distance to fault indicating the fault as occurring outside the section range (outside the range: from 0 to 1.0 p.u.) or/and the calculated fault resistance of negative value, has to be surely rejected.

The second step of the selection is used when the first step is not sufficient. In the second step, the remote source impedances (behind the terminals B and C – if it is considered that the fault locator is installed at the station A (Fig. 7.35)), are calculated for different subroutines [124]. Having calculated the impedances of the sources behind the remote buses B, C – in relation to both subroutines (SUB_B, SUB_C), first it is checked in which quadrant of the complex plane they are placed. The actual equivalent source impedance is of the form of an *RL* branch, i.e., is considered as placed in the I quadrant of the complex impedance plane. If the calculated source impedance lies outside the I quadrant, then this subroutine is false and has to be rejected. Otherwise, if at least two subroutines (out of three) still remain, then the selection has to be continued. The particular subroutine can be rejected also if the calculated value of the remote source differs from the actual impedance. For this rare fault case a certain knowledge of the actual equivalent sources behind the line terminals is necessary. This is not a serious limitation since for the other fault-location algorithms that utilize incomplete three-end measurements [23, 117, 192, 213], the impedances of the sources are involved even in calculation of the distance to fault.

7.7.3 Use of Two-end Measurements

The demand and importance of developing fault locators for three-terminal lines under the availability of two-end synchronized measurements of voltages and currents (Fig. 7.38) has been introduced and justified in [192]. This is of practical importance since the measurement infrastructure of the tapped line is usually rather poor, or there are no communication channels for sending measurements from the end of the tapped line.

It is considered in the fault-location methods presented in [115] and [192] that phasor measurement units from two line ends (PMU_A, PMU_B) provide the fault locator (FL) with three-phase phasors of voltage and current ($\underline{U}_A$, $\underline{I}_A$, $\underline{U}_B$, $\underline{I}_B$) measured synchronously (Fig. 7.38). Fault location is based on using three subroutines, designed for locating particular faults (FA, FB, FC) and selecting the valid subroutine.

Comparison of the voltage at the tee point T, obtained by analytical transfer of signals measured at the end A (according to (7.97)) and at the end B (according to

(7.95)) allows preliminary selection of the faulted-line section. If these voltages match each other, then there is a fault FC on the tapped line. Otherwise, there is a fault FA or FB on the main route of the line (sections: A–T and T–B). Distinguishing between the fault FA and FB can be performed on the basis of the calculated distance to fault [115, 192].

Figure 7.39 shows a three-terminal line network with a fault on the section A–T for considering the subroutine SUB_A to be derived according to the approach presented in [192].

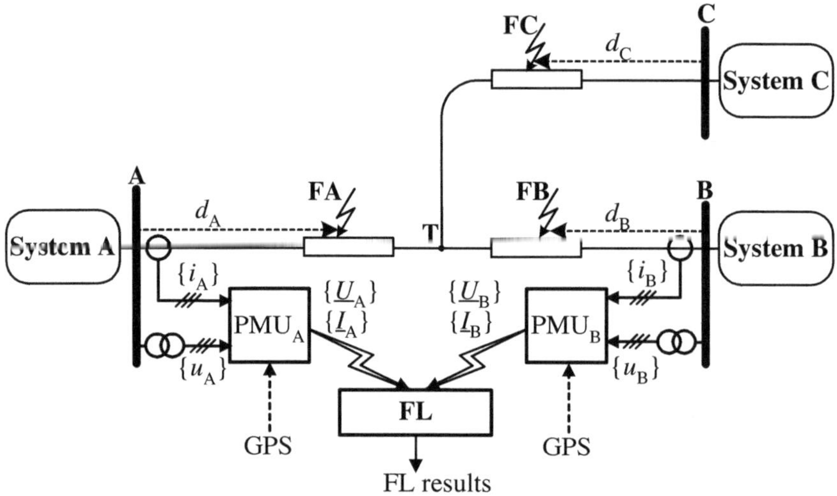

Fig. 7.38 Fault location on three-terminal line using two-end synchronized measurements

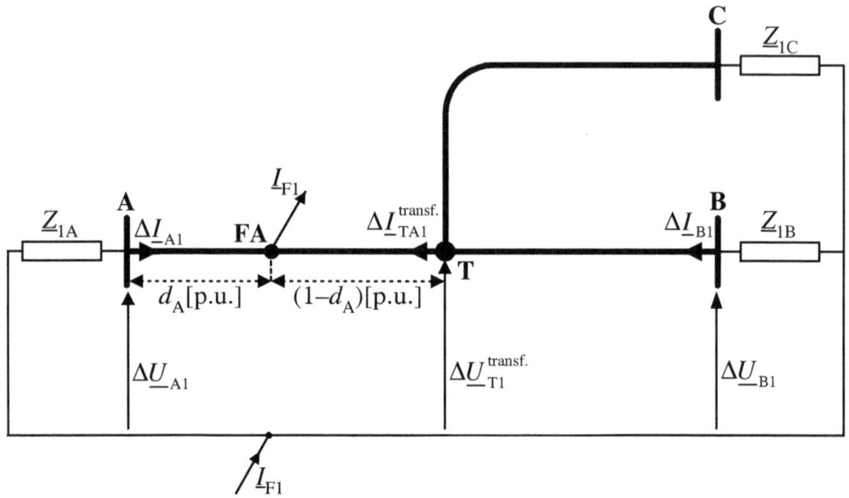

Fig. 7.39 Three-terminal line network – illustration for subroutine SUB_A utilizing incremental quantities

The subroutine SUB_B is derived analogously to the subroutine SUB_A, for which the derivation follows. It was proposed in [192] to apply an accurate fault-location algorithm (7.8), which applies complete two-end synchronized measurements for locating faults on a two-terminal line. Lin *et al.* [192] proposed to apply (7.8) in order to locate faults (FA) on the section A–T using as the input signals:

- incremental positive-sequence components of signals from the side A: $\Delta\underline{U}_{A1}$, $\Delta\underline{I}_{A1}$; and
- incremental positive-sequence components of signals transferred to the tee point T: $\Delta\underline{U}_{T1}^{\text{transf.}}$ (obtained according to (7.95)), and $\Delta\underline{I}_{TA1}^{\text{transf.}}$ derived using a current division theory as the following function:

$$\Delta\underline{I}_{TA1}^{\text{transf.}} = f(\Delta\underline{I}_{B1}, \underline{\gamma}_{1LB}, \underline{\gamma}_{1LC}, \underline{Z}_{c1LB}, \underline{Z}_{c1LC}, \ell_{LB}, \ell_{LC}, \underline{Z}_{1B}, \underline{Z}_{1C}) \tag{7.103}$$

Thus, the required current (7.103) is dependent on the measured incremental current: $\Delta\underline{I}_{B1}$, the parameters of the line section T–B: $\underline{\gamma}_{1LB}, \underline{Z}_{c1LB}, \ell_{LB}$, the line section T–C: $\underline{\gamma}_{1LC}, \underline{Z}_{c1LC}, \ell_{LC}$ and the source impedances behind the buses B, C: $\underline{Z}_{1B}, \underline{Z}_{1C}$. Inaccuracy in providing these source impedances is a source of certain additional fault-location errors. In order to assure the highest possible accuracy of fault location, a new approach, not requiring these source impedances as the input parameters, was proposed in [115]. This algorithm applies two-end voltages and one-end current and is based on the generalized fault-loop model (7.76) and its final form (7.80), derived with strict consideration being given to the distributed-parameter line model. The formula (7.80) applied to a two-terminal line requires two-end voltages ($\underline{U}_A$, $\underline{U}_B$) and one-end current ($\underline{I}_A$). When applying it to the subroutine SUB_A for locating faults FA on the section A–T of a three-terminal line (Fig. 7.38, Fig. 7.39) the following input signals are considered:

- signals measured at the bus A: $\underline{U}_A$, $\underline{I}_A$; and
- voltage at the tee point T: $\underline{U}_T^{\text{transf.}}$, obtained after analytic transfer of signals from the bus B with use of (7.95).

Analogously, the formula (7.80) can be applied to the subroutine SUB_B for locating faults FB on the section T–B.

Use of the fault-location algorithm (7.80) for locating faults (faults FA, FB in Fig. 7.38) on the main route of a three-terminal line [115] is advantageous since the source impedances are not involved for this algorithm, in contrast to the method introduced in [192].

Faults on the line section T–C (fault FC in Fig. 7.38) are located using the subroutine SUB_C, for which an illustration is given in Fig. 7.40. The signals measured at the buses A, B are transferred analytically to the tee point T, applying the distributed-parameter line section model, performing it separately for each symmetrical component (positive-sequence: $i = 1$, negative-sequence: $i = 2$, zero-sequence: $i = 0$). The transferred voltage: $\underline{U}_{\text{T1}}^{\text{transf.}}$, $\underline{U}_{\text{T2}}^{\text{transf.}}$, $\underline{U}_{\text{T0}}^{\text{transf.}}$, and the sums of transferred currents: $(\underline{I}_{\text{TA1}}^{\text{transf.}} + \underline{I}_{\text{TB1}}^{\text{transf.}})$, $(\underline{I}_{\text{TA2}}^{\text{transf.}} + \underline{I}_{\text{TB2}}^{\text{transf.}})$, $(\underline{I}_{\text{TA0}}^{\text{transf.}} + \underline{I}_{\text{TB0}}^{\text{transf.}})$ are the input signals for the subroutine SUB_C. Thus, we deal with the one-end voltage and current in this case. Two alternative formulations of the subroutine SUB_C have been presented in [115, 192].

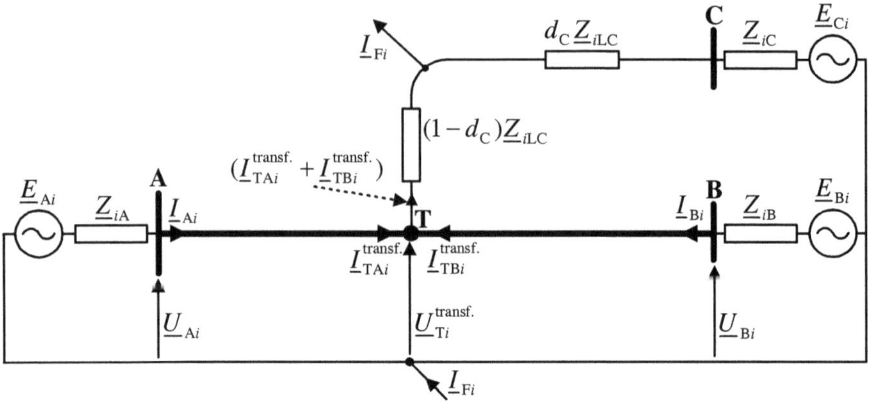

Fig. 7.40 Three-terminal line network – illustration for subroutine SUB_C designed for locating faults on line section T–C

7.7.4 Use of Minimal Measurements

In [117], the fault-location algorithm for three-terminal lines using the limited measurements, as compared to the earlier known approaches, has been presented. In particular, only the one-end measurements of three-phase current and voltage are utilized as the fault-locator input signals (Fig. 7.41). Additionally, information on amplitudes of pre-fault currents, i.e., amplitudes of pre-fault positive-sequence currents under the assumption of complete symmetry, from the remaining sections of the line has to be provided as the input data. Moreover, detailed impedance data of the network (for both the line sections and equivalent sources behind the line terminals) has to be provided as the fault-locator input data. The method is based on the symmetrical-components approach applied for formulating all three subroutines designed for locating faults on the respective line sections.

The final step in the fault-location algorithm relies on selecting a valid subroutine, i.e., on indicating which of the subroutines yields results consistent with the actual distance to fault and fault resistance. The subroutine that yields distance to fault outside its line section, and/or negative fault resistance, is surely false and has to be rejected. If it is not so, other criteria have to be utilized. In the study carried out the following criteria quantities were utilized:

- total fault currents in faulted phases (ought to correspond to the measured currents);
- amplitudes of the total fault current in healthy phases (ought to be close to zero); and
- boundary conditions for faults involving earth.

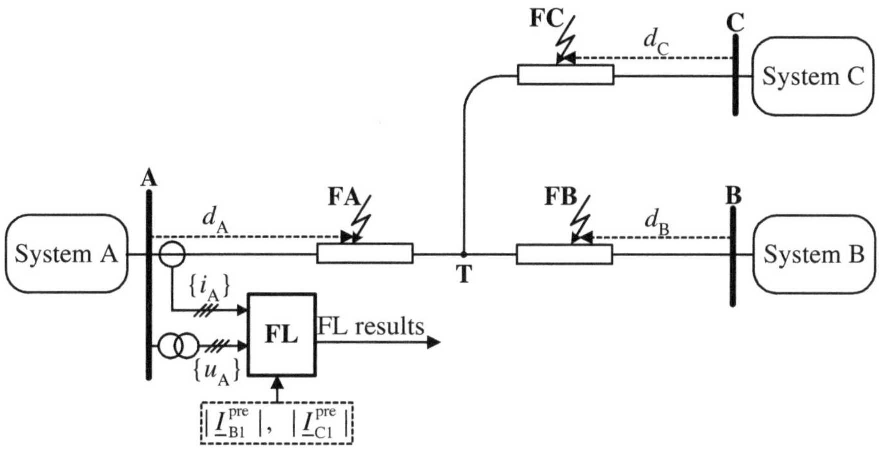

Fig. 7.41 Fault location on three-terminal line with use of minimal measurements

7.8 Fault Location on Multi-terminal and Tapped Lines

In Chap. 2, the basic models of multi-terminal and tapped lines are introduced. The networks presented there are limited to the three-terminal line case (Figs. 2.7 and 2.8) and the network with one tapped line (Fig. 2.9). However, in real power networks one can find lines with numbers of terminals higher than three and also lines with more than one tap for feeding the load. Use of the fault-location algorithms, which were presented in Sect. 7.7, is limited to a three-terminal line or to a line with a single tap. Therefore, those algorithms require certain modifications before being applied to multi-terminal lines.

In Fig. 7.42, a scheme of the multi-terminal line is shown. It has four terminals with power generation, and thus not less than three, as required in the definition of multi-terminal lines introduced in [350]. In the case of five terminals there are three tap points (T1, T2, T3) to which the tapped lines (T1–C, T3–E) are connected with the sources of power generation, and the tapped line (T2-D) feeding the load. Distributed generation included into distributed systems changes the power flow, normally radial in distribution systems, to a multi-directional one [81]. For the five-terminal lines there are four hypothetical locations of faults on the main route of the line (A–T1–T2–T3–B) and three hypothetical locations on the tapped lines (T1–C, T2–D, T3–E).

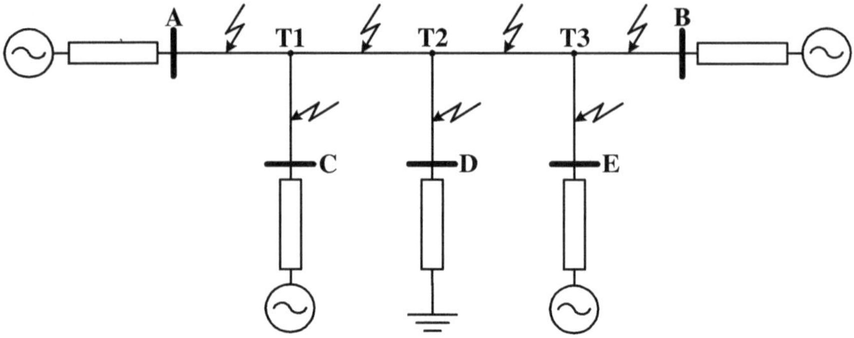

Fig. 7.42 Fault location on a five-terminal line

Figure 7.43 shows an example of the tapped line with a source of generation at the bus A and alternatively also at the bus B [3]. Thus, generation is at two of the terminals, which is in compliance with the definition of the tapped line introduced in [81]. There are also four tapped lines feeding the loads, which are usually balanced three-phase ones, but also a single-phase load can be met in distribution systems [33].

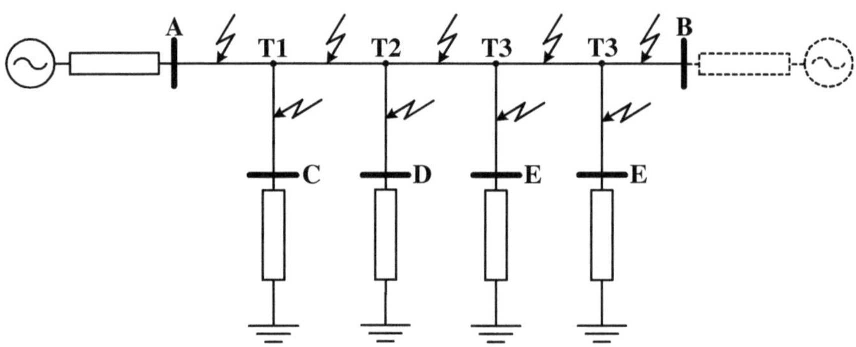

Fig. 7.43 Fault location on line A–B with four tapped lines feeding loads

A variety of algorithms for locating faults on multi-terminal and tapped lines have been developed so far. For example, fault location on a multi-terminal line has been reported in [218]. This method uses magnitudes of differential current at all terminals for locating faults on multi-terminal two parallel lines. The method is based on a three-terminal fault-location algorithm and an equivalent conversion from the n-terminal to the three-terminal system is applied.

In [1] use of synchronized three-phase voltages and currents at all terminals for fault location is considered. The algorithm using voltage differentials at terminals to gradually reduce a multi-terminal line into a two-terminal line containing the faulted section is applied there. Then, a reactive power-based method is used to locate the fault.

An approach from [68] uses synchronized current inputs from all terminals and uses two different methods to locate the fault on parallel double-circuit multi-terminal transmission lines:

- impedance relay method (employs an impedance calculation); and
- current diversion ratio method.

In [310], an analysis and evaluation of multi-terminal fault location have been presented. The technique proposed there uses the voltage from two terminals and the current data from all terminals of the line. The measurements are considered as performed synchronously with use of the GPS.

In [23] a fault-location method for a multi-terminal transmission line using synchronized voltage measurements was introduced. It contains a simple new procedure for identifying the faulted-line section first. Then, to exactly locate the fault on this section, a method is described that uses the synchronized voltage measurements at all terminals. The main advantage of this method [23] is that the current-transformer errors in the current measurements can be avoided. The method assumes that the source impedances are available. The method readily lends itself to untransposed lines and is free from the pre-fault conditions.

An interactive approach to fault location on distribution overhead lines with load taps is presented in [3]. However, the approach can be applied to transmission networks as well. The cited paper presents a novel technique in single-ended fault location for overhead systems based on the concept of incremental (pure fault) components of voltages and currents rather than total quantities. The incremental fault-path current at the assumed fault point is determined. In order to find the actual fault point, the assumed fault position is shifted in an interactive fashion in such a way as to obtain the minimum value of the fault-path currents from healthy phases. It is shown that the fault locator is highly insensitive to variations in source impedances (both the local and remote) and to the presence of taps with variable loads.

8 Fault Location in Distribution Networks

Abstract The objective of this chapter is to review basic fault-location algorithms used in distribution lines. First, it gives a review of the standard fault-location technique, namely impedance-based methods used in distribution networks. Then it gives a short review of the non-standard fault-location techniques, such as high-frequency-related methods, traveling wave and wavelet transform, for distribution systems.

8.1 Introduction

Fault location in distribution medium-voltage (MV) networks has been a subject of interest to utility engineers and researchers [44, 79, 261]. Information on accurate fault location available just after the fault helps utility personnel to expedite service restoration and to make adequate reconfiguration of the network for reducing outage time and operating costs. Therefore, more efficient methods for fault location, supply restoration and high-quality customer service, are highly required [3, 263].

Fault location in distribution networks (DN) creates new problems compared with the same task in HV and EHV transmission lines. In HV and EHV networks each transmission line may be equipped with a dedicated fault locator (FL). In such a case, the FL algorithm is a numerical procedure that converts voltage and current, given in a digital form, into a single number being a distance to fault. In contrast to the transmission lines, the distribution networks are usually non-homogeneous, with branches and loads along the line, which make the fault location difficult [79, 261].

Distribution systems, also called medium-voltage (MV) systems deliver the electrical energy to final step-down transformers at a voltage below 132 kV (this value may be different in particular systems). The MV networks provide suppliers direct to large customers, but the vast majority of consumers are connected at low voltage via distribution transformers (MV/LV) [73, 87, 327].

The scheme of a typical distribution system is presented in Fig. 2.33 (Chap. 2). The MV networks are supplied from HV/MV transforming substations. The electrical energy is then transported via a series of underground-cable and overhead-line circuits to the customers. In some cases a dispersed generation (DG) can be connected to the network. The particular line sections are of length from a few hundred meters to some dozens of kilometres. The distribution voltages in a specific service territory are likely similar because it is easier and more cost effective to stock spare parts when the system voltages are consistent. Although due to different loads and the network development, separate feeders are made up of sections with different technical data: varying cable cross-sections and OH line parameters. The configuration layout and complexity of distribution networks vary widely depending on the application.

In DN networks, FLs are usually assumed to be of a centralized type, i.e., they measure the quantities common for the whole substation (busbar voltages and transformer currents), which makes accurate fault location more difficult. Four fundamental factors contribute to this:

- If a current of a faulty line is not directly available to the FL, a certain error is introduced when assuming the transformer current during a fault to be a current of the faulty line; moreover, it is not possible to compensate accurately for the pre-fault load current of the faulty line.
- MV lines may be multi-terminal and/or contain loops that creates well-known problems for one-end fault location as, in general, there is no indication on a single fault position (few alternatives appear as possible).
- In the case of a MV line, there are often loads located between the fault point and the busbar; since the loads change and are unknown to the FL it is difficult to compensate for them.
- MV lines are mainly cable feeders that creates additional problems with adequate representation of the equivalent scheme, when compared with overhead lines.

8.2 Faults in Distribution System

Practical fault-location methods based on the use of current and voltage phasors have been implemented in the feeder-protection devices. The reliability and accuracy of substation-based fault locators is questionable because of numerous factors.

The factors that are mostly influenced by the type of neutral grounding are discussed in Sect. 2.5. The principle of DN neutral grounding must be considered when a fault-location method is used. The way the neutral is connected to the ground determines the behavior of a power system during a single phase-to-ground fault. The phase-to-phase faults, practically, are not influenced by the used method of the network neutral grounding. Some general related problems and used techniques that can reduce errors will be discussed in the following sections [106].

8.2.1 Solidly Grounded Networks

The idea of impedance-based methods used for fault location in solidly grounded DN is very similar to that applied for transmission lines. This type of DN networks are the most common in North America. The ungrounded network will require grounding transformers, which are distributed around the network, resulting in fewer sources of ground current. The net effect of this and of the latter two types of grounding is to reduce the magnitude of the ground current [14]. This can be seen as an increase in zero-sequence source impedance. Techniques to locate faults on these networks will not differ from those for solidly grounded networks except for the reduced accuracy of the computational techniques.

The unique topography of solidly grounded distribution systems presents many challenges for fault locators at the substation. These include:

1. Conductor sizes change, thus making impedance calculations non-linear. One possible solution is to model the line and create a nomograph.
2. Multiple feeder taps and laterals.
3. Inaccurate models/system data and dynamic configurations can affect results.
4. Effects of fault impedance can be significant.
5. Less energy to establish clean arcs.
6. Evolving fault characteristics and magnitudes can fool a relay's ability to select the correct fault type.

In the case of a phase-to-ground fault, the reactance algorithm using the zero-sequence voltage and the zero-sequence current of the faulted feeder may be used to compensate for the effect of the tapped loads with wye-delta distribution transformers. Another approach is to use the capacitive current of the feeder (important if a large portion of the feeder consists of cable). One can also improve the fault location for ground faults by using the ground or residual current instead of the faulted phase current.

Another method uses an on-line calculation of the source impedance (behind the unit) and the load impedance to compensate for the tapped loads (in addition to a complete compensation for the reactance effect – which is discussed in Chap. 6) in a radial distribution network. The load impedance can be estimated using the pre-fault data (Fig. 8.1):

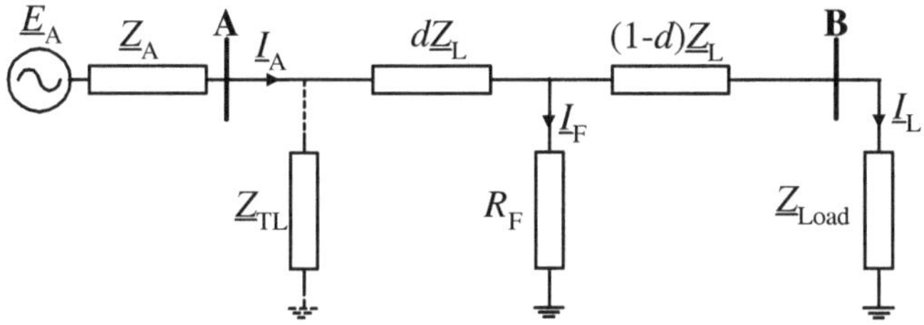

Fig. 8.1 Equivalent scheme of faulted radial network

$$\underline{Z}_{\text{Load}} = \frac{\underline{U}_{\text{A1}}^{\text{pre}}}{\underline{I}_{\text{A1}}^{\text{pre}}} - \underline{Z}_{\text{1L}} \tag{8.1}$$

where: $\underline{U}_{\text{A1}}^{\text{pre}}$, $\underline{I}_{\text{A1}}^{\text{pre}}$ are pre-fault positive-sequence voltage and current and $\underline{Z}_{\text{1L}}$ is the positive-sequence line impedance.

Compensation for tapped loads can be done when the impedance $\underline{Z}_{\text{TL}}$ is known (e.g., from pre-fault loads). Then, instead of measured current $\underline{I}_{\text{A}}$ the compensated current is used:

$$\underline{I}_{\text{A}}^{'} = \underline{I}_{\text{A}} - \frac{\underline{U}_{\text{A1}}}{\underline{Z}_{\text{TL}}} \tag{8.2}$$

Source impedance ($\underline{Z}_{\text{A}}$) behind the fault locator is calculated from the fault and pre-fault voltages and currents at the terminal as shown in (8.3)

$$\underline{Z}_{\text{A}} = -\frac{\Delta \underline{U}_{\text{A}}}{\Delta \underline{I}_{\text{A}}} \tag{8.3}$$

where: $\Delta\underline{U}_{\text{A}}=\underline{U}_{\text{A}} - \underline{U}_{\text{A}}^{\text{pre}}$, $\Delta\underline{I}_{\text{A}}=\underline{I}_{\text{A}} - \underline{I}_{\text{A}}^{\text{pre}}$.

Since the radial system is considered, by calculating $\underline{Z}_{\text{1L}}$ and $\underline{Z}_{\text{A}}$ the current distribution factor $\underline{k}_{\text{F}}$ as described in (6.5) is accurately calculated. Therefore, this method (using only local measurements) can compensate completely for the errors caused by the reactance effect. Compensation for tapped loads allows this method to provide accurate results for a variety of fault cases on distribution feeders, although the accuracy may degrade for faults.

Traditional Methodologies

Often, faults are located without the use of any measurements. Location is simply made through physical indications, field methods, and brute force methods such as:

- restoration through switching;
- restoration through recloser operation;
- indication through fuse and fault-locator operation;
- downed wires, customer calls, maps;
- relay targets;
- DC thumping of underground circuits; and
- smelling burnt cables.

Observant Technologies

The ability to collect data further improves upon the success of locating faults through such techniques as:

- local detection with communications feedback;
- "fault was upstream or downstream";
- intelligent metering with modem, cellular-based devices, LOCATE-type devices;
- detectors sending feedback through SCADA;
- Satellite-based location or carrier packages.
- choice of passing on information to an "interpreter" (i.e., a person) or to an "interpretive device" (i.e., a substation HMI or DMS); and
- fault recorders.

Intelligent Technologies

Due to complexity, intelligent technologies are being applied for distribution fault location. These include:

- artificial intelligence, neural networks to "learn" system characteristics; and
- distributed intelligence in substations to pull together information for best guess, confidence factors.

Although evolving, competition may arise among observant technologies dispersed throughout the system, but integrated together, and advanced intelligent technologies. These are briefly discussed in Chap. 9.

8.2.2 Locating Faults on Compensated Distribution Networks

The FL in high-resistance ground faults networks is a very complicated task. At present the fault location is determined by dividing the faulted feeder into sections by a disconnector and closing the substation circuit breaker against the fault until the faulty line section is found. This is time consuming, since the operation of disconnectors requires a patrol moving in the terrain [221]. The other indication and location methods proposed in the literature, are based on the direct measurement of the basic components of the currents and voltages, on analyzing their variations or their harmonic components with different methods, or on mixed versions of these methods. The difficulty with the accurate location of ground faults in high-impedance grounded networks is that the fundamental-frequency fault currents are often small compared to the load currents, even in the case of very low fault resistances [95].

In the promising methods it is assumed that the zero-sequence currents and the neutral to earth voltage are available as complex values, so the magnitude and the phase angle of their fundamental signal must be calculated from the measured signals [221, 248]. The algorithms are based on determining the total phase-to-ground admittance as seen from the substation and the unbalance of each feeder of a HV/MV substation.

The determination of the phase-to-ground admittance (for zero-sequence equivalent circuit) of the whole system makes it possible to calculate the ground fault current much more accurately than by using rough tabular values of the phase-to-ground susceptances from the network data base. The change in the unbalance of the feeder is a sensitive indicator with highly resistive ground faults. The unbalance factor $\underline{k}$ is a normalized complex parameter indicating the phase and the magnitude of the asymmetry of the three phases to ground. Its value depends on the topology and the nature of the system (overhead lines, underground cables).

After the fault indication, the fault location can be determined by connecting the faulted feeder into a closed ring with some healthy feeder usually by a remote-controlled switching device. In a compensated network this can be made without outage to customers. Details of the algorithm are presented in [221].

The calculation of the zero-sequence impedances can be sometimes inaccurate, especially in the case of underground cables.

8.2.3 Locating Faults on Underground and Paralleled Cables

Underground Cables

Locating faults on underground transmission or distribution lines is difficult to achieve using microprocessor relays. First, the zero-sequence impedance changes due to the distributed shunt capacitance along the entire line/cable (Fig. 8.2). The capacitance can change depending on the system voltage, stored charge, different zero-sequence return paths, and other factors. The effect can be likened to infeed from unpredictable sources along an entire transmission line.

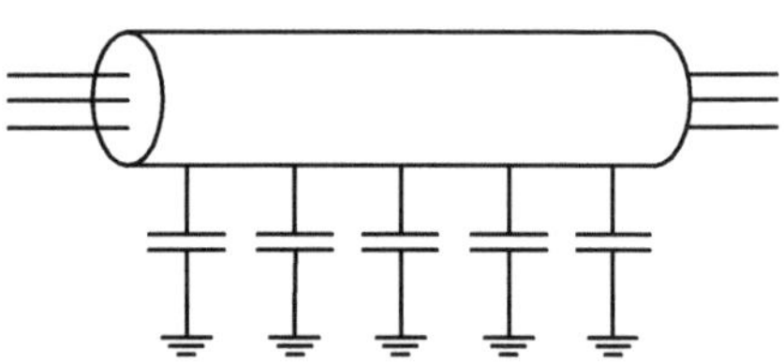

Fig. 8.2 Distributed capacitance along underground cable

Moreover, the ohmic value of the cable is usually small, so a minute error in the impedance model results in large errors in predicting fault location.

In practice, utilities use some form of pre-location followed by one or more methods to pinpoint the location of the cable fault. The methods used depend primarily on voltage class (transmission or distribution) and installation (conduit or direct burial). There are generally two categories of fault locating: terminal methods and tracer methods. These methods of fault locating usually take place after a fault has occurred and has been cleared by protective relaying.

Terminal methods measure electrical quantities at one or both ends of the cable circuit. Terminal methods are usually used to “pre-locate” the fault. The most popular terminal methods include “bridge” techniques, like the Murray-loop, and radar/low- or high-voltage pulse methods [15].

Tracer methods are most useful for “pinpointing” the location of the fault, and usually require walking the cable route in the field. The most popular tracer methods are acoustic (“thumper”) techniques and earth gradient (applying a source and measuring return current) methods, although many other methods are used. Recent developments in using the transient voltage and current waves during a fault have improved pinpointing cable fault location on distribution circuits. The method measures the time between transient peaks to compute a distance to fault [328].

Paralleled Cable Circuits

Frequently, two conductors in parallel are used instead of one larger conductor. In the case of insulated cables, a cable failure is characterized as a line-to-ground fault on only one of the conductors. The line impedance in this case is not a linear function of the distance to the fault. In the case of a radial circuit supplied from one side only, the impedance is related to the distance to the fault as

$$\underline{Z} = f(d) = d(2-d) \tag{8.4}$$

where: $\underline{Z}$ is the normalized impedance to the fault (= 1 at the end of the line); d is the distance to the fault (= 1 at the end of the line), which represents a parabolic relation.

In the usual case of sources at both ends of the line, the situation resembles a three-terminal line.

8.3 Review of the Standard Fault-location Techniques for Distribution Systems

In this section, reviews of selected standard fault-location techniques (phasor-based algorithms) proposed for distribution systems are presented [265].

8.3.1 Method Based on Lumped-parameters Network Model

The method presented in ref. [223] is based on the idea of fault location applied for short transmission lines [62, 174], with all loads, including tapped lines, represented by a lumped-parameter impedance model placed behind the fault. This way of compensating for tapped loads is accurate as tapped load impedances are much larger than the feeder impedance.

The method is based on calculating both source and load impedances based on pre-fault and fault voltages and currents measured at the substation. The load impedance is calculated as in (8.1) while the source impedance behind the fault locator is calculated as in (8.3).

To avoid inaccurate calculation of source impedance for a small difference between fault and pre-fault values, a negative-sequence network is used for unbalanced faults. The basic relation is obtained from the following equation for measured fault-loop impedance:

$$\underline{Z}_{\text{meas}} = \frac{\underline{U}_{\text{A}}}{\underline{I}_{\text{A}}} = dZ_{1\text{L}} + R_{\text{F}}\frac{\underline{I}_{\text{F}}}{\underline{I}_{\text{A}}} \tag{8.5}$$

From this equation one can obtain the quadratic relation for distance to fault:

$$d^2 - d\underline{k}_1 + \underline{k}_2 - \underline{k}_3 R_{\text{F}} = 0 \tag{8.6}$$

where: $\underline{k}_1 = \frac{\underline{U}_{\text{A}}}{\underline{I}_{\text{A}}\underline{Z}_{1\text{L}}} + \frac{\underline{Z}_{\text{Load}}}{\underline{Z}_{1\text{L}}} + 1$, $\underline{k}_2 = \frac{\underline{U}_{\text{A}}}{\underline{I}_{\text{A}}\underline{Z}_{1\text{L}}}\left(\frac{\underline{Z}_{\text{Load}}}{\underline{Z}_{1\text{L}}} + 1\right)$,

$\underline{k}_3 = \frac{\Delta\underline{I}_{\text{A}}}{\underline{I}_{\text{A}}\underline{Z}_{1\text{L}}}\left(\frac{\underline{Z}_{\text{A}} + \underline{Z}_{\text{Load}}}{\underline{Z}_{1\text{L}}} + 1\right)$.

The complex equation (8.6) has two unknowns d and R_{F}. By separating this equation into real and imaginary parts, the value of d can be obtained after elimination of R_{F} using:

$$d = \frac{-b - \sqrt{b^2 - 4ac}}{2a} \tag{8.7}$$

where: a=1, $b = -\left(\text{Re}(\underline{k}_1) - \frac{\text{Im}(\underline{k}_1)\cdot\text{Re}(\underline{k}_3)}{\text{Im}(\underline{k}_3)}\right)$ $c = \text{Re}(\underline{k}_2) - \frac{\text{Im}(\underline{k}_2)\cdot\text{Re}(\underline{k}_3)}{\text{Im}(\underline{k}_3)}$.

The fault type is considered by including the adequate voltages and currents. Unlike other methods using only local data [223], this method is not affected if the fault current at a fault locator is not in phase with current at the fault. In conclusion, immunity to effects of load current and fault resistance is achieved. Compensation for tapped loads enables this method to provide accurate results, although, for heavily tapped feeders, the accuracy may degrade toward the end of the feeder.

The method was tested on an EMTP [56] model of a typical distribution tapped network with approximately 200 fault cases by varying fault locations (main feeder and taps), fault resistances and fault-inception angle. Sample results are shown in Table 8.1, where *d* (%) is calculated fault location.

Table 8.1 Fault-location accuracy versus fault resistance

Fault resistance (Ω)	**Distance to fault (%)**	
	50%	**90%**
R_F=10 Ω	d=49.8 %	d=88.0 %
R_F=50 Ω	d=49.2 %	d=82.0 %

Additional means are required to distinguish if the fault is on the tap or on the feeder.

8.3.2 Technique with Two-port Network Section Representation

Reference [45] suggested a technique that uses the fundamental-frequency voltages and currents measured at a line terminal before and during the fault. The fault-location technique is described by considering a single-phase-to-ground fault on a radial system, shown in Fig. 8.3. The selected system consists of an equivalent source G, the line between nodes M and N and laterals. Loads are tapped at several nodes and conductors of different types are used on this circuit.

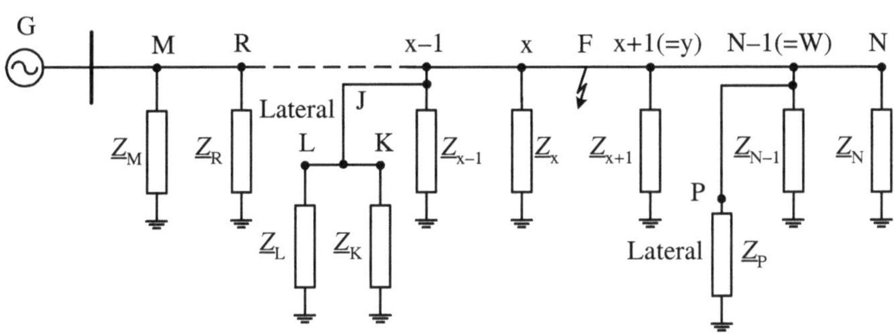

Fig. 8.3 The single line diagram of a radial line experiencing a fault at F

Generally, the shown impedances represent any type of load. The fault-location technique consists of the following six steps.

1. Choosing an apparent faulted section. A preliminary estimate of the location of the fault is made, say between nodes x and x+1 (=y). Line parameters, the type of fault and phasors of the sequence voltages and currents are used to obtain this estimate.
2. Calculation of an equivalent radial system. All laterals between node M and the apparent location of the fault are ignored and the loads on a lateral are considered to be present at the node to which the lateral is connected.

3. Load modeling. The effects of the loads are considered by compensating for their currents. Static response-type models are used for all loads up to node x and also for a consolidated load at the remote end. For a load at node, say R, this model is described by

$$\underline{Y}_{\text{R}} = \left(G_{\text{R}} \left| \underline{U}_{\text{R}} \right|^{n_{\text{P}}-2} + \text{j} B_{\text{R}} \left| \underline{U}_{\text{R}} \right|^{n_{\text{Q}}-2} \right) \tag{8.8}$$

where, $\underline{U}_{\text{R}}$ is the voltage at node R, $\underline{Y}_{\text{R}}$ is the load admittance, G_{R} and B_{R} are constants proportional to the conductance and susceptance, respectively, and n_{P} and n_{Q} are the response constants for the active and reactive components of the load. The constants G_{R} and B_{R} are estimated from the pre-fault load voltages and currents and, appropriate values of n_{P} and n_{Q}. These constants and voltages are used to estimate load admittances and sequence currents during the fault.

4. Calculation of voltages and currents at the fault and remote end. The sequence voltages and currents at node F during the fault are estimated by assuming that all loads beyond node x are consolidated into a single load at N, as shown in Fig. 8.4.

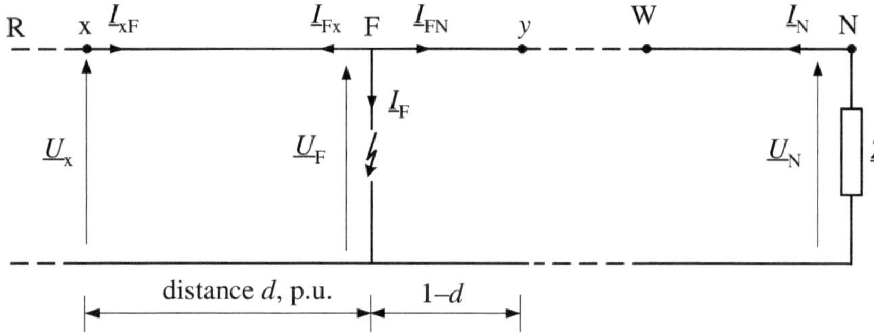

Fig. 8.4 Fault voltages and currents at nodes F and N

The relation between voltages and currents at nodes F and x can be described by using a two-port network model [45]:

$$\begin{bmatrix} \underline{U}_{\text{F}} \\ \underline{I}_{\text{Fx}} \end{bmatrix} = \begin{bmatrix} 1 & d \cdot \underline{B}_{\text{xy}} \\ d \cdot \underline{C}_{\text{xy}} & -1 \end{bmatrix} \begin{bmatrix} \underline{U}_{\text{x}} \\ \underline{I}_{\text{xF}} \end{bmatrix} \tag{8.9}$$

where: d is the per unit distance to node F from node x.

The sequence voltages and currents at nodes N and F during the fault are related by a similar equation:

$$\begin{bmatrix} \underline{U}_{\mathrm{N}} \\ \underline{I}_{\mathrm{N}} \end{bmatrix} = \begin{bmatrix} 1 & (1-d)\underline{B}_{\mathrm{xy}} \\ (1-d)\underline{C}_{\mathrm{xy}} & -1 \end{bmatrix} \begin{bmatrix} \underline{A}_e & \underline{B}_e \\ \underline{C}_e & \underline{D}_e \end{bmatrix} \begin{bmatrix} \underline{U}_{\mathrm{F}} \\ \underline{I}_{\mathrm{FN}} \end{bmatrix} \tag{8.10}$$

where: $\underline{A}_e, \underline{B}_e, \underline{C}_e$ and $\underline{D}_e$ are the equivalent constants of the cascaded sections between nodes $x+1(=y)$ and N.

The currents at node F keep to the rule:

$$\underline{I}_{\mathrm{FN}} = -\underline{I}_{\mathrm{Fx}} - \underline{I}_{\mathrm{F}} \tag{8.11}$$

The following equation is obtained by further substitutions from (8.9) and (8.11) and from an equation involving pre-fault voltages and currents in (8.10) and, neglecting the second- and higher-order terms in *d*:

$$\begin{bmatrix} \underline{U}_{\mathrm{N}} \\ \underline{I}_{\mathrm{F}} \end{bmatrix} = \frac{1}{\underline{K}_v + d \cdot \underline{K}_w} \begin{bmatrix} \underline{K}_m + d \cdot \underline{K}_n & d \cdot \underline{K}_p \\ \underline{K}_q + d \cdot \underline{K}_r & \underline{K}_v + d \cdot \underline{K}_u \end{bmatrix} \begin{bmatrix} \underline{U}_{\mathrm{x}} \\ \underline{I}_{\mathrm{xF}} \end{bmatrix} \tag{8.12}$$

where: $\underline{K}_m, \underline{K}_n, \underline{K}_p, \underline{K}_q, \underline{K}_r, \underline{K}_u, \underline{K}_v$ and $\underline{K}_w$ are complex parameters and are computed using $\underline{Y}_{\mathrm{N}}$, $\underline{B}_{\mathrm{xy}}$, $\underline{C}_{\mathrm{xy}}$, $\underline{A}_e$, $\underline{B}_e$, $\underline{C}_e$ and $\underline{D}_e$, respectively.

The sequence voltages at nodes F and N and sequence currents at node F are obtained by using (8.9) and (8.12) and various parameters.

5. Estimating the location of the fault. The distance to the fault node F from node x, *d*, expressed as a fraction of the length from node x and node x+1, is estimated from the voltage–current relationships at the fault and the resistive nature of the fault impedance. The voltage–current relationship at the fault, for an a-phase-to-ground fault:

$$\frac{\underline{U}_{0\mathrm{F}} + \underline{U}_{1\mathrm{F}} + \underline{U}_{2\mathrm{F}}}{\underline{I}_{0\mathrm{F}} + \underline{I}_{1\mathrm{F}} + \underline{I}_{2\mathrm{F}}} = R_{\mathrm{F}} \tag{8.13}$$

where: $\left(\dfrac{\underline{U}_{1\mathrm{F}} - \underline{U}_{2\mathrm{F}}}{\underline{I}_{1\mathrm{F}} - \underline{I}_{2\mathrm{F}}} \right) = R_{\mathrm{F}}$ and $\underline{I}_{0\mathrm{F}}, \underline{I}_{1\mathrm{F}}, \underline{I}_{2\mathrm{F}}$ are zero-, positive- and negative-sequence voltage and current phasors at fault M and, R_{F} is the fault impedance.

Equating imaginary parts of both sides of (8.13), substituting the sequence voltages and currents at fault from (8.9) and (8.12), neglecting the second- and higher-order terms in *d* and rationalizing, the following equation is obtained:

$$\mathrm{Im}\left(\frac{\underline{K}_{\mathrm{A}} + d \cdot \underline{K}_{\mathrm{B}}}{\underline{K}_{\mathrm{C}} + d \cdot \underline{K}_{\mathrm{D}}}\right) = 0 \tag{8.14}$$

where: $\underline{K}_{\mathrm{A}}$, $\underline{K}_{\mathrm{B}}$, $\underline{K}_{\mathrm{C}}$ and $\underline{K}_{\mathrm{D}}$ are complex parameters.

The complex parameters $\underline{K}_{\mathrm{A}} - \underline{K}_{\mathrm{D}}$ are expressed in real and imaginary parts as $\underline{K}_{\mathrm{A}} = K_{\mathrm{AR}} + \mathrm{j}K_{\mathrm{AI}}$ and so on and, substituted in (8.14). Rationalizing the resulting equation, neglecting higher-order terms in d and rearranging the following equation is obtained:

$$d = \frac{K_{\mathrm{AR}}K_{\mathrm{CI}} - K_{\mathrm{AI}}K_{\mathrm{CR}}}{(K_{\mathrm{CR}}K_{\mathrm{BI}} - K_{\mathrm{CI}}K_{\mathrm{BR}}) + (K_{\mathrm{DR}}K_{\mathrm{AI}} - K_{\mathrm{DI}}K_{\mathrm{AR}})} \tag{8.15}$$

An iterative solution of d is obtained using the pre-fault admittance of the consolidated load at node N and some of the above mentioned equations. Two more estimates of the distance are obtained by considering that the fault is either located between node x–1 and node x or is located between nodes x+1 and x+2, respectively. The most plausible solution is selected and the location of the fault from the relay location, node M, is estimated.

6. Converting multiple estimates to single estimate. The fault-location technique could provide multiple estimates if the line has ‘laterals’. The number of estimates, for a fault, depends on the system configuration and the location of the fault. Software-based fault indicators, like those commercially available, are developed for this purpose. They detect downstream faults irrespective of their location. Information from the fault indicators is combined with multiple estimates, to arrive at a single estimate for the location of a fault.

Test Results

The fault-location technique described above was tested using simulated fault data. The authors have performed the simulations using the PSCAD/EMTDC simulation tool. A 37-km long, 25-kV radial circuit, was selected for the studies [45]. The fault locations estimated by the proposed technique for single-phase-to-ground faults, using fault resistances of 5.0 Ω and 50.0 Ω, indicate that the proposed technique is more accurate than the reactive component method. The results also show that, for a 5.0-Ω fault resistance, maximum error by the proposed technique for single-phase-to-ground faults is less than 1.7-%. For a 50-Ω fault resistance, the maximum error is less than 2.2-%.

8.3.3 Algorithm Based on Fault-recorder Data

Reference [278] suggested an algorithm that uses the fundamental-frequency voltages and currents measured at a line terminal before and during the fault. Current can be also measured at a supplying transformer if only one centralized type of disturbance fault recorder (DFR) is installed at the substation. A distance to fault is estimated based on the topology principle. The proposed method is devoted to estimating the location of faults on radial MV system, which can include many intermediate load taps. In the method, the non-homogeneity of the feeder sections is also taken into account. Distribution utility MV networks are used as an example.

Algorithm for Calculating the Fault Impedance

In the proposed method [278] the calculation of fault location consists of two steps. First, the fault-loop impedance is calculated by utilizing the measured voltages and currents obtained before and during the fault. Second, the impedance along the feeder is calculated by assuming the faults at each successive section. By comparing the measured impedance with the calculated feeder impedance, an indication of the fault location can be obtained.

Measurements at the Faulty Feeder

In this case the busbar voltage and the feeder currents are available (Fig. 8.5). As far as only one-end supplied radial networks are considered, the positive-sequence fault-loop impedance is calculated according to well-known equations depending on the type of fault by using the bus voltage and the faulty feeder current.

Measurements at the Substation Level

Let us consider a radial network, where a faulty feeder, such as for example, k, has the pre-fault equivalent impedance $\underline{Z}_k$. The remaining parallel connected feeders are represented by an equivalent branch with the impedance $\underline{Z}_{\mathrm{L}k}$. Both $\underline{Z}_k$ and $\underline{Z}_{\mathrm{L}k}$ are assumed to be the positive-sequence impedances. Measurements are available only at the substation level: the bus voltage and the supplied current (Fig. 8.5). The aim of the analysis is to determine the fault positive-sequence impedance of feeder k, $\underline{Z}_{\mathrm{F}k}$, under the assumption that the equivalent impedance $\underline{Z}_{\mathrm{L}k}$ remains unchanged during a fault. The following equation is valid for the pre-fault state:

$$\underline{Z}^{\mathrm{pre}} = \frac{\underline{U}^{\mathrm{pre}}}{\underline{I}^{\mathrm{pre}}} = \frac{\underline{Z}_k \underline{Z}_{\mathrm{L}k}}{\underline{Z}_k + \underline{Z}_{\mathrm{L}k}} \tag{8.16}$$

where: $\underline{U}^{\mathrm{pre}}$, $\underline{I}^{\mathrm{pre}}$ are pre-fault voltage and current phasors, respectively.

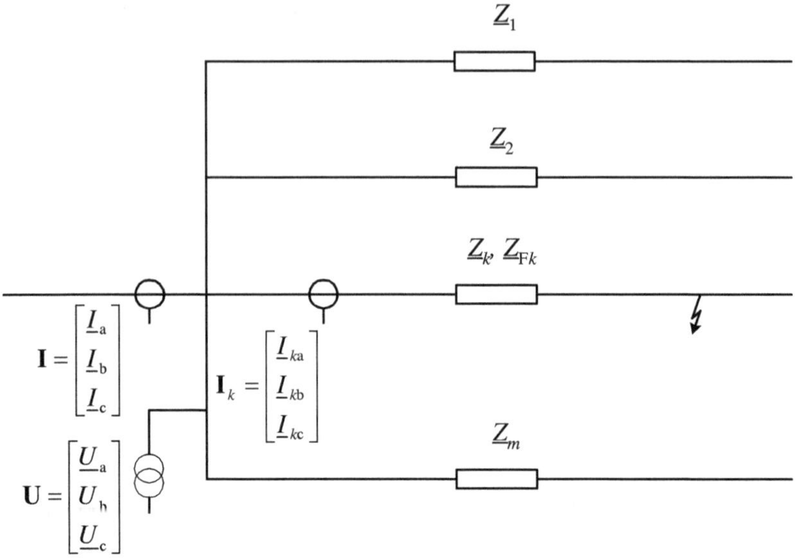

Fig. 8.5 Diagram of the network: current measurements are taken in the faulty feeder or centrally in the supplying feeder

Two fault cases are considered, namely:

1. Phase-phase fault loop: the positive-sequence impedance seen from the substation is obtained from the equation:

$$\underline{Z}_F = \frac{\underline{U}_{ph\text{-}ph}}{\underline{I}_{ph\text{-}ph}} = \frac{\underline{Z}_{Fk}\underline{Z}_{Lk}}{\underline{Z}_{Fk} + \underline{Z}_{Lk}} \tag{8.17}$$

where: $\underline{U}_{ph\text{-}ph}$, $\underline{I}_{ph\text{-}ph}$ – phase-phase fault-loop voltage and current taken at the substation, respectively.
Combining (8.16) and (8.17) yields:

$$\underline{Z}_{Fk} = \frac{\underline{Z}_F \underline{Z}^{pre}}{\underline{Z}^{pre} - \underline{Z}_F \underline{k}_{zk}} \tag{8.18}$$

$$\text{where: } \underline{k}_{zk} = \frac{\underline{Z}^{pre}}{\underline{Z}_{Lk}} = \frac{\underline{S}_{Lk}}{\underline{S}_{\Sigma}}, \tag{8.19}$$

$\underline{S}_{Lk}$ – the power in the faulty line in pre-fault condition, $\underline{S}_{\Sigma}$ – the power in all the lines in pre-fault conditions.
Combining (8.16) and (8.19) one also obtains:

$$\underline{k}_{zk} = \frac{\underline{Z}_k}{\underline{Z}_k + \underline{Z}_{Lk}} \tag{8.20}$$

The coefficient $\underline{k}_{zk}$ for each line is estimated on the basis of the pre-fault steady-state conditions. From (8.18) one can calculate the fault-loop impedance using the measurements from the substation. Dividing numerator and denominator of (8.18) by $\underline{Z}^{\text{pre}}$ and substituting (8.19) for $\underline{Z}_F$, (8.20) can be rewritten in a more convenient form:

$$\underline{Z}_{Fk} = \frac{\underline{U}_{\text{ph-ph}}}{\underline{I}_{\text{ph-ph}} - \underline{k}_{zk}\dfrac{\underline{U}_{\text{ph-ph}}}{\underline{Z}^{\text{pre}}}} \tag{8.21}$$

This is illustrated in Fig. 8.6. In a pre-fault condition the considered feeder k has an equivalent impedance $\underline{Z}_k$. After fault inception this impedance is divided among two impedances: $\underline{Z}_{ko}$ and $\underline{Z}_{Fk}$. The last one determines a distance to fault.

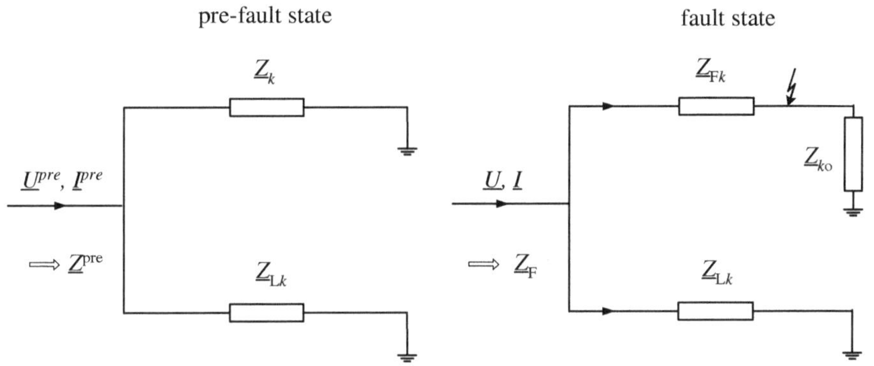

Fig. 8.6 Pre- and fault states of the faulted network

2. Phase-ground faulted loop (a phase-to-ground fault): the positive-sequence fault-loop impedance is calculated in a classical way. One can observe that as only a single phase-to-ground fault is considered, the zero-sequence current measured in the substation contains the faulty feeder current $\underline{I}_{kN}$ and zero-sequence current flows through capacitance of the healthy feeders. Knowing the voltage and current measurements at the substation, and network parameters the fault-loop impedance can be established in a similar way as for measurements from the feeder. The final expression takes the form:

$$\underline{Z}_{Fk} = \frac{\underline{Z}_g \underline{Z}^{pre}}{\underline{Z}^{pre} - \underline{Z}_g \underline{k}_{zk}\left(1 - \frac{\underline{U}_0}{\underline{U}_{ph}}\right)} \tag{8.22}$$

where: $\underline{Z}_g = \frac{\underline{U}_{ph}}{\underline{I}_{ph} + \underline{k}_{kN}\underline{I}_{kN}}$, (8.23)

$\underline{U}_0 = (\underline{U}_a + \underline{U}_b + \underline{U}_c)/3$.

The above equations define fault-loop impedance for phase-to-ground fault in terms of positive-sequence impedance.

Estimation of Distance to Fault

Based on the measured fault-loop impedance and the cable parameters it is possible to estimate the distance to a fault. The algorithm for phase-to-phase fault is discussed below in detail.

Algorithm for Phase-to-phase Fault

Let us consider the equivalent positive-sequence scheme of the fault loop. The shunt elements represent loads at successive nodes while the cable impedance is represented by the series elements. Defining an equivalent fault-loop impedance as seen from *i*th node to the fault point one obtains the following recursive form (Fig. 8.7).

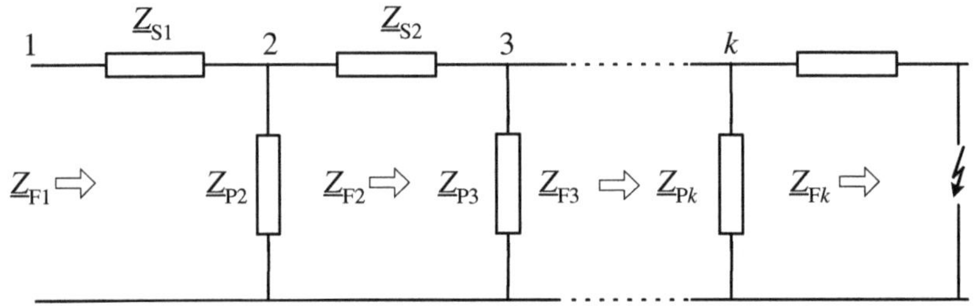

Fig. 8.7 Equivalent positive-sequence diagram of the faulty cable

$$\underline{Z}_{Fi} = \frac{\underline{Z}_{Pi}(\underline{Z}_{Fi-1} - \underline{Z}_{Si-1})}{\underline{Z}_{Pi} - \underline{Z}_{Fi-1} + \underline{Z}_{Si-1}} = R_{Fi} + jX_{Fi} \tag{8.24}$$

In the above equation $\underline{Z}_{Si-1}$ represents the cable segment impedance, while $\underline{Z}_{Pi}$ relates to the load impedance and/or equivalent impedance of the branches connected to the node. The value of this impedance is estimated from the steady-state condition of the network. One can see that the impedance obtained in the following steps tends toward zero

$$\left|\underline{Z}_{\mathrm{F}i-1}\right| > \left|\underline{Z}_{\mathrm{F}i}\right| \tag{8.25}$$

and the impedance of the faulty section is

$$\underline{Z}_{\mathrm{F}k} = l_{\mathrm{F}k-1}\underline{Z}_{\mathrm{S}k-1} + R_{\mathrm{F}} \tag{8.26}$$

where: $l_{\mathrm{F}k}$ – p.u. distance from node k to the fault point (total length of the faulty segment is assumed to be 1), $\underline{Z}_{\mathrm{S}k-1}$ – impedance of the cable segment between nodes k–1 and k ($\underline{Z}_{\mathrm{S}k-1} = R_{\mathrm{S}k-1} + \mathrm{j}X_{\mathrm{S}k-1}$), R_{F} – fault resistance.

Example 8.1 The considered 10-kV substation [278] is supplied from the 150-kV system (Fig. 8.8). The cable network is operated in a radial way. Measurements of current are available at the supplying transformer or at the feeders. Cable shield is considered as grounded only at the load points.

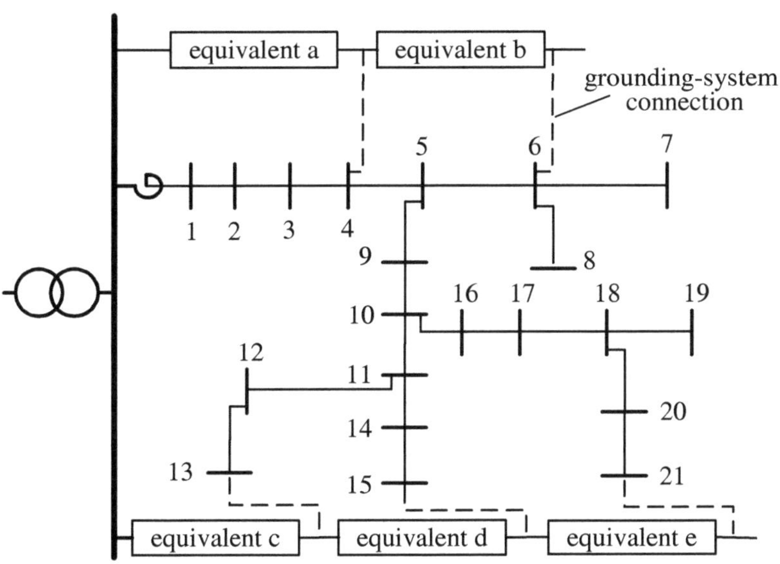

Fig. 8.8 Equivalent scheme of the analyzed network; dotted lines are for grounding-system connection

In more detailed analysis, the equivalent grounding scheme should also include the substation grounding (Fig. 8.9). The similar situation is also typical for three-phase four-wire MV systems.

The method is further developed for the distribution systems with more branches and are reported in references [280, 281]. The simulation results based on the developed methods are described here. For a distance to fault calculation each feeder should be represented by the detailed scheme with adequate line and load models. In the cable networks the grounding system has a different structure from the feeders (open cables may have connected grounding circuits), which should also be represented in the model. This requires representing all feeders connected in a given substation in the general simulation model. However, for proper post-fault transient analysis some simplifications can be introduced: the supplying system is described by steady-state parameters; the analyzed feeder is represented in detail; the all other feeders are represented by equivalent schemes by reproducing only the grounding-system connections.

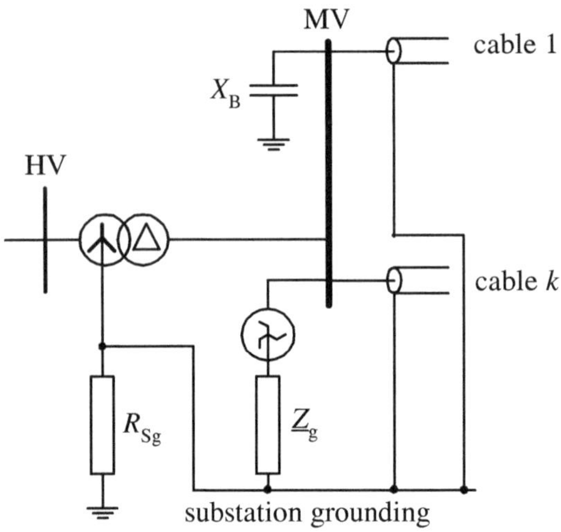

Fig. 8.9 Equivalent scheme for substation and MV network grounding

An example of the analyzed network is presented in Fig. 8.8. Cable sections are represented by appropriate π-schemes, while loads and equivalent circuits are reflected by a RL or RLC scheme. The ATP-EMTP model [56] of the analyzed network has been extensively used for investigation of the proposed fault-location algorithm. The MV network consists of 16 feeders that, except for the particular analyzed feeder, are represented by their equivalent schemes. Let us consider an example of an a–g (phase a to ground) fault at node 17 in the analyzed feeder (actual distance to fault – 8138 m, Fig. 8.8) with an assumed fault resistance $R_F = 0.1\,\Omega$. Phase voltages and currents observed at the substation during a–g and 3-phase faults are presented in Figs. 8.10 and 8.11, respectively.

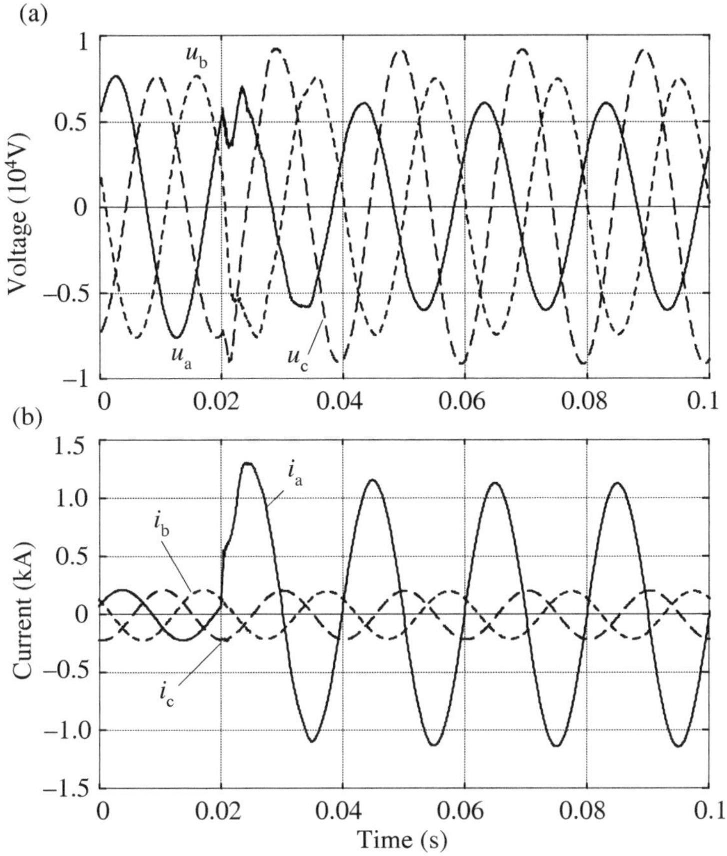

Fig. 8.10 Phase voltages (a) and currents and (b) during an a–g fault at node 17

Results of distance to fault estimation for 3-phase and phase-to-ground fault are gathered in Table 8.2. The results obtained are quite good.

Table 8.2 Results of distance to fault estimation

Type of fault	**Fault resistance** (Ω)	**Obtained result** (m)	**Error** (m)
3–phase	0.1	8145	+7
	2.0	8135	–3
a–g	0.1	8131	–7
	2.0	8081	–57
The actual distance to fault: 8138 m			

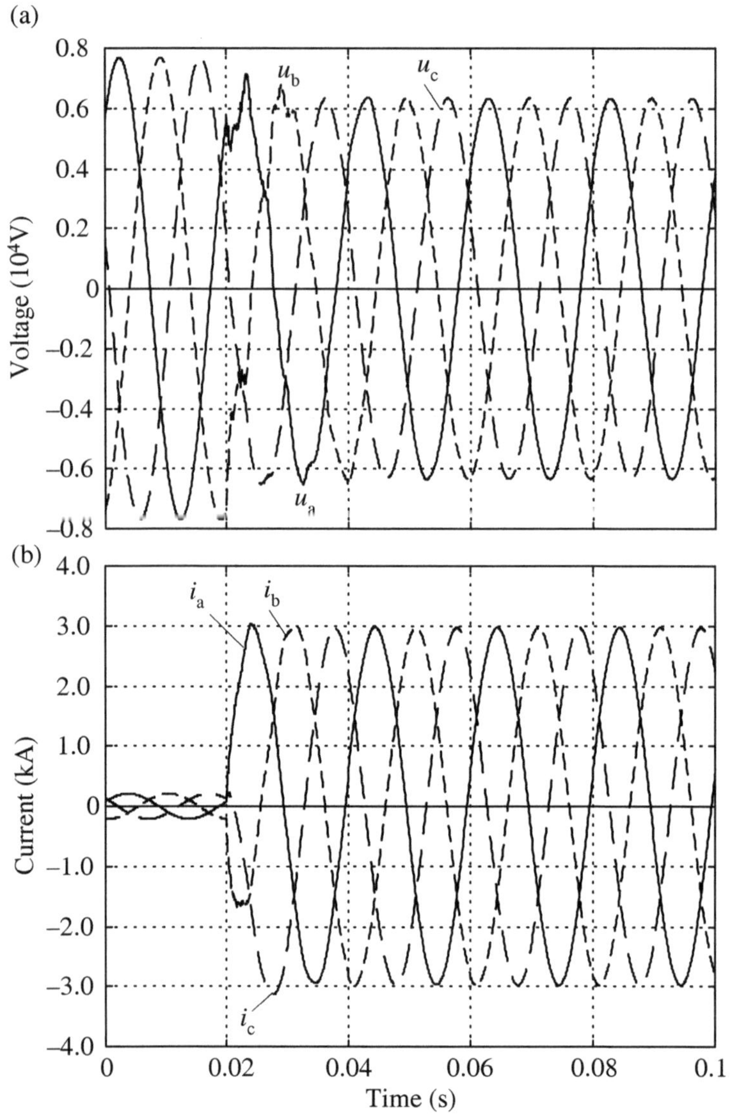

Fig. 8.11 Phase voltages (a) and currents (b) during 3-phase fault at node 17

8.3.4 General Form of Impedance-based Algorithms

There are many different fault-location algorithms that are based on fault-loop impedance calculation. The main problem arising from use of such methods in DN is connected with the specific characteristics of this object. We have seen that in contrast to the transmission lines, the distribution networks are usually non-homogeneous, with branches and loads along the line, which makes the fault location difficult.

In DN networks, FLs are usually assumed to be of a centralized type, i.e., they measure the quantities common to the whole substation (busbar voltages and supplied transformer currents), which makes accurate fault location more difficult. Four fundamental factors contribute to this:

- if a current of a faulty line is not directly available to the FL, a certain error is introduced when assuming the transformer current during a fault to be a current of the faulty line; moreover, it is not possible to compensate accurately for the pre-fault load current of the faulty line;
- DN feeders may be multi-terminal and/or contain loops that creates well-known problems for one-end fault location as, in general, there is no indication on a single fault position (few alternatives appear possible);
- in the case of a DN feeder, there are often loads located between the fault point and the busbar; since the loads change and are unknown to the FL it is difficult to compensate for them; and
- DN feeders are mainly cable feeders that creates additional problems with adequate representation of the equivalent scheme, when comparing with overhead lines.

Different methods proposed to overcome the above-mentioned difficulties. Generally, these algorithms are realized by utilizing the following three-step procedure for fault location in distribution networks (Fig. 8.12).

- First, the equivalent positive- and zero-sequence impedance of the network is computed in a pre-fault steady state for all nodes of the network based on existing topology, loads and feeder parameters.
- Then, after the fault detection, the specific fault-loop parameters are calculated depending on the fault type (phase-to-phase or phase-to-ground) and the place of measurements (at the supplying transformer or at the faulty feeder) that allows fault place determination.
- In the case of a few possible place determinations an adequate procedure should be applied for selecting the most likely result.

The question taken up in the last step is an important issue when dealing with fault location in distribution feeders. It is expected that an effective algorithmic tool can be applied to filter out inappropriate solutions, and provide the real fault-location indication at the end of the processing. In some cases, additional information can help here, such as for example, post-fault breaker position. Many authors propose to apply artificial intelligence methods to resolve the problem [162].

As was seen in the previous sections, the distance to fault is usually determined as the sum of distance to a preceding node and distance inside the considered section (cable).

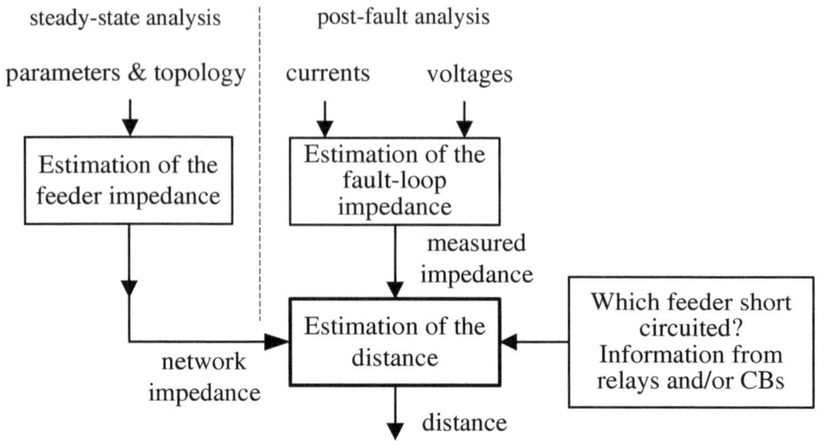

Fig. 8.12 Basic block diagram of the impedance-based FL algorithm

To improve the fault-location estimation, it is important to reduce errors caused by fault-location algorithms. The algorithms will deliver more accurate results if more information about the system is available. Below, we describe in brief one of the proposed extensions to improve the basic impedance-based methods.

Reference [289] proposes and discusses an extended fault-location formulation to be used in general distribution systems. The method is based on the apparent impedance calculation and fundamental quantities. Furthermore, the method considers the specific characteristics of distribution systems, being capable of locating faults in systems with intermediate loads and laterals/sublaterals.

The load-profile variation is also considered in this formulation, taking into account only local measurements. This characteristic normally is not included in the impedance-based fault-location methods. The formulation is also suitable for large distribution systems and different fault types. The following subsections describe the proposed extensions.

General Feeders

Distribution systems are typically composed by a main feeder, laterals, sublaterals, and end loads. Laterals are main feeder branches not always composed by three-phase connections. Figure 8.13 illustrates such a system.

To consider each lateral, the proposed method calculates equivalent systems to each possible power-flow path (PPFP), resulting in n equivalent radial systems, where n is the number of laterals. In the presented system, illustrated in Fig. 8.13, for example, there are a total of 6 PPFPs, and the node groups (1, 2, 3, 4, 5, 6, 7) and (1, 2, 3, 4, 5, 9, 10, 16, 17, 18, 19) comprise two of them. The equivalent systems are obtained by the transformation of lines and loads outside the path being analyzed into constant impedances along the radial system. Since the method analyzes the system in the first fault cycle, this assumption is a reasonable approximation.

The consideration of equivalent impedances has already been proposed in different algorithms [278, 281, 289]. This consideration was based in a systematic computation of parallel and series impedances, representing lines and loads. This type of approach, however, is not well suited for large distribution systems, especially in systems with a high degree of coupling between phases. Instead, a power-flow algorithm could be used. The advantage of using this type of approach is that three-phase power-flow algorithms are well known, have great precision, and are easily applied in general distribution systems.

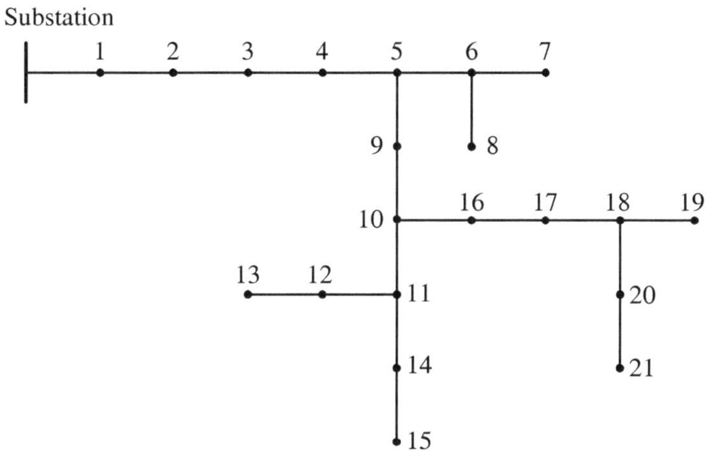

Fig. 8.13 A general distribution feeder

The proposed formulation applies a ladder-technique-based three-phase power flow [156], which is an iterative process developed for radial distribution systems applications. The implemented power flow considers the distribution system non-linearities and asymmetrical phase coupling. Therefore, the proposed fault-location method can be applied to each equivalent system. The procedure is proposed as follows [289]:

1. Run a three-phase power-flow algorithm, such as the one described in [156], considering the pre-fault conditions of the system.
2. Calculate the equivalent impedances $\underline{\underline{Z}}_{e(p-q)}$ in each node by using:

$$\underline{\underline{Z}}_{e(p-q)} = \frac{\underline{U}_p}{\underline{I}_{p-q}} \tag{8.27}$$

3. Determine the PPFP.
4. Select one PPFP and determine the nodes with laterals. As an example, in path (1, 2, 3, 4, 5, 6, 7), only nodes 5 and 6 have laterals.

5. For each node with lateral, determine an equivalent load, considering only the previously calculated equivalent impedances $\underline{Z}_e$ outside the path being analyzed. As an example, node 5 in the path (1, 2, 3, 4, 5, 6, 7) would have an equivalent impedance given by $\underline{Z}_{e(5-9)}$ and, similarly, node 6 by $\underline{Z}_{e(6-8)}$ (Fig. 8.13).
6. For each node with lateral and loads, conduct the parallel between loads and the equivalent load determined in previous step. This is the final equivalent load for the analyzed nodes in the path.
7. Go back and follow step 4 until all of the n equivalent systems have been determined.
8. Execute the fault-location algorithm for each equivalent system. A total of n fault locations will be determined.
9. Determine the correct fault location and path by using a separate fault-section-determination algorithm.

Load Profile

The load demand in electric power distribution systems is dependent on the consumer type (residential, industrial, or commercial) and may vary with many factors [156]. Also, impedance-based fault-location methodologies are dependent on the system loading during the fault period, as one of the required inputs is the load in each node of the system. If the system has a different load profile during the algorithm execution, this will lead to errors. To overcome this limitation, due to the time-varying load characteristics, a procedure to update the load value is proposed.

The direct method of a load compensation can utilize measurements in each load node. In practical distribution systems, however, such measurements are hardly available. Therefore, the proposed extension is based on local measurements, available at the substation terminals. The proposed procedure is described as follows.

1. A standard equivalent impedance is defined as the equivalent impedance seen at the substation terminals, obtained by using the local voltages and currents measured when the system is operating in a previously defined normal operation load profile, and is given by:

$$\underline{Z}_{\text{Load}m} = \frac{\underline{U}_{\text{S}_\text{m}}}{\underline{I}_{\text{S}_\text{m}}} \tag{8.28}$$

where:
$\underline{Z}_{\text{Load}m}$ – equivalent impedance in phase m; m – considered phase a, b or c;
$\underline{U}_{\text{S}_\text{m}}$ – load profile voltage at terminal S, phase m;
$\underline{I}_{\text{S}_\text{m}}$ – load profile current at terminal S, phase m.

2. When acquiring the faulted voltage and current data, the measured equivalent impedance seen from the substation is calculated by using the pre-fault data. The measured impedance represents the system's present operation state and is given by:

$$\underline{Z}_{\mathrm{Load}m}^{\mathrm{pre}} = \frac{\underline{U}_{\mathrm{S_m}}^{\mathrm{pre}}}{\underline{I}_{\mathrm{S_m}}^{\mathrm{pre}}} \tag{8.29}$$

where:

$\underline{Z}_{\mathrm{Load}m}^{\mathrm{pre}}$ – equivalent pre-fault impedance in phase m;

$\underline{U}_{\mathrm{S_m}}^{\mathrm{pre}}$ – pre-fault load profile voltage at terminal S, phase m;

$\underline{I}_{\mathrm{S_m}}^{\mathrm{pre}}$ – pre-fault load profile current at terminal S, phase m.

3. The measured equivalent impedance $\underline{Z}_{\mathrm{Load}m}^{\mathrm{pre}}$ is compared with the standard equivalent impedance $\underline{Z}_{\mathrm{Load}m}$. A load variation rate is obtained by:

$$\underline{\Delta}_{\mathrm{Load}m} = \frac{\underline{Z}_{\mathrm{Load}m}^{\mathrm{pre}} - \underline{Z}_{\mathrm{Load}m}}{\underline{Z}_{\mathrm{Load}m}} \tag{8.30}$$

4. The load matrices used in the power flow and in the proposed fault-location formulation are updated by multiplying them by the obtained load variation and considering an equal percentage change for all loads.
5. A three-phase power flow is executed to obtain the new equivalent systems.
6. The proposed fault-location technique is executed with the updated load information.

The described procedure uniformly considers changes for all loads. However, due to the stochastic nature of the time-varying load, each spot (or distributed) load can have different characteristics. If the load pattern is known for each load group, the load-estimation procedure can use this information to weigh the load changes measured at the substation differently for each load group.

Simulation results were provided in reference [289], to validate the formulation and illustrate its accuracy on fault location, showing the method's robustness and potential for real-time applications. The formulation was implemented as an embedded system and is used as an analysis tool at the considered operation center.

8.4. Review of Non-standard Fault-location Techniques for Distribution Systems

Faults in distribution systems effect power-system reliability, security and quality. Accurate fault location minimizes the time needed to repair damage, restore power and reduce costs. Fault-location techniques can be classified into two fundamental categories: traditional fault-location techniques that are based on fundamental frequency currents and voltages, and methods based on traveling wave and high frequency, which are transient-based algorithms. The application of traditional fault-location techniques (phasor-based algorithms) that uses fundamental voltages and currents at line terminals for distribution lines with tapped loads is difficult. In recent years they have been used in conjunction with the concept of traveling wave and wavelet transform [197–199, 229]. In this section, the reviews of selected non-standard (transient-based algorithms) proposed for distribution systems are presented.

Reference [197] described an improved fault-location technique based on high-frequency transient signals produced during a fault. First, the faulted signals are transformed into the modal domain; second, the modal signals are transformed into wavelet coefficients. Then, the location of the fault is calculated based on the traveling-wave information provided by the wavelet coefficients. For the transmission systems the method can be used for single-ended and synchronized two-ended recording of fault transients. It is independent of the fault impedance and is suitable for mutually coupled tower geometries as well as series-capacitor-compensated lines. For radial distribution systems with multiple laterals where the measurements are available only at the substation, the method allows the accurate determination of the fault location.

Another possible approach for addressing the fault-location problem in distribution systems is the use of fault-generated traveling waves [315]. This method provides accurate results even though it is associated with high implementation costs and poses difficulties when traveling waves are analyzed from one terminal only, due to the lateral derivations and loads that reflect the traveling waves similarly to a fault. Recently, traveling-wave methods have been reborn as an alternative for fault location. This is mostly due to the increasing desire for faster, more accurate fault location, coupled with improvements in data acquisition, GPS time synchronization and communication systems [312].

Reference [229] described the application of wavelet transform as an accurate fault-location technique for distribution lines with tapped loads. When a fault occurs, a transient wave travels from the fault point to substation busbars and load terminals. The time taken for the fault-generated transient wave to arrive at busbars or load terminals depends on the distance traveled and the velocity of the traveling wave. Fault-transient detectors based on the concept of digital wavelet transform (DWT) are installed at substation busbars and load terminals to capture the time when a fault-generated transient wave arrives. The time recorded at the busbars and terminals is synchronized with a global positioning system (GPS) clock. From the time recorded and the topological structure, an accurate fault location is achieved. The fault-location technique is validated by ATP-EMTP [56] simulation on an 11-kV distribution line with seven tapped loads.

Reference [198] described a single-ended fault-location technique for radial distribution feeders based on the information provided by the measurements and the feeder configuration. The proposed method first identifies the fault path, based on the traveling-wave information provided by the high-frequency components of the recorded fault-transient signals. It then calculates the exact fault location along the identified path based on the power-frequency signals. The multiphase transient signals are first decomposed into their modal components. Then, the modal signals are decomposed into their wavelet components and the corresponding wavelet coefficients are obtained. These wavelet coefficients are used to extract the relevant signal features that are subsequently used to identify the branch or path where the fault is located. Finally, the fault distance from the main substation is calculated using the information provided by the power frequency signal.

Reference [312] described an advanced fault indicator based on the fault-generated high-frequency noise signal (FI-HF), which is mounted between the earth wire and tower. Fault-generated high-frequency noise signal is captured by a specially designed 'earth trap' and stack tuner. The principle of FI-HF proposed is that the earth trap circuit rejects or stops the specified narrow band of frequencies where the stack tuner filters them to produce an output signal. When a fault occurs, noise frequencies traveling from the fault strike stack tuner, which is very close to the fault. This stack tuner filters the selected band of frequencies from the wide-band noise frequencies. After striking the closest stack tuner, the noise frequencies then strike the earth trap circuit. All noise frequencies are allowed to pass through the trap circuit except those frequencies that are filtered by the fault-side stack tuner, which has received a considerable level attenuation by the earth trap circuit. As a result, all other stack tuner far away from the fault point can only receive a selected low-level and trapped narrow band of frequencies. Thus, other stack tuners could only filter a low level of those frequencies to be an output signal. As the stack tuner close to the fault point has a high level of filtered signal compared to the low-level output of other stack tuners, the two highest outputs, can clearly indicate that the fault is between towers.

A technique developed and described in [21] concerns determining the fault location on distribution overhead lines and cable systems. It relies on detecting fault-generated high-frequency signals. In the scheme, a high-speed sampling system is used to capture the fault-generated high-frequency transients. To achieve the high accuracy inherent in the system, higher sampling rates are used.

8.4.1 DMS-based Fault Location

Reference [128] presents a fault-location method, that has been developed as part of distribution management system (DMS) development. The basic idea of DMS has been to develop a new information system based on integration of network information system and distribution automation. From the fault-location point of view, the basic idea means that only the existing device and data should be used instead of developing a method that requires more accurate network modeling and special devices in the network level. This basis is of great importance from the practical point of view because the method can be taken into use without major investments.

The principle of distance-based fault location is very simple: finding the similarity of calculated and measured fault current. The database of network information system includes the data needed for fault-current analysis of the distribution network, which was originally carried out as part of off-line engineering analysis. This network modeling and fault-current calculation method are also part of DMS and together with real-time topology information it provides the basis for fault location. The measured fault current can be obtained for microprocessor-based relays, which are quite common nowadays. Even one relay with such a capability is sufficient if it is located in the incoming bay of the substation. As a result, the algorithm finds one or several possible faulted-line sections based on distance.

The DMS provides an excellent environment for further processing, because the distance-based fault-location result is not explicit. The information of possible fault detectors and even the terrain and weather conditions can be taken into account. In the further processing fuzzy logic is applied. An important part of fault location is the user interface of the DMS, which provides a geographic view to the network and fault-location results. In addition to the fault location the DMS provides a whole concept of fault management, including for example restoration and fault reporting. The method has been used in several utilities for many years and the experience is very good. The accuracy is very good and by using the method the mean outage time of customers has been reduced significantly. The method is now part of commercial DMS products and is so far limited to short-circuit faults.

8.4.2 Artificial Intelligent Fault-location Techniques for Distribution Systems

Artificial intelligent (AI)-based techniques have the potential advantage over conventional techniques in significantly improving the accuracy in fault location. There has been a large number of research activities in the universities as well as research institutions to determine alternative methods to using conventional techniques for accurate location of faults both in transmission lines and distribution lines. The AI-based techniques proposed for the solution of fault location for distribution systems are briefly described in Chap. 9.

9 Artificial Intelligence Application

Abstract This chapter reviews some of the commonly used techniques with respect to fault location. In the following sections, we give introductions to basic techniques numbered among artificial intelligent (AI) methods. Next, some application examples to power-system automation and fault-location problems are presented. The attached examples confirm that these methods have been highly successful in addressing the considered tasks.

9.1 Introduction

Microprocessor technology brings unquestionable improvements to substation automation, in general, and particularly, the fault location: criteria signals are estimated in a shorter time; input signals are filtered out more precisely; it is easy to apply sophisticated corrections; the hardware is standardized and may communicate with other control systems; it is easy to use synchronizing measurements and so on. All this, however, did not make a major breakthrough in power-system protection, monitoring and control as far as security, dependability and speed of operation are considered. The key reason behind this is that the principles used for measurement and decision making (used criteria) blindly reproduce the criteria known for decades.

The considered tasks, however, may be supported with new techniques based on pattern-recognition algorithms or a decision-making problem that can help in increasing the accuracy of estimation. This observation directly leads to AI application in fault location [6, 60, 275]. Practically, it includes the artificial neural network approach (pattern recognition), as well as the expert system and fuzzy-logic methods (decision making) [127, 138, 168, 313, 321, 325].

In a broad perspective, intelligent systems underlie what is called *soft computing*. In traditional hard computing, the prime objectives of the computations are precision and certainty. However, in soft computing, the precision and certainty carry a cost. Therefore, it is realistic to consider the integration of computation, reasoning, and decision making as various partners united in order to provide a framework for the tradeoff between precision and uncertainty.

Accuracy limitation of fault-location problems is connected with signal processing and measurement. The traditionally used methods are based on Fourier transformation and other similar techniques. As is seen in Chap. 4, the traditional measurement algorithms are derived under assumption that the observed process is in a steady-state condition. However, such a postulate is far from reality in some important cases, such as for example, in analysis of post-fault transients in isolated (ungrounded) or compensated MV networks. It is shown in Chap. 4 that the promising technique proposed for investigation of such non-stationary signals is connected with wavelet transform (WT). In some of the proposed methods this technique does not, however, deliver direct information on distance to fault. The final result should be deduced from indirect particular quantities. The outcome may be produced with some form of soft processing with, for example, artificial neural network (ANN) application [31, 243].

The major reason for searching for a new solutions is that an accurate location of a fault can reduce the time required for restoration of power supply to customers. A very high degree of accuracy is thus required that cannot be obtained using conventional techniques. This is due to the wide variations in both system and fault conditions that occur in practice and these in turn can have a significant influence on the degree of accuracy achievable with conventional fault-location techniques. Moreover, some specific object configuration, such as: multi-ended lines [1, 80, 218, 310, 319], transmission system combined with overhead lines and cables [139], branching networks [229, 263], participation of non-linearities (MOVs) [10, 32, 210, 227, 241, 274, 285], resonant character of distribution systems with isolated or compensated neutral grounding [59, 96, 128, 179, 342, 345] and others, limiting application of conventional methods.

AI-based techniques have the potential advantage over conventional techniques in significantly improving the accuracy in fault location. There have been a large number of research activities in the universities as well as research institutions to determine alternative methods to using conventional techniques for accurate location of faults both in transmission and distribution lines. This chapter gives a general introduction to the techniques and shows some examples of the proposed solution.

9.2 Artificial Intelligence Methods

AI is a subfield of computer science that investigates how the thought and action of human beings can be mimicked by machines [351]. Both the numeric, non-numeric and symbolic computations are included in the area of AI. The mimicking of intelligence includes not only the ability to make rational decisions, but also to deal with missing data, adapt to existing situations and improve itself in the long time horizon based on the accumulated experience.

Three major families of AI techniques are considered to be applied in modern power-system automation and control [60, 153, 275, 325]:

- expert system techniques (XPSs);
- artificial neural networks (ANNs); and
- fuzzy-logic systems (FLog).

The XPS, ANN and FLog approaches have their own advantages and limitations. Table 9.1 compares the basic features of these AI methods.

Table 9.1 Comparison of AI methods in power system automation

Feature	**Approach**		
	XPS	**ANNs**	**FLog**
Knowledge used	Expert knowledge in the form of rules, objects, frames, etc.	Information extracted from the training set of cases.	Expert knowledge in the form of protection criteria.
Trouble shooting and improving a relay	Change of rules required.	Difficult – the internal signals are almost impossible to interpret.	Convenient - the internal signals are understandable and analyzable
Self-learning	Possible.	Natural.	Possible.
Handling unclear cases	Possible.	Natural.	Natural.
Robustness	Not critical and easy to ensure.	Difficult to ensure.	Not critical and easy to ensure.
Setting a relay	Convenient.	Large number of simulation required.	Convenient. Both knowledge and simulation are used.
Computations	Extensive.	Dedicated hardware.	Moderate.

In the classical considerations also the genetic algorithm (GA) is included in the family of the AI methods. GAs and similar so-called evolutionary computation techniques are suitable for optimization problems with applications including, but not limited to, design optimization, optimal parameter learning of ANNs or fuzzy basis function networks, or optimal design of control parameters. However, such methods are more frequently used for fault location, specially in MV networks for selection of the best (more reliable) solution from several possible results.

A short introduction to the above-mentioned techniques is presented in the following sections. Some typical applications are presented as well.

9.2.1 Expert Systems

An expert system (XPS) is usually defined as an interactive system that is able to manifest an expert level of knowledge in a specific domain and solve problems in this domain for judging or the formation of an opinion. The simple expert systems includes a few heuristic rules based on the expert's experience. The knowledge takes the form of so-called production rules written using the 'If... Then…' syntax (knowledge base). The system also includes the facts that generally describe the domain and the state of the problem to be solved (data base). A generic inference engine uses the facts and the rules to deduce new facts that allow the firing of other rules. The knowledge base is a collection of domain-specific knowledge and the inference system is the logic component for processing the knowledge base to solve the problem. This process continues until the base of facts is saturated and a conclusion has been reached (Fig. 9.1). To guide the reasoning and to be more efficient, these systems may incorporate some strategies known as meta-knowledge. Rule-based systems still represent the majority of the existing expert systems.

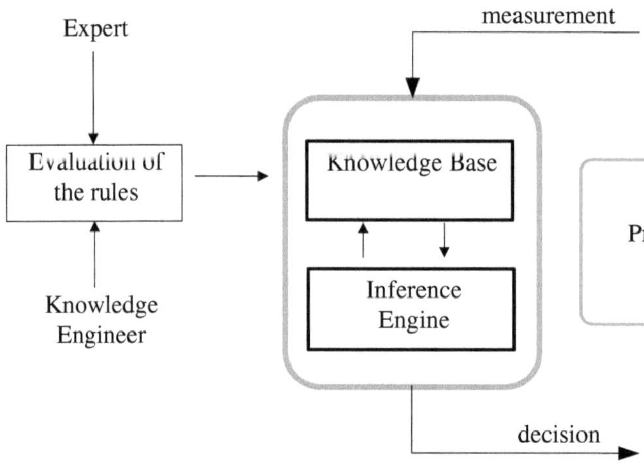

Fig. 9.1 Simplified block diagram of an expert system

There are few applications of XPS to power-system automation and control reported, but all of them solve the off-line tasks such as settings coordination, post-fault analysis or fault diagnosis [48] and classification of faults or switching events [168], and fault-section identification [311]. As yet there is no application reported of the XPS technique employed as a decision-making tool in an on-line operating protective relay. The basic reason for this is that there is no extensive rule base that describes the reasoning process applicable to protective relaying [43, 168].

9.2.2 Fuzzy Logic

Fuzzy logic (FLog), unlike the traditional two-values crisp logic, determines relations between objects and events rather by soft (fairly big – relatively small, sweet enough – almost unsweetened) than unambiguous (big – small, sweet – bitter) qualifications. So these relations are defined by description of a grade of similarity, membership or on the basis of approximation [337]. FLog provides a simple way to approach a final conclusion based upon ambiguous, vague, imprecise, noisy, or missing input information.

FLog is based on the theory of fuzzy sets that was developed for a domain in which definitions of activities and observations are 'fuzzy', in the sense that there no well-described, sharp boundaries of the set of activities or observations to which the description apply [351]. The membership in the fuzzy-sets theory, generally, is not determined as 'true – false' or numerically: '1 – 0', but rather by description of the grade of membership. In that sense the crisp logic is a special case of more general fuzzy logic. Contrary to traditional set theory where a crisp set divides the universe of discourse into two groups: members and non-members, fuzzy sets allows us to characterize various forms of moderate transition from total membership to total non-membership. This allows for step-by-step transitions from one range of operation to another.

Elements considered in fuzzy sets are linguistic objects or words, rather than numbers. The input domain is a noun, e.g., 'voltage', 'temperature', 'pressure', 'satisfaction', etc. Since the error is just the difference, it can be thought of in the same way. The fuzzy variables themselves are adjectives that modify the variable (e.g. 'extremely low' error, 'small positive' error, 'large negative' error). As a minimum, one could simply have 'good', 'neutral', and 'wrong' variables for each of the parameters. Additional rates such as 'very good' and 'very wrong' could also be added to extend the responsiveness to exceptional or very non-linear conditions, but aren't necessary in a basic system. We can see that unlike a traditional approach, where to obtain an accurate result we try to accurately measure, to accurately represent and to accurately calculate adequate variables, in the fuzzy approach we suppose some distance to the observed process and measured variables. It is interesting, and even surprising, that such an assumption frequently may lead to the goal through easier and more accurate way [183].

Returning to the fault location we may conclude that, from that point of view, there are two approaches to synthesis of adequate algorithms:

- designing using traditional crisp logic on the basis of a mathematical model of the observed process following faults: the 'objective' measurements are used in that model for determination of searching parameters; and
- designing with application of linguistic variables and with use of fuzzy sets and decision-making relations.

The other possible way is to use an intermediate solution where we use fuzzy reasoning inside the algorithm that is designed according to the known mathematical model [212, 231].

There are many circumstantial pieces of evidence supporting a fuzzy-logic-based approach for realization of different tasks in power-system automation and control, e.g.:

- fault-induced transients in electrical networks have many unknown parameters and even correct measurements could not reproduce all the characteristics in-

volved in the process; further development of the mathematical-model-based methods could not sufficiently improve the object-state identification;
- measurements during transients are inaccurate and the amount of information on the observed process is limited; therefore, application of stiff algorithms is not justified; and
- the final decision may be supported by use of many, even incomplete, criteria; fuzzy decision making is a natural and effective tool for resolving of a multi-criteria approach.

Some basic information on fuzzy logic and its application to power-system automation and control are provided below.

The general structure of the fuzzy-logic processing applied in automation and control is presented in Fig. 9.2 [275]. With respect to the power-system protection and control the adequate blocks realize the following functions [153]:

- The criteria signals are fuzzyfied in order to account for dynamic errors of the measuring. Thus, instead of real numbers, the signals are represented by fuzzy numbers.
- The thresholds for the criteria signals are also represented by fuzzy numbers to account for the lack of precision in dividing the space of the criteria signals between the tripping and blocking regions.
- The fuzzy signals are compared with the fuzzy settings. The comparison result is a fuzzy-logic variable.
- Several relaying criteria are used in parallel. The criteria are aggregated by means of formal multi-criteria decision-making algorithms.
- The tripping decision depends on multi-criteria evaluation of the status of a supervised object. In the case of a decision-making system the output signal usually should be crisp-valued: 'yes/no' (1/0) while the fuzzy reasoning delivers a fuzzy result. The block of 'defuzzyfication' is intended to translate a fuzzy decision to an adequate crisp number.

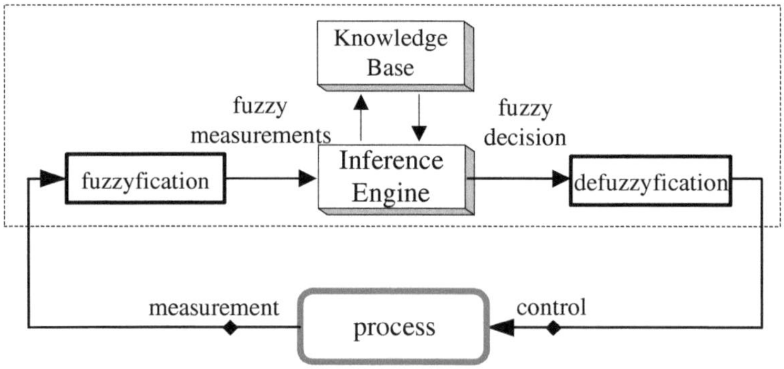

Fig. 9.2 Simplified block diagram of the fuzzy-logic approach

We can see that the structure from Fig. 9.2. is similar to the expert system (Fig. 9.1) and is often defined as the fuzzy expert system, where the knowledge base is fuzzy or imprecise in nature. Compared to the expert system, the fuzzy expert system usually has fewer rules.

Fuzzy Set

The imprecise linguistic values in a fuzzy set are formally described by using a membership function. A membership function $\mu_F(x)$ of set F is a function that assigns value, or membership degree, to every $x \in F$, μ: $X \rightarrow [0, 1]$. The set F is called a fuzzy set.

In that definition the range of possible quantitative values considered for fuzzy-set members is called the universe of discourse. A universe of discourse can be continuous or discrete. Fuzzy set F in a universe of discourse (domain) X is determined through the membership function $\mu_F(x)$, which takes the value from [0, 1]. The membership function is a measure of similarity of the element x to the fuzzy set, which is described as:

$$F = \{(x, \mu_F(x)); x \in X\} \tag{9.1}$$

Domain X defines a range of considered changing of an argument x for which the membership function $\mu_F(x)$ is defined. The membership function determines the fuzzyness of the considered set (number) and may be defined by a different way, e.g., by measurement of some statistics, the result of polling, expert judgments or others. For example, the office manager asks the employs 'what temperature in the office is too hot?' Collecting the answers (after normalizing and approximation) he obtains the function $\mu_F(x)$ as in Fig. 9.3, where x means temperature and F is related to the formulated question (fuzzy set). From that figure we can read that for temperature $x=\theta_p$ in the office is hot in the grade of 0.2. If the intention of the poll was selection of 'a better temperature' in the office, then the manager can define one crisp value K: 'it is too hot in the office if temperature $x \geq \theta_a$'. The membership function for K is indicated by the dotted line in Fig. 9.3. So the problem was reduced to two-valued logic:

$$\mu_K(x) = \begin{cases} 0 & x < \theta_a \\ 1 & x \geq \theta_a \end{cases}$$

The manager can also define another value L: 'the best temperature in the office is between θ_r and θ_s'. The membership function for L is indicated by another dotted line. It is equal to zero in all the 'universe of discourse' instead of the period: $\theta_r \leq x \leq \theta_s$, where it takes the value of 1.

In the above example we can also trace the fuzzyfication stage as in Fig. 9.2, which here is based on a polling procedure. Moreover, the final decision reproduces the defuzzyfication process: the fuzzy value is replaced by its crisp representation.

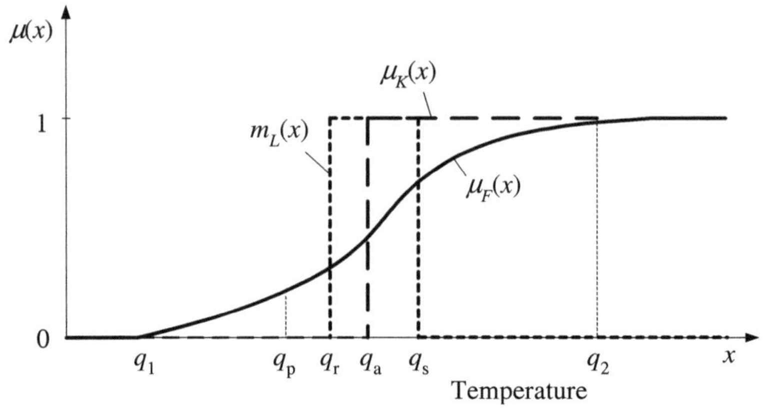

Fig. 9.3 Example of membership function of fuzzy set ($\mu_F(x)$) and a crisp ($\mu_F(x)$) set

The basic operations on fuzzy sets are defined similarly as for classical Boolean sets. If fuzzy sets A and B are defined by adequate membership functions $\mu_A(x)$ and $\mu_B(x)$ then these operations are defined as follows:

- result of fuzzy union (summation) $A \cup B$ gives the membership function of the form:

$$\mu_{A \cup B}(x) = \mu_A(x) \vee \mu_B(x) = \max\left(\mu_A(x), \mu_B(x)\right) \tag{9.2}$$

- fuzzy intersection $A \cap B$ gives the membership function:

$$\mu_{A \cap B}(x) = \mu_A(x) \wedge \mu_B(x) = \min\left(\mu_A(x), \mu_B(x)\right) \tag{9.3}$$

- fuzzy complement of set A: $\overline{A}$ is connected with the following changing of the membership function:

$$\mu_{\overline{A}}(x) = 1 - \mu_A(x) \tag{9.4}$$

It should be pointed out that the above relations may also be defined in a different way [212, 231]. Graphical illustrations of the above relations are presented in Fig. 9.4.

We can see that operations of fuzzy union and intersection give the same results as in the case of Boolean sets, however, the resulting sets are also fuzzy. The complement of a fuzzy set has a membership function that is different from zero when the original function is different from zero. That is a different situation from the result obtained for a crisp logic:

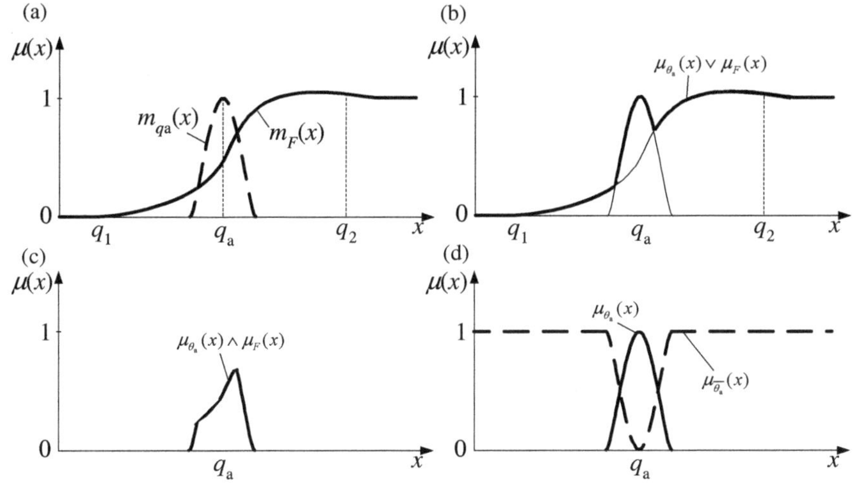

Fig. 9.4 Membership functions of fuzzy sets: (a) original, (b) result of fuzzy union, (c) result of fuzzy intersection, and (d) complement of the set θ_a

$$A \cap \overline{A} \neq 0$$
$$A \cup \overline{A} \neq V \tag{9.5}$$

where V is the universe of discourse for the set A.

Fuzzyfication

The fuzzyfication operation can be accomplished by considering the crispy input values as 'singletons' (fuzzy sets that have membership functions reduced to two values: 1 and 0 ('yes' and 'no') – as in Boolean logic) and taking the values of the set's membership function at the corresponding data value. Membership functions can attain different forms. Reasonable functions are often piecewise-linear functions, such as triangular or trapezoidal functions. A fuzzy number with a membership function of triangular form is defined as follows:

$$A = (m_A, \alpha_A, \beta_A)_{\mathrm{LR}} \tag{9.6}$$

where: m_A is medium value of fuzzy number A, for which $\mu_A(m) = 1$; α_A, β_A are adequately, left and right boundaries.

The membership function of a fuzzy number of type $\mathrm{L-R}$ is defined as:

$$\mu_A(x) = \begin{cases} \mathrm{L}\left(\dfrac{m_A - x}{\alpha_A}\right) & \text{for } x \leq m_A \\ \mathrm{R}\left(\dfrac{x - m_A}{\beta_A}\right) & \text{for } x \geq m_A \end{cases} \tag{9.7}$$

Functions L and R may be defined in different way, e.g.:

$$\mathrm{L}(x) = \mathrm{R}(x) = \max\left(0.1 - |x|^p\right)$$

$$\mathrm{L}(x) = \mathrm{R}(x) = \mathrm{e}^{|x|^p}$$

$$\mathrm{L}(x) = \mathrm{R}(x) = \frac{1}{1+|x|^p} \qquad p = 1,2,3,\ldots \tag{9.8}$$

$$\mathrm{L}(x) = \mathrm{R}(x) = \left(\frac{1}{1+|x|}\right)^p$$

A trapezoidal form of membership function is defined similarly but it has an adequate flat segment between left and right boundaries equal to one (Fig. 9.5).

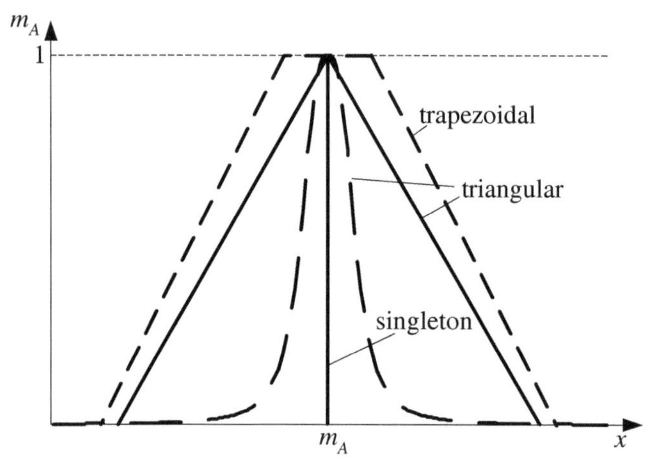

Fig. 9.5 Examples of membership functions

Fuzzy set A having the center as its only element is a fuzzy singleton. That is simply a crisp value of m_A.

In a fuzzy-logic system as in Fig. 9.2 crisp values are transformed into fuzzy sets that represent linguistic values in the process known as fuzzyfication. During this process a physical value of x is converted into fuzzy set $A(x)$ by an associating membership degree: $x \rightarrow \mu_A(x)$. The membership degree $\mu_A(x)$ represents the fuzzy equivalent of the value of x.

If the fuzzy-logic system is intended to process dynamic signals then the fuzzyfication stage may reproduce the credibility of dynamic measurements. This is illustrated if Fig. 9.6, which shows the dependence of magnitude estimate of a sinusoidal signal on actual phase and assumed noise (Fig 9.6a) and results of fuzzyfication. In steady states: the initial value X_1 and final value X_4 are measured with good accuracy, so their fuzzy representations have 'narrow' membership functions (almost crisp values) – Fig. 9.6b. In the transient period the results of measurements are inaccurate and results are represented by fuzzy numbers with wide membership functions (results are 'more fuzzy'): X_2 and X_3 in Fig. 9.6b.

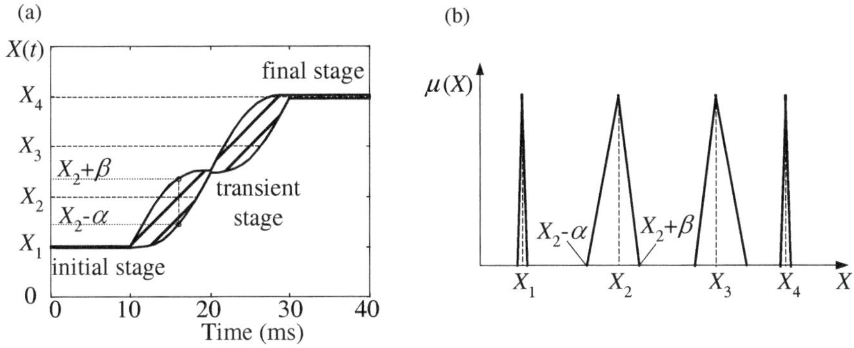

Fig. 9.6 Changes of the observed signal (a) and fuzzyfied measurements (b)

In this way, in the stage of fuzzyfication, we can account for different features of the observed process and provided measurements [147, 249].

Fuzzy Rules

The central part of a fuzzy-logic system is an inference engine (Fig. 9.2) the function of which may be defined as a 'fuzzy expert system'. It consists of a database (set of definitions and rules for data manipulation) and a rule base (establishing association and relations between propositions).

Fuzzy numbers may be adequately transformed (processed) to stress their meaning. This may be done by applying modifiers or hedges. A modifier or linguistic hedge is an operation that modifies the meaning of a term (fuzzy set, fuzzy number). For example, if fuzzy term $A(v)$ means: 'high voltage', then the variable $B(v)$: 'higher voltage' may be obtained by a 'concentration' function. This is equivalent to the following operation on a membership function $\mu_{A(v)}$: $\mu_{B(v)}=(\mu_{A(v)})^2$. Similarly, variable $C(v)$: 'less voltage' may be obtained by use of 'dilation': $\mu_{C(v)}=(\mu_{A(v)})^{1/2}$. Such operations allow formulation of complex relations between fuzzy numbers [231, 337].

A fuzzy decision is determined through consecutive realization of fuzzy rules of a form: IF A THEN B. It consists of an antecedent part (IF ..) and a consequent part (THEN ..), with antecedent parts describing causes and consequent parts describing consequences relevant for the action [337]. The above rules may have more complex form, e.g.:

$$\text{R: IF } (x_1 \text{ is } A_1) \wedge (x_2 \text{ is } A_2) \text{ THEN } (y \text{ is } B_1) \tag{9.9}$$

where membership functions $\mu_{A_1}(x_1)$, $\mu_{A_2}(x_2)$ and $\mu_{B_1}(y)$ may have different shapes so we can see that fuzzy rules relate fuzzy sets.

Structure and organizational principles of fuzzy systems may vary depending on application. Generally, we can distinguish fuzzy-control systems and decision-making systems.

Multi-criteria decision making consists in application of a number of criteria that may meet the defined condition in different degree. In that case we should define a method for processing of all particular decisions for working out the common 'better suited' decision. We can apply here the weighting aggregation method [337]. It may be described as follows.

- Each *k*th of *p* (k=1, 2, .., p) particular criterion is represented by fuzzy set Y_k that assesses the degree of fulfillment of a wanted decision, say: 'yes'. The opposite decision ('no') is characterized by fuzzy set N_k. Both fuzzy sets are described by adequate membership functions: μ_{Yk} and μ_{Nk}. Because there are only two decisions: 'yes' or 'no', it is evident that: $\mu_{Yk} = 1-\mu_{Nk}$ (more complex decision plane can be divided into an adequate number of such decision-pairs).
- Therefore, the *k*th criterion can be represented by the following fuzzy set:

$$C_k = \{\mu_{Yk} / Y, \mu_{Nk} / N\} \tag{9.10}$$

which means: the *k*th criterion is fulfilled in μ_{Yk} degree when 'yes' is concerned and in μ_{Nk} degree when 'no' is concerned.

- The weighted multi-criteria aggregation method consists in determination of the support degree represented by fuzzy number δ_Y for the output decision 'yes' as a weighted mean of particular decisions:

$$\delta_Y = w_1\mu_{Y1} + w_2\mu_{Y2} + \ldots + w_p\mu_{Yp} \tag{9.11}$$

where the weights are non-negative and:

$$\sum_{i=1}^{p} w_i = 1 \tag{9.12}$$

and similarly for support δ_N concerning the decision 'no'. If both decisions are estimated with the same weights then: $\delta_N = 1 - \delta_Y$.

Weight w_k specifies an importance of the *k*th criterion in a whole decision plane and may be determined from experience with application of experts or automatically as a result of adequate procedure (self-adjusting procedure) [337, 351]. Relation (9.11) represents an efficient solution that is performed not just by a single decision maker, but by the whole set of participants in some kind of negotiation process.

Criteria aggregated in (9.11) should have similar features in the observed process, i.e. should represent similar physical characteristics. In multi-criteria systems similar criteria can be grouped yielding branching decision structure [337].

Continuous (fuzzy) logic applied to power-system automation enables pass by of the information related to symptoms of an analyzed disturbance through a decision-making routine: the higher is the value of truth for a final signal (decision), the more evident are symptoms of the identified event. In the considered fault-location algorithms such methods are mainly used for improvement of fault detection, selection of fault type and other.

Defuzzyfication

Results of fuzzy processing are usually used in the real world as crisp values. They are obtained from fuzzy numbers in the process called defuzzyfication, which is the reverse of fuzzyfication. In other words, defuzzyfication is the process of producing a quantifiable result in fuzzy logic. Typically, a fuzzy system will have a number of rules that transform a number of variables into a 'fuzzy' result, that is, the result is described in terms of membership in fuzzy sets. For example, rules designed to decide how much temperature to apply might result in 'decrease temperature (18%), maintain temperature (36%), increase temperature (75%)'. Defuzzyfication would transform this result into a single number indicating the change in temperature.

The simplest but least useful defuzzyfication method is to choose the set with the highest membership, in this case, 'increase temperature' since it has a 75% membership, and ignore the others, and convert this 75% to some number. The problem with this approach is that it loses information. The rules that called for decreasing or maintaining temperature might as well have not been there in this case.

Translation from a fuzzy to a crisp domain is not unambiguous and in practice different methods are used. Due to the fact that fuzzy sets can be either uniform or not, symmetrical or not, bounded or not, a convenient measure of their geometrical shape is the centroid, the point within a fuzzy set designating some kind of geometrical center, e.g., its center of gravity (COG). For example, the centroid of a triangular fuzzy set is the concurrence point of its three medians. Some examples are given in Fig. 9.7.

In the first case, the output argument of considered fuzzy set takes value x^* for which membership $\mu_A(x^*)$ function has maximum:

$$\mu_A(x^*) = \max_{x \in X} \mu_A(x) \tag{9.13}$$

In that method the shape of the membership function does not influence the result. This may lead to a surprising result (Fig. 9.7a). Modification of this method consists in calculation of the final result as an average from all maximal values (Fig. 9.7b).

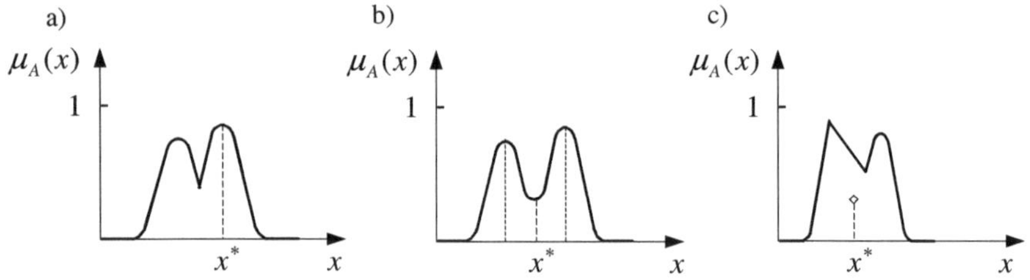

Fig. 9.7 Examples of methods applied for defuzzyfication: (a) choosing maximum value, (b) medium of maximum, and (c) center of gravity

Defuzzyfication according to COG includes calculation of the output value x^*, for which the membership function $\mu_A(x^*)$ of the considered fuzzy set *A* has center of gravity. For discrete representation of that function COG is calculated from [231]:

$$x^* = \frac{\sum_{i=1}^{M} x_i \mu_A(x_i)}{\sum_{i=1}^{M} \mu_A(x_i)} \tag{9.14}$$

where M – number of discrete representations of x.

The above approach can also be used in a fuzzy estimation procedure, where for determination of the output value the rule base method (Sugeno's method) is applied [337]. Relational estimation is based on a set of relations as in (9.9) where consequent parts (THEN ..) have the form of mathematical (not fuzzy) linear functions. The final output value is determined as in (9.14).

If the considered criteria have the sense of supporting degree for one of two choices (as in (9.10)) then the output Boolean value can be determined by comparison of a continuous logic function δ_N (for 'no' or δ_Y for 'yes' – see (9.11)) with an adequate threshold Δ. The output crisp result $\delta^\#$ is determined as:

$$\delta^\# = \begin{cases} 1 & \text{if } \delta_N > \Delta \\ 0 & \text{otherwise} \end{cases} \tag{9.15}$$

In one-criterion decision making the output signal can be determined as a result of fuzzy intersection of both membership functions. The membership function of resulting fuzzy set R(x) takes the form:

$$\mu_R(x) = \min\{\mu_Y(x), \mu_T(x)\} \tag{9.16}$$

where: $\mu_Y(x)$ – criterion membership function; $\mu_T(x)$ – threshold membership function (generally, it can be represented by fuzzy set) – Fig. 9.9.

The output crisp value can be determined by application of the maximum rule (Fig. 9.8a) or from the relation between fields of criterion membership function $\mu_Y(x)$ and the result of (9.16):

$$\delta_Y = \frac{\int\limits_{X_R} \mu_R(x)\mathrm{d}x}{\int\limits_{X_Y} \mu_Y(x)\mathrm{d}x} = \frac{F^*}{F} \tag{9.17}$$

where: F^*, F – fields determined by adequate membership functions (Fig. 9.8b).

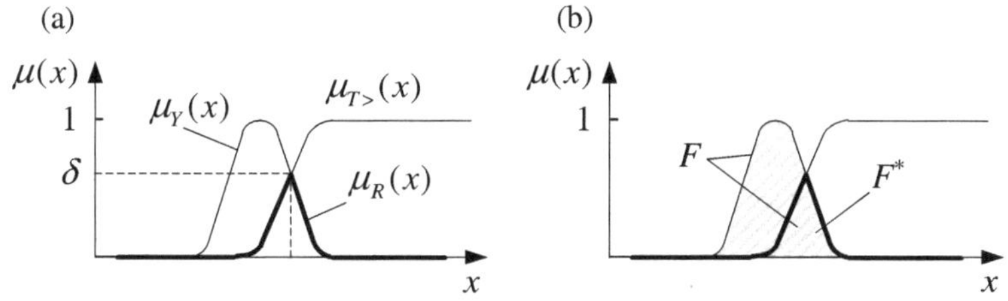

Fig. 9.8 Defuzzyfication according to: (a) maximum role, and (b) relation between fields

The final result $\delta^{\#}$ is determined by comparison of value δ_Y with a threshold Δ, as in (9.15).

There are many application examples of fuzzy-logic solution in power-system relaying and fault location. The main intention is to improve logic decisions with the use of several criteria. Fuzzy set input and output quantitative degrees of membership were based on human expert decisions and the relative cost of an incorrect decision. In the multi-criteria approach, final decisions were based on several criteria with adaptable weighting factors [147, 153]. Some examples related to fault location are presented in Sect. 9.3.1.

9.2.3 Artificial Neural Networks

The ANNs are very different from expert systems since they do not need a knowledge base to work. Instead, they have to be trained with numerous actual cases. An ANN is a set of elementary neurons that are connected together in different architectures organized in layers that are biologically inspired [36, 49]. Figure 9.9 presents the structure of a feedforward ANN known as a perceptron [94]. Each lth layer is composed of K_l numbers of neurons the inputs of which are connected to the previous layer neurons. Excitation signals are delivered to the input layer. An elementary neuron can be seen as a processor that makes a simple non-linear operation of its inputs producing its single output (Fig. 9.10). A weight (synapse) is attached to each neuron and the training enables adjusting of different weights according to the training set. The ANN are not programmed or supported by a knowledge base as are expert systems. Instead, they learn a response based on given inputs and a required output by adjusting the node weights and biases accordingly. Therefore, we can say that the ANN processing information on the base of previously 'learned' examples. From that point of view we may say that the ANNs are the most typical representatives of the 'artificial intelligent'.

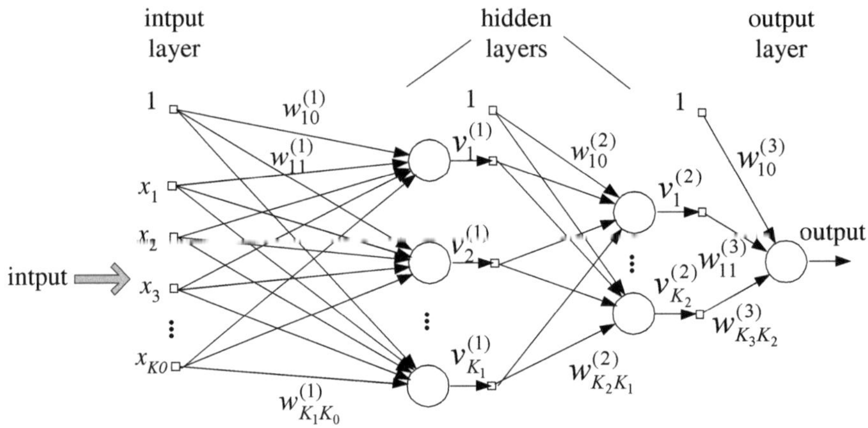

Fig. 9.9 Typical three-layer architecture of a feedforward ANN

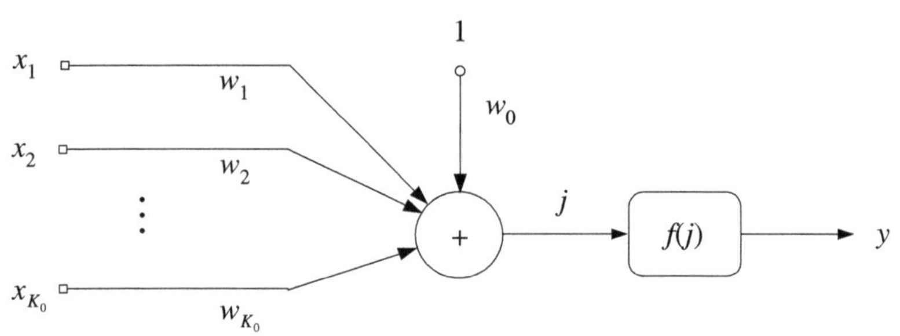

Fig. 9.10 Model of a neuron

Since ANNs can provide excellent pattern recognition, they are proposed by many researchers to perform different tasks in power-system relaying for signal processing and decision making [36, 94, 158, 183]. The common application of the ANN technique assumes:

- The ANN is fed either with non-processed samples of the input signals, or by features of those signals extracted using certain measuring algorithms.
- The sliding data widow consisting of the recent and a few historical samples of the signals, is fed to the ANN.

- The output from the ANN encodes the output decision such as tripping command, type of fault, etc.
- The training patterns exposed to the ANN cover the most important operating conditions of both internal faults and other disturbances.
- Additional pre- and post-processing may be applied.

A major problem with ANNs is that no exact guide exists for the choice of the number of hidden layers and neurons per hidden layer. On the other hand, the ability to generalize is one of the main advantages of using ANNs.

The specific feature of ANN is the extent to which it is dedicated to parallel computing. Moreover, it is designed to generalize the observed inputs and will produce a correct input–output mapping even when the input was not presented to the network in the learning process.

A very important step in ANN development was synthesizing the algorithm for adaptive learning of the network. The basic algorithm is known as the back-error-propagation method in which in the consecutive steps the neuron weights are corrected to obtain minimum global error between the actual and expected reaction of the network. Generally, this method is known as supervised learning where 'a teacher' provides a desired (target) response [94].

The neuron model (Fig. 9.10) has a central processing unit that determines an output as a function of K_0 inputs:

$$y = f(\varphi) = f\left(\sum_{i=0}^{K_0} w_i x_i\right) \tag{9.18}$$

where: $w_0 x_0$ – threshold value (polarization), $f(\varphi)$ – neuron-activation function, φ – summation output signal, y – neuron output.

Summing function in (9.18) can be represented as vector multiplication:

$$\varphi = \mathbf{w}^T \mathbf{x} \tag{9.19}$$

where: $\mathbf{w} = \begin{bmatrix} w_0 & w_1 & \ldots & w_{K_0} \end{bmatrix}$, $\mathbf{x} = \begin{bmatrix} x_0 & x_1 & \ldots & x_{K_0} \end{bmatrix}^T$.

The activation function $f(\varphi)$ defines the neuron state transition. It can take different particular forms, e.g.:

- step function (Fig. 9.11a):

$$f(\varphi) = \begin{cases} 1 & \text{if } \varphi \geq 0 \\ 0 & \text{if } \varphi < 0 \end{cases} \tag{9.20}$$

- piecewise-linear function (Fig. 9.11b):

$$f(\varphi)=\begin{cases}1 & \text{if } \varphi>1\\ -1 & \text{if } \varphi<-1\\ \varphi & \text{if } |\varphi|<1\end{cases} \tag{9.21}$$

- sigmoid unipolar function (Fig. 9.11c):

$$f(\varphi)=\frac{1}{1+\mathrm{e}^{-\beta\varphi}} \tag{9.22}$$

- sigmoid bipolar function (Fig. 9.11d):

$$f(\varphi)=\tanh(\beta\varphi)=\frac{1-\mathrm{e}^{-2\beta\varphi}}{1+\mathrm{e}^{-2\beta\varphi}} \tag{9.23}$$

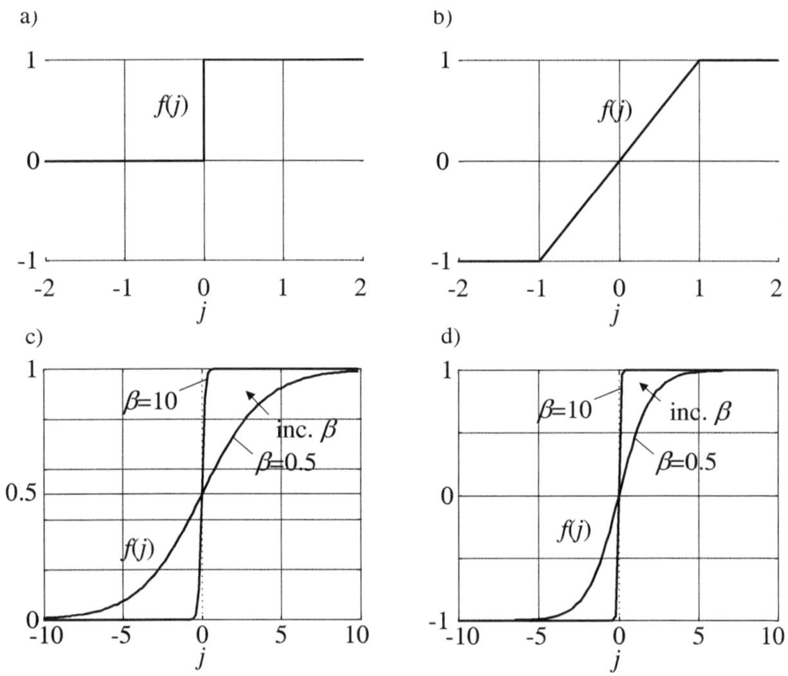

Fig. 9.11 Activation functions: (a) step function, (b) piecewise-linear, (c) unipolar, and (d) bipolar

Programming (learning) of the neuron with a given activation function consists in adequate selection of weighting coefficients w_i, such that operation (9.18) gives a desirable transformation of input vector **x** into output signal y. The described neuron model can be used to form more complex artificial neural network structures. In general, neural networks can be classified as feedforward and feedback types depending on the interconnections of the neurons.

Feedforward Networks

Figure 9.12 shows the structure of a feedforward multi-layer network with K_0 input and K_M output signals. Processing is realized in two final layers, so it is a two-layer network. The intermediate layer is called hidden because their neurons have no connections with outside information. For simplicity, the polarizing signals frequently are not indicated in such schemes.

Signal processing in the ith network layer is performed according to the following relation:

$$\mathbf{v}^{(i)} = f^{(i)}\left(\mathbf{W}^{(i)}\mathbf{q}^{(i-1)}\right) \tag{9.24}$$

where:

$\mathbf{v}^{(i)} = \left[v_1^{(i)} \quad v_2^{(i)} \quad \ldots \quad v_{K_i}^{(i)}\right]^T$ – signal vector at the output of ith layer,

$$\mathbf{W}^{(i)} = \begin{bmatrix} w_{10}^{(i)} & w_{11}^{(i)} & \cdots & w_{1K_{i-1}}^{(i)} \\ w_{20}^{(i)} & w_{21}^{(i)} & \cdots & w_{2K_{i-1}}^{(i)} \\ \vdots & \vdots & \vdots & \vdots \\ w_{K_i0}^{(i)} & w_{K_i1}^{(i)} & \cdots & w_{K_iK_{i-1}}^{(i)} \end{bmatrix}$$ – weighting matrix between (i–1)th and ith layer,

$$\mathbf{q}^{(i-1)} = \begin{cases} \mathbf{x} & \text{for } \; i = 1 \\ \begin{bmatrix} 1 \\ \mathbf{v}^{(i-1)} \end{bmatrix} & \text{for } \; i = 2,3,..,M \end{cases}$$

x vector of input signals, $f^{(i)}(.)$ – activation function of neurons at the ith layer, M – number of processing layers.

It is assumed that all neurons of the ith layer are the same. There may be more than one hidden layer. The result of the network processing is represented by the output vector:

$$\mathbf{y} = \mathbf{v}^{(M)} = \left[y_1 \quad y_2 \quad \ldots \quad y_{K_M}\right]^T \tag{9.25}$$

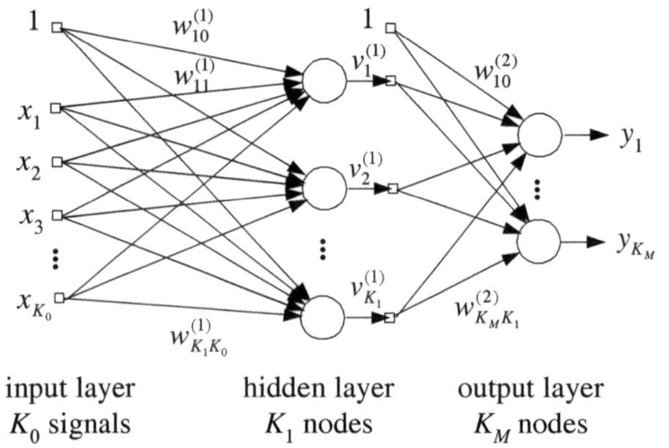

Fig. 9.12 Structure of two-layer feedforward network

At present, the majority of problems use a feedforward architecture. Such networks are widely used in prediction, filtering, information selection and decision making. Its popularity is connected with the relatively simple structure and well-defined learning algorithm [94].

Recurrent Networks

Modification of a multi-layer perceptron that leads to feedback between the output and input of the network gives a recurrent network. Such networks may have a different structure depending on the depth and type of the connections between output (from the network output or any hidden layer) and input. An example is presented in Fig. 9.13. That is a so-called real-time recurrent network. The processing layer is composed of the output and hidden neurons. Feedback is created by connection of neurons from the processing layer with input layer through the delaying lines. In the input layer there are input signals and feedback signals.

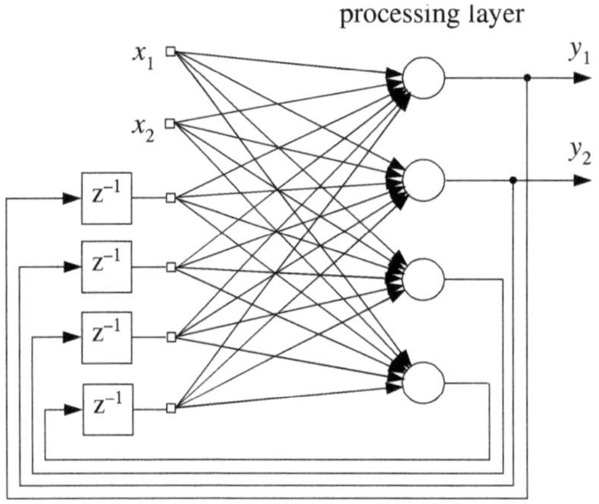

Fig. 9.13 Structure of recurrent ANN

Because of feedback such a network has his own dynamics: steady state is achieved after multiple input–output interactions. Changing of any neuron influences all the network. Therefore, such a structure is suitable for simulation of dynamic processes. Unfortunately, the negative side of the introduced feedback is the possible appearance of unstable regions in the network operation.

Strategies of Learning

The basic problem in designing of ANN lies in determination of the weights in order to achieve the desired objective. This process is usually called learning or training. Generally, there are two strategies of ANN learning: supervised or unsupervised learning.

Supervised learning consists in modification of the network weights with the intention of minimization of the difference (error) between the presented input examples and the target output values. This strategy may be considered as some form of 'teaching' in which the teacher has knowledge of the desirable input–output relations. The input signals are considered as a training vector and the desired response represents the optimum action to be performed by the network [36, 94]. The error signal is defined as the difference between the actual and desired network response.

In unsupervised learning there is no external teacher and there are no defined relations between inputs and output. The learning is performed on the basis of a set of examples where only input conditions are known. In the learning process these examples are selected with respect to some similarity principle. Such a network may be considered as a self-organizing system that is learning on a competitive principle: the output neurons compete among themselves to be activated with the result that only one neuron is on at any time (a winner-takes-all competition rule) [94]. The learning strategy depends on an ANN structure. In the case of feed-forward multi-layer networks supervised learning is used. Its structure is presented in Fig. 9.14.

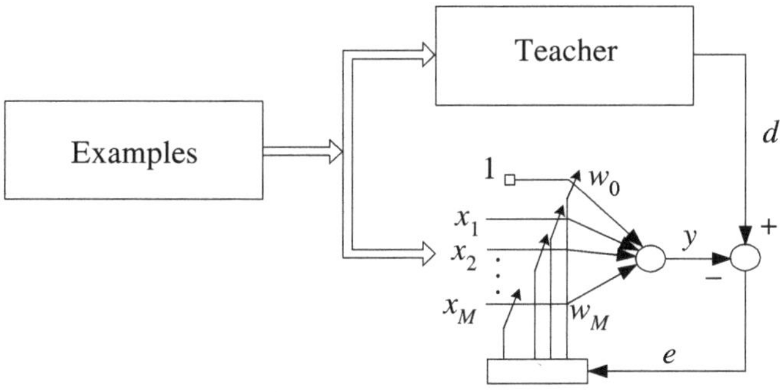

Fig. 9.14 Scheme of supervised learning

For simplicity, the considered example is connected with a one-neuron network. The set of examples used in the learning process (with known input–output relations) are collected from measurements or are obtained from simulation. In the last case it is assumed that the system model is known. Successive correction of the neuron weighting coefficients is performed according to the following rule:

$$w_{ji}(n+1) = w_{ji}(n) + \Delta w_{ji}(n) \tag{9.26}$$

where: $w_{ji}(n)$ and $w_{ji}(n+1)$ represent previous and corrected weights between the *i*th and *j*th adjoining layers; while $\Delta w_{ji}(n)$ is a correction value; *n* is the number of the iteration step.
A stable learning process is assured by such choice of corrections $\Delta w_{ji}(n)$ that the iteration calculations (9.26) are convergent.

Consider the *j*th neuron in a single-layer network. The learning process is based on minimizing errors between the actual output of the considered *j*th neuron: $y_j(n)$ and the value requested by the teacher: $d_j(n)$ (see Fig. 9.14)):

$$e_j(n) = d_j(n) - y_j(n) \tag{9.27}$$

These errors are collected in vector $\mathbf{e}(n)$. We can see that the vector $\mathbf{e}(n)$ is a function of weighting coefficients vector $\mathbf{w}(n)$ at the input to the considered layer.

The correction value in (9.26) can be calculated as follows:

$$\Delta w_{ji}(n) = \eta e_j(n) x_i(n) \tag{9.28}$$

where: x_i is the *i*th input signal; η – is the learning rate parameter determining the speed of iteration process.

The aim of learning is to obtain a minimum of error function (9.27). The adequate criterion may be formulated similarly as in LSM (see Sect. 4.3). For a network composed of *L* neurons it can be defined as:

$$S_2(\mathbf{w}) = \frac{1}{2}\sum_{j=1}^{L}(d_j - y_j)^2 \tag{9.29}$$

If the training set contains *P* learning pairs of $(\mathbf{x}(n), \mathbf{d}(n))$ with input vector $\mathbf{x}(n)$ and vector $\mathbf{d}(n)$ of desired outputs then for the *n*th iteration step of learning we have:

$$S_2(\mathbf{w}(n)) = \frac{1}{2}\sum_{n=1}^{P}\sum_{j=1}^{L}(d_j(n) - y_j(n))^2 \tag{9.30}$$

Minimization of (9.30) is a non-linear problem because of non-linear activation functions. Fortunately, there are known effective numerical algorithms for minimization of that function, which is based on the steepest-decent method [36, 94]. They are an analytical base for the back-error-propagation learning strategy. Usually, it is applied in the version known as the Levenberg–Marquard algorithm [92, 180].

The back-error-propagation algorithm consists in calculation of correction values (9.28) as functions of errors estimated from minimization of (9.30). The process is performed layer by layer across the network in the reverse direction starting from the output layer. The structure of the algorithm is presented in Fig. 9.15. Corrected weighting vectors are determined in blocks $A^{(M)}$, $A^{(M-1)}$,.., $A^{(1)}$ and errors 'propagated' to low layers are calculated in blocks $B^{(M-1)}$, $B^{(M-2)}$,.., $B^{(2)}$.

There are many kinds of different implementation of the back-error-propagation algorithm. They differ in methods of determination of new weighting coefficients, which can be successfully upgraded when errors are passed back from layer to layer, or in parallel after finalizing a given iterative step. Their modifications are also used for training of recurrent networks [36, 94].

The grade of network 'learning' can be evaluated by checking the value of corrections (9.28) in successive stages. The number of iterative steps (that is equivalent to the size of prepared learning set) needed to achieve satisfactory convergence depends on the network size, their structure, kind of investigated problem, details of used learning algorithm and so on. Sometimes, the learning set should have the size of many thousand cases.

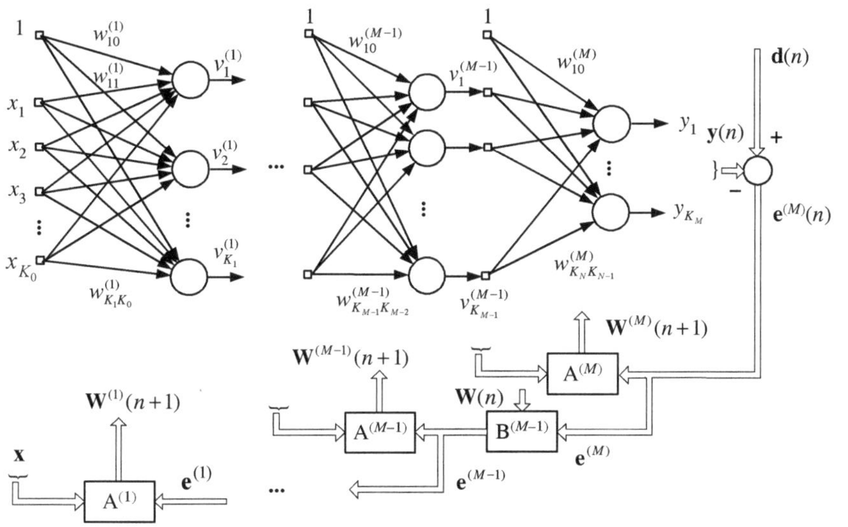

Fig. 9.15 Structure of back-error-propagation algorithm

Correctness of the chosen ANN structure and learning process can be controlled by evaluation of the network outputs on some test excitation (with known answers). A testing set of input–output pairs is a part of all learning sets of cases, which, however, was not used in the learning.

Although the back-error propagation is the most frequently used method for training feedforward ANN, it is neither the only nor necessarily the unique approach. First, the technique has no adaptive feature so all learning examples should be used every time the weighting coefficients are updated, even if a subset is not pertinent to the subject under consideration (the importance of some learning examples is not easy to determine). The method could not be easily adopted to optimize an ANN size and structure. Moreover, such an algorithm can be very slow in the vicinity of the final solution [60].

Recently, other optimization techniques, such as genetic algorithms (GA) are proposed to ANN train. The GA is based on the concept of natural selection and genetics. It is inspired by the biological process of evolution where selection, mutation and cross-over play a major role. Mathematically, GA is a parallel optimization algorithm that uses a performance criterion for evaluation and a population of possible solutions to the search for a global optimum and can handle high-dimensional, non-linear optimization problems [60]. As to the basics of GA, the reader is referred to [28, 85].

Fuzzy Neural Networks

Fusion of ANN and fuzzy-logic systems is a result of the tendency to use the better features of both techniques: capability of non-linear mapping, learning of ANN and high immunity to errors in input data and flexibility to their inaccuracy or uncertainty, which is a characteristic property of a fuzzy-logic. Therefore, the neuro-fuzzy hybrid system combines the advantages of a fuzzy-logic system, which deals with explicit knowledge that can be explained and understood, and neural networks, which deal with implicit knowledge, which can be acquired by learning. This combination is referred to as a fuzzy neural network (FNN).

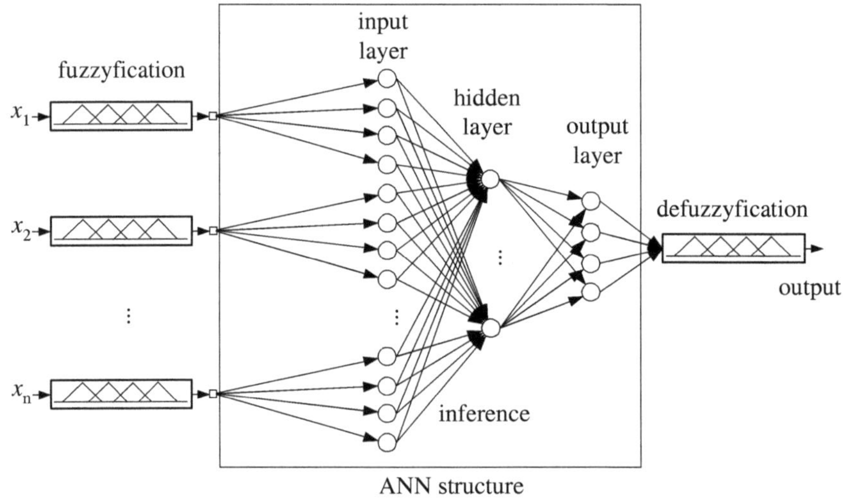

Fig. 9.16 Structure of the simple FNN

There are several approaches to integrate ANN and fuzzy logic and very often it depends on the application [183]. One of the simplest structures of FNN is presented in Fig. 9.16. In fact, that is a combination of feedforward ANN driven by fuzzy inputs. In that structure the ANN represents a fuzzy-inference system: each node in the inference layer multiplies the input signals and the output of the node becomes the result of fuzzy inputs product (fuzzy intersection). In a fuzzy-inference layer fuzzy rules are fired and the value at the end of each rule represents the initial weight of the rule, and will be adjusted to its appropriate level at the end of training. For determination of a crisp output a defuzzyfication procedure is applied. Before that, in the output layer all particular outputs are summed with adequate weights (fuzzy union operation). Training of the network can be realized with a back-propagation method.

For fault-location application the input signals represent normalized voltage and current waveforms obtained from simulation of different disturbances in the considered network (Fig. 9.17) [6].

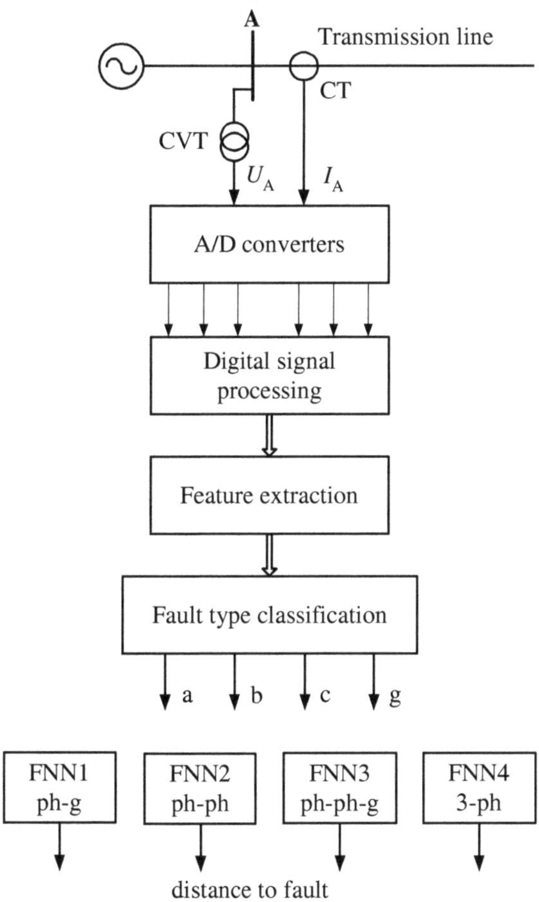

Fig. 9.17 Detailed fault-location scheme based on FNN

They are divided into groups with respect to place of fault or any other disturbances. Fuzzyfication of input signals is performed by taking into account different signal characteristics, such as for example: value of DC component (in current signals fuzzyfication), signal spectrum (classified to one of a few categories) and so on. The adequate signal characteristics are calculated in the pre-processing stage. The phase-coordinates or symmetrical-components approaches can be applied here. Designation of x_i in the input signals is related to the fault place in the training data. Details are presented in Fig. 9.17. FNNs are designed to an adequate type of fault. If the line needs to be considered as an untransposed one the FNNs should be precisely adjusted to a specific type of fault: a–g, b–g, …, a–b–c, a–b–c–g.

Recently, the neuro-fuzzy approach is becoming one of the major areas of interest because it has the benefits of neural networks as well as of fuzzy-logic systems and it removes the individual disadvantages by combining them on the common features. Such an approach is also proposed for power-system protection purposes [324].

9.2.4 Genetic Algorithms

Genetic algorithms belong to the wide family of evolutionary computation methods. They are computational optimization algorithms that are inspired by Nature's evolutionary process. They primitively emulate (with great simplicity) the rules of natural selection, which favor the stronger species and force further evolution such that they remain alive in a given environment. GAs offer the attraction that all parts of the possible space are potentially available for consideration, so the global minimum should be attained if premature convergence can be avoided. The primary focus of such algorithms is the search for the global optimal state when for any reason the analytical methods can not be used (e.g., for problems with multiple local optima, or even for problems where we do not know whether a single optimum is unique) [47, 85].

Similarly as in natural evolution, the process reproduced in GA delivers successive generations of individuals that are modified so as to approach an optimum form. Each descendant has different features from its parents, i.e., it is not a perfect but modified copy. If the new characteristics are favorable, the offspring is more likely to flourish and pass its characteristics to the next generation. However, an offspring with unfavorable characteristics is not reproduced, ending further development of the considered path. These concepts have been applied to mathematical optimization, where a population of candidate solutions develops toward an optimal point [28].

Originating from the evolutionary inspiration, GA also uses biological terminology. The calculation procedure can be tracing by calling the natural process. Characteristics of a living organism (from the point of view of its fitness with respect to an environment) is determined by a set of chromosomes placed in each organism cell. The chromosomes are made up of genes, where each gene determines a particular attribute such as hair color. The complete set of genetic material is referred to as the genome, and a particular set of gene values constitutes a genotype. The resulting set of attributes is described as the phenotype [102]. Each individual in the population of candidate solutions is graded according to its fitness. The higher the fitness of a candidate solution, the greater are its chances of reproducing and passing its characteristics to the next generation. This simplified procedure constitutes the basis for genetic algorithms.

In the computer program the process characteristics are represented by numbers. The searching algorithm optimizes the fitness function inside the search space that is defined by the set of chromosomes made up by genes. For example, in a simple case the chromosome can be composed of two genes: α, β, each defined by three binary numbers:

$$[\alpha \quad \beta] = [\alpha_0 \quad \alpha_1 \quad \alpha_2 \quad \beta_0 \quad \beta_1 \quad \beta_2], \text{ e.g.: } [4 \quad 3] = [1 \quad 0 \quad 0 \quad 0 \quad 1 \quad 1]$$

The possible values for the genes are called alleles, so there are 8 alleles for each gene in this example. Each position along the chromosome is known as a locus; there are two decimal or six binary loci in the above example. It can be seen that the investigated problem involving many variables can be represented by the chromosome with as many as needed numbers of genes or with fewer numbers of genes and adequate numbers of loci in a single gene. For N binary numbers of loci there are 2^N alleles in the chromosome. For each possible combination of genes in the search space there can be determined fitness function:

$$f(\alpha,\beta) = f(\alpha_0,\alpha_1,\alpha_2,\beta_0,\beta_1,\beta_2)$$

Each individual in the population has his specific chromosome. On the basis of this fitness function the selection process is performed: the fitter an individual, the more likely it is to be selected for further reproduction.

The optimization process is performed iteratively by selection of the individuals with higher (less) fitness function belong to the consecutive created generations. The global optimization is assured thanks to two additional mechanisms performed in one after another generation (reproduction): *cross-over* and *mutation.* These operators define the method for creation of the next-generation individuals (candidate solution) by adequate modification of its chromosomes.

Cross-over is the operation during which a new chromosome is created that typically shares many of the characteristics of its parents: mother and father. Different approaches can be applied here. For example, for a six-loci chromosome the new solutions can be generated as follows (m_i and f_i stand for: mother and father, respectively):

$$\begin{matrix}[m_0 & m_1 & m_2 & m_3 & m_4 & m_5] \\ [f_0 & f_1 & f_2 & f_3 & f_4 & f_5]\end{matrix} \mapsto \begin{matrix}[m_0 & f_1 & f_2 & f_3 & m_4 & m_5] \\ [f_0 & m_1 & m_2 & m_3 & f_4 & f_5]\end{matrix}.$$

Mutation involves changing the values of one or more loci. The classic example of a mutation operator involves a probability that an arbitrary bit in a chromosome will be changed from its original state. A common method of implementing the mutation operator involves generating a random variable for each bit in a sequence. This random variable indicates whether or not a particular bit will be modified.

In this way parents are selected for each child, and the process continues until a new population of candidate solutions of appropriate size is generated. The general scheme of the GA is presented in Fig. 9.18 [102]. GA starts with a randomly generated population (a set of solutions) and then moves from one population to another. This process continues until the stopping criteria are met. At each iteration, the new population is generated by applying various search operators.

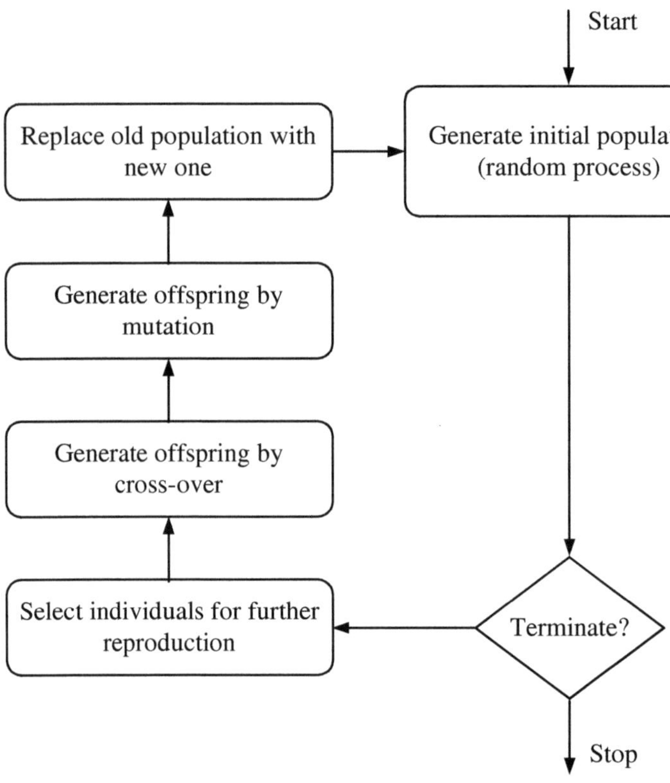

Fig. 9.18 Flow chart of basic GA

Common terminating conditions are: solution is found that satisfies minimum criteria or fixed number of generations reached.

It can be seen that in order to implement a GA, the following design decisions need to be established:

- how to use chromosomes defined by a sequence of numbers, to represent the candidate solutions;

- determination of the size of the population;
- how to define the fitness function for the consecutive population;
- how to select individuals for reproduction using fitness function;
- how to reproduce candidates, i.e., how to create a new generation of candidate solutions from the existing population that is defined by details of cross-over and mutation procedures; and
- when to stop the evolutionary process.

The chosen assumptions define the features of the algorithm.

Genetic algorithms are a very effective way of quickly finding a reasonable solution to a complex problem. Moreover, the basic principle of the method is simple and clear. However, it is important to understand that the functioning of such an algorithm does not guarantee success because it has a stochastic nature and in some applications a genetic pool may be too far from the solution. This section gives only a general view on the GA concept. For details the reader should refer to the literature [28, 102, 211] and many related materials are easily available on the Internet.

9.3 Application Examples

9.3.1 Phase Selection

Let us refer to the phase-selection method presented in Sect. 3.3.2. The structure of the algorithm is shown in Fig. 3.14. The criteria signals are formed by angles α and β. In order to speedup the measurement of the proposed criteria quantities an adequately fast algorithm should be applied. This algorithm, however, should enable a balance between speed and accuracy to be set and should not contribute to significant speeding up the final decision. The transient measuring errors may be handled by using the fuzzy representation of the criteria signals. The common core of this approach may be summarized as follows [128, 253]:

- extraction of the respective features from the available measurements;
- fuzzyfication;
- application of the specific inference system, which combines the on-line measurements with the off-line expert knowledge, formulated usually in the form of the production rules; and
- defuzzyfication and interpretation of the output.

Formally, this approach uses the expert system core, but usually requires a comparatively low number of rules, which distinguishes it from the real expert system problems. The other important feature of the considered application, in turn, is the requirement of the hard real-time operation.

In the presented approach the criteria signals representing the phase-shift angles (3.21) and (3.22) are transformed into the fuzzy numbers [153, 253]. The fuzzy signal $\alpha(k)$ is understood as a series of normalized fuzzy numbers instead of the actual crisp number. In addition, the membership function $\mu(\alpha(k))$ for the criteria quantity $\alpha(k)$ is approximated with the L-R (left-right) standard approximation. Thus, the criteria quantity $\alpha(k)$ is defined by the following three parameters (9.9) (k represents discrete time):

- the mean value, M(k);
- the left bound, L(k); and
- the right bound, R(k).

For the application considered here it is proposed to use the trapezoidal form of the membership function. An example of the membership function $\mu(\alpha(k))$ for the criteria signal $\alpha(k)$ is shown in Fig. 9.19. The presented functions are related with the following three different situations: B – a fault is indicated but the considered phase is not involved in a fault, Z – the respective phase is involved in a fault, N – no fault is detected. One can observe in Fig. 9.19 that for the limit value of the criteria quantity, which is specific for the considered fault type $\alpha = \alpha_{\text{lim}}$ (as depicted in Fig. 3.13), it is not clear whether the given phase is involved in a fault or not. Consequently, both membership functions for the numbers Z and B are greater than zero.

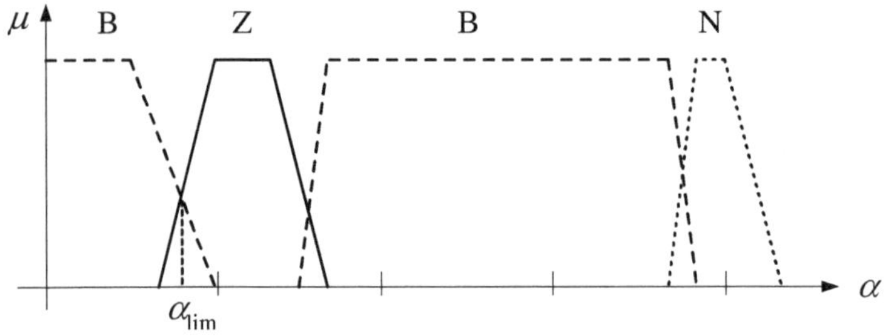

Fig. 9.19 Example of membership functions $\mu(\alpha(k))$ for the criteria quantity $\alpha(k)$

The heart of the whole fuzzy-logic process is made on a basis of fuzzy reasoning. It evaluates the overall truth grade of a set of rules and describes the system in a 'fuzzy' way. The final decision is worked out from consecutive checking of a chain of adequate rules of the form

$$\text{(IF (P1) AND (P2)) THEN (Q1)} \tag{9.31}$$

The results of the checking are also placed in a fuzzy domain. The block of a fuzzy inference represents the heart of the whole fuzzy-logic process. There is the inference from a set of fuzzy rules involving fuzzyfication of the conditions of the rule, and then propagating the membership values of the conditions to the outcome of the applied rules.

The final operation in a fuzzy-logic decision-making process is to represent the result of the fuzzy reasoning as a crisp value $\alpha(k)$ [153, 337]. In the presented algorithm the weighting-center method was chosen for defuzzyfication.

The above-described fuzzy-logic decision-making scheme has been used for determining a fault type. A decision is taken upon the fuzzy-logic rules. The inference system input vector has the form:

$$\boldsymbol{x} = \begin{bmatrix} \alpha \\ \beta \end{bmatrix} \tag{9.32}$$

where α, β are determined as in (3.21) and (3.22).

Crisp values of the vector α, β undergo transformation into fuzzy values described by the fuzzy set. Illustration of a decision-making process is shown in Fig. 9.20. The threshold element is used to reduce oscillations in the output fuzzy signal. A median filter with short data window, i.e., of 3 samples width, reduces the output instability.

For determination of different fault types the input signals represented by the vector $\boldsymbol{x}$, defined in (9.32), are simultaneously processed in nine separate blocks, having the structure as depicted in Fig. 9.20. Finally, the output signals z_i, where 'i' corresponds to the specification of the faulted phases and possible involvement of the ground, clearly determines the fault type. It should be pointed out that each type of fault is detected in the separate block; so, generally, different forms of the membership functions can be applied in the particular blocks.

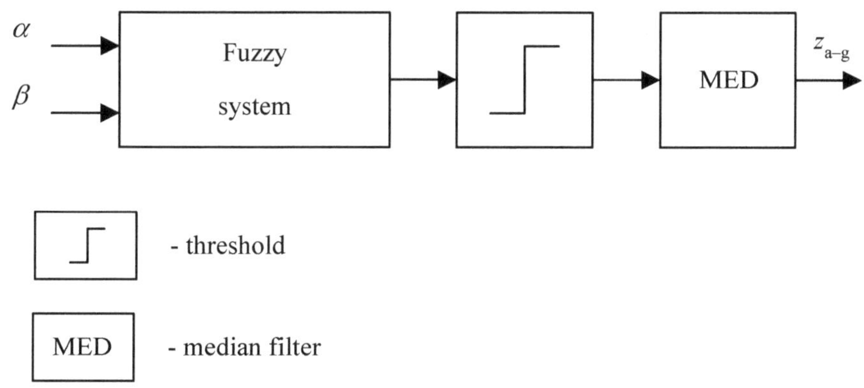

Fig. 9.20 Fuzzy inference system for phase a to ground fault (a–g)

For illustration of the method let us consider an example of the fault-classification task in a 220-kV, 200-km transmission line (terminated at buses A and B), considered here as the protected object [253]. This line is considered as affected by different short circuits: line-to-ground (ph–g), line-to-line (ph–ph), line-to-line-to-ground (ph–ph–g) faults. A database of the measured line currents was built up for various types of faults, applied at different locations, under different fault-inception angles and different source-to-line impedance ratios. Fault classification has been considered as performed at both line ends A and B.

Let us consider a phase b to earth fault (b–g) with fault resistance of 25 Ω, fault-inception time $t_{flt} = 0.067$ s, applied at 120 km from bus A. Figure 9.21a presents the waveforms of the criteria quantities: α – phase shift angle between the positive- and negative-sequence phasors, β – phase shift angle between the negative- and zero-sequence phasors.

One observes extensive changes during the transient period, in particular for the quantity β. Because of the input uncertainty, the fuzzy-system output initially takes the low level (L). This is seen in Fig. 9.21b where the output signals (flags) from the fuzzy inference blocks (z_{a-g}, z_{b-g}, z_{c-g}) are visualized. After some transient period the flag z_{b-g} is set permanently at the high level (H), giving the correct result (fault type: b–g) after 7 ms from the fault inception.

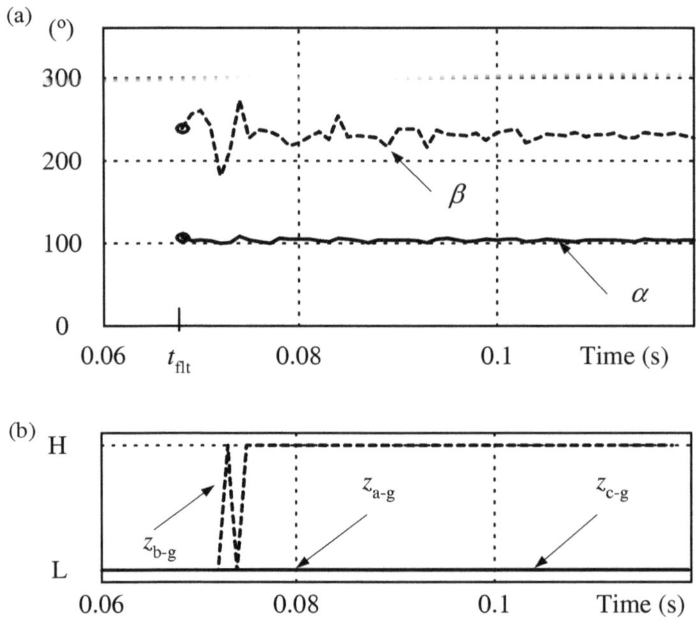

Fig. 9.21 Example of fault classification: (a) criteria quantities α, β, and (b) flags for phase-to-ground faults

The provided experiments and obtained results confirm good characteristics of the developed algorithm. There were no incorrect decisions on fault classification for many tested fault cases [253].

The fuzzy-set approach may be also used as a decision-supporting tool within other multi-criteria schemes. The usual designing program for establishing of such algorithms rely on performing the following steps [66]:

1. Removing the DC components from the currents.
2. Converting the phase currents into symmetrical component representation.

3. Simulation-based determination of the relations between: – positive- and negative-sequence currents, – zero- and positive-sequence currents, – positive- and rated line currents.
4. Establishment of the multi-rule base.
5. Fuzzy reasoning aimed at recognition of a fault type with the credibility factor for the performed recognition.

Such an approach may be recommended for classification of faults in series-compensated lines [150]. It is worth noting that the first step has to be performed only for faults in front of series capacitors. This is so because for faults behind series capacitors DC components do not appear in fault currents at all.

The fault-classification problem may also be considered as suitable task for solution by use of artificial neural networks. Such an approach was recently proposed and analyzed by many authors [43, 196, 239]. The idea of using ANNs must be supported by an efficient software and appropriate hardware, usually including specialized chips. On the other hand, application of the pattern-recognition concept in practice requires preparation of a set of patterns covering the variety of situations to be analyzed. The latter task can be done in a simulative way. The patterns may be waveforms of phase currents and voltages or/and calculated secondary values obtained from the simulation of the power-system events to be recognized by the ANN.

The main idea of the fault-type estimation is based on analyzing of phase and zero-sequence voltage and current using ANNs. The neural fault-type estimator (NFTE) consists of 4 neural networks: three recurrent nets for particular phase-fault detection and the fourth feedforward one for fault to ground recognition (Fig. 9.22) [196]. The NFTE uses feature vectors formed by voltage and current trials. The ANNs are free-layer networks with activation functions of both hidden layers of hyperbolic tangent type and linear functions in output neurons.

The nets work in parallel indicating faulted phases (the nets ANN–ph) and eventually fault to earth events (the net ANN–g). Phase detectors have recurrent structure while a phase-to-ground detector is composed of a simple feedforward net. Changes in outputs of particular ANN–ph classifiers from –1 to 1 indicate fault detection in scanned phase and changes in output of fault to ground detector inform about faults as follows: a–g, b–g, c–g, a–b–g, a–c–g or b–c–g. The decision threshold in both detectors equal to zero has been chosen [196]. It can be seen from Fig. 9.22 that the phase detector makes a decision on the basis of the following input vector (phase a):

$$P^{\mathrm{a}}(n) = \left[| u_n^{\mathrm{a}} |; | u_{n-2}^{\mathrm{a}} |; \ldots ; | u_{n-2*N_{Ph}+2}^{\mathrm{a}} |;\ | i_n^{\mathrm{a}} |; | i_{n-2}^{\mathrm{a}} |; \ldots ; | i_{n-2*N_{Ph}+2}^{\mathrm{a}} |; I_n^{\mathrm{bc}}; y_{n-1}^{\mathrm{a}}\right] \tag{9.33}$$

where u_n^{a}, i_n^{a} – nth sample of voltage and current at phase a, respectively; N_{ph}, N_{g} – number of samples in input vector of phase- and earth-detection net, respectively, and similarly for other phases.

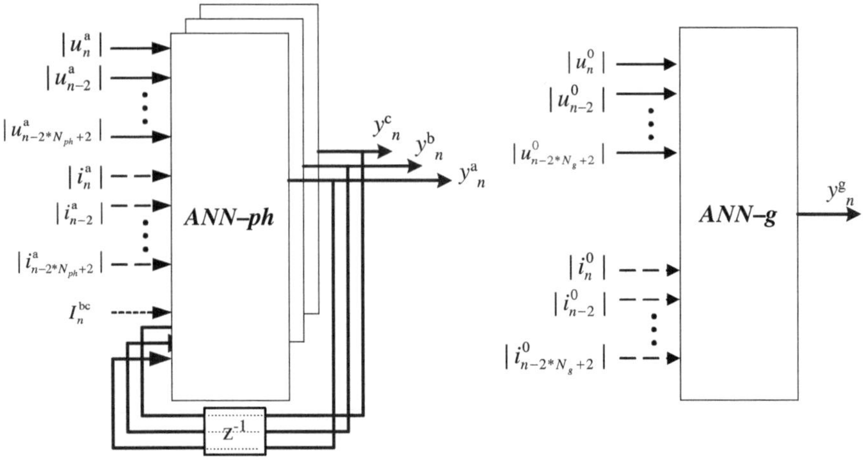

$$I_n^{bc} = \max(|i_n^b|,\ldots,|i_{n-2*N_{ph}+1}^b|,|i_n^c|,\ldots,|i_{n-2*N_{ph}+1}^c|)$$

Fig. 9.22 Structure of neural fault-type estimator

Introduction of the signal I_n^{bc} (and I_n^{ac}, I_n^{ab} for remaining phases – Fig. 9.22) causes each of the ANN–phs to make a decision based on information from its assigned voltage and current trails and condensed information on current levels in the adjacent phases. Such a choice of input signals makes it possible to avoid partitioning object and separating classifiers from all three phase. The excitation vector for the earth detector has the following form:

$$P^0(n) = \left\| u_n^0 |;| u_{n-2}^0 |;\ldots;| u_{n-2*N_{ph}+2}^0 |;\ | i_n^0 |;| i_{n-2}^0 |;\ldots;| i_{n-2*N_{ph}+2}^0 |\right\| \tag{9.34}$$

Taking the samples with indexes *n*, *n*–2, …, only, reduces a number of samples introduced to the patterns without shortening of the sliding window width (when N_{ph}=5 a half of voltage and current wave period is in the observation zone). Moreover, introducing only absolute values of voltage and current samples reduces the number of different patterns to be analyzed [196].

Faults a–b–c and a–b–c–g are considered as the same fault type (a–b–c). Since all simulated a–b–c–g faults were symmetrical ones they can not be discriminated from a–b–c type faults. Such faults do not produce voltage and current zero-sequence components and can not be recognized as earth faults. Moreover, the value of the zero-sequence current in the case of faults to earth close to symmetrical ones may be less than the value of the measured the zero-sequence current appearing during disturbances caused by the CTs saturation.

9.3.2 Fault-direction Detection

As presented in Sect. 3.3.3 a directional element is based on the impedance calculation from incremental values of voltage and current (3.23). This function can be successfully realized by ANN. The proposed [295] ANN-based direction discriminator does not explicitly use the phase or impedance information as the basis of its decisions. It learns from experience gained during the training and recognizes the hidden relationships that exist in the patterns observed during the learning phase. Such a discriminator uses various attributes to reach the decision and tends to emulate the conventional pattern-classification problem. An equation of the boundary describing the classification is embedded in the three-layer feedforward neural network (number of used neurons: 12-4-1) by training through the use of an appropriate learning algorithm and suitable training data. The discriminator uses instantaneous values of the line voltages and line currents to make decisions.

The discriminator was tested extensively by using simulated and field-recorded data. The obtained test results show that the proposed discriminator is able to determine the direction of the fault three samples after the inception of the fault. The performance is not affected by the direction of the fault, the location of the fault, phases involved, presence of fault resistance, initial power-flow conditions and variations in source impedances. The inventors claim that the ANN-based discriminator is fast, robust and accurate [295].

9.3.3 Fault Location Using ANN

Thanks to the basic features of the ANNs, such as the capability of non-linear mapping, parallel processing and learning these tools show a great enhancement in the accuracy of fault location in comparison to other techniques, especially in the case if the explicit formulation of the solution algorithm is not possible, e.g., in multi-ended transmission or distribution networks with laterals [5, 7, 59]. Although the general accuracy of the analytical fault-locating methods may be satisfactory, they suffer some drawbacks because of [138, 178, 222]:

- Lack of uncertainty of information about the line parameters, particularly regarding a zero-sequence impedance.
- Load-flow unbalance.
- Unknown fault resistance and load current, which to a large extent cause the fault location errors.
- Assumption of the quality of fault resistances in the case of two phase-to-ground faults (single phase-to-ground and phase-to-phase faults are caused by a single fault resistance, thus in these cases analytical algorithms are highly accurate).

Therefore, it is vital to reduce errors of the fault-location methods by securing the accuracy of information about the system or by the use of new decision-making algorithms.

Multilayer Perceptrons Application

The ANNs proposed for fault location differ with the used ANN configuration, learning method and applied input information. Input data set depends on a considered type of fault, which can be determined according to a method presented in Sect. 9.3.1. For example, for one-end measurement algorithm the following set of input data is usually used for a–b, a–b–g or a–b–c faults [195]:
$|\underline{U}_a|$, $|\underline{U}_b|$, $|\underline{U}_c|$ – magnitude of phase voltages,
$|\underline{I}_a|$, $|\underline{I}_b|$, $|\underline{I}_c|$ – magnitude of phase currents,
$|\Delta\underline{U}_{1a}|$, $|\Delta\underline{I}_{1a}|$ – magnitudes of incremental (post-pre-fault) positive-sequence voltage and current with respect to phase a,
$\angle\underline{U}_{ba}$, $\angle\underline{U}_{ca}$ – angle of phase-phase voltages,
$\angle\underline{I}_a$, $\angle\underline{I}_{ba}$, $\angle\underline{I}_{ca}$ – angle of phase-phase currents,
$\angle\Delta\underline{U}_{1a}$, $\angle\Delta\underline{I}_{1a}$ – angle of incremental values,
$|\underline{Z}_A|$, $\angle\underline{Z}_A$, $|\underline{Z}_B|$, $\angle\underline{Z}_B$ – magnitude and angle of system impedances at both sides of the line.

The used ANN structure is presented in Fig. 9.23. If the same ANN is used for distance to fault estimation during fault with participation of the phase b or c, the sequence of the input set should be adequately changed. For example, for faults with participation of the phase c the input set will have the following sequence: $|\underline{U}_c|$, $|\underline{U}_a|$, $|\underline{U}_b|$, $|\underline{I}_c|$, $|\underline{I}_a|$, $|\underline{I}_b|$, $|\Delta\underline{U}_{1c}|$, $|\Delta\underline{I}_{1c}|$, $\angle\underline{U}_{ac}$, $\angle\underline{U}_{bc}$, $\angle\underline{I}_c$, $\angle\underline{I}_{ac}$, $\angle\underline{I}_{bc}$, $\angle\Delta\underline{U}_{1c}$, $\angle\Delta\underline{I}_{1c}$, $|\underline{Z}_A|$, $\angle\underline{Z}_A$, $|\underline{Z}_B|$, $\angle\underline{Z}_B$. The system impedances (last four values) do not depend on a type of fault and are not changed.

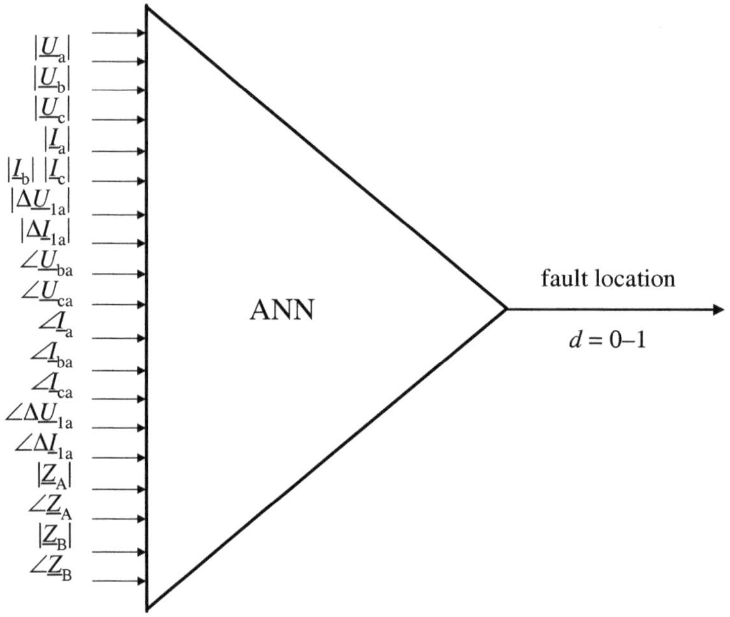

Fig. 9.23 ANN input structure for distance to fault estimation

The algorithm was tested for a 400-kV system with line length ℓ=200 km [195]. The performance of the algorithm was evaluated by varying the ANN size and firing rules of neurons were used. As a result, the two-layer ANN was chosen with 11 non-linear neurons in the hidden layer and one linear output neuron: 11N–1L that showed the best performance. The obtained maximum and mean errors made by neural and conventional impedance-based locators with respect to the actual fault distance are shown in Fig. 9.24. It was considered for a–b–g faults with the fault model as in Fig. 9.25. The fault resistance was randomly selected from the following set of values: R_F=0 Ω, 10.0 Ω and 50.0 Ω.

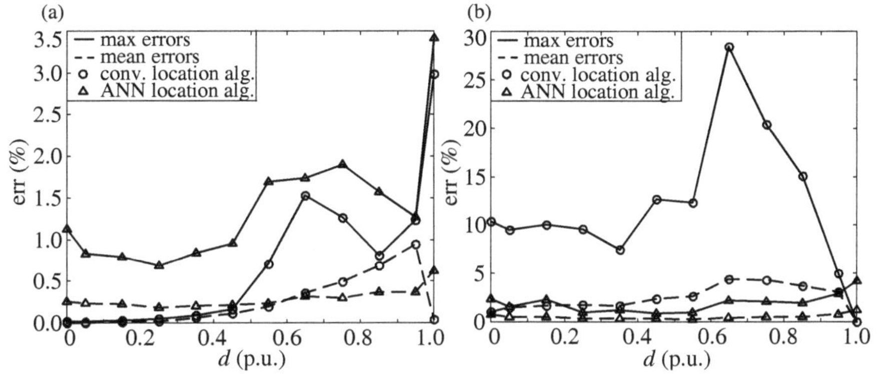

Fig. 9.24 Errors of ANN and classical fault-location algorithms for a–b–g faults with the fault resistance: R_F=0.0 Ω (a) and for R_F=0.0, 10.0 and 50.0 Ω (b)

Figure 9.24a presents the results obtained under the assumption that both fault resistances were the same: R_{Fa}= R_{Fb} while Fig. 9.24b contains results for cases when resistances R_{Fa} and R_{Fb} were randomly selected. The fault place was changed in the range of: 0.0–1.0 p.u.

Fig. 9.25 a–b–g fault model

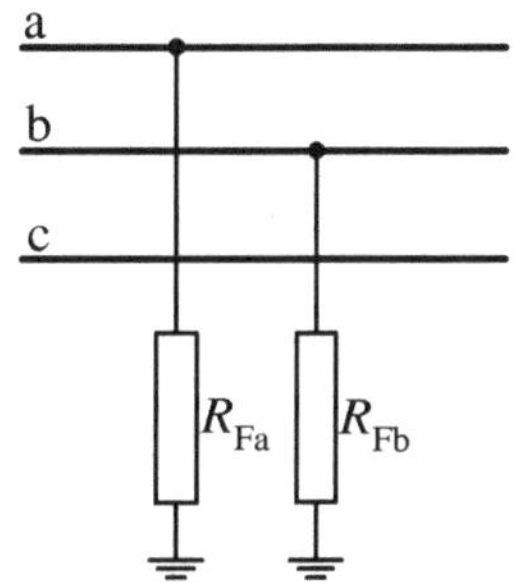

The errors are calculated according to the formula:

$$\text{err} = \frac{|d_c - d_a|}{\ell} \cdot 100\% \tag{9.35}$$

where: d_c, d_a are the computed and the actual distances, respectively.

The results obtained from the ANN's estimator is compared with that obtained by use of a classical one-end impedance-based algorithm. One can see that for equal fault resistances (symmetrical fault) the traditional estimator delivers quite good results. However, the neural estimator is more robust for faults with unsymmetrical fault resistances and arcing faults that, in general, is the more realistic case [206, 288, 305].

Fault Location in Series-compensated Line

In the case of series-compensated lines, the performance of protection systems and fault locators depend on the status of the capacitor and the transient response of the capacitor protection circuit. The MOV starts conducting immediately after the instantaneous voltage across the capacitor exceeds a certain voltage level U_{REF}. The U–I characteristic of a MOV is approximated by the single exponential model as given by (2.52). The value of q is typically chosen between 30 and 50. The MOV is a resistive device that absorbs energy and is protected from overheating by an overload-protection scheme that calculates the energy developed through the MOV and initiates protective gap flashover (by-passing the MOV). In addition, a high-current function is provided to speed up by-passing for severe internal faults if the current is greater than I_{max}.

Application of digital technology allows for modifications to be made on-line to improve the network protection and control in the presence of the controllable and non-controllable devices. The appropriate on-line modeling of the non-linear electrical behavior of such devices can be achieved either by the simulation of the devices' behavior directly or by artificial intelligence technology [222].

Artificial intelligent tools can be used to improve fault detection and fault location for the series-compensated lines [228]. In both references, the ANN has been trained with a number of EMTP [56] cases that simulate different fault conditions in a selected network. The ANN models a response of the capacitor installation and the selected network to these conditions. The designed ANNs have been trained and tested with the same power network. This approach is sensitive to a specific installation and surrounding power network conditions. The real-world application on the ANN technique will depend on its ability to generalize and provide a robust response to the various network conditions.

New techniques to improve the fault-location calculation for series-compensated lines can be used for on-line calculation of the voltage across a non-linear capacitor installation. The proposed scheme requires only local voltage and current at the bus and is independent of the series capacitor installation, operation of the capacitor protection, and the surrounding power-system elements [222].

The capacitor protection must be modeled in the equivalent scheme to accurately transfer the voltage across the capacitor. The voltage and current relationship for the capacitor is defined by

$$u_x(t) = \frac{1}{SC} \int_{-\infty}^{t} i_C(t)\,\mathrm{d}t \tag{9.36}$$

where $u_x(t)$ is the voltage across the capacitor and $i_C(t)$ is the current through the capacitor (Fig. 2.27a). The current $i_C(t)$ is the difference between the line current $i(t)$ and the current $i_{MOV}(t)$ through the MOV. By using the modeled *U–I* characteristic, it can be shown that the voltage and current relationship at the capacitor can be given by the ordinary differential equation (see (2.52) and Fig. 2.27) [222]

$$\frac{\mathrm{d}u_x(t)}{\mathrm{d}t} = \frac{i(t)}{SC} - \frac{p}{SC}\left(\frac{u_x(t)}{U_{REF}}\right)^q \tag{9.37}$$

This voltage is used to calculate the line-side voltage of the capacitor by summing the measured bus voltage and the voltage drop across the capacitor.

The ANN techniques used estimate the voltage across the capacitor in the time domain. This voltage is applied to calculate the line-side voltage of the capacitor by summing the measured bus voltage and the voltage across the capacitor. The calculated line-side voltage can be used with any of the conventional relays or fault locators.

The proposed scheme is simple, fast, accurate and reliable. It requires only local voltage and current at the bus. It is well suited for implementation in a digital fault locator or a digital distance relay to improve their performance. For the fault-location application, the scheme needs to be combined with the existing fault-location techniques [227].

Fault Location on Ungrounded or Compensated Distribution Networks

We have seen in Chap. 8 that the fault location in distribution networks shows known problems because of the network branching, laterals and dynamical change of loads along the network sections. The problem is made more difficult when the network neutral is isolated or compensated (when using a Petersen coil). In such networks a fault current during the phase-to-ground fault considerably depends on the network capacitance and has a resonant character. The fault-loop impedance is relatively high, which results in low fault current. The fundamental component fault current and voltage phasors do not contain information on distance to fault. In such a case the classical impedance-based methods usually give wrong results [59].

In such a case the phasor measurement utilizing a high-frequency voltage and current transients is proposed instead. The fault-loop parameters are estimated with use of a differential-equation line model (see Chap. 4) or wavelet transformation [27]. A more interesting alternative to these solutions is the use of ANNs. As proposed in [59] a multilayer perceptron has a structure similar as in Fig. 9.23 with the number of hidden neurons from the range 10–50. The inputs are composed of voltage and current samples taken from data window of 5 ms with a sampling frequency of 5 kHz. All input samples are scaled to the range [–1, 1]. The authors of this work have compared the ANN estimator with other conventional methods and have obtained more satisfactory results for faults with low fault resistance. Conventional algorithms worked better with higher fault resistances (around 50 Ω).

9.3.4 Wavelet-transform Application

It was shown in Sect. 4.3 that the WT is a recently developed mathematical tool for signal processing. It has attracted much interest in the field of speech and image processing since then. The possibilities of application in protection relaying and fault location have been discussed recently [2, 27, 166, 199, 241]. Use of WT to realize measurements as part of classical protection or fault-location functions needs a caution because it is not directly related to such values as: phasor, impedance, magnitude, harmonics and similar characteristics known for FT. Unlike Fourier analysis, the WT provides time information and temporary frequency divided into a few scales (details). Therefore, proposed direct applications are concentrated on completion of a fault-detection [166], phase-classification [241, 339] and fault-location [2, 199, 229] algorithms.

Fault Location in Transmission Lines

Compared to traditional Fourier analysis, which relies on a single basis function, wavelet analysis applies basis functions of a rather wide functional form. The basic concept in wavelet transform is to select an appropriate wavelet (mother transformation function) and then perform analysis using a family of shifted and dilated variations of this wavelet. A main feature of wavelets is the oscillating and fast decaying behavior that comes with the location in time and frequency. Therefore, WT is a suitable tool for analyzing fast changes with good resolution in time and frequency. In fact, the efficient analysis of high frequencies and short-duration events allows a sure and rapid knowledge of the fault.

The idea of the method is illustrated in Fig. 9.26 [283]. Measurement of arriving times of waves at the terminal buses, together with knowing the velocity of these waves v, allows determination of a distance to fault. Thus, the problem in fact boils down to proper detection of arriving waves, to distinguish the type of these waves (direct waves or reflected waves) and to capture times of their arriving at terminal buses. All these tasks can be directly realized by using wavelet transforms (time location of wave) and artificial neural networks (distinguishing kinds of wave), as is presented below.

When a fault occurs at fault point F (36 km from A) traveling waves are generated appear as a disturbance superposed on the power frequency voltage signals that are recorded by the relays mounted at both sides of the line. Processing these voltage signals using a digital wavelet transform (using in this case Daubechies4 mother wavelet) reveals their travel times between the point fault F and relay locations (Fig. 9.27): disturbance will be seen at bus A in time t_A after the fault time, similarly a wave will reach bus B in time t_B. Both these times are proportional to distances between the fault point and two ends of the line. However, they are not useful in themselves, because it is impossible to directly measure these times (fault time is unknown).

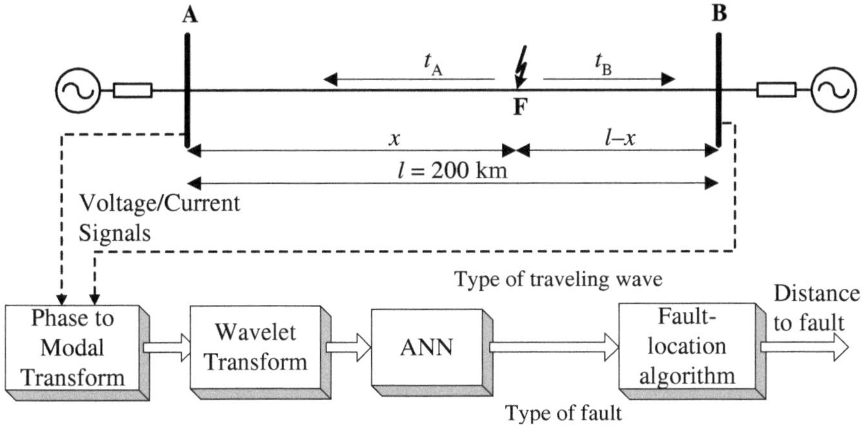

Fig. 9.26 Illustration of a distorted voltage analysis by using a wavelet transform

Much more useful is the well-measurable difference of these times: $\Delta t = t_A - t_B$ (synchronized recording is required). A wavelet transform performs decomposition of incoming voltage signals. Analyzing the obtained wavelet coefficients at different scales (especially at low levels) it is easy to determine the times of occurrence traveling waves in voltage signals – wavelet coefficients become non-zeros (peaks can be observed on waveform) – in particular the very first received waves that directly correspond to times t_A and t_B. As was said earlier, using the knowledge of Δt time and the velocity of traveling wave v, the distance x to point F can be calculated as follow:

$$x = \frac{l - v\Delta t}{2} \tag{9.38}$$

In the example of Fig. 9.26, velocity is $v = 1.175 \times 10^5$ km/s, the first peak at bus A occurs at $t_A = 29.81$ ms, the first peak at bus B at $t_B = 29.90$ ms, so distance x to the fault point F is $x = 35.96$ km. The presented example is very simple. It utilizes two-end synchronized recording, and concerns a 3-phase balanced fault. In situations like the above the proper identification and time localization of waves is not problematical at all. The fault-location algorithm is simple, too. But in general, e.g., in the case of single-ended recording, or different types of faults, these tasks are much more complicated. The general fault-location algorithm is described in detail in [199], but it assumes that all the needed times of occurrence of waves are well known as well as their types and sources. In fact, the main problem is not to detect the traveling wave in voltages. The correct selected wavelet transform can manage this task very well. Even more, a wavelet transform will indicate not only traveling waves caused by internal faults what is desirable (directed waves as well as reflected waves), but also slight disturbances that can come from the outside part of power system, which is very undesirable. Thus, the real problem is how to distinguish between direct traveling waves caused by a fault (including high-impedance faults), traveling waves reflected from the fault, traveling waves reflected from the remote end of the line, and other disturbances. The solution of this issue use of an artificial neural network. The ANN based on wavelet decomposition coefficients as well as on a voltage waveform, performs a task of filtering disturbances contained in the incoming voltage signal. First, ANN "decides" if the present disturbance belongs to a traveling wave class or not, and if it does, which phases or ground are involved in the fault. That information, plus the times of appropriate waves obtained by performing a wavelet transform, is indispensable to the fault-location algorithm.

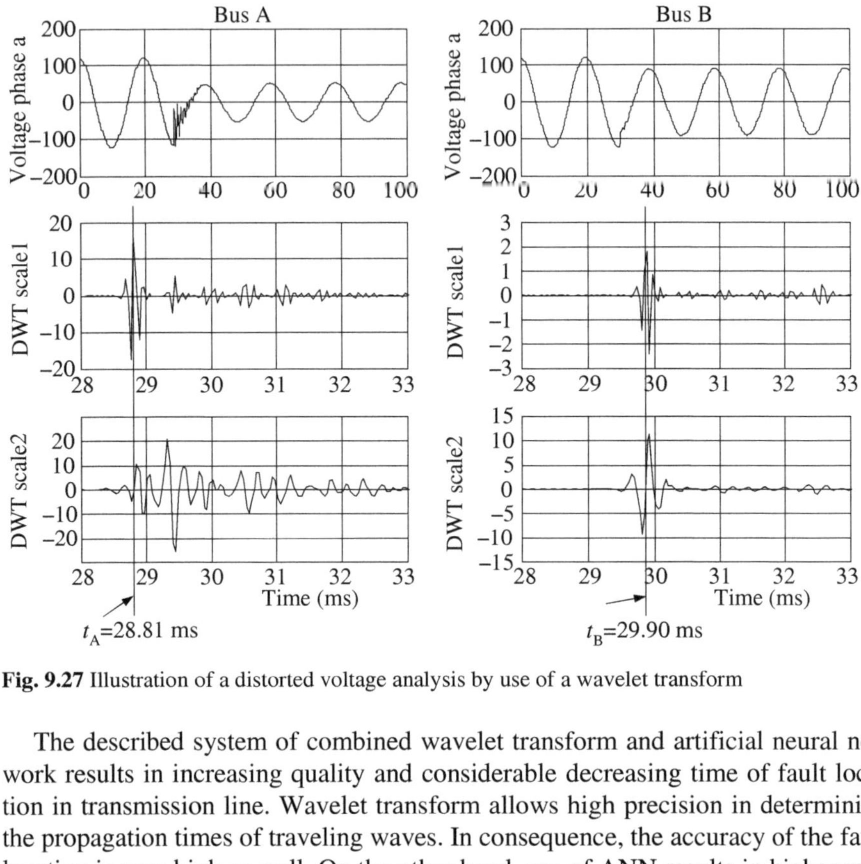

Fig. 9.27 Illustration of a distorted voltage analysis by use of a wavelet transform

The described system of combined wavelet transform and artificial neural network results in increasing quality and considerable decreasing time of fault location in transmission line. Wavelet transform allows high precision in determining the propagation times of traveling waves. In consequence, the accuracy of the fault location is very high as well. On the other hand, use of ANN results in higher reliability of the algorithm to fault location, because the system with ANN is much more selective to traveling waves and much more resistant to disturbances that are present in a power system. The time response of such a system is also very good because it is only slightly longer than the time required to propagate waves from fault point to relay.

Test results have demonstrated that this approach can be used for accurate determination of the fault-inception time, fault-type classification and distance to fault estimation. It is seen that more advanced use of WT in power-system automation needs application of additional tools for recognition and identification of information delivered by this method. That's why the WT method is frequently joined with an another technique. For example, the good time-resolution feature of WT may be connected with ANN or FNN (fuzzy neural network) techniques [35, 139, 241, 339]. The general scheme of the signal processing includes the following steps:

1. Voltage and current signals conditioning; in this process a DC component in the current waveform is removed.
2. Voltage and current signals wavelet decomposition: determination of details (D) and approximations (A).
3. Rough estimation of fault place.
4. Application of FNN to accurate fault location.

Rejection of DC from the current waveform is needed because the WT is not suitable for non-oscillation components. In step 2 it is assumed that a discrete WT is applied, however CWT methods may also be considered [27]. The final result is obtained from FNN application [139].

Fault Location in Distribution Networks

Traveling-wave methods for distribution network and cable fault location have a long history [70, 71, 328, 346]. The justification for using such techniques is explained by the domination of high-frequency post-fault current and voltage transients, especially when ungrounded or compensated networks are considered. The wavelet transform is employed to extract these transients and the transient characteristics are used to implement fault location.

The straightforward application of such an algorithm is based on utilizing the wavelet transformation with a complex-valued mother function [27, 98, 247]. In the method reported in [95] the following mother wavelet is used:

$$\psi(t) = \left(1 + \sigma|t| + \frac{\sigma^2}{2}t^2\right)\mathrm{e}^{-\sigma|t|}\mathrm{e}^{\mathrm{j}\omega_0 t} \tag{9.39}$$

where the coefficient σ is set: $\sigma = 2\pi/\sqrt{3}$ to bring the frequency characteristic of the mother wavelet to zero for $\omega = 0$ (admissible condition) [27].

The Fourier transform of (9.39) is as follows:

$$F\{\psi(t)\} = \Psi(\mathrm{j}\omega) = \frac{2\sigma^3\left(3\sigma^2 - (\omega-\omega_0)^2\right)}{\left(\sigma^2 - (\omega-\omega_0)^2\right)^3} \tag{9.40}$$

Additionally, the basic frequency is set to: $\omega_0 = 2\pi$, which gives the normalizing coefficient to the frequency: $1/p = f$ (4.141). For these settings the mother function and its frequency characteristic looks like these in Fig. 9.28).

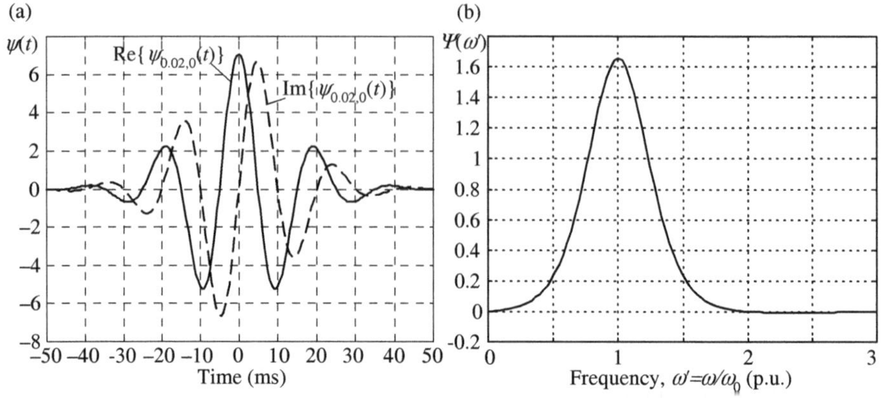

Fig. 9.28 Mother function of the considered wavelet (a) and its Fourier transform (b)

We can see that the considered wavelet has a period equal to 20 ms and the magnitude characteristic has a good frequency feature: it represents a band-pass filter with the center at 50 Hz ($\omega = \omega_0$) while the gain is equal to $6/\sigma$. It can be seen that for analyzing the 50-Hz signal component, the used data window length should last a few tens of milliseconds. Thanks to the features offered by wavelets the higher frequency components can be measured with the same resolution and within adequately shorter window length. This property can be very useful in analysis of the high-frequency transients in which the signal energy is concentrated during faults in distribution networks.

WT can also be applied for distance to fault calculation in a similar way as the FT. Consider a signal $x(t)$ sampled with the frequency $f_s = 1/T_s$. for which the CWT is performed according to (4.142) with the wavelet as in (9.38). For the previously assumed parameters and chosen frequency f we obtain [27]:

$$W_x(kT, f) = T\sqrt{f}\sum_{n=0}^{M-1} x(nT)\psi^*\big(f\cdot(nT - kT)\big) \tag{9.41}$$

where: M – number of samples in data window; note that in this case: $f = 1/p$.

The transformation (9.41) can be applied for ground–fault current and the voltage measured at the substation giving adequate current and voltage wavelets: $W_i(kT,f)$ and $W_u(kT,f)$ which are complex-valued functions. The fault distance can be estimated as proportional to the fault–loop inductance calculated as follows:

$$L_f = \frac{1}{2\pi f}\,\mathrm{Im}\left(\frac{W_u(kT,f)}{W_i(kT,f)}\right) \tag{9.42}$$

The constant frequency f for calculation of the above wavelets is chosen for obtaining the maximum of current wavelet with different time translation kT. The wavelet transform in (9.42) is similar to use of a special pair of filters for voltage and current components calculation.

9.3.5 Genetic Algorithms Application

The special difficulties of the fault location take place in the case of branching networks with non-homogeneous sections. Such a condition may be met in distribution as well as in transmission networks. The fault location based on the voltage and current phasors yields more than one location that could not be properly selected [348].

Different methods are proposed for resolving the selection problem. The general approach is based on application of an algorithm for searching for some similarity between the recorded post-fault voltage and current transients and stored signals related to the known (previously recorded or simulated) faults. The stored signal can be in the form of the recorded time-variable phasors or simply waveforms. The selection procedures are based on the fact that in the complex network its response to a fault depends, among other things, on the place of fault. It is usually assumed that the type of fault can be determined before starting the selection procedure. The general algorithm is thus composed of the following steps.

1. determine the type of fault;
2. estimate a distance to fault and set of possible fault places; and
3. select most likely result.

In the second step different methods can be used: differential-equation-based, impedance-based, traveling-wave or some other algorithm. The selection procedure decided on the effectiveness of the hole algorithm. In some of the proposed methods both the second and third steps are connected, creating the main procedure.

A promising selection procedure was formulated in the language of an optimization problem where the criterion function is related to matching the post-fault recorded file with the stored set of process files. This problem is suitable for genetic algorithms application as an optimization tool.

The most advanced algorithm of this type is presented in [162]. It is assumed that in the analyzed network area there are defined numbers of digital fault recorders (DFRs) but only limited data are available ('sparse data'). This is because some of them may not be triggered during a fault or in some network nodes DFRs may not be installed.

The authors utilize the phasors for matching. The matching degree can be represented by a value obtained from the following cost function.

$$f(d,R_f)=\sum_{k=1}^{N_u} r_{ku}\left|\underline{U}_{ks}-\underline{U}_{kr}\right|+\sum_{k=1}^{N_i} r_{ki}\left|\underline{I}_{ks}-\underline{I}_{kr}\right| \tag{9.43}$$

where:

R_f, d, – the fault resistance and fault location,

r_{ki}, r_{ku} – weights for the errors of the currents and voltages, respectively,

$\underline{U}_{kr}$, $\underline{U}_{ks}$ – the post-fault voltage phasors obtained from the short-circuit studies and from recorded waveforms, respectively,

N_i, N_u – the number of the selected voltage and current phasors, respectively.

The cost function $f(d, R_f)$ has a non-negative value and will be zero when the phasors obtained from the simulation studies exactly match those obtained from the recorded waveforms. The best fault-location estimation is achieved when the cost function is at a minimum. Therefore, the problem of fault-location estimation is actually the optimization problem. Since there may be several local minimum and maximum points, it is difficult to use the gradient-based method to find the global minimum. In fact, this is a good problem for application of a GA. This requires formulation of the fitness function (see Sect. 9.2.4), which in contrast to (9.43) should take the maximum value for the best matching. The following fitness function was proposed [162]:

$$g(d,R_f)=C_{\max}-f(d,R_f) \tag{9.44}$$

where: $C_{\max}$ is the maximum value of $f(d, R_f)$ in the current population.

The algorithm is suitable for the situation in which the conventional algorithms cannot be applied because some recorded data is not available. However, a system model including the static parameters and topology is required.

The searching process is run according to the GA methodology: at the beginning, the initial population is generated randomly. Then posing the fault according to the initial population, the short-circuit study is carried out to obtain the simulated during-fault phasors and further calculate the fitness value for each individual. The next generation is produced by applying the three steps as described above. The process is repeated until the best match is found.

The similar problems with selection between a few results also arise in the case of the fault location in distribution networks. They also can be formulated in the category of optimization problem and resolved by using GA.

AI techniques are in the process of fast development and evolution. The practical experiences of using different structures of ANNs and fuzzy-reasoning methods in real faults have shown that they are applicable and worthy of consideration in solving fault-location and protection problems. They can be even more effective when assisted by modern optimization and signal processing methods.

References

1. Abe M, Otsuzuki N, Emura T et al. (1995) Development of a new fault location system for multi-terminal single transmission lines. IEEE Trans on Power Deliv 10(1):159–168
2. Abur A, Magnago FH (2000) Use of time delays between modal components in wavelet based fault location. Int J Electr Power and Energy Syst 22(6):397–403
3. Aggarwal RK, Aslan Y, Johns AT (1997) An interactive approach to fault location on overhead distribution lines with load taps. In: Proc of 6th Int Conf on Developments in Power System Protection – DPSP, IEE CP434 pp 184–187
4. Aggarwal RK, Coury DV, Johns AT et al. (1993) A practical approach to accurate fault location on extra high voltage teed feeders. IEEE Trans on Power Deliv 8(3):874–883
5. Aggarwal RK, Coury DV, Johns AT et al. (1993) Computer-aided design and testing of an accurate fault location for EHV teed feeders. In: Proc of 5th Int Conf on Developments in Power System Protection – DPSP, IEE CP368 pp 60–64
6. Aggarwal RK, Joorabian M, Song YH (1997) Fuzzy neural network approach to accurate transmission line fault location. Int J of Eng Intell Syst for Electr Eng and Commun 5(4):251–258
7. Aggarwal RK, Xuan QY, Dunn RW et al. (1999) A novel fault classification technique for double-circuit lines based on a combined unsupervised/supervised neural network. IEEE Trans on Power Deliv 14(4):1250–1256
8. Ahn S-P, Kim C-H, Aggarwal RK et al. (2001) An alternative approach to adaptive single pole auto-reclosing in high voltage transmission systems based on variable dead time control. IEEE Trans on Power Deliv 16(4):676–686
9. Akke M, Thorp JT (1998) Some improvements in the three-phase differential equation algorithm for fast transmission line protection. IEEE Trans on Power Deliv 13(1):66–72
10. Al-Dabbagh M, Kapuduwage SK (2005) Using instantaneous values for estimating fault locations on series compensated transmission lines. Electr Power Syst Res 76(1–3):25–32
11. Andersson PM (1999) Power System Protection. IEEE Press, McGraw-Hill, New York
12. Apostolov A, Duffy C, Tholomier D (1999) Advanced methods for faulted phase selection and directional detection in numerical distance relays. In: Proc of 52-Annual Conference for Protective Relay Engineers, Texas A&M University, 16 p
13. Aurangzeb M, Crossley PA, Gale P (2001) Fault location using high frequency travelling waves measured at a single location of a transmission line. In: Proc of 7th Int Conf on Developments in Power System Protection – DPSP, IEE CP479 pp 403–406
14. Baldwin T, Renovich F Jr, Saunders LF (2001) Fault locating in ungrounded and high-resistance grounded systems. IEEE Trans on Ind Appl 37(4):1152–1159
15. Bascom EC III, Von Dollen DW, Ng HW (1994) Computerized underground cable fault location expertise. In: Proc of IEEE Power Engineering Society Transmission and Distribution Conference, pp 376–382
16. Begovic M, Novosel D, Djokic B (2008) Issues related to the implementation of synchrophasor measurements. In: Proc of 41st Annual Hawaii International Conference on System Sciences, pp 160–164

17. Benmouyal G (2001) A combined directional and faulted phase selector element based on incremental quantities. IEEE Trans on Power Deliv 16(4):478–484
18. Benmouyal G (1995) Removal of DC-offset in current waveforms using digital mimic filtering. IEEE Trans Power Deliv 10(2):621–630
19. Bewley LV (1963) Travelling Waves on Transmission Systems. John Wiley & Sons, Inc., New York
20. Bo ZQ, Johns AT, Aggarwal RK (1995) A new non-unit protection scheme based on fault generated high frequency current travelling waves. In: Proc of Int Conf on Advance in Power System Control, Operation & Management, APSCOM-95, Hong Kong pp 422–427
21. Bo ZQ, Weller G, Redfern MA (1999) Accurate fault location technique for distribution system using fault-generated high frequency transient voltage signals. IEE Proc – Gener Transm Distrib 146(1):73–79
22. Bradley DA, Gray CB, O'Kelly D (1978) Transient compensation of current transformers. IEEE Trans on PAS 97(4):1264–1271
23. Brahma SM (2005) Fault location scheme for a multi-terminal transmission line using synchronized voltage measurements. IEEE Trans on Power Deliv 20(2):1325–1331
24. Brahma SM, Girgis AA (2004) Fault location on a transmission line using synchronized voltage measurements. IEEE Trans on Power Deliv 19(4):1619–1622
25. Brandao Faria JA (2000) Application of Clarke's transformation to the modal analysis of asymmetrical single-circuit three-phase line configurations. Eur Trans on Electr Power 10(4):225–231
26. Bunyagul T, Crossley PA (2001) Design and evaluation of an overcurrent relay suitable for operation with measurement current transformers. In: Proc of 7th Int Conf on Developments in Power System Protection – DPSP, IEE CP479 pp 201–204
27. Chaari O, Meunier M, Brouaye F (1996) Wavelets: A new tool for the resonant grounded power distribution system relaying. IEEE Trans on Power Deliv 11(3):1301–1308
28. Chambers L (2000) The Practical Handbook of Genetic Algorithms Application. Chapman & Hall/CRC, Boca Raton
29. Chen C-S, Liu C-W, Jiang J-A (2002) A new adaptive PMU based protection scheme for transposed/untransposed parallel transmission lines. IEEE Trans on Power Deliv 17(2):395–404
30. Chen W-K (2003) The Circuits and Filters Handbook. CRC Press LLC, Boca Raton
31. Chen Z, Maun J-C (2000) Artificial neural network approach to single-ended fault locator for transmission lines. IEEE Trans on Power Syst 15(1):370–375
32. Cheong WJ, Aggarwal RK (2004) A novel fault location technique based on current signals only for thyristor controlled series compensated transmission lines using wavelet analysis and self organising map neural networks. In: Proc of 8ht Int Conf on Developments in Power System Protection – DPSP, IEE CP500 vol 1 pp 224–227
33. Choi M-S, Lee S-J, Lee D-S et al. (2004) A new fault location algorithm using direct circuit analysis for distribution systems. IEEE Trans on Power Deliv 19(1):35–41
34. Chua LO, Lin P-M (1975) Computer-aided Analysis of Electronic Circuits. Prentice-Hall, Inc., Englewood Cliffs
35. Chunju F, Li KK, Chan WL (2007) Application of wavelet fuzzy neural network in locating single line to ground fault (SLG) in distribution lines. Int J Electr Power and Energy Syst 29(6):497–503
36. Cichocki A, Unbehauen R (1993) Neural Networks for Optimization And signal Processing. John Wiley & Sons, Inc., New York
37. Collatz L (1960) Numerical Treatment of Differential Equations. Springer Verlag, New York
38. Cook V (1986) Fundamental aspects of fault location algorithms used in distance protection. IEE Proc C 133(6):359–368
39. Cook V (1985) Analysis of Distance Protection. Research Studies Press Ltd., John Wiley & Sons, Inc., New York
40. Crossley PA, McLaren PG (1983) Distance protection based on travelling waves. IEEE Trans on PAS, 102(9):2971–2983

41. Crossley PA, Southern E (1998) The impact of the global positioning system (GPS) on protection & control. In: Proc of 11th Int Conf on Power System Protection–PSP, Bled, pp 1–5
42. Dalcastagne AL, Filho SN, Zurn HH et al. (2008) An iterative two-terminal fault-location method based on unsynchronized phasors. IEEE Trans on Power Deliv 23(4):2318–2329
43. Dalstein T, Kulicke B (1995) Neutral network approach to fault classification for high-speed protective relaying, IEEE Trans on Power Deliv 10(2):1002–1011
44. Das R, Novosel D (2000) Review of fault location techniques for transmission and subtransmission lines. In: Proc of 54th Annual Georgia Tech Protective Relaying Conference (CD–ROM)
45. Das R, Sachdev MS, Sidhu TS (2000) A fault locator for radial subtransmission and distribution lines. In: Proc of IEEE Power Engineering Society Summer Meeting, Seattle, paper 0-7803-6423-6/00
46. Dash PK, Pradhan AK, Panda G et al. (2000) Adaptive relay setting for flexible AC transmission systems (FACTS). IEEE Trans on Power Deliv 15(1):38–43
47. da Silva APA, Falcao DM (2008) Fundamentals of genetic algorithms. In: Lee KY, El-Sharkawi MA (ed) Modern Heuristic Optimization Techniques. Theory and Applications to Power Systems. IEEE Press, Wiley-Interscience, New Jersey
48. Dillon TS, Laughton MA (1990) Expert System Application in Power Systems. Prentice Hall, New York
49. Dillon TS , Niebur D (1996) Neural Networks Applications in Power Systems. CRL Publishing Ltd., London
50. Director SW, Rohrer RA (1972) Introduction to System Theory. McGraw-Hill, New York
51. Djuric MB, Radojevic ZM, Terzija VV (1999) Time domain solution of fault distance estimation and arcing fault detection on overhead lines. IEEE Trans on Power Deliv 14(1):60–65
52. Djuric MB, Radojevic ZM, Terzija VV (1998) Distance protection and fault location utilizing only phase current phasors. IEEE Trans on Power Deliv 13(4):1020–1026
53. Djuric MB, Terzija VV (1995) A new approach to the arcing faults detection for autoreclosure in transmission systems. IEEE Trans on Power Deliv 10(4):1793–1798
54. Djuric MB, Terzija VV, Radojevic ZM (1998) Overhead lines fault location and arc voltage estimation numerical algorithm derived in time domain. Electr Eng, 81(1):45–53
55. Djuric MB, Terzija VV, Skoljev I. (1993) Transmission line arcing faults recognition from the voltage signal In: Proc of 11th Power System Computations Conference – PSCC, pp 1037–1040
56. Dommel HW (1986) Electro-Magnetic Transients Program (EMTP), BPA, Portland, Oregon, USA
57. Dommel HW (1969) Digital computer solution of electromagnetic transients in single- and multi-phase networks. IEEE Trans on PAS, 88(4):338–396
58. Dudurych I, Rosolowski E (1999) Analysis of overvoltages in overhead ground wires of EHV power transmission line shield under single phase-to-ground faults. Electr Power Syst Res 53(2):105–111
59. Eberl G, Hanninen S, Lehtonen M et al. (2000) Comparison of artificial neural networks and conventional algorithms in ground fault distance computation. In: Proc of Power Engineering Society Winter Meeting, vol 3 pp 1991–1996
60. El-Sharkawi M, Niebur D (1996) A tutorial course on artificial neural network with applications to power systems. IEEE PES, 96 TP 112-0, New York
61. Erezzaghi MEL, Crossley PA, Elferes R (2004) Design and evaluation of an adaptive distance protection scheme suitable for series compensated transmission feeders. In: Proc of 8th Int Conf on Developments in Power System Protection – DPSP, IEE CP500 vol 2 pp 453–456
62. Eriksson L, Saha MM, Rockefeller GD (1985) An accurate fault locator with compensation for apparent reactance in the fault resistance resulting from remote-end infeed. IEEE Trans on PAS 104(2):424–436

63. Evrenosoglu CY, Abur A (2005) Travelling wave based fault location for teed circuits. IEEE Trans on Power Deliv 20(2):1115–1121
64. Evrenosoglu CY, Abur A (2003) Fault location for teed circuits with mutually coupled lines and series capacitors. In: Proc of IEEE Bologna Power Tech Conference, (CD–ROM), paper IEEE 03EX719C
65. Fardanesh B, Zelingher S, Meliopolos APS et al. (1998) Multifunctional synchronized measurement network. IEEE Comput Appl in Power 11(1):26–30
66. Ferrero A, Sangiovanni S, Zapitelli E (1995) A fuzzy-set approach to fault-type identification in digital relaying. IEEE Trans on Power Deliv 10(1):169–175
67. Funabashi T, Otoguro H, Mizuma Y et al. (2001) Influence of fault arc characteristics on the accuracy of digital fault locators. IEEE Trans on Power Deliv 16(2):195–199
68. Funabashi T, Otoguro H, Mizuma Y et al. (2000) Digital fault location for parallel double-circuit multi-terminal transmission lines. IEEE Trans on Power Deliv 15(2):531–537
69. Funk AT, Malik OP (2000) Impedance estimation including ground fault resistance error correction for distance protection. Int J Electr Power and Energy Syst 22(1):59–66
70. Gale PF, Crossley PA (1998) Fault location on overhead power lines using travelling waves. In: Proc of 11th Int Conf on Power System Protection – PSP'98, Bled, pp 89–94
71. Gale PF, Crossley PA, Bingyin X et al. (1993) Fault location based on travelling waves. In: Proc of 5th Int Conf on Developments in Power System Protection – DPSP, IEE CP368 pp 54–59
72. Gangadharan PK, Sidhu TS, Finlayson GJ (2007) Current transformer dimensioning for numerical protection relays. IEEE Trans on Power Deliv 22(1):108–115
73. Gers JM, Holmes EJ (1998) Protection of Electricity Distribution Networks. The Institution of Electrical Engineers, London
74. Ghassemi F, Goddarzi J, Johns AT (1998) Method to improve digital distance relay impedance measurement when used in SC lines protected by a metal oxide varistor. IEE Proc – Gener Transm Distrib 145(4):403–408
75. Ghassemi F, Johns AT, Goddarzi J (1997) A method for eliminating the effect of MOV operation on digital distance relays when used in series compensated lines. In: Proc of 32nd Universities Power Engineering Conference – UPEC, Manchester, pp 113–116
76. Gilany MI, Malik OP, Hope GS (1992) A digital protection technique for parallel transmission lines using a single relay at each end. IEEE Trans on Power Deliv 7(1):118–123
77. Gilbert DM, Morrison IF (1995) A statistical method for detection of power system faults. In: Proc of Int Conf on Power System Transients – IPST, Lisbon, pp 288–293
78. Girgis AA, Brown RG (1981) Application of Kalman filtering in computer relaying. IEEE Trans on PAS, 100(7):3387–3397
79. Girgis AA, Fallon CM, Lubkeman DL (1993) A fault location technique for rural distribution feeders. IEEE Trans on Ind Appl, 29(6):1170–1175
80. Girgis AA, Hart DG, Peterson WL (1992) A new fault location technique for two- and three-terminal lines. IEEE Trans on Power Deliv 7(1):98–107
81. Girgis AA, King D (2003) Fault location in distribution feeders in the presence of distributed generation. In: Proc of CIGRE Study Committee B5 Colloquium, Sydney, paper 217
82. Girgis AA, Makram EB (1988) Application of adaptive Kalman filtering in fault classification distance protection and fault location using microprocessors. IEEE Trans on Power Syst 3(1):301–309
83. Girgis AA, Sallam AA, El-din AK (1998) An adaptive protection scheme for advanced series compensated (ASC) transmission lines. IEEE Trans on Power Deliv 13(2):414–420
84. Glover JD, Sarma M (1994) Power System Analysis and Design. PWS Publishing Company, Boston
85. Goldberg DE (1989) Genetic Algorithms in Search, Optimization and Machine Learning. Addison Wesley, Reading
86. Goldsworthy DL (1987) A linearized model for MOV-protected series capacitors. IEEE Trans on Power Syst 2(4):953–958
87. Gonen T (1986) Electric Power Distribution System Engineering. McGraw-Hill, New York

88. Gopalakrishnan A, Kezunovic M, Mckenna SM et al. (2000) Fault location using the distributed parameter transmission line model. IEEE Trans on Power Deliv 15(4):1169–1174
89. Gopalakrishnan A, Kezunovic M, Mckenna SM et al. (1999) Extension to fault location algorithm based on synchronized sampling. In: Proc of CIGRE – Study Committee 34 Colloquium and Meeting, Florence, paper 216
90. Gracia J, Mazon A, Zamora I (2005) Best ANN structures for fault location in single- and double-circuit transmission lines. IEEE Trans on Power Deliv 20(4):2389–2395
91. Gu JC, Shen KY, Yu SL et al. (2006) Removal of dc offset and subsynchronous resonance in current signals for series compensated transmission lines using a novel Fourier filter algorithm. Electr Power Syst Res 76(5):327–335
92. Hagan MT, Menhaj M (1994) Training feedforward networks with the Marquart algorithm. IEEE Trans on Neural Net, 5(6):989–993
93. Harmeet SK, Raghuveer T (1998) A new design distance to fault locator for power transmission lines. In: Proc of Int Conf on Modern Trends in the Protection Schemes of Electric Power Apparatus and Systems, New Delhi, paper VI-7–11
94. Haykin S (1994) Neural Networks. A Comprehensive Foundation. Macmillan Collage Publishing Company, Inc., New York
95. Hänninen S (2001) Single phase earth faults in high impedance grounded networks. Characteristics, indication and location. PhD Dissertation, Helsinki University of Technology
96. Hänninen S, Lehtonen M (1998) Characteristics of earth faults in electrical distribution networks with high impedance earthing. Electr Power Syst Res 44(3):155–161
97. Hänninen S, Lehtonen M, Hakola T (2002) Earth faults and related disturbances in distribution networks. IEE Proc – Gener Transm Distrib 149(3):283–288
98. Hänninen S, Lehtonen M, Hakola T, Rantanen R (1999) Comparison of wavelet and differential equation algorithms in earth fault distance computation. In: Proc of 13th Power Systems Computations Conference, Trondheim, vol 2, pp 801–807
99. He B, Li Y, Bo ZQ (2006) An adaptive distance relay based on transient error estimation of CVT. IEEE Trans on Power Deliv 21(4):1856–1861
100. Herrmann HJ (2001) Design numerical relays with regard to CT saturation. In: Proc of CIGRE 2001 S.C. 34 Colloquium, Sibiu, CP301 7 p
101. Hess-Nielsen N, Wickerhauser MV (1996) Wavelets and time-frequency analysis. Proc of the IEEE, 84(4):523–540
102. Hopgood AA (2001) Intelligent Systems for Engineers and Scientists. CRC Press LLC, Boca Raton
103. Hosemann G, Steigerwald HM (1993) Modal saturation detector for digital differential protection. IEEE Trans on Power Deliv 8(3):933–940
104. Ibe AO, Cory BJ (1986) A travelling wave-based fault locator for two- and three-terminal networks. IEEE Trans on Power Deliv 1(2):283–288
105. IEEE Power System Relaying Committee Working Group Report (1994) Synchronized sampling and phasor measurements for relaying and control. IEEE Trans on Power Deliv 9(1):442–449
106. IEEE Std C37.114 (2005) IEEE guide for determining fault location on AC transmission and distribution lines. IEEE Power Engineering Society Publ., New York
107. IEEE Std. C37.118 (2006) IEEE standard for synchrophasors for power systems. IEEE Power Engineering Society Publ., New York
108. Isaksson I (1988) Digital protective relaying through recursive least squares identification. IEE Proc – Gener Transm Distrib 135(5):441–449
109. Izykowski J (2008) Fault location on power transmission lines, The Technical University of Wroclaw Press, Wroclaw, Poland
110. Izykowski J, Balcerek P, Kawecki R (2002) Location of single phase-to-ground faults combined with broken conductor failure in single power transmission lines. In: Proc of Int Symposium on Modern Electric Power Systems – MEPS, Wroclaw, pp 313–318
111. Izykowski J, Kasztenny B, Rosolowski E, Saha MM (1998) Fundamental frequency equivalenting of series capacitors equipped with MOVs under fault conditions of a series-

compensated line. In: Proc of 8th Int Symposium on Short-Circuit Currents in Power Systems, Brussels, pp 13–18

112. Izykowski J, Kasztenny B, Rosolowski E, et al. (1998) Dynamic compensation of capacitive voltage transformers. IEEE Trans on Power Deliv 13(1):116–122
113. Izykowski J, Kawecki R (2001) Location of faults in partially parallel transmission networks. In: Proc of IEEE Porto Power Tech Conference, Porto, vol III, paper EDT4-230
114. Izykowski J, Kawecki R, Rosolowski E (2002) Accurate location of faults in parallel transmission lines under availability of measurements from one circuit only. In: Proc of Power Systems Computation Conference –PSCC'02 (CD ROM), Sevilla, paper 6
115. Izykowski J, Molag R, Bozek M (2006) Fault location on three-terminal overhead lines with using two-terminal synchronized voltage and current phasors. In: Proc of Int Symposium: Modern Electric Power Systems – MEPS, Wroclaw, pp 306–311
116. Izykowski J, Molag R, Rosolowski E (2006) Accurate location of faults on power transmission lines with use of two-end unsynchronized measurements. IEEE Trans on Power Deliv 21(2):627–633
117. Izykowski J, Molag R, Rosolowski E (2005) Fault location in three-terminal line with use of limited measurements. In: Proc of Int Conf 2005 IEEE St. Petersburg Power Tech (CD–ROM)
118. Izykowski J, Rosolowski E, Molag R et al. (2008) Accurate location of faults on three-terminal line with use of three-end unsynchronised measurements In: Proc of 16th Power Systems Computation Conference – PSCC, Glasgow (CD–ROM)
119. Izykowski J, Rosolowski E, Saha MM (2008) A new fault location algorithm for use with current differential protective relays of two-terminal line. In: Proc of 16th Power Systems Computation Conference – PSCC, Glasgow (CD–ROM)
120. Izykowski J, Rosolowski E, Saha MM (2007) Post-fault analysis of operation of distance protective relays of power transmission lines. IEEE Trans on Power Deliv 22(1):74–81
121. Izykowski J, Rosolowski E, Saha MM (2004) Locating faults in parallel transmission lines under availability of complete measurements at one end. IEE Proc – Gener Transm Distrib 151(2):268–273
122. Izykowski J, Rosolowski E, Saha MM (2003) Adaptive digital distance algorithm for parallel transmission lines. In: Proc of IEEE Power Tech, Bologna, (CD–ROM)
123. Izykowski J, Rosolowski E, Saha MM, Balcerek P (2005) Accurate algorithm for locating faults in power transmission lines under saturation of current transformers. In: Proc of Power Systems Computation Conference – PSCC, Liege, (CD–ROM)
124. Izykowski J, Rosolowski E, Saha MM et al. (2007) A fault location method for application with current differential relays of three-terminal lines. IEEE Trans on Power Deliv 22(4):2099–2107
125. Jamali S, Ghezeljeh A (2004) Fault location on transmission line using high frequency travelling waves. In: Proc of 8th Int Conf on Developments in Power System Protection – DPSP, IEE CP 500 vol 1 pp 220–223
126. Jayasinghe JASB, Aggarwal RK, Johns AT et al. (1998) A novel non-unit protection for series compensated EHV transmission lines based on fault generated high frequency voltage signals. IEEE Trans on Power Deliv 13(2):405–413
127. Järventausta P, Verho P, Karenlampi M et al. (1997) AI methods for fault location in MV networks. Int J of Eng Intell Syst for Electr Eng and Commun 5(4):265–271
128. Järventausta P, Verho P, Partanen J (1994) Using fuzzy sets to model the uncertainty in the fault location of distribution feeders. IEEE Trans on Power Deliv 9(2):954–960
129. Jeyasurya B (1997) Accurate distance relaying with error compensation. Electr Mach and Power Syst 25(1):29–40
130. Jeyasurya B, Bath CA (1989) An accurate algorithm for transmission line fault location using digital relay measurements. Electr Mach and Power Syst 16(1):25–34
131. Jiale S, Song GB, Xu QQ et al. (2006) Time-domain fault location algorithm for parallel transmission lines using unsynchronized currents. Int J Electr Power and Energy Syst 28(4):253–260

132. Jiang J-A, Lin Y-H, Yang J-Z et al. (2000) An adaptive PMU based fault detection/location technique for transmission lines - Part II: PMU implementation and performance evaluation. IEEE Trans on Power Deliv 15(4):1136–1146
133. Jiang J-A, Yang J-Z , Lin Y-H et al. (2000) An adaptive PMU based fault detection/location technique for transmission lines - Part I: Theory and algorithms. IEEE Trans on Power Deliv 15(2):486–493
134. Johns AT, Aggarwal RK, Song YH (1994) Improved techniques for modelling fault arcs on faulted EHV transmission systems. IEE Proc – Gener Transm Distrib 141(2):148–154
135. Johns AT, Jamali S (1990) Accurate fault location technique for power transmission lines. IEE Proc C 137(6):395–402
136. Johns AT, Moore PJ, Whittard R (1995) New technique for the accurate location of earth faults on transmission systems. IEE Proc – Gener Transm Distrib 142(2):119–127
137. Johns AT, Salman SK (1995) Digital Protection for Power Systems. IEE Power Series 15, Peter Peregrinus Ltd., London
138. Joorabian M, Taleghani ASL SMA, Aggarwal RK (2004) Accurate fault locator for EHV transmission lines based on radial basis function neural networks. Electr Power Syst Res 71(3):195–202
139. Jung CK, Kim KH, Lee JB et al. (2007) Wavelet and neuro-fuzzy based fault location for combined transmission systems. Int J Electr Power and Energy Syst 29(6):445–454
140. Kang S-H, Lee S-J, Kwon Y-J (2001) A fault location algorithm for parallel transmission line with a teed circuit. In: Proc of IEEE PES Summer Meeting, pp 921–926
141. Kang Y-C, Lim U-J, Kang S-H et al. (2004) Compensation of the distortion in the secondary current caused by saturation and remanence in a CT. IEEE Trans on Power Deliv 19(4):1642- 1649
142. Kang Y-C, Kang S-H, Park J-K et al. (1996) Development and hardware implementation of a compensating algorithm for the secondary current of current transformers. IEE Proc – Electr Power Appl, 43(1):41–49
143. Kang Y-C, Ok S-H, Kang S-H (2004) A CT saturation detection algorithm. IEEE Trans on Power Deliv 19(1):78–85
144. Kang Y-C, Ok S-H, Kang S-H et al. (2004) Design and evaluation of an algorithm for detecting current transformer saturation. IEE Proc – Gener Transm Distrib 151(1):27–35
145. Kang Y-C, Park J-K, Kang S-H et al. (1997) An algorithm for compensating secondary currents of current transformers. IEEE Trans on Power Deliv 12(1):116–124
146. Kasztenny B (2001) Distance protection of series compensated lines – problems and solutions. In: Proc of 28th Annual western Protective Relay Conference, Spokane (CD–ROM)
147. Kasztenny B (1995) Digital relaying in power system using fuzzy sets theory. Arch of Control Syst 4(1–2):113–127
148. Kasztenny B, Cambell B, Mazereeuw J (2000) Phase selection for single-pole tripping-weak infeed conditions and cross-country faults. In: Proc of 27th Annual Western Protective Relay Conference, Spokane (CD–ROM)
149. Kasztenny B, Hatziadoniu C, Funk A (1999) VSI-based series compensation scheme for transmission lines. Eur Trans on Electr Power 9(2):101–108
150. Kasztenny B, Izykowski J, Rosolowski E, Saha MM (1997) Algorithms for protection of series compensated lines – the comparative analysis. In: Proc of 32nd Universities Power Engineering Conference – UPEC, Manchester, vol 1, pp 299–302
151. Kasztenny B, Premerlani W, Adamiak M (2008) Synchrophasor Algorithm Allowing Seamless Integration with Today's Relays. In: Proc of 9th Int Conf on Developments in Power System Protection – DPSP, IET CP536 pp 724–729
152. Kasztenny B, Rosolowski E, Lukowicz M, Izykowski J (1997) Current related relaying algorithms immune to saturation of current transformers. In: Proc of 6th Int Conf on Developments in Power System Protection, – DPSP, IEE CP434 pp 365–369
153. Kasztenny B, Rosolowski E, Saha MM, Hilstrom B (1997) Fuzzy sets and logic in power system protection. Int J of Eng Intell Syst for Electr Eng and Commun 5(4):193–203

154. Kasztenny B, Sharples D, Asaro V (2001) Distance relays and capacitive voltage transformers - balancing speed and transient overreach. In: Proc of 55th Annual Georgia Tech Protective Relaying Conference (CD–ROM)
155. Kawady T, Stenzel J (2003) A practical fault location approach for double circuit transmission lines using single end data. IEEE Trans on Power Deliv 18(4):1166–1173
156. Kersting WH (2002) Distribution System Modeling and Analysis. CRC Press, Boca Raton
157. Kezunovic M (1999) Fault location. In: Wiley Encyclopedia of Electrical and Electronics Terminology, vol 7, John Wiley & Sons, Inc., New York, pp 276–285
158. Kezunovic M (1997) A survey of neural net applications to protective relaying and fault analysis. Int J of Eng Intell Syst for Electr Eng and Commun 5(4):185–192
159. Kezunovic M, Aganagic M, Mckernns S et al. (1995) Computing responses of series compensation capacitors with MOV protection in real time. IEEE Trans on Power Deliv 10(1):244–251
160. Kezunovic M, Kojovic LA, Abur A et al. (1994) Experimental evaluation of EMTP-based current transformer models for protective relay transient study. Trans on Power Deliv 9(1):405–413
161. Kezunovic M, Kojovic LA, Skendzic V et al. (1992) Digital models of coupling capacitor voltage transformers for protective relay transient studies. IEEE Trans on Power Deliv 7(4):1927–1935
162. Kezunovic M, Luo S, Sevcik DR (2002) A novel method for transmission network fault location using genetic algorithms and sparse field recordings. In: Proc of PES IEEE Summer Meeting (CD–ROM)
163. Kezunovic M, Mrkic J, Perunicic B (1994) An accurate fault location algorithm using synchronized sampling. Electr Power Syst Res 29(3):161–169
164. Kezunovic M, Perunicic B (1996) Automated transmission line fault analysis using synchronized sampling at two ends. IEEE Trans on Power Syst 11(1):441–447
165. Khorashadi-Zadeh H, Aghaebrahimi MR (2005) A novel approach for decreasing CVT transients in distance protection using artificial neural network. IEICE T Inf Syst E88D(7):1630–1637
166. Kim CH, Aggarwal RK (2001) Wavelet transforms in power systems, Part 2 Examples of application to actual power system transients. Tutorial: Wavelet transforms. Power Eng J 15(4):193–202
167. Kim CH, Aggarwal RK (2000) Wavelet transforms in power systems, Part 1 Tutorial: Wavelet transforms. Power Eng J 14(2):81–87
168. Kim CJ, Russel BD (1989) Classification of faults and switching events by inductive reasoning and Expert System methodology. IEEE Trans on Power Deliv 4(3):1631–1637
169. Kizilcay M, Koch K-H (1994) Numerical fault arc simulation based on power arc test. Eur Trans on Electr Power 4(3):177–185
170. Kizilcay M, Pniok T (1991) Digital simulation of fault arcs in power systems. Eur Trans on Electr Power 1(1):55–60
171. Koglin HJ, Lobos T (1996) Recognition of arcing faults on transmission lines using a neural network. In: Proc of 12th Power System Computations Conference - PSCC, Dresden pp 688–693
172. Kojovic LA (2002) PCB Rogowski coils benefit relay protection. IEEE Comp Appl in Power 15(3):50–53
173. Kojovic LA, Bishop MT, Skendzic V (2003) Coiled for protection. IEEE Power and Energy Mag 1(3):43–48
174. Korejwo E, Synal B, Trojak J (1980) Short HV transmission lines problems. In: Proc of 2nd Int Conf on Development in Power System Protection, IEE CP185, London pp 196–200
175. Kumai T, Nakabayashi H, Hirata Y et al. (2002) Field trial of optical current transformer using optical fiber as Faraday sensor. In: Proc of IEEE Power Engineering Society Summer Meeting vol 2, pp 920–925

176. Kwon WH, Lee GW, Park YM et al. (1991) High impedance fault detection utilizing incremental variance of normalized even order harmonic power. IEEE Trans on Power Deliv 6(2):557–564
177. Lee CJ, Park JB, Shin JR et al. (2006) A new two-terminal numerical algorithm for fault location, distance protection, and arcing fault recognition. IEEE Trans on Power Syst 21(3):1460–1462
178. Lee S-J, Choi M-S, Kang S-H et al. (2004) An intelligent and efficient fault location and diagnosis scheme for radial distribution systems. IEEE Trans. on Power Deliv 19(2):521–532
179. Lehtonen M, Matsinen A, Antila E et al. (2000) An advanced model for automatic fault management in distribution networks. In: Proc of IEEE Power Engineering Society Winter Meeting, vol 3, pp 900–1904
180. Lera G, Pinzolas M (2002) Neighborhood based Levenberg-Marquardt algorithm for neural network training. IEEE Trans on Neural Net, 13(5):1200–1203
181. Lewis LJ (1951) Travelling wave relations applicable to power system fault locators. AIEE Trans 70(2):1671–1680
182. Li F, Li Y, Aggarwal RK (2002) Combined wavelet transform and regression technique for secondary current compensation of current transformers. IEE Proc – Gener Transm Distrib 149(4):497–503
183. Li HX., Chen PCL, Huang HP (2001) Fuzzy Neural Intelligent Systems. Mathematical Foundation and the Applications in Engineering. CRC Press, Boca Raton
184. Liao Y (2008) Fault location for single-circuit line based on bus-impedance matrix utilizing voltage measurements. IEEE Trans Power Deliv 23(2):609–617
185. Liao Y (2007) Unsynchronized fault location based on distributed parameter line model. Electr Power Comp and Syst 35(9):1061–1077
186. Liao Y (2006) Fault location utilizing unsynchronized voltage measurements during fault. Electr Power Comp and Syst 34(12):1283–1293
187. Liao Y, Elangovan S (2006) Unsynchronised two-terminal transmission-line fault-location without using line parameters. IEE Proc – Gener Transm Distrib 153(6):639–643
188. Liao Y, Kang N (2009) Fault-location algorithms without utilizing line parameters based on the distributed parameter line model. IEEE Trans Power Deliv 24(2):579–584
189. Liao Y, Kezunovic M (2007) Optimal estimate of transmission line fault location considering measurement errors. IEEE Trans Power Deliv 22(3):1335–1341
190. Lin Y-H, Liu C-W, Chen C-S (2004) A new PMU-based fault detection/location technique for transmission lines with consideration of arcing fault discrimination-part I: theory and algorithms. IEEE Trans on Power Deliv 19(4):1587–1593
191. Lin Y-H, Liu C-W, Chen C-S (2004) A new PMU-based fault detection/location technique for transmission lines with consideration of arcing fault discrimination–Part II: Performance Evaluation. IEEE Trans on Power Deliv 19(4):1594–1601
192. Lin Y-H, Liu C-W, Yu C-S (2002) A new fault locator for three-terminal transmission lines–using two-terminal synchronized voltage and current phasors. IEEE Trans on Power Deliv 17(2):452–459
193. Lobos T (1992) Fast estimation of symmetrical components in real time. IEE Proc – Gener Transm Distrib 139(1):27–30
194. Lucas JR, McLaren PG (1991) A computationally efficient MOV model for series compensation studies. IEEE Trans on Power Deliv 6(4):1491–1497
195. Lukowicz M (2004) New fault location method based on artificial neural networks. In: Proc of 14th Int Conf on Power System Protection – PSP, Bled, pp 290–295
196. Lukowicz M, Rosolowski E (1998) Fault type classification in high voltage power systems using artificial neural networks. In: Proc of 11th Int Conf on Power System Protection – PSP, Bled, pp 141–146
197. Magnago FH, Abur A (1999) Advanced techniques for transmission and distribution system fault location. In: Proc of CIGRE – Study Committee 34 Colloquium and Meeting, Florence, paper 215

198. Magnago FH, Abur A (1999) A new fault location technique for radial distribution systems based on high frequency signals. In: Proc of IEEE PES Summer Meeting, pp 426–431
199. Magnago FH, Abur A (1998) Fault location using wavelets. IEEE Trans on Power Deliv 13(4):1475–1480
200. Marti JR (1982) Accurate modeling of frequency-dependent transmission lines in electromagnetic transient simulations. IEEE Transactions on PAS, 101(1):147–155
201. Marusic A, Cavlovic M (1998) Application of fuzzy logic to digital fault locator algorithm. In: Proc of 11th Int Conf on Power System Protection – PSP, Bled, pp 71–76
202. MATLAB user's guide (2000) The Math Works Inc., Natick
203. Maun J-C, Philppot L, Coemans J et al. (1995) Power system modelling for the design of advanced fault locators and line protections. In: Proc of the IEEE/KTH Stockholm Power Tech Conference, Stockholm, paper SPT IC 12–01–0381:394–399
204. Mazon AJ, Minambres JF, Zorrozua MB et al. (1995) New method of fault location on double-circuit two-terminal transmission lines. Electr Power Syst Res 35(3):213–219
205. Mazon AJ, Zamora I, Gracia J et al. (2001) Fault location system on double circuit two-terminal transmission lines based on ANNs, In: Proc of IEEE Porto Power Tech Conference, Porto, paper 0-7803-7139-9/01
206. Mazon AJ, Zamora I, Minambres JF et al. (2000) A new approach to fault location in two-terminal transmission lines using artificial neural networks. Electr Power Syst Res 56(3):261–266
207. McKenna SM, Hamai D, Kezunovic M et al. (1997) Transmission line modeling requirements for testing new fault location algorithms using digital simulators. In: Proc of 2nd Int Conf on Digital Power System Simulators – ICDS, Montreal, pp 63–69
208. McLaren PG, March G (1989) A saturation detector and its application to differential current protection. In: Proc of 4th Int Conf on Developments in Power System Protection – DPSP, IEE CP302 pp 15–19
209. McLaren PG, Swift GW, Zhang Z et al. (1995) A new positive sequence directional element for numerical distance relays. IEEE Trans on Power Deliv 10(2):666 – 675
210. Megahed AI, Moussa AM, Bayoumy AE (2006) Usage of wavelet transform in the protection of series-compensated transmission lines. IEEE Trans on Power Deliv 21(3):1213–1221
211. Michalewicz Z (1996) Genetic Algorithms + Data Structure = Evolution Programs. Springer-Verlag, New York
212. Mielczarski W (1998): Fuzzy Logic Techniques in Power Systems. Springer-Verlag, Heidelberg
213. Minambres JF, Zamora I, Mazon AJ et al. (1996) A new technique, based on voltages, for fault location on three-terminal transmission lines. Electr Power Syst Res 37(2):143–151
214. Moore PJ, Bo ZQ, Aggarwal R (2005) Digital distance protection for composite circuit applications. IEE Proc – Gener Transm Distrib 152(2):283–290
215. Moore PJ, Johns AT (1992) Performance of adaptive distance protection under high resistance earth faults. In: Proc of CIGRE: paper 34–203, 7 p
216. Moore PJ, Whittard R, Johns AT (1995) A novel earth fault location technique utilizing single ended measurements. In: Proc of IEEE/KTH Stockholm Power Tech Conference, Stockholm, paper SPT IC 12–03–0515:406–410
217. Musierowicz K, Bandurski W (2002) Estimation of protection's criterion values during arc distortion, Int J Electr Power and Energy Sys 24(9):737–742
218. Nagasawa T, Abe M, Otsuzuki N, et al. (1992) Development of a new fault location algorithm for multi-terminal two parallel transmission lines. IEEE Trans on Power Deliv 7(3):1516–1529
219. Network Protection & Automation Guide (2002) T&D Energy Automation & Information, Alstom, France
220. Nie YX, Yin XG, Li KC et al. (2002) Development of an optical voltage transformer for power protection relays. Electr Power Syst Res 61(2):127–131

221. Nikander A, Järventausta P (1998) Methods for earth fault identification and distance estimation in a compensated medium voltage distribution network. In: Proc of Int Conf on Energy Management and Power Delivery – EMPD, Singapore, pp 595–600
222. Novosel D, Bachmann B, Hart DG et al. (1996) Algorithms for locating faults on series compensated lines using neural network and deterministic methods. IEEE Trans on Power Deliv 11(4):1728–1736
223. Novosel D, Hart D, Hu Y et al. (1998) System for locating faults and estimating fault resistance in distribution networks with tapped loads. US Patent number 5,839,093
224. Novosel D, Hart DG, Saha MM et al. (1994) Optimal fault location for transmission systems. ABB Rev (8):20–27
225. Novosel D, Hart DG, Udren E et al. (1996) Unsynchronized two-terminal fault location estimation. IEEE Trans on Power Deliv 11(1):130–138
226. Novosel D, Hart DG, Udren E et al. (1995) Fault location using digital relay data. IEEE Comp Appl in Power 8(3):45–50
227. Novosel D, Phadke A, Saha MM et al. (1997) Problems and solutions for microprocessor protection of series compensated lines In: Proc of 6th Int Conf on Developments in Power System Protection – DPSP, IEE CP434 pp 18–23
228. Novosel D, Saha MM (1997) Locating faults on series compensated lines using intelligent methods. Int J of Eng Intell Syst for Electr Eng and Commun 5(4):259–264
229. Nouri H, Wang C, Davies T (2001) An accurate fault location technique for distribution lines with tapped loads using wavelet transform. In: Proc of IEEE Porto Power Tech Conference, Porto, paper 0-7803-7139-9/01 vol 3, 4 p
230. Oppenheim AV, Schafer R., Buck JR (1999) Discrete-time Signal Processing. Prentice-Hall, New Jersey
231. Pedrycz W (1989) Fuzzy Control and Fuzzy Systems. John Willey & Sons, Inc., New York
232. Pereira CEM, Zanetta LC Jr (2005) Optimization algorithm for fault location in transmission lines considering current transformers saturation. IEEE Trans on Power Deliv 20(2):603–608
233. Pereira CEM, Zanetta LC Jr (2004) Fault location in transmission lines using one-terminal post-fault voltage data. IEEE Trans on Power Deliv 19(2):570–575
234. Peterson W, Novosel D, Hart DG et al. (1999) Tapping IED data to find transmission faults. IEEE Comp Appl in Power 12(2):36–42
235. Phadke AG (1998) Synchronized phasor measurements for protection and local control. In: Proc of CIGRE Session, Paris, paper 34–106
236. Phadke AG, Thorp JS (2008) Synchronized Phasor Measurements and Their Applications. Springer, New York
237. Phadke AG, Thorp JS (1988) Computer Relaying for Power Systems. Research Studies Press Ltd., John Wiley & Sons, Inc., New York
238. Pilch Z, Winkler W (1985) Phase-comparison and distance protection performance during simultaneous double faults. In: Proc of 3rd Int Conf on Developments in Power System Protection – DPSP, CP249 pp 42–45
239. Poeltl A, Frohlich K (1999) Two new methods for very fast fault type detection by means of parameter fitting and artificial neural networks. IEEE Trans on Power Deliv 14(4):1269–1275
240. Pollock DSG (1999) A Handbook of Time-series Analysis. Signal Processing and Dynamics. Academic Press, London
241. Pradhan AK, Routray A, Pati S et al. (2004) Wavelet fuzzy combined approach for fault classification of a series-compensated transmission line. IEEE Trans on Power Deliv 19(4):1–7
242. Premerlani W, Kasztenny B, Adamiak M (2008) Development and implementation of a synchrophasor estimator capable of measurements under dynamic conditions. IEEE Trans on Power Deliv 23(1):109–123
243. Purushothama GK, Narendranath AU, Thukaram D et al. (2001) ANN applications in fault locators. Electr Power and Energy Syst 23(6):491–506

244. Radojevic ZM, Terzija VV (2006) Two-stage numerical algorithm for distance protection. fault location and arcing faults recognition. Electr Eng, 88(4):289–295
245. Richards GG, Tan OT (1983) Fault location for transmission lines with current-transformer saturation. IEE Proc C 130(1):22–27
246. Rioul O, Vetterli M (1991) Wavelets and signal processing. IEEE Signal Process Mag 8(4):14–38
247. Robertson DC, Camps OI, Mayer JS, Gish WB (1996) Wavelets and electromagnetic power system transients. IEEE Trans on Power Deliv 11(2):1050–1058
248. Roman H, Druml G (1999) Distance location of earth faults in compensated medium voltage networks. In: Proc of 15th Int Conf and Exhibition on Electricity Distribution – CIRED (CD–ROM)
249. Rosolowski E (2002) Digital signal processing for power system protection and control. Academic Publisher EXIT, Warsaw (in Polish)
250. Rosolowski E, Izykowski J, Kasztenny B (2001) Adaptive measuring algorithm suppressing a decaying DC component for digital protective relays. Electr Power Syst Res. 60(2):99–105
251. Rosolowski E, Izykowski J, Kasztenny B, Saha MM (1999) Differential equation based impedance measurement for series-compensated lines. In: Proc of IEEE Power Tech Conference, Budapest, paper BPT99–316–16 (CD–ROM)
252. Rosolowski E, Izykowski J, Kasztenny B, Saha MM (1998) Fault loop impedance analysis for a transmission line with series capacitors and their overvoltage protection. In: Proc of 10th Int Conf on Power System Protection – PSP, Bled, pp 21–26
253. Rosolowski E, Izykowski J, Lisik I (2004) Selection of faulted phases in power transmission line with fuzzy logic application. Acta Technica CSAV (49): 47–58
254. Rosolowski E, Izykowski J, Saha MM (2005) Differential equation based fault location algorithm for series-compensated transmission line. In: Proc of Power Systems Computation Conference – PSCC, Liege (CD–ROM)
255. Rosolowski E, Izykowski J, Saha MM et al. (2009) Accurate transmission line fault location using two-terminal data without time synchronization. Prz Elektrotech (6):170–174
256. Rosolowski E, Kasztenny B, Izykowski J, Saha MM (1996) Comparative analysis of impedance algorithms for series compensated lines. In: Proc of 10th Int Conf on Power System Protection – PSP, Bled, pp 21–26
257. Rosolowski E, Michalik M (1994) Fast identification of symmetrical components by use of a state observer. IEE Proc – Gener Transm Distrib 141(6):617–622
258. Sachdev MS (1997) Advancement in microprocessor based protection and communication. IEEE Tutorial Course, IEEE PES, IEEE Catalog Number: 97TP120-0
259. Sachdev MS, Agarwal R (1988) A technique for estimating line fault locations from digital distance relay measurements. IEEE Trans on Power Deliv 3(1):121–129
260. Sachdev MS, Agarwal R (1985) Accurate fault location estimates from digital impedance relay measurements. In: Proc of 3rd Int Conf on Developments in Power System Protection – DPSP, IEE CP249 pp 180–184
261. Sachdev MS, Das R, Sidhu TS (1997) Determining locations of faults in distribution systems. In: Proc of 6th Int Conf on Developments in Power System Protection – DPSP, IEE CP434 pp 188–191
262. Sachdev MS, Nagpal M (1991) A recursive least error squares algorithm for power system relaying and measurement application. IEEE Trans on Power Deliv 6(3):1008–1015
263. Sachdev MS, Sidhu TS, Das R (1999) A fault location technique for radial subtransmission lines with tapped loads. In: Proc of CIGRE Colloquium SC34, Florence, paper 203
264. Sadeh J, Hadjsaid N, Ranjbar AM et al. (2000) Accurate fault location algorithm for series compensated transmission lines. IEEE Trans on Power Deliv 15(3):1027–1033
265. Saha MM, Das R, Verho P, Novosel D (2002) Review of fault location techniques for distribution systems. In: Proc of Power Systems and Communications Infrastructures for the Future Conference, Beijing (CD–ROM) 6 p

266. Saha MM, Izykowski J, Kasztenny B, Rosolowski E (1998) Improving protective relaying by on-line dynamic compensation of capacitive voltage transformers. In: Proc of 33rd Universities Power Engineering Conference – UPEC, Edinburgh, vol 1 pp 336–339
267. Saha MM, Izykowski J, Kasztenny B, Rosolowski E (1998) Relaying algorithms for protection of series-compensated lines. In: Proc of Int Conf on Modern Trends in the Protection Schemes of Electric Power Apparatus and Systems, New Delhi, paper V-50–61
268. Saha MM, Izykowski J, Kasztenny B, Rosolowski E (1997) Modeling fault conditions for protection of series compensated lines. In: Proc of Int Conf on Power Systems Transients – IPST, Seattle, pp 383–388
269. Saha MM, Izykowski J, Lukowicz M, Rosolowski E (2001) Application of ANN methods for instrument transformer correction in transmission line protection. In: Proc of 7th Int Conf on Developments in Power System Protection – DPSP, IEE CP479 pp 303–306
270. Saha MM, Izykowski J, Rosolowski E (2004) A method of fault location based on measurements from impedance relays at the line ends. In: Proc of 8th Int Conf on Developments in Power System Protection – DPSP, IEE CP500 pp 176–179
271. Saha MM, Izykowski J, Rosolowski E (2004) A two-end method of fault location immune to saturation of current transformers. In: Proc of 8th Int Conf on Developments in Power System Protection – DPSP, IEE CP500 pp 172–175
272. Saha MM, Izykowski J, Rosolowski E, Palki BS (2004) Fault location technique for a transmission line compensated with series capacitors at both ends. In: Proc of 3rd Int Conf on Power System Protection and Automation, New Delhi, pp I/92–I/101
273. Saha MM, Izykowski J, Rosolowski E et al. (2008) Accurate location of faults on series-compensated lines with use of two-end unsynchronised measurements. In: Proc of the 9th Int Conf on Developments in Power System Protection – DPSP, IEE CP536 pp 338–343
274. Saha MM, Izykowski J, Rosolowski E, Kasztenny B (1999) A new accurate fault locating algorithm for series compensated lines. IEEE Trans on Power Deliv 14(3):789–797
275. Saha MM, Kasztenny BZ (1997) AI methods in power system protection. Int J of Eng Intell Syst for Electr Eng and Commun 5(4):183–184
276. Saha MM, Kasztenny B, Izykowski J, Rosolowski E (1999) A novel fault locating technique for transmission lines with series-compensation. In: Proc of Power Systems Computations Conference – PSCC, Trondheim, pp 679–685
277. Saha MM, Kasztenny B, Rosolowski E, Izykowski J (2001) First zone algorithm for protection of series compensated lines. IEEE Trans on Power Deliv 16(2):200–207
278. Saha MM, Provoost F, Rosolowski E (2001) Fault location method for MV cable network. In: Proc of 7th Int Conf on Developments in Power System Protection – DPSP, IEE CP476 pp 323–326
279. Saha MM, Rosolowski E, Izykowski J (2006) Differential equation algorithm for locating faults on parallel series-compensated lines. In: Proc of 3rd Int Symposium: Modern Electric Power Systems – MEPS, Wroclaw pp 289–295
280. Saha MM, Rosolowski E, Izykowski J (2005) ATP-EMTP investigation for fault location in medium voltage networks. In: Proc of Int Power System Transients Conference–IPST, Montreal (CD_ROM), paper IPST05 – 220, 6 p
281. Saha MM, Rosolowski E, Izykowski J (2004) A new fault location algorithm for distribution networks. In: Proc of 8th Int Conf on Developments in Power System Protection – DPSP, IEE CP500 pp 168–171
282. Saha MM, Rosolowski E, Izykowski J (2003) ATP-EMTP investigation of a new distance protection principle for series compensated lines. In: Proc of Int Conf on Power Systems Transients–IPST, New Orlean, paper 5b-1
283. Saha MM, Rosolowski E, Izykowski J, Pierz P (2002) Artificial intelligent and wavelet techniques application to power system protection. Indian J of Power and River Valley Dev – Power-2002 Annu Rev 52(9–10):196–203
284. Saha MM, Wikstrom K, Izykowski J, Rosolowski E (2000) Fault location in uncompensated and series-compensated parallel lines. In: Proc of 2000 IEEE PES Winter Meeting, Singapore (CD–ROM)

285. Saha MM, Wikstrom K, Izykowski J, Rosolowski E (2001) New concept for fault location in series-compensated parallel lines. In: Proc of 2001 IEEE PES Winter Meeting, Columbus (CD–ROM)
286. Saha MM, Wikstrom K, Izykowski J, Rosolowski E (2001) New fault location algorithm for parallel lines. In: Proc of 7th Int Conf on Developments in Power System Protection – DPSP, IEE CP476 pp 407–410
287. Saha MM, Wikstrom K, Lindahl S (1997) A new approach to fast distance protection with adaptive features. In: Proc of 6th Int Conf on Developments in Power System Protection – DPSP, IEE CP434 pp 32–35
288. Salat R, Osowski S (2004) Accurate fault location in the power transmission line using support vector machine approach. IEEE Trans on Power Syst 19(2):979–986
289. Salim RH, Resener M, Filomena AD et al. (2009) Extended Fault-Location Formulation for Power Distribution Systems. IEEE Trans on Power Deliv 24(2):508–516
290. Sanderson JVH, Pereira PS (1989) A device for the detection of CT saturation. In: Proc of 4th Int Conf on Developments in Power System Protection – DPSP, IEE CP302 pp 20–23
291. Sant MT, Paithankar YG (1979) Online digital fault locator for overhead transmission line. IEE Proc C 126(11):1181–1185
292. Schweitzer EO III (1990) A review of impedance-based fault locating experience. In: Proc of 14th Annual Iowa–Nebraska System Protection Seminar, Omaha, 1–31 p
293. Sheng LB, Elangovan S (1999) A fault location method for parallel transmission lines. Int J Electr Power and Energy Syst 21(4):253–259
294. Shengfang L, Chunju F, Weiyong Y et al. (2004) A new phase measurement unit (PMU) based fault location algorithm for double circuit lines. In: Proc of 8th Int Conf on Developments in Power System Protection – DPSP, IEE CP500 vol 1 pp 188–191
295. Sidhu TS, Singh H, Sachdev MS (1995) Design, implementation and testing of an artificial neural-network based fault direction discriminator for protecting transmission-lines. IEEE Trans on Power Deliv 10(2):697–706
296. Silva M, Oleskovicz M, Coury DV (2004) A fault locator for transmission lines using travelling waves and wavelet transform theory. In: Proc of 8th Int Conf on Developments in Power System Protection – DPSP, IEE CP500 pp 212–215
297. Silveira EG, Pereira C (2007) Transmission line fault location using two-terminal data without time synchronization. IEEE Trans on Power Syst 22(1):498–499
298. Smith SW (1999) The Scientist and Engineer's Guide to Digital Signal Processing. California Technical Publishing, San Diego
299. Song GB, Suonan J, Xu QQ et al. (2005) Parallel transmission lines fault location algorithm based on differential component net. IEEE Trans on Power Deliv 20(4):2396–2406
300. Song YH, Aggarwal RK, Johns AT (1995) Digital simulation of fault arcs on long-distance compensated transmission systems with particular reference to adaptive autoreclosure. Eur Trans on Electr Power 5(5):315–324
301. Srinivasan K, St-Jacques A (1989) A new fault location algorithm for radial transmission systems with loads. IEEE Trans on Power Deliv 4(3):1676–1682
302. Steinmetz CP (1893) Complex quantities and their use in electrical engineering. In: Proc of the International Electrical Congress, Chicago, AIEE Proc pp 33–74
303. Stringfield TW, Marihat DJ, Stevens RF (1957) Fault location methods for overhead lines. Trans of the AIEE, Part III, PAS 76(3):518–530
304. Styvaktakis E, Bollen MHJ, Gu IYH (1999) A fault location technique for two and three terminal lines using high frequency fault clearing transients. In: Proc of IEEE Power Tech Conference, Budapest, paper BPT99–410–22
305. Sultan AF, Swift GW, Fedirchuk DJ (1992) Detection of high impedance arcing faults using a multilayer perceptron. IEEE Trans on Power Deliv 7(4):1871–1877
306. Suonan JL, Song GB, Zhang JK et al. (2005) A novel short-window algorithm of phasor computation for EHV line protection. Electr Power Syst Res 73(1):1–7

307. Szafran J, Rosolowski E (1987) Fast signal recognition for power system protection, In: Proc of 2nd Int Symposium on Measurement of Electrical Quantities – IMEKO-TC4, Warsaw, pp 253–260
308. Tag Eldin EM, Gilany MI, Abdelaziz MM et al. (2006) An accurate fault location scheme for connected aged cable lines in double-fed systems. Electr Eng 88(5):431–439
309. Takagi T, Yamakosi Y, Yamura M et al. (1981) Development of new type fault locator using the one–terminal voltage and current data. IEEE Trans on PAS, 101(8):2892–2898
310. Takani H, Kurosawa Y, Im Ai S et al. (2004) Analysis and evaluation of multi terminal fault location using actual fault data. In: Proc of 8th Int Conf on Developments in Power System Protection – DPSP, IEE CP500 vol 1 pp 208–211
311. Tan JC, Crossley PA, Kirschen D et al. (1999) Fault section identification on a transmission network using action factors and expert system technology. In: Proc of 13th Power Systems Computations Conference – PSCC, Trondheim, pp 820–826
312. Tang Y, Wang HF, Aggarwal RK et al. (2000) Fault indicators in transmission and distribution systems In: Proc of Int Conf on Electric Utility Deregulation and Restructuring and Power Technologies – DRPT, pp 238–243
313. Tawfik MM, Morcos MM (2001) ANN-based techniques for estimating fault location on transmission lines using Prony method. IEEE Trans on Power Deliv 16(2):219–224
314. Terzija VV, Koglin H-J (2001) A new approach to arc resistance calculation. Electr Eng, 83(4):187–192
315. Thomas DWP, Carvalho RJ, Pereira ET (2003) Fault Location in Distribution Systems Based on Traveling Waves. In: Proc of IEEE Bologna Power Tech Conference (CD–ROM) 6 p.
316. Torrence C, Compo GP (1998) A practical guide to wavelet analysis. Bull of the Am Meteorol Soc 79(1):61–78
317. Tziouvaras DA, Benmouyal G, Roberts J et al. (2001) The effect of conventional instrument transformer transients on numerical relay elements. In: Proc of CIGRE S.C. 34 Colloquium, Sibiu, CP308, 5 p
318. Tziouvaras DA, McLaren P, Alexander G (2000) Mathematical models for current, voltage, and coupling capacitor voltage transformers. IEEE Trans on Power Deliv 15(1):62–72
319. Tziouvaras DA, Roberts J, Benmouyal G (2001) New multi-ended fault location design for two- or three-terminal lines. In: Proc of 7th Int Conf on Developments in Power System Protection – DPSP, IEE CP476 pp 395–398
320. Tziouvaras DA, Roberts J, Benmouyal G (1999) New multi-ended fault location design for two- or three-terminal lines. In: Proc of CIGRE – Study Committee 34 Colloquium and Meeting, Florence, paper 213
321. Vasilic S, Kezunovic M (2002) An improved neural network algorithm for classifying the transmission line faults. In: IEEE PES Winter Meeting, New York, 6 p
322. Vaughan M, Moore PJ (1999) A digital signal processing technique utilising VLF radio spectra for the detection of power system arcing faults. In: Proc of 13th Power Systems Computations Conference – PSCC, Trondheim, pp 693–699
323. Wang C, Dou CX, Li XB et al. (2007) A WAMS/PMU-based fault location technique. Electr Power Syst Res 77(8):936–945
324. Wang H, Keerthipala WWL (1998) Fuzzy-neuro approach to fault classification for transmission line protection. IEEE Trans on Power Deliv 13(4):1093–1104
325. Warwick K, Ekwue A, Aggarwal R. (1997): Artificial Intelligence Techniques in Power Systems. (ed) The Institution of Electrical Engineers, London
326. Welfonder T, Leitloff V, Fenillet R et al. (2000) Location strategies and evaluation of detection algorithms for earth faults in compensated MV distribution systems. IEEE Trans on Power Deliv 15(4):1121–1128
327. Whitaker JC (1999) AC Power Systems Handbook, CRC Press LLC, Boca Raton
328. Wiggins CM (1994) A novel concept for URD cable fault location. IEEE Trans Power Deliv 9(1):591–597

329. Wilson RE, Zevenbergen GA, Mah D et al. (1999) Calculation of transmission line parameters from synchronized measurements. Electr Mach and Power Syst 27(12):1269–1278
330. Wiszniewski A (1983) Accurate fault impedance locating algorithm. IEE Proc - Part C 130(6):311–315
331. Wiszniewski A (1990) Digital high-speed calculation of the distorted signal fundamental component. IEE Proc C 137(1):19–24
332. Wiszniewski A (1985) Fault location correction of errors due to current transformers. In: Proc of 3rd Int Conf on Developments in Power System Protection – DPSP, CP249 pp 185–187
333. Wiszniewski A, Izykowski J (1975) Influence of ferroresonance suppression circuits upon the transient response of capacitive voltage transformers. In: Proc of Int Conf on Developments in Power System Protection – DPSP, IEE CP125 pp 182–188
334. Wiszniewski A, Szafran J (1989) Distance algorithm immune to saturation of current transformers. In: Proc of 4th Int Conf on Developments in Power System Protection – DPSP, London, IEE CP302 pp 196–199
335. Wright A, Christopoulos C (1993) Electrical Power System Protection. Chapman & Hall, London
336. Xia YQ (1994) Development and implementation of a variable window algorithm for high-speed and accurate digital distance protection. IEE Proc – Gener Transm Distrib 141(4):383–389
337. Yager RR, Filev DP (1994) Essentials of Fuzzy Modeling and Control. John Wiley & Sons, Inc., New York
338. Young P (1984) Recursive Estimation and Time-series Analysis. Springer-Verlag, Berlin
339. Youssef OAS (2004) Combined fuzzy-logic wavelet-based fault classification technique for power system relaying, IEEE Trans on Power Deliv 19(2):582–589
340. Yu C-S, Liu C-W, Yu S-L et al. (2002) A new PMU-based fault location algorithm for series compensated lines. IEEE Trans on Power Deliv 17(1):33–46
341. Zamora I, Minambres JF, Mazon AJ et al. (1996) Fault location on two-terminal transmission lines based on voltages. IEE Proc – Gener Transm Distrib 143(1):1–6
342. Zhang Q, Zhang Y, Song W et al. (1998) Transmission line fault location for single-phase-to-earth fault on non-direct-ground neutral system. IEEE Trans on Power Deliv 13(4):1086–1092
343. Zhang Q, Zhang Y, Song W et al. (1999) Transmission line fault location for phase-to-earth fault using one-terminal data. IEE Proc – Gener Transm Distrib 146(2):121–124
344. Zhang Q, Zhang Y, Song W et al. (1999) Fault location of two-parallel transmission line for nonearth fault using one-terminal data. IEEE Trans on Power Deliv 14(3):863–867
345. Zhang Y, Zhang Q, Song W et al. (2000) Transmission line fault location for double phase-to-earth fault on non-direct-ground neutral system. IEEE Trans on Power Deliv 15(2):520–524
346. Zhao W, Song Y-H, Chen W-R (2001) Improved GPS travelling wave fault locator for power cables by using wavelet analysis. Int J Electr Power and Energy Syst 23(5):403–411
347. Zhenyu H, Kasztenny B. Madani V et al. (2008) Performance evaluation of phasor measurement systems. In: Proc of IEEE Power and Energy Society General Meeting – Conversion and Delivery of Electrical Energy in the 21st Century, (CD–ROM) 7 p
348. Zhu J, Lubkeman DL, Girgis AA (1997) Automated fault location and fault diagnosis on electrical power distribution feeders. IEEE Trans on Power Deliv 12(2):801–809
349. Ziegler G (2006) Numerical Distance Protection. Principles and Applications. Siemens AG, Publicis MCD Verlag, Erlangen
350. Ziegler G (1990) Application guide on protection of complex transmission network configuration. In: CIGRE SC34-WG04, Paris, 137 p
351. Zimmermann H-J, Zadeh LA, Gaines BR (1984) Fuzzy Sets and Decision Analysis. North-Holland, Amsterdam
352. Zimmerman K, Costello D (2006) Impedance-based fault location experience. In: Proc of IEEE Rural Electric Power Conference, 16 p

Index

MIX
Papier aus verantwortungsvollen Quellen
Paper from responsible sources
FSC® C105338
FSC
www.fsc.org

If you have any concerns about our products,
you can contact us on
ProductSafety@springernature.com

In case Publisher is established outside the EU,
the EU authorized representative is:
Springer Nature Customer Service Center GmbH
Europaplatz 3, 69115 Heidelberg, Germany

Printed by Libri Plureos GmbH
in Hamburg, Germany